Dinosaurs
A Concise Natural History
FOURTH EDITION

The ideal textbook for nonscience majors, this lively and engaging introduction encourages students to ask questions, assess data critically and think like a scientist. Building on the success of previous editions, *Dinosaurs* has been reorganized and extensively rewritten in response to instructor and student feedback. This edition has been thoroughly updated to include new discoveries in the field, such as the toothed bird specimens found in China and recent discoveries of dinosaur soft anatomy. Illustrations by leading paleontological illustrator John Sibbick and new, carefully chosen photographs, clearly show how dinosaurs looked, lived and their role in Earth history. Making science accessible and relevant through clear explanations and extensive illustrations, the text guides students through the dinosaur groups, emphasizing scientific concepts rather than presenting endless facts. Grounded in the common language of modern evolutionary biology – phylogenetic systematics – students learn to think about dinosaurs the way that professional paleontologists do.

DAVID E. FASTOVSKY is Professor in the Department of Geosciences at the University of Rhode Island. His interest in dinosaurs began in his early years when he read about a paleontologist's adventures in the Gobi Desert early in the twentieth century. Dinosaurs won out years later when he had the tough decision of choosing between a career in music or paleontology. He has since carried out fieldwork all over the world. He is known as a dynamic teacher as well as a respected researcher on the environments in which dinosaurs roamed, as well as their extinction.

DAVID B. WEISHAMPEL is a Professor Emeritus at the Center for Functional Anatomy and Evolution at Johns Hopkins University School of Medicine. His research focuses on dinosaur evolution and how dinosaurs function, and he is particularly interested in herbivorous dinosaurs and the dinosaur record of Europe. He is senior editor of *The Dinosauria* and has contributed to a number of popular publications, including acting as consultant to Michael Crichton in the writing of *The Lost World*, the inspiration for Steven Spielberg's film *Jurassic Park*. He was recently honored in an International Symposium on duck-billed dinosaurs, dedicated to him and his research.

JOHN SIBBICK has been creating illustrations of extinct life forms and their environments for over 30 years, producing numerous books on dinosaurs, as well as pterosaurs, and general books on prehistoric life. His work has appeared in scientific magazines, television documentaries and museums, and featured on a set of stamps depicting dinosaurs and other prehistoric reptiles for the United Kingdom's Royal Mail.

Gideon Mantell (1790–1852), the "father" of
modern dinosaur paleontology.

Dinosaurs
A Concise Natural History

FOURTH EDITION

DAVID E. FASTOVSKY
University of Rhode Island

DAVID B. WEISHAMPEL
The Johns Hopkins University School of Medicine

With illustrations by **JOHN SIBBICK**

CAMBRIDGE
UNIVERSITY PRESS

Shaftesbury Road, Cambridge CB2 8EA, United Kingdom

One Liberty Plaza, 20th Floor, New York, NY 10006, USA

477 Williamstown Road, Port Melbourne, VIC 3207, Australia

314–321, 3rd Floor, Plot 3, Splendor Forum, Jasola District Centre, New Delhi – 110025, India

103 Penang Road, #05–06/07, Visioncrest Commercial, Singapore 238467

Cambridge University Press is part of Cambridge University Press & Assessment,
a department of the University of Cambridge.

We share the University's mission to contribute to society through the pursuit of
education, learning and research at the highest international levels of excellence.

www.cambridge.org
Information on this title: www.cambridge.org/9781108475945

DOI: 10.1017/9781108567565

First published 2009
Second edition 2012
Third edition 2016
Fourth edition 2021
Reprinted 2022

Printed in the United Kingdom by TJ Books Limited, Padstow Cornwall

A catalogue record for this publication is available from the British Library

ISBN 978-1-108-47594-5 Hardback
ISBN 978-1-108-46929-6 Paperback

Additional resources for this publication at www.cambridge.org/dinosaurs4

To **Lesley**, **Naomi**, and **Marieke**,
my family.

To poor **Robert**, because. . .

To **Sarah** and **Amy**.
Thanks for continuing to remind your dad
that there are things other than dinosaurs!

TABLE OF CONTENTS

PREFACE TO THE FOURTH EDITION

Bigger, Better, and Badder

About 40 years ago, a number of dinosaur specialists inaugurated a revolution in our understanding of dinosaurs. Somehow dinosaurs had languished scientifically although they were big draws in museums, on cereal boxes, and in horror flicks. Perhaps it was because they were thought to be cold-blooded, stupid, and extinct because of it. But that revolution – the "Dinosaur Renaissance" (it was just that, a rebirth) – galvanized paleontologists into what can only be thought of as the greatest makeover in history. A shot in the arm from the movie *Jurassic Park*, itself a product of the Dinosaur Renaissance, and dinosaurs got hot (literally!).

In the 40 years since the Dinosaur Renaissance began, we have learned that they were not crocodile cold-blooded; that they were not slow or stupid; that they did not all go extinct (modern birds are dinosaurs), that many kept nests and raised their babies through dinosaur adolescence; that they lived from Antarctica to the Sahara, and everywhere in between; that they were colorful – and that we can know some of those colors(!); that the sexes showed off to each other; and that when the dinosaurs went extinct, they were killed in the most dramatic and shocking way (un)imaginable: by an asteroid impact with Earth. Dinosaurs got bigger, better, and badder. No wonder people became interested in them!

With so many advances in our understanding of dinosaurs coming so fast, we needed to update our book for our readers. To do this every chapter has been thoroughly revised, a new chapter has been added, and three chapters have been completely rewritten. And with the revisions come all kinds of spectacular new photographs, drawn from all over the world, augmenting the collection of beautiful photographs and drawings commissioned especially for this book that it already had. Most importantly, the book is now up to date. We hope you find the book as rich an experience as we have in writing it.

To the Student

How to Get the Most out of this Book

Dinosaurs: A Concise Natural History Fourth Edition (DCNH IV) is designed to be used with an unusually broad range of levels, from the very inexperienced to the crazy into it.

Organization

It is in its organizational design that DCNH IV is unique. The key is that while each chapter explores each subject in increasing detail, it is *not* necessary to push to the end of each chapter before proceeding to the next; users have considerable latitude regarding how deeply they delve into each subject. The most interested readers will take a more comprehensive approach in each chapter; those not wishing to overload on Dinosauria need not explore the full range of subject matter encompassed within each chapter.

The book is divided into four consecutive parts, designed to be read sequentially.

Part I – Introductory background scaffolding, including collecting, time, phylogeny, and the position of dinosaurs within the vertebrate biota. Some basic details about plate tectonics and evolutionary biology are also provided; because these are not built into the chapters, they are not necessary to get the flow.

Parts II and III – These are the core of the book. You need to know who dinosaurs are, before you can learn about what they did. Each of the dinosaur group-centered chapters is laid out with parallel organization:

(1) basic (and brief) taxonomic context;
(2) paleobiology of the dinosaur group; followed by
(3) a more detailed evolutionary treatment.

This organization is key, because it allows you to choose how deeply you wish to explore each group. Students with less interest in the details of which dinosaur is related to which other dinosaur (dinosaur systematics) need only go, in each chapter, as far as the sections on dinosaur paleobiology, stopping short of the detailed sections on dinosaur systematics (uniformly entitled in each chapter, "The Evolution of…," signaling the systematic complexity to follow).

Students with greater interest can go deeper into the "Evolution of…" sections, exploring the more detailed cladograms, even assessing, should they so choose, the nature of the diagnostic characters provided for each group. *Again, the book is designed to allow readers to choose the level at which they wish to engage with this material.*

Part IV – Part IV is more synthetic, and includes paleobiological, and macroevolutionary aspects of dinosaur paleontology. The chapters on the paleobiology (Chapter 13), warm-bloodedness (Chapter 14), Mesozoic (Chapter 15), and the dinosaur extinction (Chapter 17) are uniquely comprehensive coverages of these difficult topics; and, as in all chapters throughout this book, they are supported by a significant series of carefully chosen original citations from the primary scientific literature.

Chapter 16 is a history of ideas in dinosaur paleontology. History only has resonance when one knows something about the subject, and so we have put this chapter near the end of the book; that way, when you encounter it, you will remember those ideas from the preceding chapters.

This chapter is *not* about names and dates; rather, it is about the development of ideas – and the people who developed them. Today, there are more active dinosaur paleontologists than have ever before, and so in this chapter we also try to introduce you to a few of them.

Textbooks are expensive, and we mean for our students to get the most out of their investment! We hope that you find this book rewarding, and that we can successfully convey some of the excitement and wonder that all professional paleontologists experience in their careers.

To the Instructor

Dinosaurs: A Concise Natural History is designed to introduce first- and second-year university students, many commonly seeking to fulfill general science requirements, to the logic of scientific inquiry and to concepts in natural history and evolutionary biology. The perspectives and methods introduced through dinosaurs have a relevance that extends far beyond the dinosaurs, teaching scientific logic and critical thinking. The approach has been successful for around 40 years, and new discoveries and interpretations now merit this fourth edition. Professional paleontologists, including even dinosaur specialists, will find in it a comprehensive overview of the group, with many of the key issues highlighted.

In its preparation, Cambridge University Press again devoted considerable energy to obtaining extensive feedback from the many instructors who had had experience teaching from previous editions. The thoughtful, detailed, and, in many cases, comprehensive, answers obtained for this fourth edition were particularly useful in determining the ways in which this edition could be strengthened as a teaching tool. Accordingly, we have responded to virtually all suggestions and recommendations. The care that veteran instructors have put into their answers has surely enriched our book; we are most grateful!

A Unique Conceptual Approach

Names, dates, places, and features are available everywhere these days. But litanies of names, dates, and places is not science; the *creative* synthesis of these data is far more important and, fortunately, far more interesting. The goal of this book is to help students achieve that synthesis.

Reflecting its field, DCNH IV is organized through the lens of phylogenetic systematics. This approach allows students to understand dinosaurs as professional paleontologists do. To have had an entire class in dinosaurs, and yet be insensible to the underlying phylogenetic connections among these (and all) organisms is indefensible; it would be akin to studying biology without evolution. The cladograms used in this book are drawn in a way that highlights the evolutionary relationships they depict, ensuring that both the methods and conclusions of phylogenetic systematics remain accessible.

Part I introduces the fundamental intellectual tools of the trade, including collecting, geological time, the logic of phylogenetic systematics, and enough basic tetrapod anatomy to get the ball rolling. Parts II and III cover, respectively, Saurischia and Ornithischia. The chapters within Parts II and III cover the major groups within Dinosauria, treating them in terms of behavior, lifestyle, and finally evolution. The central role of birds as living dinosaurs is developed through Theropoda I, II, and III. The normally prominent status accorded to *Archaeopteryx* has here been diminished, since the astounding Liaoning fossil discoveries in the past 25 years have undermined the uniqueness of *Archeopteryx* as a transition to birds. Reflecting this, we've augmented the section on Mesozoic bird evolution, supported by some extraordinary photographs of fossils from the Jehol avian biota, generously donated by Dr. Luis Chiappe and Ms. Stephanie Abramowicz, both of the Natural History Museum of Los Angeles. Ornithischians are treated in Chapters 10 to 12, culminating in Ornithopoda, a group that remains phylogenetically somewhat fraught. By the time students reach this chapter, however, they will be in a position to understand, appreciate, and assimilate some of the uncertainty.

Part IV covers the aspects of the paleobiology of Dinosauria, from their biology (Chapter 13), to their metabolism (Chapter 14), to the great rhythms that drove their evolution (and coevolution; Chapter 15), to a fully updated discussion of their extinction (Chapter 17). Chapter 16, the penultimate chapter, is devoted to a history of dinosaur paleontology. Although commonly introduced at the beginning of dinosaur books, our history chapter (Chapter 16) – a history of *ideas* – is placed toward the end, so that the thinking that currently drives the field can be understood in context. We believe that the history of dinosaur paleontology is much more resonant when students already know something about the fossils being hunted and the ideas being developed. Finally, the book ends, like the dinosaurs themselves, with a discussion of the great Cretaceous–Paleogene mass extinction. Here we might say, as so many have, that Earth then entered the Age of Mammals, but, paradoxically, we'll try to persuade readers that we're still in the "Age of Dinosaurs."

We would cheat our readers if we left out accounts of the dinosaur specialists, whose colorful personalities and legendary exploits make up the lore of dinosaur paleontology; so we've included a few of their stories as well (Chapter 16). The fourth edition also highlights Generation X and even a few Millennial paleontologists in the hope that our readers might see something of themselves in these accomplished young professionals.

Finally, as in all previous editions, any errors that appear in this work are entirely Dave's fault.

Features

DCNH IV is designed to help instructors to teach and to help students learn.

- The book continues to be richly illustrated with especially commissioned art by John Sibbick, one of the world's foremost illustrators of dinosaurs. We have also dramatically increased the number of photographs and, in this new edition, obtained many replacements as well. Cambridge University Press now prints the book in color which, we believe, increases the impact of its contents.
- As always, the chapters are arranged so that they present the material in order of increasing complexity and sophistication, building the confidence of the student early on, and extending their sophistication gradually as they progress through the book.
- The tone of the text is light, lively, and readable, engaging readers in the science, and dispelling the apprehension many students acquire when they pick up a science textbook.
- Objectives at the beginning of each chapter help students to grasp chapter goals.
- Boxes scattered throughout the book present a range of ancillary topics, from dinosaur poetry, to extinction cartoons, to how bird lungs work, to colorful accounts of unconventional, outlandish, and extraordinary people and places.
- A comprehensive series of "Topic Questions," to be used as study guides, are located at the end of each chapter. The questions probe successively deeper levels of understanding, and students who can answer all of the "Topic Questions" will have a good grasp of the material. Variants of these questions can serve as excellent templates for examination questions.
- A Glossary ties definitions of key terms into the pages where the terms are used.
- There are two indices: an Index of subjects and an Index of genera, which includes English translations of all dinosaur names.
- Appendices are included in certain chapters to introduce material that students may need in order to understand chapter concepts, such as the chemistry necessary to understand radioactive decay, plate tectonics, the morphology of modern birds, and the basic principles of evolution by natural selection (Darwinian evolution).

Online resources to help you deliver your dinosaur course include:

- electronic files of the figures and images within the book;
- lecture slides in PowerPoint with text and figures to help you to structure your course; and
- solutions to the questions in the text for instructors.

All resources are available to instructors at www.cambridge.org/dinosaurs4.

PART I
Remembrance of Things Past

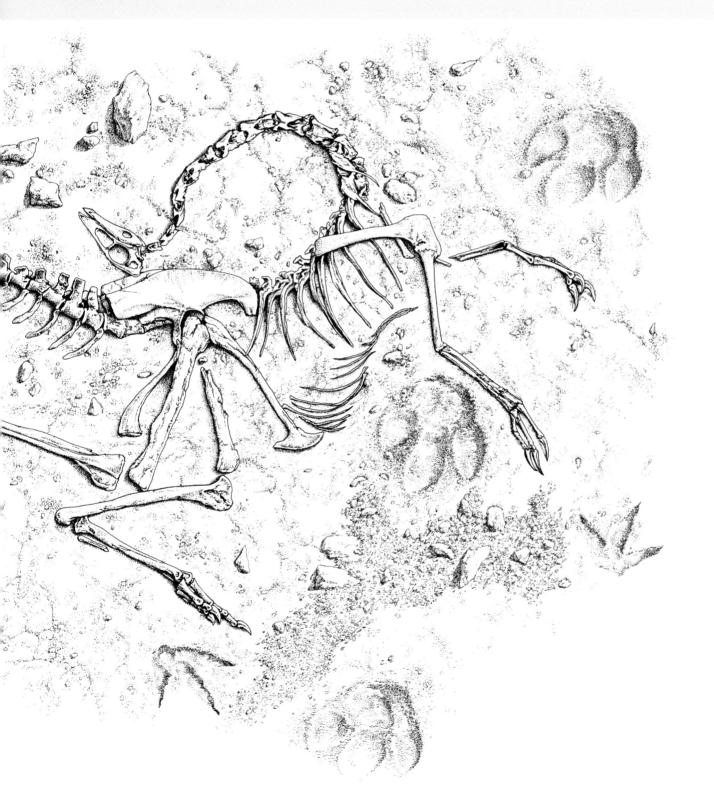

Dinosaur Tales

This book is a tale of dinosaurs; who they were, what they did, and how they did it. But, more significantly, it is also a tale of natural history. Dinosaurs enrich our concept of the biosphere, the three-dimensional layer of life that encircles the Earth. Our biosphere has a 3.8 *billion*-year history; we, and all other life around us, are merely Earth's latest tenants. To not know the history of life is to not know our organic connections to the biosphere and ultimately to our Earth. To know the history of life is certainly one way to begin to answer the perennial question, "Why are we here?". Dinosaurs have significant lessons to impart in this regard, because as we learn who dinosaurs really are, we can better understand who *we* really are.

Since the first dinosaur bones were identified in the early nineteenth century, both the total number of dinosaurs known, and the rate of finding them, has spiraled almost beyond anybody's expectations. By one estimate, new dinosaur species are reported at a rate of three per month, many coming from previously incompletely explored regions including China and Mongolia, Chile and Argentina, and, most recently, Saharan and sub-Saharan Africa (Figure I.1). We haven't done the math, but we're certain that there are far more trained paleontologists studying the fossil records than there ever were before.

Moreover, it is clear that dinosaurs are no longer your parents' (or is it their parents'?!) cast-iron clunkers. Their relationships to other "reptiles" – and even to us – are not what we thought; their lifestyles are not what we thought; they look very different from what we'd imagined; indeed, almost everything about them has changed since the late 1960s and, thankfully, they're a whole lot more interesting now. If you like dinosaurs, it's a great time to be alive: we're definitely living in a golden age of dinosaur discovery.

The Word "Dinosaur" in this Book

The term "dinosaur" (*deinos* – terrible; *sauros* – lizard) was invented in 1842 by the English naturalist Sir Richard Owen (see Box 16.2) to describe a few fossil bones of large, extinct "reptiles." With modifications (for example, not all dinosaurs are "large"), the name has proven resilient. It has become clear in the past 25 years, however, that not all dinosaurs are extinct; in fact, almost all specialists now agree that *birds are living dinosaurs*. The (technically correct) term nonavian

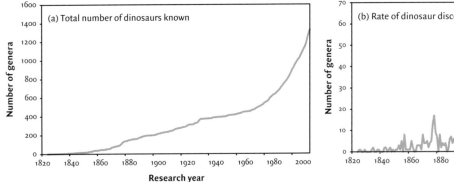
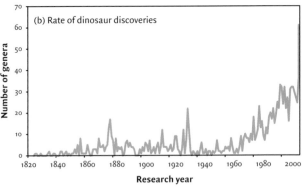

Figure I.1. Paleontologist Michael Benton's compilation of dinosaur discoveries from 1822 to 2014; we've extended his estimates to 2019. (a) Total number of dinosaurs known; (b) rate of dinosaur discoveries. There were spikes in the rate of dinosaur discoveries in the late 1880s and in the early to mid twentieth century, but all of that pales in comparison to what's happening now.

dinosaurs to specify *all* dinosaurs except birds is a mouthful, so we prefer to use the term "dinosaurs" as shorthand for "nonavian dinosaurs." The distinction between nonavian dinosaurs and all dinosaurs will be most relevant only when we discuss the origin of birds and their early evolution in Chapters 6, 7, and 8; there, we'll take care to avoid confusing terminology.

Science

Ours is also a tale of science, which is, in turn, all about imagination and creativity, with a little help from data. Creativity is the currency of science; integrity is the currency of the scientist. That is, the value of the science is surely dependent upon the creativity of the scientist; but it is equally dependent upon his or her integrity, because without personal integrity a scientist has nothing of value.

In the following pages, we hope to build a sense of the intellectual richness of science; a quality that philosopher of science Karl Popper called the "logic of scientific discovery." And so if this is a story about science, we'd better articulate what we mean by the word "science."

Science = Testing Hypotheses

Science is the business of constructing testable hypotheses and then testing them. An example of a simple scientific hypothesis is: "The Sun will rise tomorrow." This hypothesis makes specific predictions. Most importantly, the hypothesis that the Sun will rise tomorrow is testable; that is, it makes a prediction that can be assessed. The test is relatively straightforward: we wait until tomorrow morning and either the Sun rises or it doesn't.

Notice that here we're not looking for things that *support* our hypothesis; rather, we're looking for things that *test* it. If you were designing a race car, you would not drive it around the block and say, "See, it runs!". Rather, you would subject it to the grueling conditions that it might experience on a race track, to test whether or not the car was up the job. So it is with a scientific hypothesis: it needs to be *tested* and, like the car, continue to run (e.g., fail to be falsified) under the most extreme set of conditions.[1]

So what's *not* science? Important questions that you can't get at using science might be "Is there a God?", "Does she love me?", and "Why don't I like hairy men?". In *Music Man*, Marian "the librarian" Paroo asks "What makes Beethoven great?". She won't learn it from science. Some questions are far better suited to science than others.

The "Proof" is in... the Test!

In our example of a scientific hypothesis (above), if the Sun does not rise, the statement has been *falsified*, or demonstrated to be not correct, and the hypothesis can be rejected. On the other hand, if the Sun rises, the statement has *failed falsification*, and the hypothesis cannot be rejected. For a variety of relatively sophisticated philosophical reasons, scientists do not usually claim that they have *proven* a statement to be true; rather, the statement has simply been tested and not falsified. It turns out that, in a philosophical sense, it is very difficult to "prove" anything. For this reason, scientists rarely use words like "prove" or "true" when discussing their work.

---

[1] Notice our use of the word hypothesis. It is not some idea or conjecture, as it is used colloquially. Rather, in science, a hypothesis is a (potentially complex) explanation of a (potentially complex) phenomenon, to be tested and, if provisionally accepted, to resist falsification.

But we *can* test hypotheses, and one of the basic tenets of science is that it consists of hypotheses that have predictions which can be tested. We will see many examples of hypotheses in the coming chapters; to be valid, all must involve testable predictions. Without testability, it may be very interesting; it may be very exciting, it may be very important, but it is not science.

Science in the Popular Media

"Science" is everywhere in the popular media, from Animal Planet to *National Geographic*, to the Weather Channel. That's good; our lives are increasingly intersected by science, and the more savvy we are scientifically, the better the decisions that we can make. But the problem is that not all of the "science" that one sees in the popular media is very scientific!

Among scientists, work is recognized and accepted only after it is peer-reviewed; that is, when it has been vetted and carefully scrutinized by other scientists qualified to evaluate the work. Very commonly, the work goes through multiple cycles of review and revision by the authors in response to the critiques of professionals in the field. At that time – and only at that time – it can be published. This is the time-honored – and time-proven – way of producing the very highest-quality scientific research.

In the popular media, the goals are somewhat different. In movies, on television, and on the radio, the goals are generally entertainment as well as immediacy: publicizing the newest discoveries as quickly as possible. Both of these have very little to do with high-quality science. In the case of immediacy, the rush to publicize can mean that half-baked theories and interpretations are popularized long before they have seen anything like peer review. And when peer review finally comes, the results can be (and have been) devastating.

The goal of entertainment can be equally destructive, in terms of the science. For example, television and radio reports commonly present "the other side" of a particular scientific issue, even when the vast preponderance of scientists take a particular position and the "other side" has very little legitimacy. People can always be found who will voice heterodox opinions, even when their views are plainly way outside of the mainstream of peer-reviewed science. It's always *entertaining* to present controversy.

In paleontology, the quest for entertainment brings a variety of excesses, beyond mere controversy. Size sells; teeth and claws sell; asteroids sell; extinction sells. But ultimately, we hope in this book to show that the real riches – and entertainment – lie in a balanced, undistorted treatment of dinosaurs: a more magnificent, fascinating group of creatures could hardly have been invented, and the ongoing scientific process of revealing who they were and what they did is compelling drama.

The web, a popular and extremely important source of information, is driven by a variety of sometimes-contradictory imperatives (knowledge; entertainment; retail), and our relationship to it is more nuanced. Accessibility, ease, and breadth are its great strengths; reliability and depth can be, but, as always, care about sources is required. Much really good, peer-reviewed science (and lots of other information!) is available on the web; indeed, most major peer-reviewed journals have a significant web presence; some can be found *only* on the web. Information from the websites of professional societies (for example, the Society of Vertebrate Paleontology) is reliable in the way that the peer-reviewed journals published by such societies are reliable. Academic institutions tend to have reliable web pages; although personal web pages, even in an academic context, generally reflect the opinions of their authors, idiosyncratic or not. Wikipedia – a stupendous tool for research and, in our view, a stunning contribution to the availability of human knowledge – reflects the web: it is often very accurate, but it is not without idiosyncratic entries. In short, the web is a powerful tool, to be sure; but *caveat emptor*: Not all sources of information are equivalently authoritative. It is ultimately the longer, more considered process of peer review – of vetting by other specialists – that ensures that the best science trumps "fake news."

SELECTED READINGS

Fastovsky, D. E., Huang, Y., Hsu J., *et al.* 2004. Shape of dinosaur richness. *Geology*, **32**, 877–880. doi: 10.1130/G20695.1.

Starrfelt, J. and Liow, L.-H. 2016. How many dinosaur species were there? Fossil bias and true richness estimated using a Poisson sampling model. *Philosophical Transactions of the Royal Society B*, **371**, 20150219. http://dx.doi.org/10.1098/rstb.2015.0219

Wang, S. and Dodson, P. 2006. Estimating the diversity of dinosaurs. *Proceedings of the National Academy of Sciences*, **103**, 13601–13605. www.pnas.org_cgi_doi_10.1073_pnas.0606028103

Chapter

1 To Catch a Dinosaur

"Fossils are our planet's memory. Our Earth has written its autobiography in rocks and fossils."
- Kirk R. Johnson, Sant Director, National Museum of Natural History (Smithsonian).

WHAT'S IN THIS CHAPTER

This chapter contains an introduction to:

- How once-living organisms get preserved as fossils.
- How dinosaur fossils are found and collected from field sites.
- How they are readied for study and public viewing.

Preservation and Fossils

Fossils, the buried remains of organic life, are just about the only evidence we have of the kinds of living creatures that existed on Earth in the past. Fossils are how we know that there were creatures such as dinosaurs. And even that fact is near-miraculous: most organisms don't get fossilized, and undoubtedly only a few of the innumerable dinosaurs that must have once lived have been preserved.

There are many types of fossils: body fossils, which commonly involve some hard part of the animal (e.g., bones and teeth); trace fossils, which are impressions such as animal tracks; and fossils such as skin impressions that are a mixture of both. Until the very latest part of the twentieth century, paleontologists thought that dinosaur soft tissue – muscles, blood vessels, organs, skin, fatty layers, etc. – were long gone. We assumed that the hard parts would not be as easily degraded over time as the soft tissues. Yet, while this assumption was generally true, new techniques and determined study have revealed all kinds of unexpected preservation, including tissues, cells, and molecules, for example, the discovery of actual red blood cells and connective tissues from *Tyrannosaurus* (see Box 7.1). Fair warning: fossils are no longer just about old bones!

Making Body Fossils

Before Burial

Consider what might happen to a dinosaur – or any land-dwelling vertebrate – after it dies.

Carcasses are commonly disarticulated (dismembered), first by predators and then by scavengers, ranging from mammals and birds to beetles and bacteria. Some bones might be stripped clean of meat and left to bleach in the sun. Others might get carried off and gnawed. Sometimes the disarticulated remains are trampled by herds of animals, breaking and separating them further. Plants growing in soils produce acids, such as humic acid, that dissolve bone. But ultimately, the nose knows that the heavy lifting in the world of decomposition is done by bacteria that feast on rotting flesh. If the animal isn't disarticulated right away, it is not uncommon for a carcass to bloat, as feasting bacteria produce gases that inflate the dead body. After a bit, the carcass will likely deflate (sometimes explosively), and then dry out, leaving bones, tissues, ligaments, tendons, and skin like jerky, crusty and inflexible (Figure 1.1).

Figure 1.1. A horse carcass, lying, drying on a grassy plain. It is literally skin and bones, and if buried, it will be on its way to becoming fossilized.

Burial

Sooner or later bones are either destroyed or buried. If they aren't gnawed and digested as somebody's lunch, their destruction can come from weathering, which means that the minerals in the bones break down and the bones wash away. But the game gets interesting for paleontologists when weathering is stopped by rapid burial. At this point, they (the bones, not the paleontologists) become fossils. Figure 1.2 shows two of the many paths that bones might take toward fossilization.

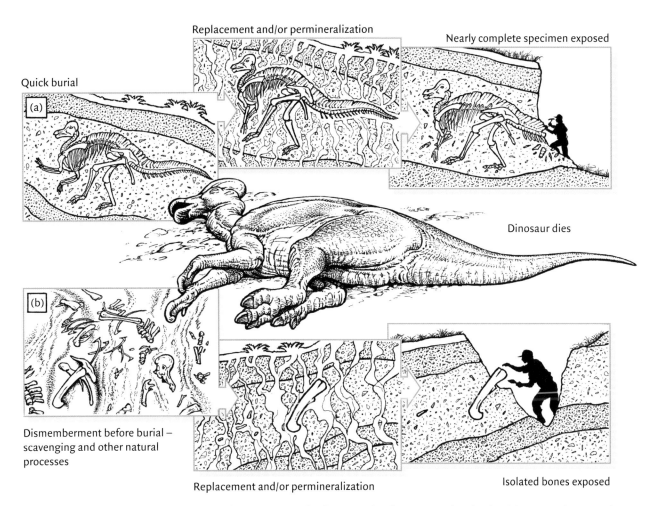

Replacement and/or permineralization

Nearly complete specimen exposed

Quick burial

(a)

Dinosaur dies

(b)

Dismemberment before burial –
scavenging and other natural
processes

Replacement and/or permineralization

Isolated bones exposed

Figure 1.2. Two endpoint processes of fossilization. In both cases, the first step is the death of the animal. Some decomposition occurs at the surface. In the upper sequence (a), the animal dies, the carcass undergoes quick burial, followed by bacterial decomposition underground, and permineralization and/or replacement. Finally, perhaps millions of years later, there is exposure. Under these conditions, when the fossil is exhumed, it is largely complete and the bones articulated (connected). This kind of preservation yields bones in the best condition. In the lower sequence (b), the carcass is dismembered on the surface by scavengers and perhaps trampled and distributed over the region by these organisms. The remains may then be carried or washed into a river channel and buried, replaced, and/or permineralized, eventually to be exposed perhaps millions of years later. Under these conditions, when the fossil is exhumed, it is disarticulated, fragmented, and the fossil bones may show water wear and/or the gnaw marks of ancient scavengers. Different conditions of fossil preservation tell us something about what happened to the animals after death.

After Burial

The burial environment is not chemically static. The *bad* news is that bone is made out of calcium–sodium hydroxyapatite, a mineral that when buried and subjected to concentrated ion-rich fluids deep in the Earth's crust, can easily and rapidly degrade (leaving no fossil to discover). The *good* news is that under the right chemical conditions, those same burial fluids develop a chemical environment that causes the original bone to be **replaced** by such minerals as fluorapatite, among others, ultimately producing a much more dissolution- and weathering-resistant product.

Replacement occurs on a spectrum. On one end, if no fluids are present throughout the history of burial to provoke a chemical reaction (a time interval that could be measured in millions of years), the bone may sit there unreplaced, which is to say that 100 percent of the original bone mineralogy remains. This, however, is rare in very old fossils.

On the other end of the spectrum, the chemistry of the burial environment over the same (or even smaller) amounts of time may produce complete replacement with no original bone material left. Given burial deep in the Earth's crust lasting millions of years, however, most dinosaur bones are altered (replaced) to a greater or lesser degree. Replacement tends to be progressively greater in the case of older fossils; and because the last nonbird dinosaur lived some 66.1 millions of years ago, some degree of replacement is generally the norm. This seems to have been the case with the recently discovered ornithopod dinosaur *Fostoria*, whose original bone was partially replaced with nothing less than opal (Figure 1.3)!

Fidelity to the shape and detail of the original bone occurs on a spectrum too. The replacement can be a crude replica of the original; however, replacement can also produce magnificent natural forgeries: chemically and texturally a **rock**, but retaining the exact shape and original delicate features, so as to appear indistinguishable from the bone it once was.

Since bones are porous, the spaces once occupied by soft tissues such as blood vessels, connective tissues, and nerves, easily fill up with minerals. This is called **permineralization** (Figure 1.4). Most fossil bones undergo a combination of some degree of replacement and permineralization.

Figure 1.3. A toe bone from the Australian dinosaur *Fostoria*, whose original bones were replaced with opal during burial.

Figure 1.4. Permineralized bone from the Jurassic-aged Morrison Formation, Utah, USA. The fossilized bone is now a solid piece of rock.

The preceding description explains how isolated fossil bones, or even part of a single animal, might come to be preserved. But along with these isolated finds, on rare and lucky days, we can sometimes come upon **bonebeds**, rich finds with hundreds – even thousands – of bones preserved. Sometimes, these bonebeds are **monospecific**, that is, they contain fossils of mostly one species,

Figure 1.5. Theropod dinosaur footprint from the Early Jurassic Moenave Formation, northeastern Arizona, USA. Human foot for scale.

and one can't help but wonder if these reflect some kind of gregarious, or herding behavior captured in the fossil record.

Trace Fossils

The single most important type of dinosaur fossil, other than the bones themselves, are trace fossils. Dinosaur trace fossils (sometimes also called ichnofossils [*ichnos* – track or trace]), come as footprints or as complete trackways. Figure 1.5 shows a mold, or impression, of a dinosaur footprint. We also find casts, which are made up of material filling up the mold. Thus a cast of a dinosaur footprint is a three-dimensional object that formed inside the impression (or mold).

In the past 40 years the importance of ichnofossils has been recognized. Ichnofossils have been used to show that dinosaurs walked erect, to reveal the position of the foot, and to reconstruct the speeds at which dinosaurs traveled. Trackways tell remarkable stories, such as that fateful day 150 or so million years ago when a large theropod dinosaur, perhaps an ornithomimid (Chapter 6) was attacked by a pack of smaller theropods (Figure 1.6).

Other Fossils

For want of a more imaginative name, we'll lump the various other kinds of fossils under "Other." Sometimes the fossilized feces of dinosaurs and other vertebrates are found. Called coprolites, these (in some cases impressive) relics can give an intestine's-eye view of dinosaurian diets. Likewise, as we shall see later in this book, fossilized eggs and skin impressions have been found; nests are known (Figure 1.7); molecules; cells; stomach stones (gastroliths; see Figure 6.18); and soft tissue; the list is really as long as there are parts of a dinosaur.

Figure 1.6. Tracks of a medium-sized theropod dinosaur among those of a pack of smaller theropods; from Shar-tsav, Gobi Desert, Mongolia. Our drawing suggests an interpretation consistent with the evidence: a *National Geographic* moment in the Late Cretaceous when a pack of *Velociraptor* attacked a single *Gallimimus*.

Bagging a Dinosaur

Finding Fossils

So, if the fossils are buried, how is it that we find them? The answer is that they're no longer buried when we find them, because the natural processes of the rock cycle have exhumed them, that is, brought the strata in which they are entombed back to the Earth's surface. Still, the act of finding them can be providential gift: if at the moment and place that some fossil-bearing sedimentary rocks are exhumed and erode, a paleontologist happens to be looking for fossils, the fossil *may* be observed and *may* be collected. Who knows how many times, throughout their 160+ million-year existence on Earth, dinosaurs stepped on exposed, weathering fossils of earlier dinosaurs, now surely lost to eternity (Figure 1.8)? Clearly then, choosing the place to look for fossils is a key part of the enterprise, as we discuss below.

It's a tenuous connection: the vagaries of fossil preservation, the chance of geological exposure, and the chance of their discovery by an ambitious paleontologist.

Figure 1.7. Fossil burrow of the dinosaur *Oryctodromeus*. Careful study of the sedimentary context of this dinosaur revealed the burrow.

Collecting

The romance of dinosaurs is bound up with collecting: exotic locales, heroic field conditions, and the manly extraction of gargantuan, fierce beasts (see Chapter 15). But ultimately dinosaur collecting is a process that draws upon good planning, a strong geological background, and a bit of luck. The steps are:

(1) *planning* (including field logistics and the science of where to look);
(2) **prospecting**; that is, hunting for fossils;
(3) **collecting**, which means getting the fossils out of whichever (usually remote) locale they are situated; and
(4) **preparing** and **curating** them; that is, getting them ready for viewing, and incorporating them into museum collections.

These steps involve different skills and sometimes different specialists.

Planning

Collecting dinosaur fossils should not be undertaken lightly. Dinosaur bones are – even in the richest sites – quite rare, and the moment they are disturbed the loss of important information becomes a concern. For this reason, most professional paleontologists have advanced degrees – often a PhD in

Figure 1.8. A pair of *Parasaurolophus* walking over some exposed fossilized bones of an earlier dinosaur that are weathering out of the cliff. Fragments of the fossilized bone have fallen at the dinosaurs' feet.

the geological or biological sciences – but before actually leading an expedition themselves, *all* have acquired many years of experience both in the logistical as well as the scientific ends of fieldwork.

Logistics of an Expedition

You can't collect much of a dinosaur by yourself. Maybe an isolated bone or tooth here and there, but anything like a skeleton? Not really. You'll need a team: the rarity of dinosaurs (lots of eyes to look for them); their size and complexity (the time that it takes to jacket all the pieces); and the physical requirements of collecting them (the mechanics of jacketing and then their transport), requires more than one person (OK, you can be in charge!). The logistical end of an expedition involves keeping one's team fed, watered, healthy, and happy in remote places where, in many cases, these commodities don't come easily. Sun, heat, dust, bugs, lack of amenities, a limited diet, and isolation from the "real world" all conspire to wear down even the most enthusiastic. It's the Great Outdoors, true, but

Figure 1.9. Preparation for one of the American Museum of Natural History's 1920s expeditions to the Gobi Desert. In the intervening 100 years, nobody has found a way to get around hauling the basic necessities into the field. The vehicles are newer, but the basic planning and preparation processes are the same.

it's no camping catalog! Add to these, language problems when you are working in other countries and limited access to medical facilities in the event of an accident involving either you or one of your crew, and the potential for disaster increases dramatically.

Work in many localities requires complete self-sufficiency: fuel, water, food, all gear for the maintenance of daily life – as well as all the maps and equipment necessary to successfully carry out the science and safely retrieve heavy, yet delicate, dinosaur bones. This takes some planning and experience; you and your crew's lives may depend upon it (Figure 1.9). You have to know what you are doing.

Fossils generally, and dinosaurs in particular, are not renewable resources, which means that collecting a dinosaur is a one-shot deal: it must be done right, because we will never be afforded another chance to collect that specimen again. Any information that is lost – any piece of it that is damaged – may be lost or damaged forever. For this reason, there are many regulations associated with collecting vertebrate fossils.

The most basic are the collection permits required for work on public lands. Obtaining the permits requires advanced planning because the agencies in charge of issuing the permits reasonably require detailed accounts of your plans before the process can go forward.

One important part of the permit-obtaining process, especially in the case of dinosaur fossils (which tend to be large and heavy), is the eventual location of the fossils. Who gets them? Does that person or place have the proper resources – or even the space – to store, preserve, and make them accessible to scientists and the general public? How is all this to be accomplished? Most of the truly great collections and many of the most important dinosaur fossils are housed in major museums, such as the American Museum of Natural History (New York), the Field Museum (Chicago), the Yale Peabody Museum (New Haven, CT), Tyrrell Museum (Alberta), the Smithsonian (Washington, DC), the Natural History Museum (London), the Musée National d'Histoire Naturelle (Paris), and many others around the world. These institutions have the resources required for the care of important specimens and the data associated with them.

Work on private lands can be as tricky as work on public lands. Ownership of specimens can be contentious, and when the specimens are potentially worth millions of dollars, a simple handshake or a verbal agreement may not be nearly enough. That was certainly the case for the notorious *Tyrannosaurus rex* specimen now known as "Sue" (Box 1.1).

Box 1.1 A Dino Named "Sue"

"Sue" (Figure B1.1) is a spectacular fossil of a very large *T. rex* – and a notorious object lesson in the kinds of misunderstandings and problems that can arise when collecting dinosaurs. Paleontologist Peter Larson, cofounder and President of the Black Hills Institute of Geological Research (BHIGR; a company that collects, prepares, and sells fossils and casts), believed that he had obtained permission from South Dakota rancher Maurice Williams for the rights to collect (and then, presumably, prepare and sell) fossils he and his crews might find on Mr. Williams' land. They agreed that BHIGR would pay Mr. Williams for any fossils that he found. In July, 1990, Sue Hendrickson, a volunteer BHIGR crew member, discovered the dinosaur (hence the name).

It was *very* rare – a *T. rex*: somewhat jumbled, but clearly large, semi-articulated, well-preserved, and obviously extremely valuable. Larson and his crews devoted considerable resources (ultimately, around $209 000) and time to collecting the many parts of the specimen (the skull alone was almost 2 m long!), starting with shaving almost 10 m off of the top of the hill in which it had been found so that they could reach the bones. By the end of the field season, they had brought a very large chunk of the specimen back to their laboratory in South Dakota, where preparation began. In accordance with Mr. Larson's understanding of the original agreement, a check for $5000 was issued to Mr. Williams for rights to the specimen. Talks were presented at national scientific meetings; publications were planned; preparation continued apace; so far, it seemed as if it was an exciting specimen, but not a socio-political phenomenon.

Then the patient went septic. Mr. Williams claimed that he had not given BHIGR rights to the specimen; only that they were allowed to remove the fossil and prepare it. Complicating the issue, because Mr. Williams is a member of the Sioux Tribe, the Sioux Nation claimed that the fossil belonged to it. And as if that weren't difficult enough to sort out, the US Federal Government claimed that Mr. William's land was held in trust by the United States, and therefore the fossil belonged to the United States as represented by the Department of the Interior!

In 1992, without warning, the Feds moved in: guns, FBI agents, police.[a] Somewhat unceremoniously (by all accounts), the FBI searched the place, and the US National Guard forcibly removed Sue's remains from the BHIGR, storing them at the University of South Dakota School of Mines and Technology, a well-known facility able to handle a specimen of this magnitude – and equally importantly, presumably, neutral ground.

Then the courts took over. The dinosaur languished. After a rather lengthy trial, the court returned the specimen to Mr. Williams, who sought to auction it through the famous auction house of Sotheby's. In the meantime, Mr. Larson served jail time for permit infractions, unrelated to Sue; however, there are those who claim that the charges and subsequent sentencing were in effect political.

The value of the specimen was undoubted, and the concern was that it would leave the United States and end up in a private collection somewhere where it could not be appreciated. So, an unholy alliance formed between the Field Museum of Natural History in Chicago, the California State University System, Disney Parks, McDonalds, the Ronald McDonald House, and private donors, who all pitched in to see the fossil displayed at the Field Museum, which won the bid for a total output of about $8.3 million.

Preparation was carried out both at the Field Museum and at Disneyworld in Orlando, Florida, where thousands of visitors to both venues saw the bones being freed of matrix, cast, and readied for mounting. Sue is now the star attraction of the Field Museum. All's well that ends well, right?

The "Sue" story is perhaps the most egregious example of the kinds of things that can go awry in paleontological collecting. By most accounts it ended well: the dinosaur is safe, intact – estimates are that it is an almost unbelievable 80% complete – and on display where it is properly maintained and optimally appreciated. On the other hand, maybe things didn't end so well: Sue changed the landscape of dinosaur paleontology, because now everybody recognizes that dinosaur fossils, formerly thought to be valueless, can potentially be sold for millions of dollars. Collecting dinosaurs has become an expensive business, and many scientists no

---

[a] Larson quotes an FBI agent as saying, "This can be real easy, or real hard, depending on whether or not you are willing to cooperate." We favor Robert De Niro for the role.

Box 1.1 *cont'd*

longer have the resources to stay in the game. And how clear that became on October 6, 2020, when "Stan," a second spectacular *T. Rex* specimen, sold at Christie's auction house for $31.8 million . . . to an anonymous bidder!

Figure B1.1. The mighty *T. rex* specimen "Sue," collected on a ranch in South Dakota, and battled over by the ranch owner, the collector, the Sioux Nation, and the United States Department of the Interior. It ultimately landed in the Field Museum of Natural History, in Chicago, IL, having been auctioned for a record $8.3M.

Work overseas – and paleontology can involve a lot of foreign travel, no matter where you live – requires a whole new level of administrative preparation. All of the problems described above are compounded by language barriers, by the necessity to obtain visas along with permits, by the logistics of preparing a field expedition in a foreign country, and by the necessity of arranging for the eventual disposition of the fossils. What country, after all, would gladly see its fossil resources dug up and exported elsewhere? It's a delicate balance, sometimes requiring the skills of a diplomat.

Science of Where to Look

All of that care expended upon all those logistics is meaningless unless our planning involves science as well. Paleontologists don't just go to weird places and grab bones. If they did, they'd lose, forever, essential information bearing upon four major problems:

(1) What kind of environment was it in which the dinosaur was preserved?
(2) Where did it live (geographically)?

(3) When it did it live?

(4) How did it die?

Oryctodromeus, a small burrowing, herbivorous dinosaur (see Figure 1.7), is a perfect example of the importance of geological context. Here was an animal found fossilized in its own burrow. Had the important geological context not been properly interpreted, the burrow would not have been recognized and this animal's unusual behavior (for a dinosaur, at least) would have gone unappreciated.

So before even collecting the fossil, the locality – the area in which the fossil or fossils occur – must be mapped geologically, in a way that records the most information possible about the setting in which the fossil was found. This kind of information requires specialized geological study of the paleoenvironments, that is the ancient environments represented by the rocks in which the fossils are found, as well as the geological context above and below the fossil. Usually this is accomplished by geological mapping and by detailed study of the sedimentary geology of the locality. Interpreting the ancient environment in which the bones came to rest commonly involves teaming up with sedimentologists – geoscientists with specialized knowledge of sedimentary rocks and the ancient environments that they preserve. This kind of teamwork allows paleontologists to develop the most complete picture of the fossils and the conditions in which they lived and died.[1]

A question commonly asked, is "How do you know where the dinosaur fossils are?". The simplest answer is, "We don't." There is no secret, magic formula for finding dinosaurs, unless it is long, hard hours of preexpedition library time and careful assessment of potentially productive – that is, fossil-bearing – regions. On the other hand, a well-educated guess rooted in knowing something about the kinds of environment in which dinosaurs lived can greatly increase the odds of finding fossils.

Some basic criteria help the success of the search. These are:

(1) Right rocks: the rocks must be sedimentary.

(2) Right time: the rocks must be of the right age.

(3) Living on land: the rocks must be terrestrial.

Right Rocks

Sedimentary rocks have the best potential to preserve dinosaur fossils. Indeed, sedimentary rocks form in, and represent, sedimentary environments, many of them places where dinosaurs lived and died. Dinosaurs are known from other types of rocks, but their fossils are most likely found in sedimentary rocks.

Right Time

Dinosaurs first appeared during the Late Triassic Period and went extinct at the end of the Cretaceous Period. If the rocks you search were *not* deposited sometime between the Late Triassic and the latest Cretaceous, you won't find dinosaurs. Older and younger rocks will yield amazing fossil creatures, but not dinosaurs.

Living on Land

Dinosaurs were terrestrial, that is, nonmarine beasts through and through, which means that their bones will generally be found in rocks that preserve the remnants of ancient river systems, deserts, and deltas. However, dinosaur remains are also known from lake deposits and from near-shore marine deposits.

---

[1] The study of all that happens to an organism after its death is called "taphonomy," and is a specialized field combining sedimentology and paleontology. Understanding the taphonomy of a fossil is the best way to know whether the animal actually lived in the environment in which its fossils were found, or whether its carcass was just deposited there after death.

Many of the richest fossil localities in the world are in areas with considerable rock exposure, such as badlands, rugged landscapes that are deeply sculpted by rivers. Fossil localities are common in deserts: plant cover on the rocks is low, and the dry air slows down the rates of weathering so that, once a fossil is exposed, it isn't chemically destroyed or washed away. Paleontologists, therefore, don't often find themselves in the jungle looking for fossils; the weathering rates are too high and the rocks are covered by vegetation. The chances of finding fossils are best in relatively dry regions. Still, not all dinosaur material by any means has been found in deserts. As long as the three criteria above are met, there is a possibility of finding dinosaur fossils, and that's usually reason enough for going in and taking a look.

Because fossils are a nonrenewable resource, collecting them should be treated with the utmost circumspection. Poor planning, indifference to the significance of fossils, lack of training, and ignorance when retrieving them from the ground, will at best lose important data, and at worst place you and your team's lives in jeopardy.

Prospecting

Once we've planned properly, and we've got ourselves, our team, and our equipment to outcrops that match the criteria above, and once we have developed a concept of their geological context, what then? Simply put, we drop our eyes and start searching for bone weathering out of the rock. So, when people talk about going on a "dinosaur dig," it is a misnomer. Nobody digs into sediment to find bones, although you might dig a bit to excavate them; bones are found because they are spotted weathering out of sedimentary rocks. If we're lucky and/or have a good eye, we'll spot something. Some field paleontologists just find more and better specimens than others, and a "feel" borne of experience (and perhaps some inherent talent) is surely involved in finding bone, as well as an experienced eye and often luck (Figure 1.10). Ace Los Angeles County Museum field paleontologist Harley Garbani used to say that the fossils "whistled" to him. But in the end, the more enthusiastic eyes you can bring to an outcrop, the more likely you are to find whatever is there.

Collecting

Collecting is the arena in paleontology in which finesse meets brute force. Delicacy is required in preparing the fossils for transport; raw power is required for lifting blocks of bone and matrix (the rock which surrounds the bone) – commonly weighing many hundreds of pounds – out of the ground

Figure 1.10. Field paleontologist extraordinaire René Hernandez-Rivera bags another dinosaur: duck-billed dinosaur vertebrae from the Late Cretaceous of Baja California, México. His very successful track record of finding spectacular fossils suggests that he has a basic affinity for this kind of work. Photo courtesy of Maria Luisa Chavarría.

and back to civilization. Because of their size and delicacy, most dinosaur fossils are encased in a rigid **jacket**, or protective covering. Figure 1.11 shows how this is done.

Transport out of the field can be difficult, depending upon the size and weight of the jackets. A small jacket (soccer-ball size) can be carried out easily enough by a few dedicated, beefy people;

(a) A fossil is found sticking out of the ground; now it needs to be cleaned off so that its extent can be assessed. Exposing bone can be done with a variety of tools, from small shovels, to dental picks, to fine brushes. As the bone is exposed, it is "glued"; that is, impregnated with a fluid hardener that soaks into the fossil and then hardens.

(b) The **pedestal**. When the surface of the bone is exposed, the rock around it is then scraped away. For small fossils, this can be quite painless; however, for large fossils, this can mean taking off the face of a small hill. Anyway, this process continues until the bone (or bones) is sitting on a pedestal, a pillar of matrix underneath the fossil.

(c) Toilet paper cushion. Padding is placed around the fossil to cushion it. The most cost-effective cushions are made from wet toilet paper patted onto the fossil. It takes a lot of toilet paper: for example, a 1 m thigh bone (femur) could take upward of one roll. On the other hand, this is not a step where we should cut corners, because returning from the field with a shattered specimen, or one in which the plaster jacket is stuck firmly to the fossil bone, is not so good.

(d) Plaster jacket. Jackets are made of strips of burlap cloth soaked in plaster, and then applied to the toilet-paper-covered specimen. A bowl of plaster is made up, and then precut, rolled strips of burlap are soaked in it and then unrolled onto the specimen and the pedestal.

Figure 1.11.

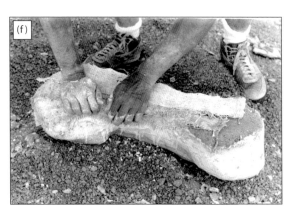

(e) Turning the specimen. After the plaster jacket is hardened, the bottom of the pedestal is undercut, and the specimen is **turned**; that is, separated at the base of the pedestal from the surrounding rock and turned over. This is a delicate step in which the quality of the jacket is tested.

(f) The top jacket. More plaster and burlap are then applied to the open (former) bottom of the jacket, now its top. At this point, the fossil is fully encased in the plaster-and-burlap jacket, and is ready for transport from the field.

Figure 1.11. (*cont.*)

but large jackets can require braces, hoists, winches, cranes, flatbed trucks, front-end loaders, helicopters, and even freight cars (as well as the dedicated people!).

Back at the Ranch: Preparing and Curating

Once the fossil dinosaur bone is out of the field and back where it can be studied, the jacket must be cut open, and the fossil prepared, or freed from the matrix. This runs from simple brushing, to scraping with dental needles, to sophisticated treatments such as acid removal of the rock matrix. These techniques are generally carried out in a preparation laboratory (or prep lab; Figure 1.12) by preparators, professionals whose job it is to free the fossil from its rock matrix and stabilize it for study and display. The work is very long, difficult, requires great skill, considerable knowledge, and real patience. When the bone is harder than the matrix, the work can go smoothly; but when the bone is softer than the matrix, the work can be extremely taxing. Fine details require special attention (not to say skills!), and much of the work of preparation must be done under a large, illuminated magnifying lens. Every paleontologist knows it: preparators are the unsung heroes of paleontology.

People expect to see free-standing displays in a museum. But, while mounts of real fossil bone are attractive, they are also extraordinarily heavy (remember, those are rocks!), time-consuming and costly to put up, and the steel frames that support the bones, as well as the process of mounting them, can be destructive to the fossil. Moreover, mounted specimens commonly undergo damage over time; slight shifts in the mounts because of the extraordinary weights of the fossil bones, or vibrations in the buildings in which the bones are housed, or museum patrons lifting apparently "insignificant" bits all diminish the quality of mounted specimens. In addition, when the specimens are assembled and mounted, they can be hard to examine for study.

Many museums, therefore, cast the bones in fiberglass and other resins, and display the casts. When well done, such displays are visually virtually indistinguishable from the originals, even to a professional. With their light weight, and the possibility of internal frames, they can be spectacular and dynamic (Figure 1.13) in a way that is not possible if the fossils are used in the display.

Figure 1.12. Scenes from a prep lab, in this case, that of the Museum of Northern Arizona in Flagstaff, USA. (a) Foreground: a large plaster jacket containing the theropod dinosaur *Coelophysis*. Background: sand tables for stabilizing specimen fragments (to be glued), open jackets, storage shelves, and grinding equipment. (b) The jacket shown in (a). The *Coelophysis* specimen is visible in the foreground. The arms and hands are to the left; the pelvis, legs, and tail are visible to the right. (c) A dinosaur skull (*Pentaceratops*) laid out for study. In the background are the large bays in which specimens are stored.

Leaving the bones disarticulated, properly curated, and available for study maximizes returns on the very substantial investments that are involved in collecting dinosaur remains. Paleontology is carried out in large part by public support, and mounted casts give the public the best value for money.

Stickershock

Dinosaurs are expensive. Here is a list of the variables whose expense must be assessed before people get to see a fossil on display:

- How many people are in the crew;
- Difficulty of travel to the field area;
- How many weeks the crew spends in the field;
- How much material is collected;
- The cost of transport to a preparation laboratory;
- The time for preparation; and
- The nature of the display, including, potentially, the casting of missing bones to make a complete mount.

In the case of a small dinosaur, maybe the plaster jackets can be completed in a couple of days of fieldwork. In the case of a really big dinosaur, it could take up to five field seasons (years) for your

Figure 1.13. A spectacular mount of the sauropod *Barosaurus* and the theropod *Allosaurus*. This mount is made of fiberglass and epoxy resin, cast from the bones of the original specimens. A dynamic pose like this would not have been possible using the original fossil bones.

team to completely jacket all the pieces and get them out of the field; and that's before preparation, casting, and mounting, all of which would more than double the time of the fieldwork. No wonder dinosaurs are perceived as valuable by private collectors and, given the right dinosaur, attract prices in the millions of dollars.

SUMMARY

Fossils, the buried remains of organic life, are divided into body fossils, trace fossils, and a grab bag of other types of fossils. The body fossils include bones, shells, and other organic remains; trace fossils consist of tracks, trackways, and other impressions in the form of molds and casts. "Other" includes everything (and anything) else.

Fossilization is a process that occurs after the organism dies. It consists of burial, and commonly involves a variety of types of partial or complete replacement, in which the original organic and mineral material of the once-living organism is naturally replaced by other minerals while buried.

Obtaining fossils, particularly dinosaur fossils, requires rigorous training and preparation, along with educated guessing. Five steps are involved: planning, prospecting, collecting, and laboratory preparation and curation. The planning ranges from choosing where to look, to getting the legal permission to carry out the study, to outfitting an expedition properly to safely meet its goals. The prospecting requires a well-trained, experienced eye. Collecting is a process designed to bring delicate fossils safely back to where they can be prepared, which involves cleaning, reconstruction, and protection. Curation makes it possible for fossils to be safely stored on the long term, and for them to be accessible to researchers and to an interested public alike.

Dinosaurs are costly.

SELECTED READINGS

Behrensmeyer, A. K. and Hill, A. P. (eds.) 1980. *Fossils in the Making.* University of Chicago Press, Chicago, IL, 338 pages.

Cvancara, A. M. 1990. *Sleuthing Fossils: The Art of Investigating Past Life.* John Wiley and Sons, New York, 203 pages.

Farlow, J. O. 2018. *Noah's Ravens: Interpreting the Makers of Tridactyl Dinosaur Footprints.* Indiana University Press, Bloomington, 643 pages.

Farlow, J. O., Chapman, R. E., Breithaupt, B., and Matthews, N. 2012. The scientific study of dinosaur footprints. In Brett-Surman, M. K., Holtz, T. R. Jr., Farlow, J. O., and Walters, B. (eds.) *The Complete Dinosaur.* Indiana University Press, Bloomington, pp. 713–759.

Gillette, D. D. and Lockley, M. G. (eds.) 1989. *Dinosaur Tracks and Traces.* Cambridge University Press, New York, 454 pages.

Horner, J. R. and Gorman, J. 1988. *Digging Dinosaurs.* Workman Publishing, New York, 210 pages.

Kielan-Jaworowska, S. 1969. *Hunting for Dinosaurs.* Maple Press Co. Inc., York, PA, 177 pages.

Larson, P. and Donnan, K. 2004. *Rex Appeal: The Amazing Story of Sue, the Dinosaur That Changed Science, the Law, and My Life.* Invisible Cities Press Llc, Monpelier, Vermont, 424 pages.

Lockley, M. G. and Hunt, A. P. 1995. *Dinosaur Tracks and Other Footprints of the Western United States.* Columbia University Press, New York, 338 pages.

Martin, R. E. 1999. *Taphonomy: A Process Approach.* Cambridge University Press, Cambridge, 524 pages.

Rogers, R. and Eberth, D. A. 2008. *Bonebeds: Genesis, Analysis, and Paleobiological Significance.* University of Chicago Press, Chicago, 512 pages.

Sternberg, C. H. 1985 (originally published 1917). *Hunting Dinosaurs in the Badlands of the Red Deer River, Alberta, Canada.* NeWest Press, Edmonton, Alberta, 235 pages.

TOPIC QUESTIONS

1. Define: fossil, dinosaur, replacement, preparation, coprolite, biosphere, body fossil, molds, casts, and ichnofossils.
2. What kind of training does it take to be a paleontologist?

3. What steps are involved in the collection of dinosaur bones?
4. What criteria maximize the likelihood of finding dinosaur fossils?
5. What kinds of paleoenvironments are most likely to preserve dinosaur bones? Why?
6. Why is it that, generally, the older the fossil, the less likely it is that original bone material will be preserved?
7. Why is understanding the geological context so important when collecting dinosaurs?
8. Why do dinosaur fossils tend to be preserved better in deserts than elsewhere?
9. What kinds of conditions might be required to find a fully articulated dinosaur fossil?
10. What kinds of geological activities are important to carry out before a fossil is extracted from the ground? Why?
11. If you were to look for a previously undiscovered dinosaur, (a) what paleoenvironmental features in the search area would you look for? (b) What must the ages of the rocks be? (c) What type of rocks should they be? (d) Where on Earth would you go to find a place with these qualities?

2 Dinosaur Days

WHAT'S IN THIS CHAPTER

This chapter contains introductions to:

- Deep (geological) time.
- Moving continents.
- Global climate change (it's not just for the twenty-first century!).

When Did Dinosaurs Live (and How Do We Know)?

Fossils, including dinosaur remains, are found in layers of rock that are commonly called strata (singular, stratum). The study of strata, called stratigraphy, is the geological specialty that tells us how old or young particular strata, and the fossils contained within them, are. Thus, stratigraphy is a technique by which we can learn the age of dinosaur fossils. Stratigraphy is divided into geochronology (*geo* – Earth; *chronos* – time), which gives geologists the numerical age of a rock; lithostratigraphy (*lithos* – rock), and biostratigraphy (*bios* – organisms), both of which give the age of something in relation to something else (older; younger, or same age). Applying these tools, we can learn something about when dinosaurs lived, a question that is as fundamental for dinosaur (pre) history as it is for human history.

Geologists generally signify time in two ways: numerically, in numbers of years before present, and by reference to blocks of time with special names (like "Triassic" and "Mesozoic"). For example, we say that the Earth was formed 4.6 billion *years before present*, meaning that it was formed 4.6 billion *years ago* and is thus 4.6 billion years old. Unfortunately, determining the precise age in years of a particular rock or fossil is not always easy, or even possible. For this reason, geologists have divided time into blocks of varying lengths, and rocks and fossils can be referred to as belonging in these blocks of time, depending upon how exactly the age of the rock or fossil can be estimated. For example, you might not know that a fossil was 92.3 million years old, but you might be able to determine that it was within the interval of time known as the Late Cretaceous, meaning that its age is somewhere between 100.5 and 66.1 million years old, dates about which you have more information.

We start our discussion intuitively, by talking about the age in years, or the numerical age. Later we will address the division of time into blocks of varying lengths.

Geochronology

The Ages of the Ages

Geoscientists are happiest when they can learn the numerical age of a rock or fossil; that is, its age in years before present. Ages in years before present are reckoned from the decay of unstable isotopes found in certain minerals (see Appendix 2.1 for a brief review of the chemistry behind isotopes). Unstable isotopes spontaneously decay from an energy state that is not stable to one that is more stable (that is, they "want" to change from a higher-energy state to a lower-energy molecular configuration). The change from unstable to stable thus releases energy (as promised!), and occurs over short or long amounts of time, depending upon the isotope. The basic decay reaction for all unstable isotopes runs as follows:

$$[\text{Unstable "parent" isotope} \rightarrow \text{stable daughter isotope}$$
$$+ \text{ nuclear emissions} + \text{energy (generally, heat)}]$$

Here is an example: The element potassium (K; atomic number 19; atomic weight 39). Potassium has a relatively rare unstable isotope, ^{40}K, which has 19 protons and 21 neutrons. When ^{40}K decays, a proton in the nucleus of the atom captures an electron (the process is called, unsurprisingly, "electron capture"), and in so doing, it makes a neutron, which decreases the atomic number from 19 to 18 (a proton is lost when it became a neutron). So now we're no longer dealing with the element potassium (atomic number 19); we're dealing with argon (Ar; atomic number 18). However, the atomic weight stays at 40, since no mass has been lost (a proton was simply converted to an electron; see Table 2.1).

Table 2.1 **Protons, neutrons, atomic number, and atomic weights of isotopes of potassium and argon.**			
	Stable K isotope	Unstable K isotope	Stable Ar isotope
	^{39}K	^{40}K	^{40}Ar
Protons	19	19	18
Neutrons	20	21	22
Atomic weight	39	40	40
Atomic number	19	19	18

Following the general pattern of decay reactions (above), this reaction can be written as follows:

$$[^{40}K \text{ (unstable parent)} \longrightarrow {}^{40}Ar \text{ (stable "daughter") + energy}]$$

The *rate* of the decay reaction is the key to obtaining a numerical age. If we know:

(1) the original amount of parent isotope at the moment that the rock was formed or the animal died (before becoming a fossil);
(2) how much of the parent isotope is left; and
(3) the rate of the decay of that isotope,

we can estimate the amount of time that has elapsed. For example, suppose we know that only 30 percent of an unstable isotope remains in a particular rock. If we know the rate at which the element decayed, we can estimate the amount of time that has elapsed since the rock was formed; that is, the *age of the rock*. The age of the rock would be equal to the time that it took for 70 percent of the unstable parent isotope to decay.

It is convenient to summarize the rate of decay by a single number. That number is called the half-life, which is the amount of time that it takes for 50 percent of the atoms of an unstable isotope to decay (leaving half as much parent as was originally present). These relationships are shown in Figure 2.1.

Choosing the Right Isotope

Different unstable isotopes have different, but constant, rates of decay,[1] which means they have different half-lives. So, the idea is to pick an isotope with a half-life (representing its rate of decay) that is appropriate to the amount of time you are interested in. For example, when people think of geological dating, they commonly think "carbon-14" (^{14}C). But when it comes to dinosaurs, ^{14}C is definitely *not* the way to go. Dating dinosaur bones using ^{14}C, which has a half-life of 5730 years, would be like giving your own age in milliseconds. Dinosaurs lived tens to hundreds of millions of years ago (Ma),[2] and after so many half-lives had elapsed, there wouldn't be nearly enough isotope left to measure. Conversely, to date human remains, not likely more than several thousand years old,

---

[1] The rate at which individual molecules decay fluctuates in the short term but is statistically constant over long periods of time.
[2] Ma, from *mille annos* – a million years; we'll use the Ma notation throughout this book.

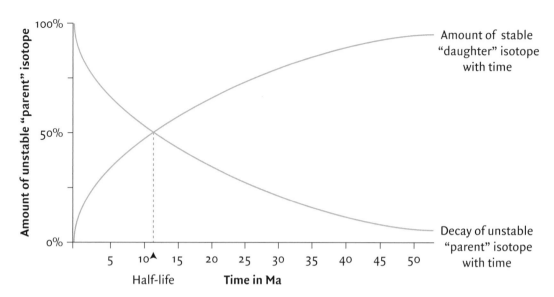

Figure 2.1. An isotopic decay curve. Knowing the amount of unstable isotope that was originally present, as well as the amount of unstable isotope now present and the rate of decay of the unstable isotope, it is possible to determine the age of a rock with that isotope in it. Suppose we found a rock with a ratio of 25 percent unstable parent: 75 percent stable daughter of a particular isotope. That would mean *two* half-lives had elapsed (half-life no. 1 = 50 percent of 100 percent parent [50 percent parent : 50 percent daughter]; half-life no. 2 = 50 percent of 50 percent parent [25 percent parent : 75 percent daughter]). The amount of time represented by two half-lives can be read on the axis marked "Time in Ma"; in this case, about 26 million years. The rock would thus be about 26 million years old.

the rubidium/strontium isotopic system (^{87}Rb/^{87}Sr), with a half-life of 48.8 billion years, would hardly be ideal. This would be a bit like timing a 100-m dash with a sundial.

Unstable isotopes are powerful dating tools, but they cannot be used directly to date dinosaur bone, because they are not generally found in the fossils themselves. Most unstable isotopes useful for dating commonly (although they are still extremely rare) form as lava cools and the minerals crystallize in the magma chamber of a volcano. The decay process begins when the unstable isotope is first formed (that is, when it crystallizes in the lava), so, the decay clock starts ticking, so to speak, from the moment the crystal formed as the lava cooled.

But no dinosaur ever lived *within* the magma chamber of a volcano (not even the *Jurassic World* superdinosaurs), so how can we get the age of the dinosaur bone when all we have is a date for some lava? The answer to that is the province of lithostratigraphy.

Lithostratigraphy

Sedimentation and Sedimentary Rocks

Sediments – sand, silt, mud, dust, and other less-familiar materials – are deposited over geographic scales of square meters (m^2) to hundreds of square kilometers (km^2) and are the direct result of sedimentation such as flowing water, wind, or explosion of a volcano, to name a few more or less common processes. Virtually every geographical location we can think of – a river, a desert, a lake, an estuary, a mountain, the bottom of the ocean, the *pampas* – has sedimentary processes peculiar to it that will produce distinctive sediments and, with time and burial, distinctive sedimentary rocks. These geological processes – most commonly, wind and water – deposit sediments in strata. The strata pile up on one another, stacking into thick sequences of sedimentary rocks (Figure 2.2).

Figure 2.2. Superposition of strata exposed by the downcutting of the Colorado River, Dead Horse Point State Park, UT, USA. The sequence of rock layers represents around 100 million years of sedimentation.

Relative Dating

It is a fact that younger sediments are deposited upon older sediments (exemplified in Figure 2.2), and yet this apparently self-evident insight is the fundamental basis of all correlations of sedimentary strata in time. Determining which of two strata is older than the other is termed **relative dating** and, while not providing the age in years before present, provides the age of one stratum *relative* to another stratum. Here, then, is part of the solution to dating dinosaur bone. Suppose that a stratum containing a dinosaur bone is sandwiched between two layers of volcanic ash. Ideally, a numerical age date could be obtained from each of the ash layers. We would know that the bone was younger than the lower layer but older than the upper layer. Depending upon how much time separates the two layers, the bone between them can be dated with greater or lesser accuracy (Figure 2.3).

But how can one tell that two geographically separated deposits were deposited at the same time if numerical ages are unknown? In this, fortunately, stratigraphers are aided by one last, extremely important tool: biostratigraphy.

Biostratigraphy

Biostratigraphy is a method of relative dating that utilizes the presence of fossil organisms. It is based upon the idea that a particular time interval can be characterized by a distinctive **assemblage**, or group, of organisms. For example, if one knows that dinosaurs lived from around 230 Ma to 66 Ma, then any rock containing an identifiable dinosaur fragment must fall within that age range. Although biostratigraphy cannot provide ages in years before present, the fact that many species of organisms have existed on Earth for 1 to 2 million-year intervals enables them to be used as powerful dating tools. For example, *Tyrannosaurus rex* lived for only around 1 million years, from 67 Ma to 66 Ma. Therefore when Sue Hendrickson found the *T. rex* "Sue" in South Dakota (Box 1.1), she knew that it could be correlated with *T. rex*-bearing sediments in Montana that have been well dated at 67 Ma to 66 Ma, or any other location where *T. rex* has been found.

Eras and Periods and Epochs, Oh My!

Geological time is hierarchical, much as our time is divided into years, months, weeks, days, hours, minutes, and seconds. We'll begin with large blocks of time called **Eras**. The Eras are, from oldest to

Figure 2.3. Bone between two dated horizons (flagged). As we know the ages of the two horizons, the age of the bone can be interpolated between them (see text).

youngest, the Paleozoic (*paleo* – ancient; *zoo* – animal), the Mesozoic (*meso* – middle), and the Cenozoic (*cenos* – new). Within each Era are smaller subdivisions (still consisting of tens of millions of years each) called Periods. We're interested in the Mesozoic Era, and so, within it, the Periods are, again from oldest to youngest, the Triassic (named for a three-fold division of sedimentary rock in Germany), the Jurassic (named for the Jura Mountains of the Franco-Swiss border), and the Cretaceous (*creta*, Latin for "chalk," named for the White [chalk] Cliffs of Dover, England). Within each Period are yet smaller subdivisions of time called Epochs (consisting of several millions of years each). Figure 2.4 shows the part of the geological time scale during most of which dinosaurs roamed the Earth.

Continents and Climates

Although the idea is just over 100 years old, most people are aware that, throughout its history, somehow the continents have moved around the Earth's surface; that the surface of the Earth has been anything but static. The process is commonly called continental drift but is in fact part of a much more significant concept: plate tectonics (see Appendix 2.2). Plate tectonics, when it was first conceived in the late 1960s, revolutionized our understanding of the Earth, because it ultimately provided explanations for phenomena as different as the rise of mountains, oceanic circulation, and even the distribution of organisms. Here, because we are interested in dinosaurs, we'll take the luxury of bypassing a mere *3.7 billion* years of continental evolution and cut to the chase – or at least to the Late Triassic, when all the continents coalesced into a single landmass, now known as Pangaea (Figure 2.5).

Late Triassic Epoch

Pangaea was the dominant land feature of the Late Triassic Earth. Like any large landmass, Pangaea had great mountain ranges; however, it was at least theoretically possible to walk on land from any place to any other. Unlike today, when the continents are largely separated by oceans, and the faunas and floras on different continents differ, Late Triassic faunas and floras all around the world were somewhat similar, presumably reflecting the connectedness of all the landmasses.

Early and Middle Jurassic Epochs

The initial break-up of Pangaea took place in the Early Jurassic. The effect was like unzipping the great supercontinent from south to north. Sediments in the eastern seaboard and Gulf Coast regions of North America, and in Venezuela and western Africa, record the opening and widening of a seaway (the "proto-Atlantic" ocean) through the Middle Jurassic. The single continent of Pangaea was beginning to reorganize into northern and southern supercontinents – Laurasia and Gondwana, respectively – although even as late as Middle Jurassic time, a land bridge is thought to have existed between the two.

Also at this time, some of the earliest epicontinental (or "epeiric") seas of the Mesozoic Era first made their appearances. Epicontinental seas are shallow marine waters that cover parts of continents. In the past, epicontinental seas were considerably more widespread than they are today, because then eustatic (or global) sea levels were higher than they are now. Epicontinental seas are a very familiar part of Earth's geologic past and, with global climate change, there is good reason to think that they are also part of its future.

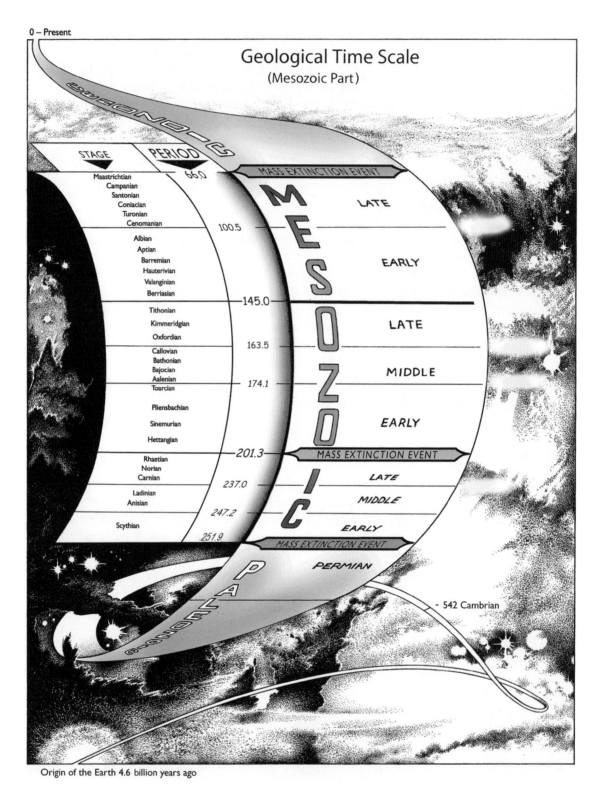

Figure 2.4. The Mesozoic part of the geological time scale, when dinosaurs roamed the Earth. The Mesozoic Era consists of the Triassic, Jurassic, and Cretaceous Periods, and although it lasted 185 million years, it constitutes only a rather tiny fraction of the expanse of the complete age of the Earth. If you compacted all of Earth time into a single year, from January 1 (the formation of the Earth) to December 31 (now), then dinosaurs were on Earth from about December 11 to December 25 (dates from International Commission on Stratigraphy (ICS); www.stratigraphy.org; v. 2020/03).

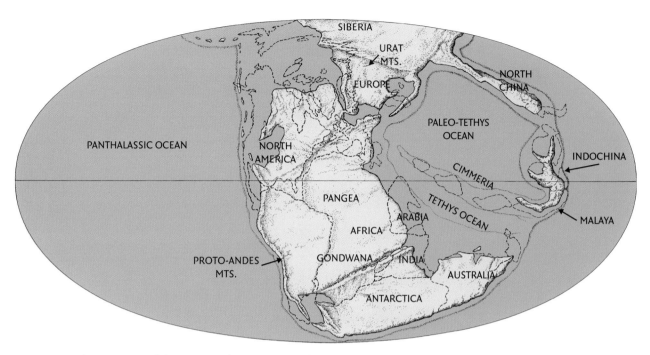

Figure 2.5. The positions of the present-day continents during the Late Triassic (237 Ma to 201 Ma). Earth was dominated by the unified landmass Pangaea.

Late Jurassic Epoch

In the Late Jurassic (Figure 2.6) and Early Cretaceous, continental separation was well underway. A broad seaway, the *Tethyan Seaway* (after the Greek titan goddess *Tethys*, a goddess of the Sea), ran between Laurasia and Gondwana.

Early Cretaceous Epoch

The first half of the Cretaceous Period was a time of active mountain-building, sea-floor spreading, high eustatic sea levels, and broad epeiric seas. The Tethys Ocean, a sea that eventually became the Atlantic Ocean, remained a dominant geographical feature, as Europe continued to separate from North America.

In Gondwana, a stable continental marriage dating back to the Early Paleozoic Era finally dissolved as the southern continent underwent rifting involving two of its largest constituents – Africa and South America – as well as two smaller constituents, India and Madagascar. While India and Madagascar were in the first bloom of unconfinement, Australia and Antarctica remained together (a union that would not end until about 50 Ma), and a land connection remained, as it *almost* does today, between South America and Antarctica.

Late Cretaceous Epoch

The global positions of continents during the Late Cretaceous would be almost familiar to us (Figure 2.7). North America became nearly isolated, connected only by a newly emergent land connection across the modern Bering Straits to the eastern Asiatic continent. Although best known

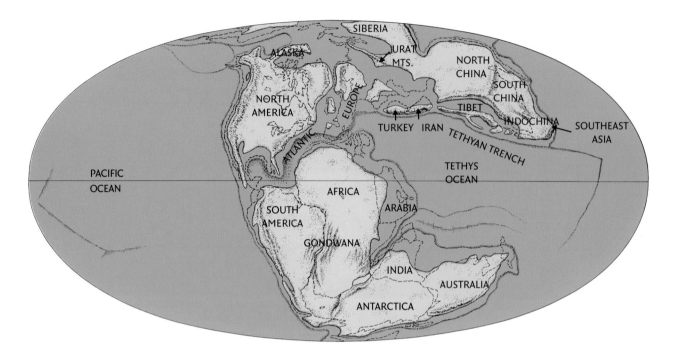

Figure 2.6. The positions of the continents during the Late Jurassic (163.5 Ma to 145.0 Ma). Pangaea has begun its dismemberment while the southern continent of Gondwana remains together.

from the last Ice Age (100 000 years ago), this land bridge has come and gone several times since the Cretaceous. Africa and South America were fully separated, the former retaining its satellite, Madagascar, and the latter retaining a land bridge to the Antarctica/Australia continent. India was by now well on its way towards its eventual collision with southern Asia.

Climates During the Time of the Dinosaurs

Earth has recorded traces that allow us to infer at least *aspects* of past climates, and indeed the flavor of the Mesozoic would be lost without some general sense of Mesozoic **paleoclimates** (ancient climates). Distributions of the landmasses as well as the amount and distribution of the oceans on the globe drastically modify temperatures, humidity, and precipitation patterns. In the following, we explore this, comparing the extreme case of the continents coalesced into the single landmass of Pangaea (see Figure 2.5) with the equally extreme (but more familiar) case of the continents widely distributed around the globe (see Figure 2.7).

Heat Retention in Continents and Oceans

Continents (land) and oceans (bodies of water) respond very differently to heat from the Sun: you can jump into a swimming pool and be comfortable long after the air and land have gotten chilly. This is due to differences in respective **heat capacity** (the amount of heat required to change the temperature of a mole of material by 1 °C): how easily the temperature of a given material can be changed. Water has a higher heat capacity than continental crust, which means that it takes more energy to change

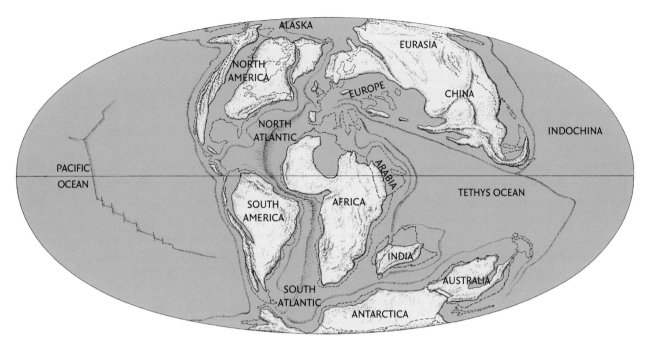

Figure 2.7. The positions of the continents during the Late Cretaceous (100.5 Ma to 66.0 Ma). The positions of the continents did not differ greatly from their present-day distributions. Note the land bridge between Asia and North America, as well as the European archipelago. By this point in time, both of the supercontinents, Gondwana and Laurasia, had disintegrated.

its temperature than it takes to change the temperature of continents. The result is that the water takes more time to heat up, and more time to cool down. This is also why temperatures at the coast are almost always milder than those found inland.

Consider how these properties of continents and oceans might modify climates at the dawn of the Mesozoic, when the continents were united into the single landmass Pangaea (see Figure 2.5). Here, continental effects – more rapid warming and cooling of continents than oceans – would have been more intense than today. Pangaea must have experienced wide temperature extremes. It would have heated up quickly and got hotter, and then cooled off more rapidly and got colder faster than modern continents, whose continental effects are mitigated by the broad, temperature-stabilizing expanses of oceans between them.

The post Late Triassic break-up of Pangaea weakened the strong continental effects. With the rise in eustatic sea level and supercontinental dismemberment, the effects of the large epeiric seas were superimposed upon the diminishing continental effects. These large bodies of water would have stabilized global temperatures, decreasing the magnitude and rapidity of the temperature fluctuations experienced on the continents during times of lower sea level.

Climates Through the Mesozoic Era

The Late Triassic and Early Jurassic were times of heat and aridity. Throughout the Triassic, the evidence is that heat and aridity increased, with a possible episode of increased humidity during the early part of the Late Triassic. They also were times of marked seasonality; that is, well-defined seasons, strongly affected by the Pangaea continental mass. By the latter two-thirds of the Jurassic, however, as well as most of the Cretaceous, Earth is thought to have been *without* polar ice or glaciers on the northern parts of the continents. This is, as yet, quite beyond our own experience; now, glaciers occur at high latitudes at both poles, and the poles themselves are covered in ice. The conclusion that there were no ice or glaciers above the Arctic and Antarctic Circles (latitudes 66.5° N and 66.5° S) is based largely upon the presence of warm climate indicators such as the fossils of warmth-loving plants and certain fish at high latitudes, and upon the absence of any evidence of continental glaciation from this time.

The absence of polar ice had an important consequence for climates: water that would have been bound up in ice and glaciers was instead in ocean basins. This in turn meant higher global sea levels, which led to extensive epeiric seas. The increased abundance of water on the continents as well as in the ocean basins had a stabilizing effect on temperatures (because it decreased continental effects), and decreased the amount of seasonality experienced on the continents.

Continental climates are enormously variable, however, and in North America, Upper Jurassic terrestrial deposits (features preserved in the rocks, such as oxidized sediments and calcium carbonate deposits) suggest that in the Late Jurassic there was seasonally arid conditions. So much for dinosaurs in steamy, swampy jungles!

Global paleoclimates in the Cretaceous are somewhat better understood than those of the preceding periods. During the first half of the Cretaceous at least, global temperatures remained warm and equable. The poles continued to be ice-free, and the first half of the Cretaceous saw far less seasonality than we see today. This means that, although equatorial temperatures were approximately equivalent to those we experience today, the temperatures at the poles were somewhat warmer. Temperatures at the Cretaceous poles have been estimated at 0 °C to 15 °C, which means that the temperature difference between the poles and the equator was only between 17 °C and 26 °C, considerably less than the ±41 °C of the modern Earth.

During its first half, the Cretaceous was subject to synergistic forces: tectonic activity, such as mountain building and sea-floor spreading, caused an increase in atmospheric CO_2 and a decrease in the volume of the ocean basins, which in turn increased the area of epeiric (inland) seas. The seas thus stabilized climates already warmed by enhanced absorption of heat in the atmosphere. Decreased ice volumes meant less sunlight reflected back into the atmosphere, which meant warmer temperatures on the Earth'ssurface.

Sound familiar? The Early to mid Cretaceous experienced the notorious "greenhouse" conditions that are currently of such concern today. Because several times in its past history, including in the mid Cretaceous, the Earth has "experimented" with greenhouse conditions, Earth history has a lot to offer to the dialog about global warming.

The last 30 million years of the Cretaceous produced a mild deterioration in the equable conditions of the mid Cretaceous. A pronounced withdrawal of the seas took place, and evidence exists of more pronounced seasonality.

SUMMARY

Determining the ages of dinosaurs is accomplished by a mixture of biostratigraphy, lithostratigraphy, and geochronology. These allow paleontologists to date rocks and fossils in relation to each other, as well as to obtain estimates of their ages in years before present. Using these techniques, geoscientists have constructed and refined a geological time scale for the entire history of the Earth. The time scale is hierarchically divided into successively more refined time intervals: Eras, Periods, and Epochs.

The earliest dinosaurs appeared during the Late Triassic, a time in which the Earth's continents were united into a single supercontinent called Pangaea. Since then, the continents have separated, moving to their present positions.

The presence of a single supercontinent had major implications for climates, which, because of that configuration, were strongly seasonal. These conditions mitigated throughout the Jurassic, although the Late Jurassic appears to have had strong seasonality, at least in North America. By mid Cretaceous time, with the separation of the continents well underway, high sea levels, melted polar ice, and high levels of atmospheric CO_2 appear to have acted synergistically to produce global warming and greenhouse conditions.

SELECTED READINGS

Barron, E. J. 1983. A warm, equable Cretaceous: the nature of the problem. *Earth Science Reviews*, 19, 305–338.

Berry, W. B. N. 1987. *Growth of a Prehistoric Time Scale Based on Organic Evolution*. Blackwell Scientific Publications, Boston, MA, 202 pages.

Crowley, T. J. and North, G. R. 1992. *Paleoclimatology*. Oxford Monographs in Geology and Geophysics, no. 18. Oxford University Press, New York, 339 pages.

Gradstein, F., Ogg, J., and Schmitz, M. (eds.) 2012. *A Geological Time Scale 2012*. Elsevier, Amsterdam, 1176 pages.

Lewis, C. 2000. *The Dating Game: One Man's Search for the Age of the Earth*. Cambridge University Press, Cambridge, 253 pages.

Robinson, P. L. 1973. Palaeoclimatology and continental drift. In Tarling, D. H. and Runcorn, S. K. (eds.) *Implications of Continental Drift to the Earth Sciences*, vol. I. NATO Advanced Study Institute, Academic Press, New York, pp. 449–474.

Ross, M. I. 1992. *Paleogeographic Information System/Mac Version 1.3: Paleomap Project Progress Report no. 9*, University of Texas at Arlington, 32 pp.

TOPIC QUESTIONS

1. Define: eustatic, biostratigraphy, stratigraphy, geochronology, half-life.
2. What is the difference between "numerical" dates and relative ages?
3. If a dinosaur bone is found between two layers dated at 90 million and 70 million years, respectively, what is the age of the dinosaur bone?
4. Referring to question no. 3, how precisely can that bone be dated?
5. What is the numerical value for the half-life shown in Figure 2.1?
6. If there were dinosaur-bearing rocks in North America and Africa, and the African dinosaur remains could be dated biostratigraphically, how could you correlate the North American deposits with the African ones?
7. Compare the Late Triassic Earth with the Late Cretaceous Earth.
8. Please name three extinct organisms that are biostratigraphic indicators, and suggest what time ranges they would be suitable for identifying.
9. *Tyrannosaurus rex*, from the Late Cretaceous Epoch of the western United States, is remarkably similar to an Asian dinosaur called *Tarbosaurus baatar*, found in sediments of approximately the same age. What is the best explanation for this?

APPENDIX 2.1 CHEMISTRY QUICK 'N DIRTY

Earth is made up of elements, as seen[3] on a Periodic Table. Many of these, such as hydrogen, oxygen, nitrogen, carbon, and iron, are familiar, while others, such as berkelium, iridium, and thorium, are less so. All elements are made up of atoms, an atom being the smallest particle of any element that still retains the properties of that element. Atoms, in turn, are made up of protons, neutrons, and yet smaller electrons, which are collectively termed subatomic ("smaller-than-atomic") particles. Protons and neutrons reside in the central core, or nucleus, of the atom. The electrons are located in a cloud surrounding the nucleus. Some electrons are more tightly bound around the nucleus and others are less tightly bound. Those that are less tightly bound are, as one might expect, more easily removed than those that are more tightly bound (Figure A2.1).

Keeping that in mind, let us further consider the subatomic particles. Protons and electrons are electrically charged; electrons have a charge of -1 and protons have a charge of $+1$. Neutrons, as their name implies, are electrically neutral and have no charge. To keep a charge balance in the atom, the number of protons (positively charged) must equal the number of electrons (negatively charged). This number – which is the same as the number of protons – is called the atomic number of the element, and is displayed to the lower left of the elemental symbol. In our example, the element potassium is identified by the letter K (its Latin name is *kalium*), and it has 19 protons. Its atomic number is thus 19, and it may be written $_{19}K$.

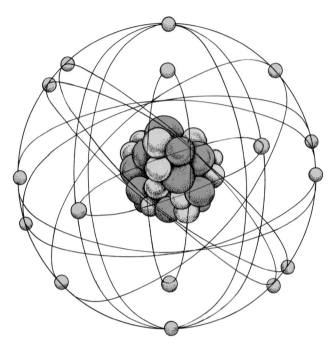

Figure A2.1. Diagram of a potassium atom. In the nucleus are 19 protons and 20 neutrons. In a cloud around the nucleus are 19 electrons, whose position relative to the nucleus is governed by their energy state.

Along with having an electrical charge, some subatomic particles also have mass. Rather than force us to work with the extremely small mass of a proton (one of them weighs about 1.67×10^{-24} g!), it is assigned a mass of 1. Neutrons have a mass of 1 as well. Because relative to protons and neutrons the masses of electrons are negligible, the mass number of an element is composed of the total number of neutrons *plus* the total number of protons. In the case of the element potassium, for example, the mass number equals the total number of neutrons (20) plus the total number of protons (19); that is, 39. This is usually written ^{39}K. Because the atomic number is always the same for a particular element, it is commonly not included when the isotope is discussed. Thus $^{39}_{19}K$ is usually abbreviated to ^{39}K.

Variations in elements that have the same atomic number but different mass numbers are called isotopes. For example, an important (for our purposes) isotope of ^{39}K is ^{40}K. Since ^{40}K is an isotope of potassium, it has the same atomic number as ^{39}K (because it has 19 electrons and 19 protons). The change in *mass* number results from additional neutrons. ^{40}K has 21 neutrons, which, with the 19 protons, increase its atomic mass to 40. Because it is potassium, of course, its atomic number remains 19.

---

[3] And heard, sung by asapSCIENCE (www.asapscience.com) to Jacques Offenbach's can-can tune from "Orpheus in the Underworld," and by Tom Lehrer to Sir Arthur Sullivan's "Major General's Song" from the album *An Evening Wasted with Tom Lehrer*. Both can be heard on YouTube.

APPENDIX 2.2 PLATE TECTONICS

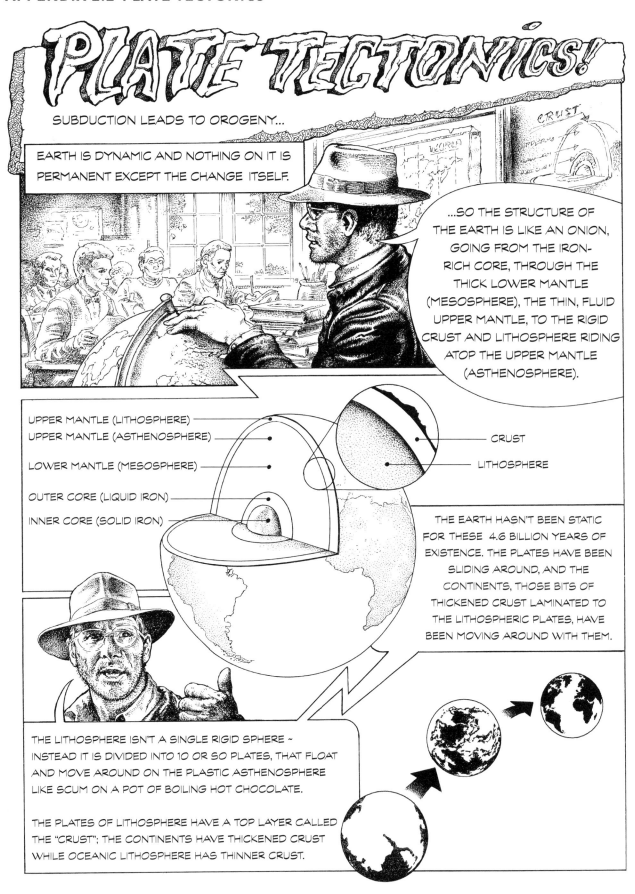

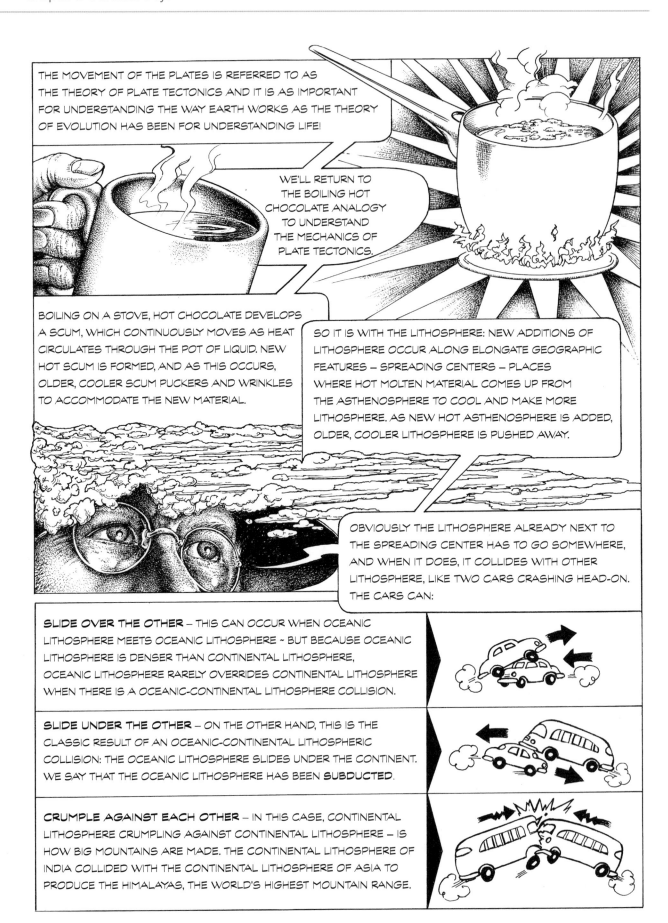

THE MOVEMENT OF THE PLATES IS REFERRED TO AS THE THEORY OF PLATE TECTONICS AND IT IS AS IMPORTANT FOR UNDERSTANDING THE WAY EARTH WORKS AS THE THEORY OF EVOLUTION HAS BEEN FOR UNDERSTANDING LIFE!

WE'LL RETURN TO THE BOILING HOT CHOCOLATE ANALOGY TO UNDERSTAND THE MECHANICS OF PLATE TECTONICS.

BOILING ON A STOVE, HOT CHOCOLATE DEVELOPS A SCUM, WHICH CONTINUOUSLY MOVES AS HEAT CIRCULATES THROUGH THE POT OF LIQUID. NEW HOT SCUM IS FORMED, AND AS THIS OCCURS, OLDER, COOLER SCUM PUCKERS AND WRINKLES TO ACCOMMODATE THE NEW MATERIAL.

SO IT IS WITH THE LITHOSPHERE: NEW ADDITIONS OF LITHOSPHERE OCCUR ALONG ELONGATE GEOGRAPHIC FEATURES – SPREADING CENTERS – PLACES WHERE HOT MOLTEN MATERIAL COMES UP FROM THE ASTHENOSPHERE TO COOL AND MAKE MORE LITHOSPHERE. AS NEW HOT ASTHENOSPHERE IS ADDED, OLDER, COOLER LITHOSPHERE IS PUSHED AWAY.

OBVIOUSLY THE LITHOSPHERE ALREADY NEXT TO THE SPREADING CENTER HAS TO GO SOMEWHERE, AND WHEN IT DOES, IT COLLIDES WITH OTHER LITHOSPHERE, LIKE TWO CARS CRASHING HEAD-ON. THE CARS CAN:

SLIDE OVER THE OTHER – THIS CAN OCCUR WHEN OCEANIC LITHOSPHERE MEETS OCEANIC LITHOSPHERE ~ BUT BECAUSE OCEANIC LITHOSPHERE IS DENSER THAN CONTINENTAL LITHOSPHERE, OCEANIC LITHOSPHERE RARELY OVERRIDES CONTINENTAL LITHOSPHERE WHEN THERE IS A OCEANIC-CONTINENTAL LITHOSPHERE COLLISION.

SLIDE UNDER THE OTHER – ON THE OTHER HAND, THIS IS THE CLASSIC RESULT OF AN OCEANIC-CONTINENTAL LITHOSPHERIC COLLISION: THE OCEANIC LITHOSPHERE SLIDES UNDER THE CONTINENT. WE SAY THAT THE OCEANIC LITHOSPHERE HAS BEEN **SUBDUCTED**.

CRUMPLE AGAINST EACH OTHER – IN THIS CASE, CONTINENTAL LITHOSPHERE CRUMPLING AGAINST CONTINENTAL LITHOSPHERE – IS HOW BIG MOUNTAINS ARE MADE. THE CONTINENTAL LITHOSPHERE OF INDIA COLLIDED WITH THE CONTINENTAL LITHOSPHERE OF ASIA TO PRODUCE THE HIMALAYAS, THE WORLD'S HIGHEST MOUNTAIN RANGE.

SO HERE'S THE COMPLETE GEOLOGICAL PACKAGE: HOT MANTLE MOVES UP, FORMING AN OCEANIC SPREADING CENTER; FAR AWAY, AT THE EDGE OF A CONTINENT, THIS CAUSES THE OLDER OCEANIC CRUST THAT FORMED EARLIER TO BE SUBDUCTED UNDER THE CONTINENT. BECAUSE PIECES OF THE SUBDUCTING SLAB ARE HOT, THEY TEND TO MELT AND RISE THROUGH THE OVERRIDING CONTINTENTAL LITHOSPHERE. WHEN THEY FINALLY REACH THE SURFACE, ALL THAT MANTLE-DERIVED HEAT ENERGY, AND GAS WANTS TO ESCAPE, AND SO VOLCANIC MOUNTAIN RANGES ARE BUILT; FOR EXAMPLE, THE ANDES IN SOUTH AMERICA AND IN NORTHWESTERN UNITED STATES, THE CASCADES.

PLATE TECTONICS REALLY EXPLAINS HOW THE WORLD *WORKS* ~ FROM WHY THE CONTINENTS HAVE MOVED THROUGH TIME, TO WHY VOLCANOES ARE WHERE THEY ARE, TO THE ORIGIN OF MOUNTAIN RANGES (TERMED "OROGENY"), TO THE REASONS FOR AND LOCATIONS OF EARTHQUAKES (MOVEMENTS IN THE CRUST).

THIS IS WHY GEOLOGISTS LIKE TO SAY, "SUBDUCTION LEADS TO OROGENY!"

Who's Related to Whom – and How Do We Know?

Who are you? Who? Who?

Peter Townsend, the Who

WHAT'S IN THIS CHAPTER

In this chapter we get cozy with the following subjects, which we revisit again and again throughout this book:

- *Phylogeny*, the evolutionary relationships of organisms.
- *Biotic evolution*, the mechanism behind the fundamental genealogical connections among organisms.
- *Phylogenetic systematics*, the method used to reconstruct patterns of evolution.
- *Cladograms*, branching diagrams that, using diagnostic characters, depict how closely (or distantly) organisms are related to each other. That depiction is a hypothesis; the cladogram is therefore a hypothesis of relationship.

Who Are You?

Identity is *the* fundamental question in all animals – including humans. Knowing who we are provides the essential context for us, and, for this reason, a favorite dramatic technique (think soap operas) is to have a key character lose some aspect of his or her identity. If you think about it, who we are governs every aspect of our behavior.

To answer "Who are you?" you could begin by asking "To whom are you related?", because *relationship*, as in human families, is the key to identity. So to understand who dinosaurs are, we need to know their relationships – to each other and to other animals.

Those relationships – who we are and where we come from – are the special province of phylogeny: the history of the descent of organisms. And here is where evolution comes in: evolution is the cause of the fundamental genealogical connection among organisms.

Organic Evolution

Biotic or organic evolution refers to *descent with modification*: the concept that organisms have changed and modified their morphology (*morph* – shape; *ology* – the study of) through each succeeding generation (see Appendix 3.1 for a brief refresher on the theory of evolution). That makes each new generation the most recent bearer of the unbroken genetic thread that connects life. Each new generation is forward-looking in that its members potentially contain new features that might be useful for whatever the organism encounters in its life; but it is also connected to its ancestors by features that it has inherited. The web of those connections, how close or distant they are – is called relationship.

Relationship and the Linnaean Classification System

Organisms are commonly classified according to the system devised by the Swedish naturalist Carolus Linnaeus (1707–1778). His hierarchical system is the very famous (or infamous!) ranking of organisms in groups of decreasing size: kingdom, phylum, class, order, family, genus, and species. Individuals are generally referred to by italicized generic (genus) and specific (species) names, for example, in the case of a famous large dinosaur, *Tyrannosaurus rex*. Any name in the hierarchy – representing a group of organisms – is considered a taxon (plural *taxa*). He correctly observed that the features that characterize life are *hierarchical*: composed of large groups, within which are smaller groups, within which are smaller groups, etc. We revisit hierarchy in greater detail below.

All classifications serve a purpose; for example, our movies are classified by both subject (Drama, Horror, Comedy, etc.) as well as by suitability for viewing (PG-13, R, etc.). The Linnaean classification has come to reflect degree of relatedness. Thus, all members of a taxon – at any level in the hierarchy – are said to be more closely related to each other than to anything else. So, for example, in the case of the taxon "Mammalia," all members (mammals) are thought to be more closely related to each other (other mammals) than to anything else.

For all its ubiquity, the Linnaean classification represents a kind of historical sleight-of-hand. Linnaeus developed his classification system well before evolution was properly understood. Linnaeus himself did not think that evolution was responsible for the Earth's diversity (in fact, he was what we would now call a "creationist"), and so his classification was developed solely on the basis of overall similarities among organisms; it had nothing to do with relationship as we understand it today. The idea of relationship, of descent with modification, quietly slipped in later, after

Darwin had published *On the Origin of Species* (1859). This was reasonable, of course: things that are more related tend to look more similar; however, overall similarity is not always the best indicator of evolutionary relationship. For this reason, although the Linnaean classification is still widely in use, many evolutionary biologists have observed that the Linnaean classification poorly accommodates a world where biotic diversity is generated by evolution. In Chapter 4, we will see some of the problems that arise.

Determining Relationships

Homology

If there are genetic relationships among organisms, then there must be genetic relationships among their parts. For example, the five "fingers" in the human "hand" and the five "toes" in the front "foot" of a lizard didn't just occur independently. They're all digits of the forelimb, a particular feature that happens to have been evolutionarily retained in these two lineages (humans and lizards). In principle and in fact, the digits on the forelimbs of lizards and humans can be traced back in time to digits in the forelimb of the common ancestor of humans and lizards. We call two anatomical structures homologous when they can, at least in theory, be traced back to a single original structure in a common ancestor (Figure 3.1). Thus we infer that the digits in the forelimbs of all mammals are homologous with those of, for example, dinosaurs. The wings of a fly, however, are not homologous with those of a bird, since they cannot be traced to a single structure on a common ancestor. Because the wings of a fly and the wings of a bird do the same thing (enable flight), they are said to be analogous.

Here is a subtle, but important distinction: the *limbs* of the pterosaur, bird, and mammal shown in Figure 3.1 are surely homologous; however, their function as *wings* is equally surely not. If wings were homologous, it would imply that wings evolved once – in the common ancestor – and all these wings are variants on the original wing. In fact, we have good reason to conclude that each of the ancestors of each of these three animals was nonflying, and that in each lineage, wings (and flight) evolved separately. Thus, wingedness in these vertebrates is analogous (Figure 3.2). even if their limbs are homologous. Evolution teaches us that there are many ways to make a wing.

Chopping Down Evolutionary Trees

We've all seen these; they begin with a pulsating, primordial slime-blob that eventually evolves into everything else as you trace the branches outward (Figure 3.3 is an example). Such trees show which descendant came from which ancestor, and when that occurred. They are common in textbooks, museum displays, and *National Geographic* articles, and deeply influence our ideas about evolution. We might call such trees evolutionary trees.

But how do we know exactly which ancestor gave rise to which descendant? After all, no human witnessed the zillions of speciation events that constitute the "tree of life."

The problem is worse than that. We learned in Chapter 1 that fossil preservation is a rare event. Is it likely that some fossil that happens to be preserved and that we happen to find, turns out to be the actual ancestor of some other that happens to be preserved and that we happen to find? The chances of this occurring are vanishingly small. Despite what one reads in the news and in the glossy magazines, the oldest *fossil* in the human family is very unlikely to be the great great great grand-daddy of all subsequent humanity.

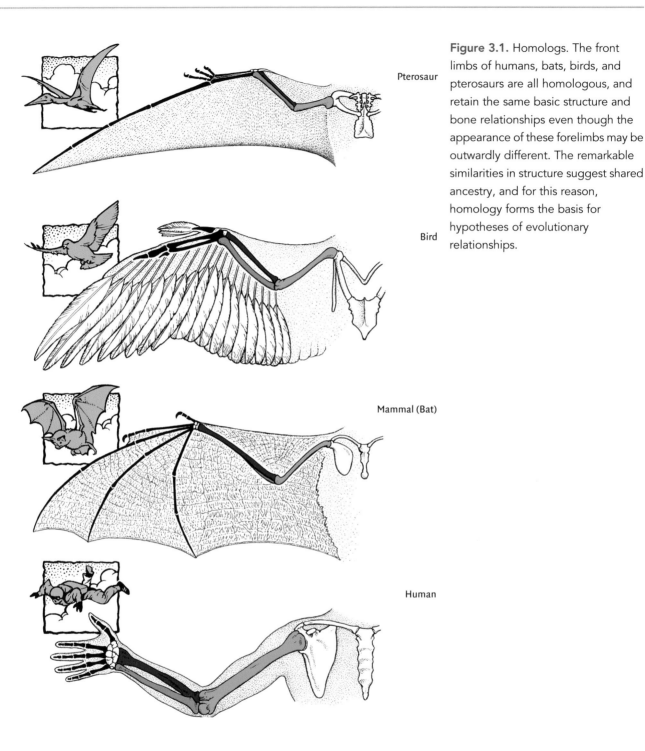

Figure 3.1. Homologs. The front limbs of humans, bats, birds, and pterosaurs are all homologous, and retain the same basic structure and bone relationships even though the appearance of these forelimbs may be outwardly different. The remarkable similarities in structure suggest shared ancestry, and for this reason, homology forms the basis for hypotheses of evolutionary relationships.

Pterosaur

Bird

Mammal (Bat)

Human

Still, that fossil *is* likely to have many features that the real great great great granddaddy possessed. And therein lies the key to recognizing who is related to whom: while we'll never find the *actual* ancestor, we have a really good chance of finding out more or less how that ancestor looked. Because evolutionary trees specify actual ancestors, and because, as we have seen, we're not likely to actually have the real ancestor in hand, we'll avoid evolutionary trees, and instead use a revolutionary method to understand who is related to whom or, thinking of it another way, the course of evolution. That method is called **phylogenetic systematics**.

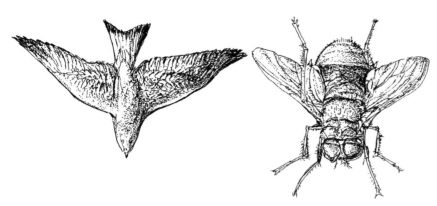

Figure 3.2. Analogs. The wings of the fly are not structurally homologous with those of vertebrates, such as the bird shown. While they have the same function (flight), they are structurally very different, and thus cannot be traced to a single structure on a common ancestor.

Phylogenetic Systematics – the Reconstruction of Phylogeny

Phylogenetic systematics was developed in the mid 1900s specifically for reconstructing the course of evolution. As we shall see later in this chapter, it is the only *scientific* means of determining relationship, and for this reason it is used by virtually all evolutionary biologists and paleontologists. And although it hasn't yet trickled down into the popular literature, the only way to really understand dinosaurs is through the lens of phylogenetic systematics. And so that's what we do in this book.

To reconstruct phylogeny, we need a way to recognize how closely two creatures are related. Superficially this is very simple: *things that are more closely related tend to share specific features.* We know this intuitively because we can see that organisms that we believe are closely related (for example, a dog and a coyote) share many similarities. Breeders have taken advantage of this for thousands of years. They've depended upon the fact that offspring look, and sometimes act, very much like their parents to obtain plants and animals with the qualities for which they select. Phylogenetic systematics is the technique by which relationships between organisms can be inferred using unique features of organisms. It depends first and foremost upon the hierarchical distribution of features in the natural world.

Hierarchy

One of Linnaeus' great insights was his recognition that all features in the natural world are organized in a **hierarchy** (Figure 3.4), which can be understood to be a successive ranking of subsets within sets. A familiar hierarchy, for example, is ranks within the military. Or, to choose a biological example, all creatures possessing fur[1] are a subset of all animals possessing a backbone, which are in turn a subset of all living organisms; these features are distributed hierarchically. *In fact, all features of living organisms are hierarchically distributed in nature*, from the possession of DNA – which is almost ubiquitous – to highly restricted features such as the possession of a brain capable of producing a written record of culture.

---

[1] A number of organisms in the world are fur-covered besides mammals; for example, bees and some spiders (like tarantulas) have a fur-like covering. But the fur on mammals is unique: it is made of unique proteins, grows in a unique way, and thus is different from these other coatings; which are analogs.

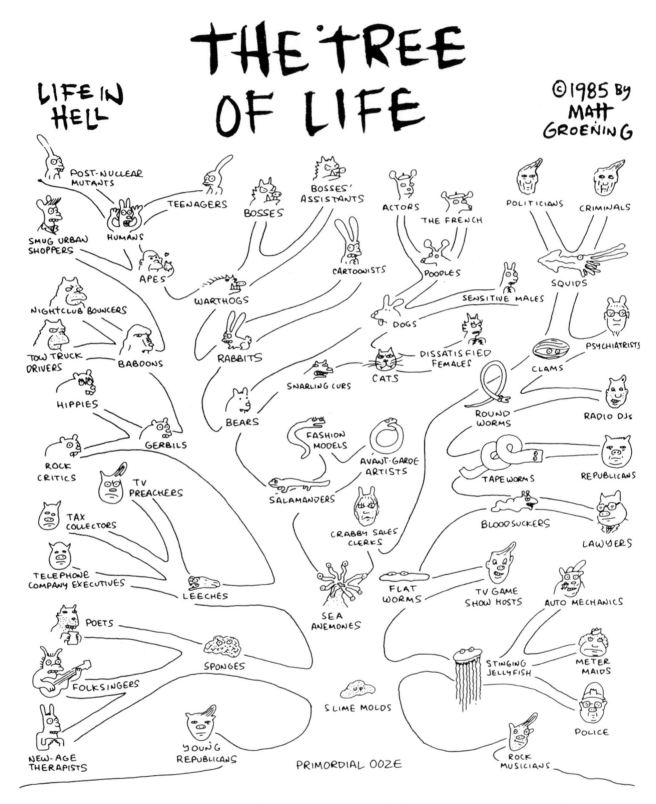

Figure 3.3. An evolutionary tree. The image of evolution as a tree goes back to Darwin (1859) and is completely familiar. From the Big Book of Hell © Matt Groening. All Rights Reserved. Reprinted by permission of Pantheon Books, a division of Random House, Inc., NY.

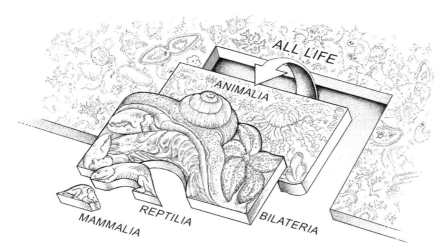

Figure 3.4. The natural hierarchy illustrated as a wooden jigsaw puzzle. The different organisms represent the larger groups to which they belong. For example, the mouse, representing Mammalia, and the lizard, representing Reptilia, together fit within the puzzle to represent Vertebrata, itself a subset of bilaterally symmetrical organisms (Bilateria), which would include invertebrates such as a lobster or a mosquito. Bilateria and other groups constitute those organisms we call Animalia (animals). Animalia, then, is a subset of all living organisms.

Always, however, unmodified or slightly modified vestiges of the original plan remain, and these provide the keys to the fundamental hierarchical relationships that reveal who's related to whom.

Characters

Identifying the features themselves is a prerequisite to establishing life's hierarchy, so we need to look more closely at what we mean by "features." Observable features of anatomy are termed characters. Unique bones or unusual morphologies would all be characters. On the other hand, "observable features" would not include what something does – or how it does it. So, for example, "bites hard" is not a character, but perhaps "big jaw muscles" would be.

Characters acquire their meaning not as a single feature on a particular organism but when their *distribution* among a selected group of organisms is considered. For example, modern birds are generally linked on the basis of having feathers. All living feathered animals are birds and all birds have feathers. Thus, not only is a penguin a bird but so are eagles, ostriches, and kiwis: they all have feathers. And if someone told us that something is a bird, we could confidently predict that it too has feathers.

Because characters are distributed hierarchically, their *position* in the hierarchy is obviously dependent upon the groups they are being used to identify. Consider again the simple example of mammalian fur. Because mammals uniquely have this type of fur, it follows that, if you wanted to tell a mammal from a nonmammal (that is, any other organism), you need only observe that the mammal is the one that has that type of fur. On the other hand, the character "possession of fur" is not useful for distinguishing a bear from a dog; both have mammalian fur. To distinguish a bear from a dog, you'd need some character that distinguishes different types of fur-bearing creatures. And there's the hierarchy: fur-bearing is more broadly distributed than the characters you might choose to distinguish the fur-bearing bear from the fur-bearing dog.

These distinctions are extremely important in establishing the hierarchy, and for this reason, characters function in two ways: as **diagnostic** characters and as **nondiagnostic** characters. The word "diagnostic" here has the same meaning as in medicine: Just as a doctor diagnoses a malady by distinctive and unique properties, so a group of organisms is diagnosed by distinctive and unique characters.

The same character may be diagnostic in one group, but nondiagnostic in a smaller subset of that group (because it is now being applied at a different position in the hierarchy). We saw that fur allowed us to tell a mammal from a nonmammal, but it can't distinguish one mammal from another: it wouldn't tell a bear from a dog.

Cladograms

Cladograms (*clados* – branch; *gramma* – letter) are branching diagrams that show hierarchies of diagnostic characters. But, as we'll see, they're not just visual aids, they're the keys to understanding who's related to whom.

To understand how a cladogram works, we begin with two familiar animals; say, a cat and a dog. A cladogram of a cat and a dog is shown in Figure 3.5.

So we're looking for diagnostic characters for these animals. Here, we choose:

(1) possession of fur;
(2) possession of a backbone; and
(3) possession of carnivorous teeth of a unique design.

On the cladogram, each branching point (bifurcation) is marked by a **node** (see Figure 3.5); and it is at the node, just before the split, that we conventionally use a **hatch mark** to list the diagnostic characters that everything connected to that node possesses. The cladogram links two separate

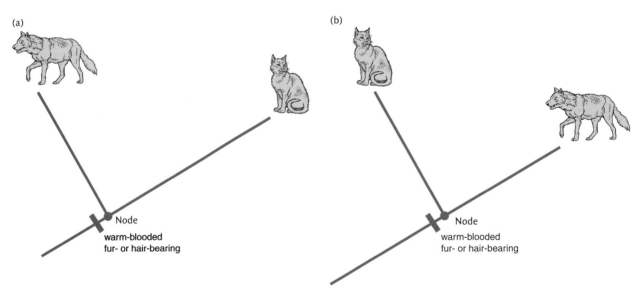

Figure 3.5. (a) A cladogram. The cat and dog are linked by the characters listed at the hatch mark (or bar), just below the **node** (dot at branching arms). The node indicates which taxa share those characters. Groups of organisms linked by shared characters at a particular node can be named; here such a name might be "mammalian carnivores." (b) The same cladogram with the dogs and cats reversed. Reversing the order of two taxa around a node does not change the cladogram's meaning, because the cladogram is only stating that both taxa share the same characters.

objects – the cat and the dog – based upon the diagnostic characters that they *share*. In short, cladograms are all about *shared diagnostic characters*.

Cladograms, of course, are not evolutionary trees; they are simple branching diagrams, showing two organisms uniquely sharing characters (see Box 3.1). So it follows that there is absolutely no difference in the order in which the two organisms sharing the characters are presented. It could go, from left to right, dog, cat (Figure 3.5a); or cat, dog (Figure 3.5b). Either way, both organisms share the character listed at the node.

The issue becomes more complicated (and more interesting) when a third animal is added to the group (Figure 3.6), in this case a monkey. Now, for the first time, because none of the three animals is identical, two of the three will have more in common with each other than either does with the third. It is in this step that the hierarchy is established. The group that contains all three animals is diagnosed by certain features shared by all three (fur and possession of a backbone). Notice that a subset containing two animals (the cat and the dog) has also been established, linked together by a character (uniquely designed carnivorous teeth) that diagnoses them as being exclusive of the third animal (the monkey).

How the characters, and even the animals, are arranged on the cladogram, is controlled by the *choice* of characters. Let's try something different:

- shortened snout
- large eyes.

Based upon these characters, the cladogram in Figure 3.7 contradicts the cladogram in Figure 3.6.

How do we choose? *The cladogram that is most likely correct is the one that doesn't change when new characters are added.* All characters that apply to these mammals support the cladogram in Figure 3.6. And we can infer from the cladogram in Figure 3.6 that a dog shares many more characters with a cat that it does with a monkey.

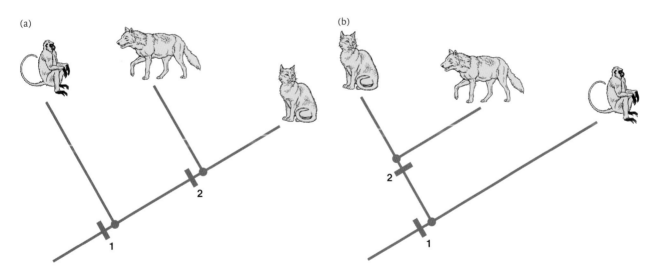

(a)   (b)

Figure 3.6. (a) One possible distribution of three mammals. Members of the group designated by node **1** are united by the possession of fur and warm-bloodedness; that group could be called Mammalia. Within the group Mammalia is a subset united by possession of carnivorous teeth (for example, "mammalian carnivores"). That subset is designated at node **2**. (b) The same cladogram, with the taxa reversed at nodes **1** and **2**. In terms of information conveyed, there is no difference between (a) and (b).

Box 3.1 What's in a Name? Cladograms versus Trees

From their outward appearance, to their etymology, to their purpose, cladograms are easily confused with trees. They look like trees; they "branch" like trees, and of course they both have something to do with the course of evolution. To make matters worse, confusion of the two has been exacerbated by idiosyncratic language like "phylogenetic trees" (which seems to mix the two) and "phylograms" (of dubious clarity); ironically, the word "tree" peppers the professional phylogenetic literature, when in fact a cladogram is what is being discussed. And in the popular literature, trees – such as those seen before the cladistic revolution of the 1970s and 1980s – still dominate evolutionary iconography. So what's in a name?

A lot, as it turns out. The differences between cladograms and trees are so fundamental that a particularly aggressive critique of systematists who used trees to reconstruct evolution, was that they were not doing "science"; a charge that, unsurprisingly, added considerable acrimony to the discussion. What scientist enjoys hearing that his/her work is not science?! Yet, for all the nastiness, there was much truth in the charge because the cladogram is testable and the tree is not. And as we have seen, testability is the hallmark of scientific hypotheses.

So what, then, is a tree? The tree is a **scenario**; a story that we tell about the history of life. If it is concordant with the data (e.g., a cladogram), then it might be correct, although because it is untestable, we'll never be able to find out. Still, humans love stories, in evolution as well as in everything else, and the word "tree" remains the most popular visible manifestation of the course of evolution – however misleading the word.

To clarify, then, in Table B3.1 is a "Cliff Notes" no-frills, "just-the-facts-ma'am" summary of some of the key differences between cladograms and evolutionary trees.

Table B3.1 Phylogenetic trees vs. evolutionary trees.

Criterion	Phylogenetic trees (cladograms)	Evolutionary trees
Diagram	Data-based hierarchical dichotomous branching diagram of *character distributions* among taxa.	Scenario about the course of evolution, specifying presumed ancestors and their descendants through time.
Ancestors and descendants	Unspecified; but the characters or traits that both the ancestors and descendants might have had are specified.	Specified.
Time	Time is unspecified in a cladogram. Older organisms may show more derived features than younger ones. When (in time) organisms first appear in the fossil record, is not a valid indicator of how highly evolved characters are; thus time of appearance has no relevance in a cladogram.	Time is on the vertical axis of a tree. When organisms first appear in the fossil record is directly related to whether they can be considered ancestral to later forms.
Nodes	The place that specifies which taxa share which characters.	The position of the inferred ancestor relative to its descendants.
Appearances in popular literature	Uh, not so much.	Very common, particularly in popular treatments of human evolution.
Science?	Yes; a **testable hypothesis** of relationship.	No, an untestable scenario of who gave rise to whom.

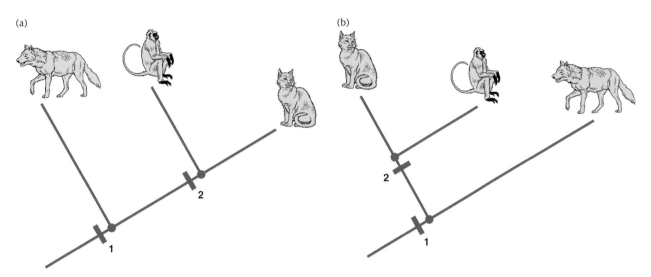

Figure 3.7. (a) An alternative distribution of three mammals. The characters selected at node **2** (see Figure 3.6) suggest that the cat and monkey share more diagnostic characters with each other than either of them does with the dog (see the text). (b) The same cladogram with the taxa reversed at nodes **1** and **2**. In terms of information conveyed, there is no difference between (a) and (b). There is no difference between the cladograms shown in Figure 3.6, nor is there any difference between the cladograms shown here. But, there is a significant difference between the cladograms shown in Figure 3.6. (both of which say that the cat and the dog are more closely related to each other than either is to the monkey), and those shown here (both of which say that the monkey and the cat are more closely related to each other than either is to the dog).

Cladograms and Organic Evolution

So how does this apply to evolution? Using character hierarchies portrayed on cladograms, we establish clades or monophyletic groups (sometimes termed "natural groups"): *groups that have evolutionary significance because the members of each group are more closely related – by genealogy – to each other than they are to any other creature*. If a group is monophyletic, it also implies that all members of that group share a more recent common ancestor with each other than with any other organism. The cladogram in Figure 3.6 suggests that the cat and the dog are more closely related to each other than either is to the monkey.

Now that we're speaking in terms of organic evolution, the specific characters that we said characterize groups can now be treated as homologous among the groups that they link. Mammalian fur once again (!) provides a convenient example. We conclude that mammals are monophyletic based upon the fact that mammals all share a unique type of fur (among many other characters). If all mammals are fur-bearing (and mammals are monophyletic), the implication is that the fur found in bears and that found in horses can in fact be traced back to fur that must have been present in the most recent common ancestor of bears and horses.

Because now we're working with the evolution of organisms (and not just hierarchies of characters), the terms derived or advanced can be used along with "diagnostic," and the term general ("nondiagnostic") is replaced by the words primitive or ancestral. "Primitive" certainly does *not* mean worse or inferior, just as "advanced" certainly does *not* mean better or superior; these refer only to how much the character has been changed by evolution. "Primitive" specifies the condition of a particular feature in the ancestor; "advanced" specifies an evolved condition of that character in its descendant. Five fingers on the human hand is indubitably primitive; in an evolutionary sense the character has been carried through evolution, unchanged, since the earliest common ancestor of all living Tetrapoda (*tetra* – four; *pod* – legs; the group of vertebrates that includes most land-dwelling vertebrates; see Chapter 4). That five fingers is primitive does not diminish their value to those of us who possess them!

Because the form of a cladogram (a branching stick diagram) is hierarchical, it is an excellent way to map the hierarchical distributions of characters in nature. *Derived characters are indicators of new monophyletic groups because, as newly evolved features, they are potentially transferable from the first organism that acquired them to all its descendants*: they therefore characterize the bifurcations at each node on the cladogram. Primitive characters – those with a more ancient history – provide no such distinction of monophyly.

Consider a simple, familiar example: possession of four limbs. Four limbs is a primitive character for the group Tetrapoda. Since all tetrapods are four-limbed (at least historically), to say that you can tell one tetrapod from another using the character "possesses four limbs" would be absurd. Considering only the character "possesses four limbs," you could not tell a dog from a frog from a chimpanzee. But now let's consider all vertebrates. The same character – "possesses four limbs" – becomes diagnostic at a lower level in the hierarchy, because some vertebrates (fish) do not have four legs while other vertebrates (all tetrapods) do, and so the character is diagnostic for Tetrapoda. It implies that Tetrapoda could be a monophyletic group (because sharing that character suggests that all its members are more closely related to each other than they are to anything else, including other vertebrates).

On any cladogram, therefore, we look for characters that mark a node in the diagram. These characters are exclusively shared by all members of the group associated with that node, and thus the cladogram is proposing that the group that shares these characters is monophyletic; e.g., that its members are more closely related to each other than they are to any other group of organisms.

Documenting the hierarchy of character distributions in nature, a cladogram shows monophyletic groups within increasingly larger monophyletic groups. In Figure 3.8, a small part of the hierarchy is shown: humans (a monophyletic group, possessing shared, derived characters) are contained in mammals (another monophyletic group possessing other shared, derived characters). Notice that the character of warm-bloodedness is primitive for *Homo sapiens*, but derived for mammals. As we have seen, features can be derived or primitive, all depending upon what part of the hierarchy one is investigating.

A cladogram need not depict every organism within a monophyletic group. If we are talking about humans and carnivores, we can put them on a cladogram and show the derived characters that diagnose them, but we might (or might not) include other mammals (for example, a gorilla). As we said, with regard to Figures 3.6 and 3.7, if the hierarchical relationships that we have

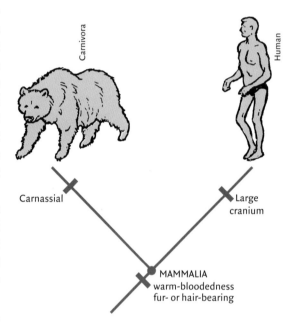

Figure 3.8. A phylogenetic tree (cladogram) showing humans within the larger group Mammalia. Mammalia is diagnosed by warm-bloodedness and possession of fur (or hair); many other characters unite the group as well. Carnivora, a group of mammals that includes bears and dogs (among others) is shown to complete the cladogram. Carnivores all uniquely share a special tooth (the carnassial) and humans all uniquely share, among many other features of the skull and skeleton, a large cranium. Note that all mammals (including humans and carnivores) are warm-blooded and have fur (or hair), but only humans have a large cranium, and only members of Carnivora have the carnassial tooth.

established are valid, the addition of other organisms into the cladogram should not alter the basic hierarchical arrangements established by the cladogram. Figure 3.9 shows the addition of one other group into the cladogram from Figure 3.8. The basic relationships established in Figure 3.8 still hold, even with the new organism added. The cladogram is likely correct.

Cladograms are Phylogenetic Trees

We identified monophyletic groups using derived characters, and that the hierarchies of characters designate hierarchies of groups. So, looking at Figure 3.9, the distribution of shared, derived characters suggests that humans and gorillas are more closely related to each other than either is to a bear. It also suggests that all three are more closely related to each other than they are to something that does not possess the derived character of bearing fur or hair.

And how does that apply to evolution? The evolution of the derived character of fur is associated with the evolution of the group Mammalia. As we currently understand their relationships, what we call "mammals" was invented when that character – among others – first evolved. And, the cladogram tells you that sometime thereafter – the cladogram does not specify when – a character that unites both humans and gorillas evolved, a character that we now recognize diagnoses a new group within Mammalia. That new group is called "Hominoidea," as it happens, and is diagnosed by lots of characters, among which are a series of specializations in the arms and trunk associated with living in trees (although not evolutionary ones!).

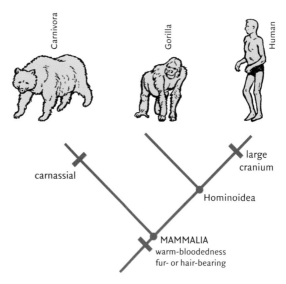

Figure 3.9. Addition of the genus *Gorilla*. The addition of gorillas to the phylogenetic tree does not alter the basic relationships outlined on the cladogram shown in Figure 3.8.

Now that the cladogram is being used as a means to reconstruct biotic evolution, it becomes a kind of "tree" itself. But it is not the familiar *evolutionary tree* seen in popular media; to distinguish it from that, we call it a **phylogenetic tree**, understanding that the rules of its construction and interpretation are those of phylogenetic systematics.[2]

And so there are fundamental differences between phylogenetic trees (e.g., cladograms) and evolutionary trees. Phylogenetic trees do not incorporate time, nor do they tell you who the ancestors were. Instead, they indicate the sequence of the evolutionary events and, more importantly, they specify the *characteristics* that the ancestor possessed. So, in the case of the phylogenetic tree shown in Figure 3.9, we aren't told who the ancestor of bears, gorillas, and humans was, but the cladogram

---

[2] Here we encounter a confusing historical problem in semantics. The metaphor of a tree was used by Darwin in *On the Origin of Species*, and ever since has been applied to evolution. During the 1970s and 1980s, however, early phylogenetic systematists (e.g., "cladists") were extremely careful to distinguish between "trees" and "cladograms." As phylogenetic systematic methods became more widely accepted, systematists began (perhaps a bit sloppily!) to use the word "tree" when they meant cladograms. Today cladists use the word "tree," by which they mean cladogram; fine for professionals, but confusing to students, who are not helped by the fact that the root of the word "cladogram" means branching!

So we compromise by using the word "tree" (referring to organic evolution) but modifying it with the words "evolutionary" (meaning old-style [pre-cladistic] systematic methods), and "phylogenetic" (meaning cladistic systematics). In short, here, *evolutionary tree* = old-style taxonomy based upon overall similarity and appearance in the fossil record; but *phylogenetic tree* = cladogram.

specifies what that earliest mammal must have been like: that it was fur-bearing (among the many other characters that diagnose Mammalia). Some important differences between evolutionary trees and phylogenetic trees (cladograms) are highlighted in Box 3.1.

Used as a tool to reconstruct evolution, then, a phylogenetic tree (cladogram) is actually a hypothesis of relationship; that is, a hypothesis about how closely (or distantly) organisms are related, and about what the sequence of the appearance of different diagnostic characters must have been. But how to test the hypothesis of relationship implicit in the cladogram?

Parsimony

As we have seen, it is possible to construct several possible cladograms which represent different potential evolutionary sequences. Which to choose? We choose using the principle of parsimony. Parsimony, a sophisticated philosophical concept first articulated by the fourteenth-century English theologian William of Ockham, states that *the explanation with the least necessary steps is probably the best one.* Why resort to complexity when simplicity is equally informative? Why suppose more steps took place when fewer can provide the same information?

Figure 3.10 shows two phylogenetic trees that are possible with birds, humans, and bats and the characters of wings, fur, feathers, and mammary glands. In part (b), the bird has to lose ancestral mammary glands and it has to replace fur with feathers. In part (a), wings must be invented by evolution twice; however, cladogram (a) is in fact the simpler of the two because it will not be complicated by the addition of more characters. In contrast, the addition of virtually any other characters that are shared by humans and bats to phylogenetic tree (b) (for example, the arrangement, shape, and number of bones, particularly those in the skull and forelimbs, the structure of the teeth, the biochemistry of each organism) requires that each of these shared characters evolved independently: once in bats and once in humans. That considerably increases the number of

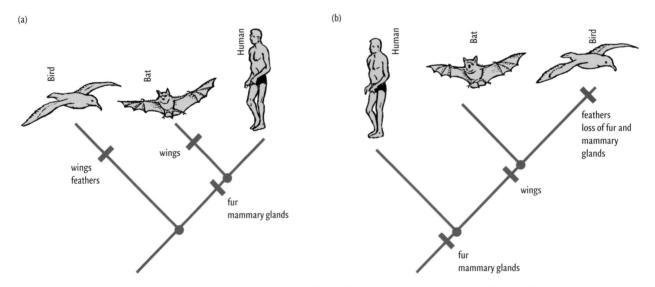

Figure 3.10. Two possible arrangements for the relationships of birds, bats, and humans. (a) The left-hand phylogenetic tree requires wings to have evolved two times; (b) the right-hand tree (cladogram) requires birds to have lost fur and mammary glands. These as well as many other characters suggest that (a) is the more parsimonious of the two cladograms.

evolutionary steps, leaving us with the conclusion that phylogenetic tree (b), the hypothesis that birds and bats are more closely related to each other than either is to a human, is a less parsimonious alternative.

Indeed, as a hypothesis about the *evolution* of these vertebrates, it is much more likely that bats and humans share a more recent common ancestor than that either does with a bird (which, obviously, is why bats and humans are classified together here as mammals). In this case, the use of shared, derived characters has led us to the most parsimonious conclusion with regard to the evolution of these three creatures.

Phylogenetic Trees (Cladograms) are Science

Cladograms are, in effect, scientific hypotheses concerning phylogenetic relationships. Like all good science, they make testable predictions about the distributions of characters in organisms, from which evolutionary relationships can be inferred. Any organism – living or extinct – can test an existing cladogram-based phylogenetic hypothesis. With living organisms, not only do we use their anatomy on the cladogram; we can also use their genetic material. Parsimony is then used to determine which cladogram most likely approximates to the course of evolution. As will become evident, cladograms are among the most powerful tools available for learning about what occurred in ages long past.

Testing a Phylogenetic Tree

If cladistic methods are really scientific, they must rise to the hallmark of science: they must be testable. So how does one test a phylogenetic tree? Here are several different kinds of tests:

(1) The phylogenetic tree is based upon characters that were observed. First we ascertain whether the characters were observed correctly – that is, do all observers of the characters in question reach the same conclusions about their interpretation? When we say that a particular character is shared, do all observers agree that the organisms we say share these characters actually do share them?

(2) The phylogenetic tree shows a hierarchical distribution of characters; is the hierarchy reconstructed correctly? That is, do all observers agree that the primitive characters are primitive, and the derived characters are derived? In other words, in (1) and here in (2), we are looking to see if there is general agreement about the characters and their relationships.

(3) Is the phylogenetic tree that has been selected the most parsimonious? This is simple to determine with three taxa and a few characters, but large data sets with tens of characters (or more) ultimately require computer power to find the most parsimonious distribution. The algorithms for determining parsimony can be quite complex, and different algorithms will lead to different arrangements of taxa on the phylogenetic tree. The methods available for making determinations of parsimony would easily fill a book bigger than this one.

(4) One good test of a phylogenetic tree occurs when a new taxon is inserted into an old tree: If the tree is robust, the new taxon will not disrupt the fundamental distribution of taxa or the shape of the tree. If it does, the tree was likely originally flawed.

These basic tests give the phylogenetic tree its power – its testability – and position it as a scientific tool that was simply unavailable to systematists before the introduction of cladograms.

Leaving Linnaeus: Introducing a Phylogenetic Classification – by Definition

We regularly encounter the groups that Linnaeus established in his classification (terms like *Canis* [dog], Reptilia, Aves [birds], etc.), but recall that Linnaeus diagnosed his groups on the basis of the overall similarity of their characters; overall similarity, as we have seen, is not a suitable basis for reconstructing evolutionary events. Moreover, properly considered, characters in and of themselves do not really encompass the idea of ancestors giving rise to descendants (e.g., evolution). They are just particular qualities either possessed, or not possessed, by an organism. So how to bring an evolutionary perspective into the names of the groups that we discuss?

In the early 1990s, cladists Jacques Gauthier (see Figure 16.12) and Kevin de Queiroz established a subtle but important distinction between *diagnosis* (the familiar characters that we introduced earlier in this chapter that are shared – or not! – by groups) and phylogenetic **definition**, which explicitly involves the process of organic evolution in the membership of a group. We can highlight a group by considering two of its members, and then define it as *all organisms stemming from the most recent common ancestor of those two organisms.* Naturally, this is most easily visualized in conjunction with a cladogram (Figure 3.11).

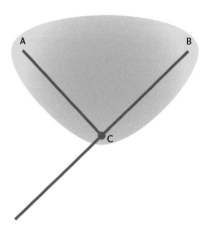

Figure 3.11. Definition. Here, two subgroups ("A" and "B") belong to a group "C." The *definition* of Group C is "the clade – the monophyletic group – that includes A and B and all the descendants of their most recent common ancestor."

Consider, for example, humans. A phylogenetic definition for the group might be "*Homo sapiens sapiens* and *Homo sapiens neandertalensis* (Neanderthals), and all members of the group stemming from their most recent common ancestor." This is a definition rooted in the idea of a biota that *evolves* (rather than is static, as Linnaeus thought), and while it does not tell us what characters to look for in humans, it effectively delimits on the cladogram what can and what cannot be considered a human. Identifying the group by its *definition*, common ancestry – the fingerprint of organic evolution – becomes fundamental to the evolutionary meaning of the name "human." Defining taxa this way distinguishes our use of *any* taxon from the nonevolutionary origin of many still-useful Linnaean terms presently in use. We will use the important concept of definition to identify all of the key groups proposed in this book; for many, there will be clear-cut, diagnostic characters, but, for a few, there will not. Our first close encounter with the brave new world of definition occurs in Chapter 4.

SUMMARY

Relationship is essential to understanding the identities of organisms. To reconstruct relationships, branching diagrams called cladograms are used. Organisms are grouped on these using the presence of shared, derived (or diagnostic) characters; the groups of organisms that result are all inferred to be more closely related to each other than to anything else; that is, they are monophyletic. The evolution of new types of organisms is represented on the cladogram by the appearance of new features, which distinguish descendants from their ancestors. These novel features are the diagnostic (or shared, derived) characters.

Cladograms differ from "trees of life" in several fundamental aspects. Cladograms are based upon shared derived characters; they do not show time; they do not show ancestors (although they specify what the ancestral condition of an organism must have been like); and, most importantly, they are testable. Falsification consists of the prediction on the cladogram that a character will be present that is not (or vice versa). It also occurs when the addition of a new organism upsets the character distributions on a cladogram.

With several hypotheses of relationship (cladograms) to choose among, the cladogram requiring the least number of steps (that is, the most parsimonious) is the preferred one.

SELECTED READINGS

Cracraft, J. and Eldredge, N. (eds.) 1981. *Phylogenetic Analysis and Paleontology*. Columbia University Press, New York, 233 pages.

Darwin, C. 2008 [1859]. *On the Origin of Species: A Facsimile of the First Edition*. Harvard University Press, Cambridge, MA, 540 pages.

Eldredge, N. and Cracraft, J. 1980. *Phylogenetic Patterns and the Evolutionary Process: Method and Theory in Comparative Biology*. Columbia University Press, New York, 349 pages.

Futuyma, D. J. 2005. *Evolution*. Sinauer Associates, Sunderland, MA, 603 pages.

Nelson, G. and Platnick, N. 1981. *Systematics and Biogeography, Cladistics and Vicariance*. Columbia University Press, New York, 567 pages.

Stanley, S. M. 1979. *Macroevolution*. W. C. Freeman and Company, San Francisco, 332 pages.

Wiley, E. O., Siegel-Causey, D., Brooks, D., and Funk, V. A. 1991. *The Compleat Cladist: A Primer of Phylogenetic Procedures*. University of Kansas Museum of Natural History Special Publication no. 19, Lawrence, KS, 158 pages.

Zimmer, C. 2014. *The Tangled Bank: An Introduction to Evolution* (2nd edn). Roberts and Company, CO, 452 pages.

TOPIC QUESTIONS

1. Define: phylogeny, morphology, homologous, analogous, a tree of life, hierarchy, characters, diagnostic, cladogram, node, derived, advanced, monophyletic groups, primitive, ancestral, parsimony, test, falsify, science.
2. Why is it that the direct ancestor of any organism isn't easy to identify?
3. If you have several possible phylogenetic trees, how do you determine which is the preferred one?
4. Construct a cladogram of something that particularly interests you (examples: guitars, music, sports, shoes, etc.). Be sure to show the diagnostic characters in their correct hierarchical locations.
5. If you don't know the ancestor, how can you hope to understand how something evolved?
6. We have said that we could "reconstruct" the course of evolution with a cladogram. How do cladograms allow us to do this?
7. How is a phylogenetic tree a "hypothesis of relationship"?
8. Explain the difference between the definition of a group, and its diagnosis.

APPENDIX 3.1 WHAT IS "ORGANIC EVOLUTION"?

Charles Darwin, via his book *On the Origin of Species* (1859), is generally regarded as the starting point of modern evolutionary theory. Yet, the most common understanding of the word "evolution," that organisms on Earth have changed through time, was *not* the point of Darwin's work. That organisms have changed through time had been well established by savvy natural historians (or natural philosophers, as they were sometimes called) well over 200 years before Darwin. Darwin's contribution was to propose the *means* by which such changes occurred. His hypothesis was constructed in the following way:

(1) Domesticated animals and plants show a wide range of variation.

(2) A wide range of variation exists among wild animals and plants as well.

(3) All living creatures are engaged in a "struggle" to survive and ultimately reproduce, and that struggle is most severe among those individuals that are most closely related.

(4) The struggle to survive in combination with the variation that exists among individuals leads to the survival and, most importantly, production of viable progeny in some variants as opposed to some others, a process that Darwin called natural selection.

(5) The viable progeny of some variants as opposed to others ensures that the characteristics of the successfully reproducing variants make it into the next generation.

(6) This process, repeated over hundreds or even thousands of generations, is *evolution by natural selection*, sometimes called "Darwinian evolution."

The variants that survived to produce viable offspring are said to be more fit than those that did not (note that this definition of fitness has no relationship to time spent in the gym). And successive generations of "fit" offspring would, in a manner analogous to breeding, eventually produce a descendant very different from its ancestor (for example, a new species). So, if higher fitness in some hypothetical lineage of organisms was conferred by having longer legs, then that lineage might show an evolutionary trend toward increasing leg length until the long-legged descendant was sufficiently different from its ancestor to be considered a different species. So Darwin's contribution to ideas about evolution, therefore, was actually the hypothesis of *evolution by natural selection*.

Because the science of genetics was not known to Darwin (having been invented, so to speak, during his lifetime), Darwin had no mechanism to explain what exactly was meant by "closely related," although he knew that in some physical way parents, for example, are closely related to their children. The explicit meaning of relationship came with the understanding of chromosomes, genes, alleles, and, some 70 or so years later, DNA.

Nonetheless, all of those ideas were, from the 1920s onward, integrated into Darwin's original hypothesis (except, of course the molecular basis of inheritance, which came somewhat later) in an intellectual movement called the "New" or "Modern Synthesis." The New Synthesis applied the then-burgeoning fields of population ecology, genetics, paleontology, and statistics to Darwinian evolution, to understanding the precise mechanisms by which particular combinations of genes (genotypes) are selected and passed on to succeeding generations. This led to the hypothesis of an unbroken genetic chain of successive changes in the physical appearance (phenotype) and behavior of organisms, to the origin of new species, as life perpetuated itself on Earth.

In the intervening years since the New Synthesis, the theory of evolution by natural selection has been refined and now incorporates very significant advances in our understanding of genetics, embryology, and molecular biology. The origin of new features is recognized as gradual in some cases, as postulated by Darwin, but also as abrupt in other cases. Sometimes changes in genotype produce new species, but sometimes they may produce phenotypic changes that are either smaller- or larger-scale than the development of merely new species. Natural selection then acts upon phenotypes in the current generation favoring, as Darwin hypothesized, the reproductive viability of some phenotypes and not others.

The fitness – or lack of it – of an organism is determined by an immense number of variables, including, but not limited to, the environment in which it evolves, the other organisms (plant and animal) with which it must interact, and ultimately the viability of its progeny. Obviously these variables are utterly unpredictable, and so evolution by natural selection, as currently understood by evolutionary biologists, does not involve the possibility of predicting future evolutionary events.

As we cannot use it to predict the future, should we consider the hypothesis of evolution by natural selection to be unscientific? Recall that science requires a hypothesis with explicit predictions that can be tested. Although the hypothesis of evolution does not predict the future, it certainly makes testable predictions. For a few examples:

- It predicts that we will find organisms that contain mixes of characters of older organisms and younger organisms occurring intermediate in time between them. (*We do.*)
- It predicts that life's diversity takes the form of many variations on a basic design, with modifications upon modifications that take us to the present. (*It does.*)
- It predicts that the biochemical building blocks of life will be present – if slightly modified – in all organisms. (*They are.*)
- Of particular relevance in this book, it predicted that creatures that mixed bird and "reptile" features would exist. (*They do*; see Chapter 8.)

Geneticist Theodosius Dobzhansky, one of the pioneers of the Evolutionary Synthesis, once wrote in *The American Biology Teacher*, "Nothing in biology makes sense except in the light of evolution." We concur.

Yet for all that there is a pesky irony. The word "evolution" strictly speaking can refer to an unfolding to a predetermined and inevitable end, such as the *evolution* of a tragedy. Because the evolution of life – organic evolution – does *not* unfold along predetermined or inevitable pathways, it is not surprising that the word "evolve" was avoided by Darwin until the very last word of the very last sentence of the very last paragraph of the very last page of *On the Origin of Species* (1859).

Chapter 4

Who Are the Dinosaurs?

WHAT'S IN THIS CHAPTER

This chapter is all about where dinosaurs fit in relation to all the other vertebrates, particularly amniotes. To understand that, we need to:

- Learn something about the course of vertebrate, especially tetrapod, evolution.
- Get on top of some basic tetrapod anatomy.
- Learn who dinosaurs are (and are not).
- Construct both a definition and a diagnosis of Dinosauria (and learn the difference!).

Finding the History of Life

In the preceding chapter, we explored the methods that scientists use to learn the identity and origin of all organisms. Now we will apply those techniques – diagnostic characters hierarchically distributed on parsimonious cladograms – to properly position dinosaurs within the biota. The history of life will unfold as we systematically encounter each bifurcation in the cladistic road, reconstructing the path of evolution until we reach Dinosauria. We'll go with Glinda the Good's admonition that "it's always best to start at the beginning."

The Beginning

Modern life is generally understood to be monophyletic. It's united by the possession of RNA, DNA, cell membranes with distinctive chemical structure, a variety of amino acids (proteins), the metabolic pathways (that is, chemical reactions) for their processing, and the ability to replicate itself (not simply grow).[1] Notice we said "modern" life – for who knows how many forms of molecular life arose, proliferated, and died out very early in Earth's history – before the thing that we now call "life" finally prevailed?

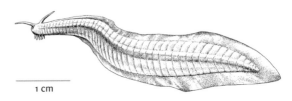

Figure 4.1. *Pikaia gracilens,* a presumed chordate from the Middle Cambrian of Canada.

While it is theoretically possible to construct a cladogram to develop the full history of modern life, we'd have to summarize about 3.8 *billion* years of biotic evolution. Instead, let's cut to the chase, to Animalia, and to a mere 510 Ma, where we first meet an early – but not the earliest – member of Chordata: *Pikaia gracilens*, a 5 cm, flattened, miniature anchovy fillet of a creature that represents an early data point in the ancestry of vertebrates (Figure 4.1).

Chordata

Pikaia reveals characters that are diagnostic of the clade to which we (and the dinosaurs) belong: Chordata ("nerve cord-bearing"). Although *Pikaia* provides an inkling about our distant relatives, what we know about the early evolution of vertebrates and their forebears among Chordata comes principally from living organisms, with some input from a few fossils. The living chordates consist of living urochordates (*uro* – tail; popularly called "sea squirts"), cephalochordates (sometimes called "lancets") and, most important for our story, vertebrates (Figure 4.2). All of these groups are united within Chordata on the basis of a distinctive suite of characters: (a) pharyngeal gills (gills in the throat region); (b) a **notochord**, the stiffening rod that runs down the back of all chordates; and (c) the familiar dorsal (back) nerve cord that, in animals with backbones (e.g., vertebrates), we call a "spinal cord." This whole suite of characters evolved only once, thus uniting these animals as a monophyletic group. We see Chordata and its diagnostic characters at the base of the cladogram in Figure 4.3. With *Pikaia* and several other forms, we – and the dinosaurs – appear to have chordate relatives as far back as the Cambrian (541 Ma to 485.4 Ma).

---

[1] This statement is not strictly true, because viruses don't have intrinsic membranes, amino acids, and metabolic pathways (they hijack the molecules and mechanisms of the cells they invade), nor do they reproduce themselves; retroviruses don't even have DNA. The origin of viruses remains shrouded in mystery; but one hypothesis is that they lost these features early in their evolutionary history.

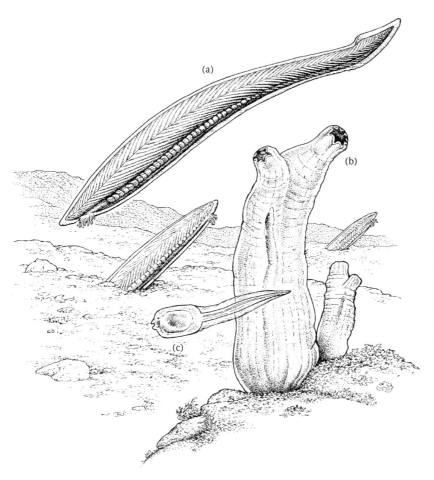

Figure 4.2. Two primitive chordates. (a) A cephalochordate (the lancet *Amphioxus*); (b) urochordate (the sea squirt *Ciona*); and (c) larval sea squirt. Cephalochordates (a) and urochordates (b) share with the vertebrates a host of derived features, including segmentation of the muscles of the body wall, separation of upper and lower nerve and blood vessel branches, and many newly evolved hormone and enzyme systems. The juvenile sea squirt (c) is free-swimming and has a notochord running down its tail. When it metamorphoses into the stationary adult, it parks on its nose and rearranges its internal and external structures.

Vertebrata

Most interesting for us, within Chordata is also found the familiar Vertebrata (*vertere* – to turn, referring to joints in the spine that bend). The diagnostic characters of vertebrates, with the exception of Chondrichyes (sharks, skates, and rays' whose ancestor is thought to have lost its bone through evolution), include a mineralized (and thus, hardened) internal skeleton, bone, divided into discrete pieces called elements,[2] and a variety of other characters (Figure 4.3).

Among Vertebrata, we're naturally interested in the subset Gnathostomata (*gnathos* – jaw; *stome* – mouth), vertebrates with true jaws (among other diagnostic features), which includes us and most of the other vertebrates that might come readily to mind. Ignoring jawless vertebrates (lampreys are a living example), we'll focus directly on the subset of gnathostomes that we call bony fish.

Within the gnathostome lineage occurs a major split in the cladogram, representing a major evolutionary branching point. On the one hand is the lineage of bony fish (Osteichthyes; *osteo* – bone; *ichthys* – fish) leading to old friends such as goldfish, tuna, and salmon. But on the other branch is Sarcopterygii (*sarco* – flesh; see Figure 4.3), a not-so-familiar, but equally friendly group

---

[2] Confusingly, in anatomy, the word "element" has a different meaning from the way it is used in chemistry – even in this book (see Chapter 2 and the Glossary).

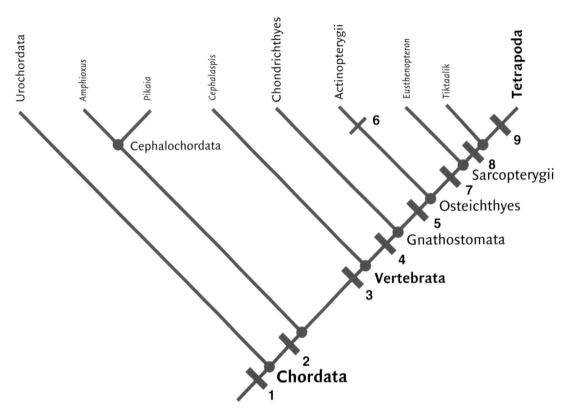

Figure 4.3. Cladogram of Chordata. Because this is a book about dinosaurs (and not all chordates), we provide diagnoses for only some of the groups on the cladogram. The characters include the following: at 1, pharyngeal gill slits, a notochord, and a nerve cord running above the notochord along its length; at 2, segmentation of the muscles of the body wall, separation of upper and lower nerve and blood vessel branches, and new hormone and enzyme systems; at 3, bone organized into elements, neural crest cells, the differentiation of the cranial nerves, the development of eyes, the presence of kidneys, new hormonal systems, and mouthparts; at 4, true jaws; at 5, bone in the endochondral skeleton; at 6, ray fins; at 7, appearance of distinctive bones in fleshy pectoral and pelvic fins (see Figure 4.4); at 8, bones acquiring a more tetrapod appearance and function; elongate rays still present in fin; at 9, skeletal features relating to mobility on land – in particular, four limbs with stable (unchanging) element patterns. Consistent with a cladistic approach, only monophyletic groups are presented on the cladogram. Some of the groups may not be familiar, but are included to fill out the cladogram. *Cephalaspis* was a primitive, jawless, bottom-dwelling, swimming vertebrate, and *Eusthenopteron* was a predaceous lobe-finned fish, bearing many characters present in the earliest tetrapods. *Tikaalik* is an extremely important fossil fish, which shows a striking intermediate morphology between lobe-finned fin morphology and the legs of modern tetrapods.

that is diagnosed by, among other things, the presence of distinctive *lobed* fins, fleshy places where the fin attaches to the body.

Although these are fish, those lobes contain bones that are recognizable as – and thus homologous with – bones in the limbs of tetrapods (Chapter 3), vertebrates with four limbs, including dinosaurs and ourselves (Figure 4.4). What's more, bones of the pelvis, vertebral column, and even bones in the skulls of lobe-finned fishes can also be recognized within tetrapods. All these diagnostic characters strongly indicate that it is here, among the lobe-fins, that the ancestry of Dinosauria – as well as our own ancestry – is to be found (Box 4.1).

(a)   (b)   (c)

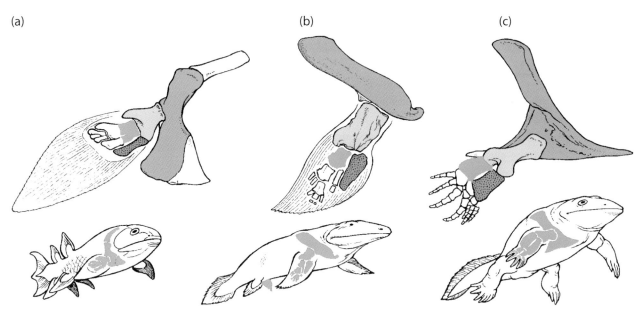

Figure 4.4. Some homologous features linking lobe-finned fish and tetrapods. (a) The fin and shoulder bones (shoulder girdle) of *Eusthenopteron*, an extinct lobe-fin fish; (b) the fin and shoulder girdle of *Tiktaalik*, an animal snugly positioned, morphologically speaking, between lobe-finned fish and tetrapods (see Figure 4.3); and (c) the front limb and shoulder girdle in early tetrapods. Because aspects of the forelimb in different early tetrapods are incomplete, the forelimb shown here is a composite prepared from two early tetrapods (*Acanthostega* and *Ichthyostega*). Key homologous bones are highlighted in both sets of drawings.

Tetrapoda

Remembering what is meant by definition (Chapter 3), modern Tetrapoda is *defined* as the clade that includes salamanders, mammals, and all the descendants of their most recent common ancestor. It is *diagnosed* by the appearance of limbs with the distinctive arrangement of bones shown in Figure 4.4. So now let's take a closer look at Tetrapoda and, because we're interested in dinosaurs, we'll try to understand the part that's generally best preserved: the skeleton.

The Tetrapod Skeleton Made Easy

Figure 4.5 shows a typical tetrapod skeleton – in this case a prosauropod dinosaur – exploded. Not surprisingly (because Tetrapoda is monophyletic), all tetrapods are built on the same basic plan: a vertebral column is sandwiched by paired forelimbs and paired hindlimbs. The limbs are attached to the vertebral column by groups of bones called girdles. At the front end is the head, composed of a skull and mandible, or lower jaw. At the back end is the tail.

Vertebral Column
The vertebral column is composed of distinct, repeated structures (the vertebrae), which consist of a lower spool (the centrum), above which, in a groove, lies the spinal cord (Figure 4.5, inset). Planted on the centrum and straddling the spinal cord is the neural arch. Various processes, that is, bumps on bone that are commonly ridge-, knob-, or blade-shaped, may stick out from each neural arch.

Box 4.1 Fish and Chips

As 1978 turned to 1979, a provocative and entertaining letter and reply were published in the scientific journal *Nature*, discussing the relationships of three gnathostomes: the salmon, the cow, and the lungfish. English paleontologist L. B. Halstead argued that, obviously, the two fish must be more closely related to each other than either is to a cow. After all, he argued, they're both *fish*! A coalition of European cladists disagreed, pointing out that, in an evolutionary sense, a lungfish is more closely related to a cow than to a salmon. In their view, if the lungfish and the salmon are both to be called "fish," then the cow must also be a fish. Can a cow be a fish?

The vast majority of vertebrates are what we call "fishes." They all make a living in either salt or fresh water and, consequently, have many features in common that relate to the business of getting around, feeding, and reproducing in a fluid environment more viscous than air. But, as it turns out, even if "fishes" describes creatures with gills and scales that swim, "fishes" is not an evolutionarily meaningful term because there are no shared, derived characters that unite all fishes that cannot also be applied to all nonfish gnathostomes. The characters that pertain to fishes are either characters present in all gnathostomes (that is, primitive in gnathostomes) or characters that evolved independently.

The cladogram in Figure B4.1.1 is universally regarded as correct for the salmon, the cow, and the lungfish. In light of what we have discussed, this cladogram might look more familiar using groups to which these creatures belong: salmon are ray-finned fish, that is, fish with long rays made of a distinct protein supporting their pectoral and pelvic fins; cows are tetrapods; and lungfishes are lobe-finned fishes. Clearly, lobe-finned fishes share more derived characters in common with tetrapods than they do with ray-finned fishes. Thus there are two clades on the cladogram:

Clade (1) lobe-finned fishes + tetrapods; and

Clade (2) lobe-finned fishes + tetrapods + ray-finned fishes.

Clade 1 is familiar as Sarcopterygia. Clade 2 occurs at the level of all fish (and the descendants of fish) and looks

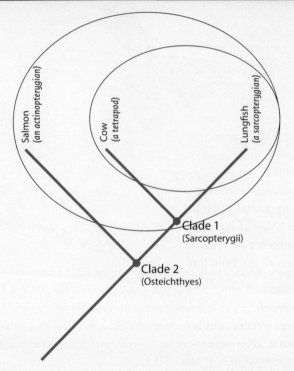

Figure B4.1.1. The cladistic relationships of a salmon, a cow, and a lungfish. The lungfish and the cow are more closely related to each other than either is to the salmon.

like part of the cladogram presented in Figure 4.3 for gnathostome relationships. If only the organisms in question are considered, the only two monophyletic groups on the cladogram must be (1) lungfish + cow; and (2) lungfish + cow + salmon (that is, representatives of the sarcopterygians and Actinopterygii, respectively).

Which are the "fishes"? Clearly the lungfish and the salmon. But the lungfish and the salmon do not in themselves form a monophyletic group unless the cow is also included. The cladogram is telling us that the term "fishes" has *phylogenetic* significance only at the level of Osteichthyes (or even below). But we can and do use the term "fishes" informally. Fish and chips will never be a burger and fries.

Halstead, L. B. 1978. The cladistic revolution – can it make the grade? *Nature*, **276**, 759–760.

Gardiner, B. G., Janvier, P., Patterson, C., *et al.* 1979. The salmon, the cow, and the lungfish: a reply. *Nature*, **277**, 175–176.

These can be for muscle and/or ligament attachment, or they can be sites against which the ends of ribs can abut. The repetition of vertebral structures, a relic of the segmented condition that is primitive for chordates, allows flexibility along the length of the animal.

Girdles

Girdles wrap around things – and the pelvic girdle and pectoral girdle are sheets of bone (or bones) that wrap part-way around – or at least, sandwich – the ribs and pelvic vertebrae (Figure 4.5). The limbs attach to the girdles for the support of the body. The pelvic girdle is the attachment site of the hindlimbs; the pectoral girdle is the attachment site of the forelimbs.

Each side of the pelvic girdle is made up of three bones: (1) a flat sheet of bone, called the ilium, which is fused to the sacrum, a block of vertebrae between the iliac blades; (2) a piece that points forward and down, called the pubis; and (3) a piece that points backward and down, called the ischium. Primitively, the three bones come together in a sculpted-out area of the pelvis called the acetabulum: the hip socket.

By contrast, the pectoral girdle consists of a flat sheet of bone, the scapula (shoulder blade), on each side of the body, attached to the outside of the ribs by ligaments and muscles. The scapula is connected with the coracoid, a shield-like bone, and together these two bones support the attachment of the front limb.

Chest

Some chest elements deserve mention. The breastbone (sternum) is generally a flat or nearly flat sheet of bone that is locked into its position on the chest by the tips of the thoracic (or chest) ribs. Paired collar bones – the clavicles – sit just above and between the paired (right and left) coracoids.

Legs and Arms

Limbs (front and back) in tetrapods show the arrangement pioneered in their sarcopterygian ancestors, an arrangement that paleontologist Neil Shubin immortalized in *Your Inner Fish* as "one bone, two bones, many bones" (compare Figures 4.4 and 4.5). All limbs, whether fore or hind, have a single upper bone connecting to a pair of lower bones ("one bone"). In a forelimb, the upper arm bone is the humerus, and the paired lower bones (forearms) are the radius and ulna ("two bones"). The joint in between is the elbow. In a hindlimb, the upper bone (thigh bone) is the femur, the joint is the knee, and the paired lower bones (shins) are the tibia and fibula ("two bones").

Beyond the paired lower bones of the limbs are the wrist and ankle bones, termed carpals and tarsals, respectively ("many bones"). The bones in the palm of the hand are called metacarpals, the corresponding bones in the foot are called metatarsals and collectively they are termed metapodials. Finally, the small bones that allow flexibility in the digits of both the hands (fingers) and the feet (toes) are called phalanges (singular phalanx). At the tip of each digit, beyond the last joint, is each ungual phalange.

Tetrapods primitively had as many as eight digits on each limb. Very early in the evolutionary history of tetrapods, this number rapidly reduced to, and stabilized at, five digits on each limb, although many groups of tetrapods (including dinosaurs, most notoriously *T. rex*) subsequently reduced that number even further (Figure 4.5).

Head

At the front end of the vertebral column of chordates are the bones of the head, composed, as we have seen, of the skull and mandible (Figure 4.6). Primitively, the skull has a distinctive arrangement: the braincase, a bone-covered box containing the brain, is located centrally and toward the back of the skull. At the back of the braincase is the occipital condyle, the knob of bone that connects the

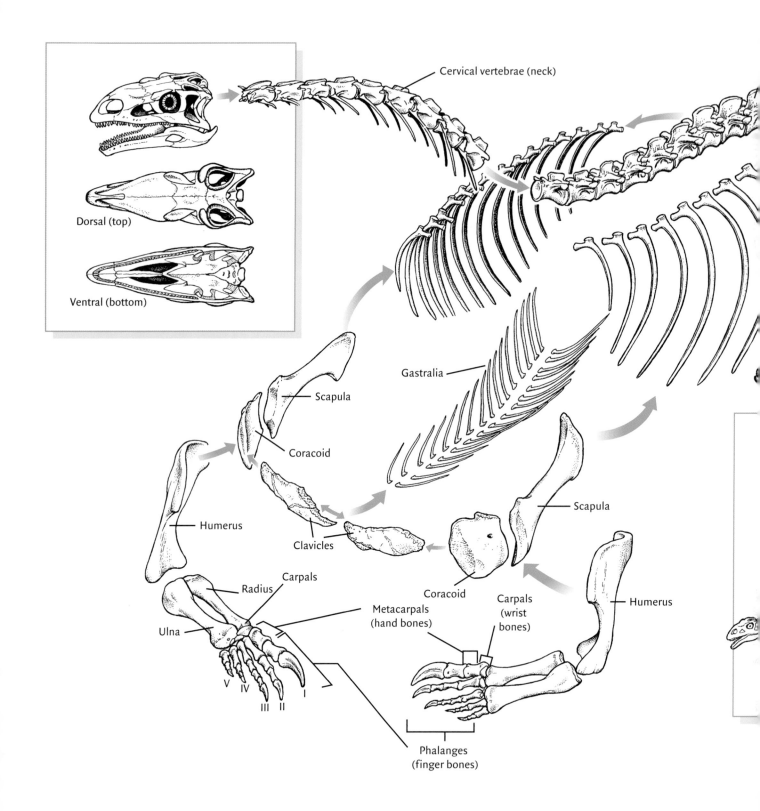

Figure 4.5. Exploded view of a tetrapod skeleton exemplified by the saurischian dinosaur *Plateosaurus*.

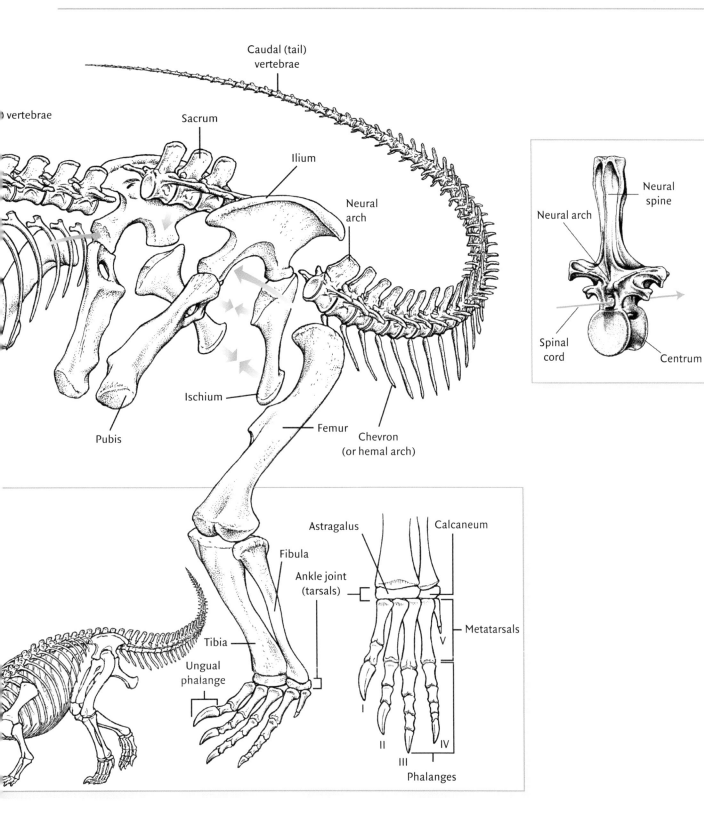

Caudal (tail) vertebrae

vertebrae

Sacrum

Ilium

Neural arch

Neural spine

Neural arch

Spinal cord

Centrum

Ischium

Pubis

Femur

Chevron (or hemal arch)

Astragalus

Calcaneum

Fibula

Ankle joint (tarsals)

Metatarsals

V

Tibia

Ungual phalange

I

II

III

IV

Phalanges

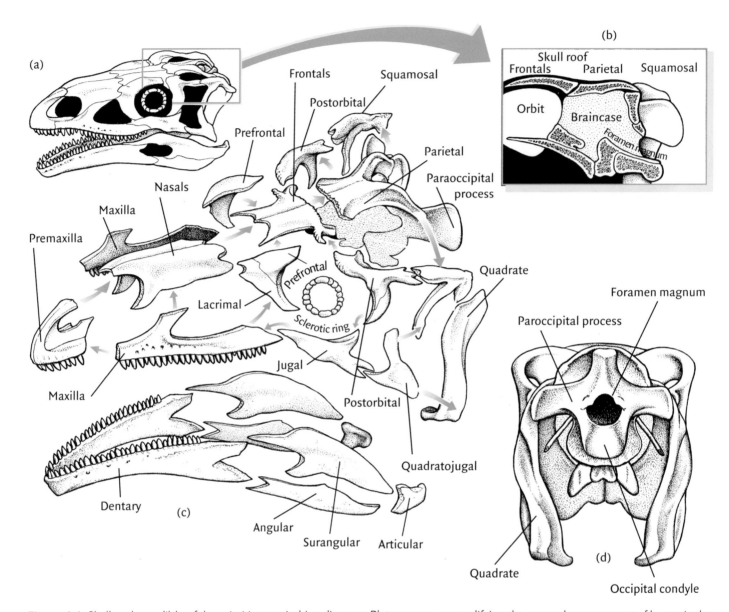

Figure 4.6. Skull and mandible of the primitive saurischian dinosaur *Plateosaurus*, exemplifying the general arrangement of bones in the skull and mandible. (a) Skull elements exploded; (b) cross-section through braincase; (c) exploded elements of mandible (lower jaw); (d) rear (occipital) view of skull.

braincase (and hence the skull) to the vertebral column. A rear-facing opening in the braincase, the **foramen magnum**, allows the spinal cord to enter the braincase to attach to the brain. Located on each side of the braincase are openings for the **stapes**, the bone that transmits sound from the **tympanic membrane** (eardrum) to the brain. Finally, covering the braincase and forming much of the upper rear part of the skull is a curved sheet of interlocking bones, the **skull roof** (inset to Figure 4.6).

The skull has two familiar pairs of openings. Located midway along each side of the skull is a large, round opening – the eye socket, or **orbit**. At the **anterior** tip of the skull is another pair of openings – the **nares** (singular, naris), or nostril openings. Finally, flooring the skull, above the mandible, is a paired series of bones, organized in a flat sheet, which forms the **palate**. Primitively, air

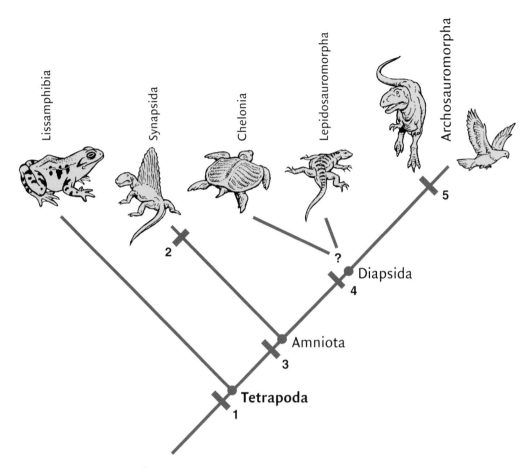

Figure 4.7. Cladogram of Tetrapoda. Derived characters include: at 1, the tetrapod skeleton (see Figure 4.5); at 2, lower temporal fenestra (see Figure 4.9); at 3, the presence of an amnion (see Figure 4.8); at 4, lower and upper temporal fenestrae (see Figure 4.9); and at 5, antorbital fenestra (see Figure 4.12). Lepidosauromorpha is a monophyletic group, the living members of which are snakes, lizards, and the tuatara. The relationships of Chelonia – turtles – within Diapsida are uncertain.

is taken in through the nares, directed through paired openings at the front of the palate (the choanae) and then, once in the mouth, it can be directed into the trachea and to the lungs.

In many vertebrates, including all mammals, a passage forms between the floor of the nasal cavity and the roof of the oral cavity (mouth), so that air breathed in through the nostrils is guided to the back of the throat, bypassing the mouth. As a result, it is possible for chewing and breathing to occur at the same time. That extra passage, a *derived* feature of the skull, is called a secondary palate, and is known in other tetrapods besides mammals, including some turtles, crocodiles, and many dinosaurs. Primitively, however, such as we see in this dinosaur (Figure 4.6), there is no secondary palate, and if food were extensively chewed in the mouth, it would quickly get mixed up with the air that is breathed in. For this reason, chewing is not a behavior of primitive tetrapods.

Within Tetrapoda

Tetrapods share a variety of derived features (Figure 4.7). We have seen many of these in the tetrapod skeleton: the distinctive morphologies of the girdles and limbs, as well as the fixed patterns of skull roofing bones. The hypothesis that all of these shared similarities evolved separately in distantly related organisms is not parsimonious; for this reason, these characters reaffirm Tetrapoda as a monophyletic group. Now we continue our journey to find out exactly where to situate dinosaurs.

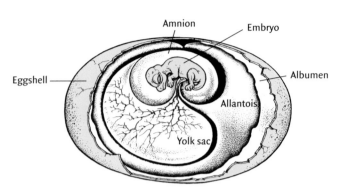

Figure 4.8. The incredible, edible (if unfertilized!), amniote egg.

Amniota

A subset of the tetrapods, Amniota is defined as the monophyletic group that includes birds, mammals, and all the descendants of their most recent common ancestor. Amniota is characterized by the invention of a special membrane for the egg-bound, developing embryo called an **amnion** (Figure 4.8, and see below). Tetrapods without an amnion – **anamniotes** – are today represented only by frogs, salamanders, and a very rare, limbless, tropical amphibian known as a caecilian (almost not worth mentioning, unless you're a caecilian, of course). If the living amphibians are any guide, the life cycles of anamniotes are intimately associated with water, as the eggs require, and likely required, an external source of moisture.

Amniotes, by contrast, are fully terrestrial, and need a means of retaining moisture within the egg. The semi-permeable amnion allows gas exchange but retains water, which permits the embryo to be continuously bathed in liquid. The evolutionary appearance of the amnion occurred in conjunction with several other features including a calcified shell, a large yolk for the nutrition of the developing embryo, and a special bladder for the management of embryonic waste. Amniotic eggs can thus be laid on land without drying out, which allowed amniotes to sever all ties with water (other than for drinking). This was a key step in the evolution of a completely terrestrial lifestyle, and is commonly associated with the advent of reptiles.

There are three great groups of amniotes – primitive amniotes, sometimes termed **anapsids** (*a* – without; *apsid* – arch), **Synapsida** (*syn* – with), and **Diapsida** (*di* – two). They're most easily distinguished by the number and position of the openings in the skull roof behind the eyes, called **temporal fenestrae** (*fenestra* – window; Figure 4.9). Our main interest is in diapsids, but we'll detour briefly to look at some basal amniotes and Synapsida.

Anapsids and Synapsida

The anapsid condition represents what is thought to have been the original morphology of the skull roof in amniotes. In these amniotes, the skull behind the eyes is completely roofed; they therefore have no temporal fenestrae. The anapsid condition is seen in many long-extinct, bulky quadrupeds that do not concern us here, and until recently was thought to persist today only in turtles. Recently, considerable work – both morphological and molecular – suggests that turtles are actually a monophyletic group of highly derived diapsids (see below). This means that there are no living anapsid amniotes. It also has major consequences for the meaning of the word Reptilia (see Box 4.2).

Synapsida is one of two great lineages of amniotic tetrapods (the other is Diapsida). All mammals (including you if, in fact, you are mammalian) are synapsids, as are a host of extinct forms, traditionally (and ironically – see Box 4.2) called "mammal-like reptiles" (Figure 4.10). The split between the earliest synapsids and the earliest representatives of the other great lineage, Diapsida

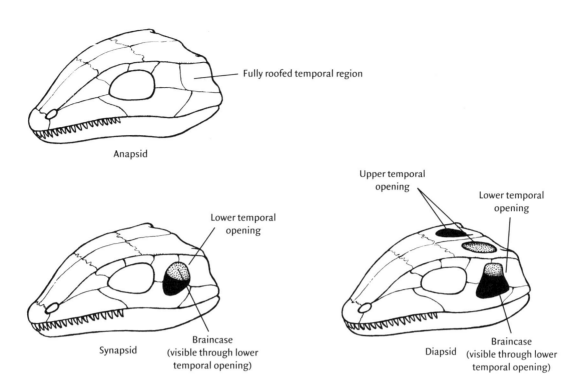

Figure 4.9. Three major skull types found in amniotes.

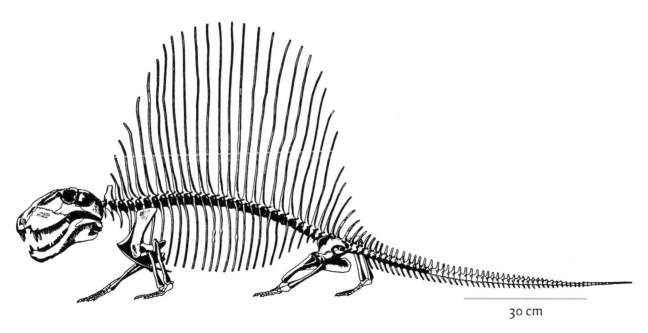

Figure 4.10. *Dimetrodon grandis*, a fin-backed synapsid from the late Paleozoic (Permian Period) of eastern Texas, USA. Not a dinosaur, despite the insistence of your breakfast-cereal box. There may not be much family resemblance, but this bad boy is distantly related to you.

Box 4.2 What, if Anything, is a "Reptile"?

Observant readers will note that the group "Reptilia" is missing from Figures 4.7 and 4.11. But aren't dinosaurs famously reptiles? To abuse Shakespeare: "The fault, dear Brutus, lies not in dinosaurs, but in reptiles."

Recall from Chapter 3 that Linnaeus developed his classification by grouping similar-looking things. His Reptilia (*reptere* – to crawl; Linnaeus coined the name) denoted a group of scaly, four-legged creatures that "crawl on [their] belly like a reptile," as the song[a] goes.

But if the classification is all about relationship, are the living reptiles – snakes, lizards, crocodilians, and the tuatara – monophyletic? Are they really more closely related to each other than they are to anything else? The cladogram in Figure 4.11 and those in Chapter 8 demonstrate that, on the basis of shared derived characters, birds are more closely related to crocodiles than crocodiles are to lizards. And that's just not possible unless a bird is a reptile. But how can a bird be a reptile?

Plainly, a modern, evolution-based Reptilia cannot be Linnaeus' traditional motley crew of crawling, scaly, non-mammalian, nonbird, nonamphibian creatures. Because crocodiles and birds are more closely related to each other than either is to snakes and lizards, a monophyletic group that includes snakes, lizards, and crocodiles (e.g., Reptilia) must also include birds.

So where does that leave reptiles? Reptilia has no diagnostic characters that are not those of Amniota, and so an obvious conclusion that one might reach is that Reptilia = Amniota. This was a common viewpoint in precladistic days, when "reptiles" were identified as a group that could lay eggs on land, through the gentle offices of an amnion. Thought about in those terms, birds, crocodilians, lizards, and dinosaurs are of course amniotes, but then so are mammals, when last we checked. This perspective left systematists with no logical alternative but to do something they were utterly unwilling to do: call mammals reptiles (a moniker applied exclusively to one's ex in those days).

But here the fossil record saves the day: If we also include *extinct* organisms in our cladogram (here we pick

[a] Jerry Leiber and Mike Stoller's "Little Egypt" (1961).

(including dinosaurs), likely occurred between 310 Ma and 320 Ma. Since then, therefore, the synapsid lineage has been evolving independently, genetically unconnected to any other group.

Synapsids are united by having a skull roof that departs from basal amniotes: the skull roof has developed a low opening behind the eye – the **lower temporal fenestra** (see Figure 4.9). Jaw muscles pass behind this opening and attach to the outside of the brain case. Synapsids are a remarkable and diverse group of amniotes and could easily fill a book just like this; indeed, this sentence is currently being read by a synapsid. But because we're interested in dinosaurs, we'll regretfully move right on past them.

Diapsida

The other great clade of amniotes is Diapsida (see Figures 4.7 and 4.9). The living diapsids include about 15,000 total species including snakes, lizards, crocodiles, the tuatara (a lizard-like reptile found only in New Zealand), and birds; extinct diapsids include dinosaurs as well as the dolphin-shaped ichthyosaurs, plesiosaurs (think Loch Ness monster), and mosasaurs, magnificent marine-adapted predatory lizards of the Cretaceous (see Figure 17.10b). Even turtles, long thought to represent the anapsid condition, are now recognized as diapsids (see above, also Box 4.2). Nobody really knows how many members of the diapsid clade have come and gone.

an arbitrary example of one such group, the small, superficially lizard-like Captorhinomorpha), then it can be seen that there exists a clade which we call Reptilia that includes some extinct anapsid amniotes, but which does not include synapids (and thus, mammals). In a phylogenetic telling, the clade closest to Reptilia – e.g., its "sister group" – is Synapsida (including Mammalia), which together form the clade Amniota Figure (Figure B4.2.1).

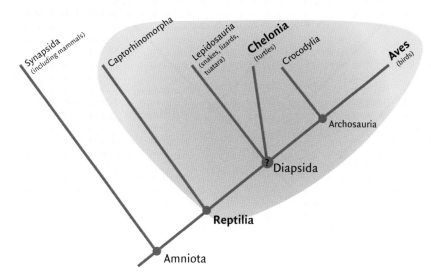

Figure B4.2.1. The inclusion in the cladogram of extinct anapsid amniotes, such as Captorhinomorpha, gives us a slightly more familiar Reptilia, in which Amniota ≠ Reptilia. Here, all reptiles are amniotes, but not all amniotes are reptiles.

Diapsida is defined as birds, lizards, and all the descendants of their most recent common ancestor. It is united by a suite of shared, derived features, including having two temporal openings in the skull roof: the upper temporal fenestra and the lower temporal fenestra (Figure 4.9). The upper and lower temporal fenestrae are thought to have provided accommodation for the bulging of contracted jaw muscles. In dinosaurs, that upper temporal fenestra may have been the location for vascularized (blood vessel-rich) soft tissue that may have supported features for display, such as, for example, the "cockscomb" on a rooster, or even for body temperature control.

There are two major clades of diapsids. The first, Lepidosauromorpha (*lepido* – scaly; *morpho* – shape), is composed of snakes and lizards (including mosasaurs) and the tuatara (among the living), as well as a number of extinct lizard-like diapsids; the second, Archosauromorpha (*archo* – ruling), brings us within striking distance of dinosaurs.

Archosauromorpha

Archosauromorpha is supported by many important, shared, derived characters (Figure 4.11). Within archosauromorphs are a series of basal members that are known mostly from the Triassic. Some bear a superficial resemblance to large lizards; others look like beefed-up crocodiles; a few even have snouts that look like reptilian pigs (see Figure 14.4).

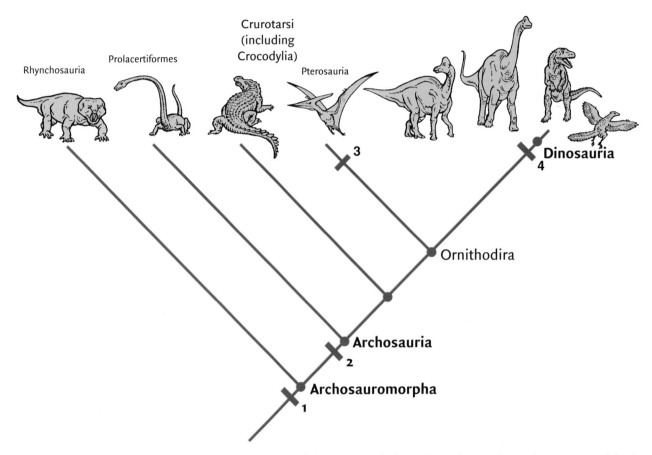

Figure 4.11. Cladogram of Archosauromorpha. Derived characters include: at 1, teeth in sockets, elongate nostril, high skull, and vertebrae not showing evidence of embryonic notochord; at 2, antorbital fenestra (see Figure 4.12), loss of teeth on palate, and new shape of articulating surface of ankle (calcaneum); at 3, a variety of extraordinary specializations for flight, including an elongate digit IV; at 4, **perforate acetabulum** (see complete list of shared derived characters of Dinosauria in Figures 4.14 and 4.15).

A subset of archosauromorphs possesses a number of significant evolutionary innovations (Figure 4.11), most notably an opening on the side of the snout, just ahead of the eye, called the **antorbital fenestra** (Figure 4.12). This is the key character that unites **Archosauria**, the group that contains crocodilians, birds, and dinosaurs. It is ironic that, for all its phylogenetic importance, the function of the antorbital fenestra is still unknown (although it may have had some relationship to maintaining salt balance in the body).

Crocodilians and their close relatives belong to a clade called **Crurotarsi** (*crus* – the joint between the ankle and fibula; *tarsi* – ankles; including some things that only *look* like crocodiles, as well as true crocodiles) about which we won't be too concerned here; dinosaurs and their close relatives constitute a clade called **Ornithodira** (*ornith* – bird; *dira* – neck; see Figure 4.11).

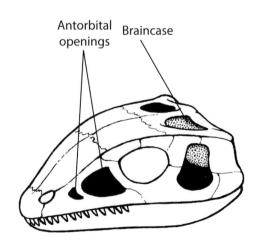

Figure 4.12. An archosaur skull with the diagnostic antorbital fenestra indicated.

Ornithodira brings us quite close to the ancestry of dinosaurs. This group is composed of two monophyletic groups, Dinosauria (*deinos* – terrible) and Pterosauria (*ptero* – winged; Figure 4.13). That pterosaurs are unapologetically Mesozoic archosaurs has led to their being called "dinosaurs"; that they had wings and flew has led some to mistake them for birds; but in fact they were something utterly different from either dinosaurs or birds. They were unique, magnificent, and now, tragically, very extinct.

Dinosaurs

This climb up the cladogram leaves us wheezing and gasping, but at long last situated at the subject of our book: Dinosauria. Dinosauria is formally defined as *Triceratops horridus*, *Passer domesticus* (sparrow; a representative derived bird), and all members of the clade descended from their most recent common ancestor (Figure 4.14).

Here, the utility of the evolutionary definition becomes clear, because as we've come to understand in the past five or so years, the earliest dinosaurs were not very different from their nondinosaur contemporaries, and as more has been understood about those contemporaries, the number of obvious diagnostic characters for the group has shrunk. As we shall see in Chapter 5, however, the evolutionary definition

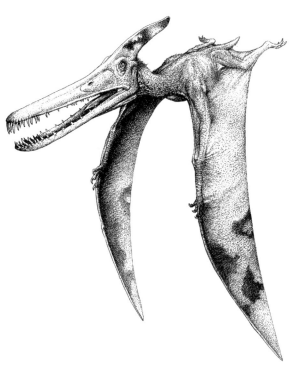

Figure 4.13. A close relative of Dinosauria: Pterosauria, as represented by a pterodactyloid pterosaur, from the Cretaceous of Asia.

precisely and effectively delimits the monophyletic group that we wish to designate as Dinosauria.

So while the nature of the transition from nondinosaur to dinosaur is the subject of the next chapter, for now, however, we highlight a single, universally accepted diagnostic character that pertains to Dinosauria: the perforate acetabulum (Figure 4.15; an opening in the hip socket of the pelvis).

Standing Tall

A feature shared by all dinosaurs, but that is *not* uniquely dinosaurian (although at one time it was thought to be) is the erect, or parasagittal stance; that is, a stance in which the plane of the legs is parallel to the vertical (mirror image) plane of the torso (see Box 4.3). This stance is also present in a few archosaurs very close to Dinosauria (see Chapter 5) but because it is truly emblematic of Dinosauria, we include it here.

In dinosaurs, an erect stance consists of a suite of anatomical features with important behavioral implications. The head of the femur (thigh bone) is oriented at approximately 90° to the shaft. The head of the femur itself is barrel-shaped (unlike the familiar ball seen in a human femur), so that motion in the thigh is restricted largely to forward and backward, within, as we've seen, a plane parallel to that of the body. The ankle joint is modified to become a single, linear articulation. This type of joint, termed a modified mesotarsal joint, allows movement of the foot only in a plane parallel to that of the body: forward and backward (Figure 4.16). This design speaks clearly for a particular behavior: efficient running locomotion. And running, of

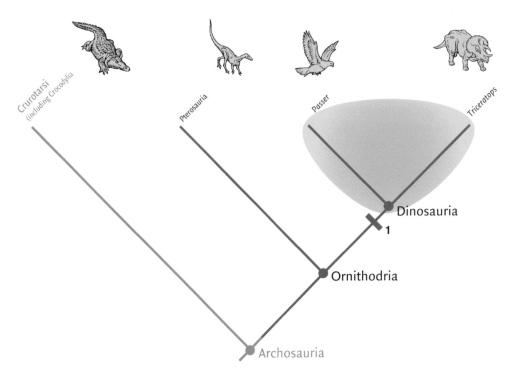

Figure 4.14. Cladogram of Ornithodira showing the phylogenetic definition of Dinosauria. Derived character at **1**, perforate acetabulum (see also Figure 4.15).

Figure 4.15. The dinosaur perforate acetabulum, elegantly modeled by a *Tyrannosaurus rex* cast in the Fukui (Japan) Prefectural Dinosaur Museum.

Box 4.3 Stance: It's Both Who You Are and What You Do

Tetrapods that are most highly adapted for land locomotion tend to have an erect, or upright, stance. This clearly maximizes the efficiency of the animal's movements on land, and it is not surprising that, for example, all mammals are characterized by an erect stance. Tetrapods such as salamanders (which are adapted for aquatic life) display a **sprawling stance**, in which the legs splay out from the body nearly horizontally. The sprawling stance seems to have been inherited from the original position of the limbs in early tetrapods, whose sinuous trunk movements (presumably inherited from swimming locomotion) aided the limbs in land locomotion.

Some tetrapods, such as crocodiles, have a semi-erect stance, in which the legs are directed at something like 45° downward from horizontal (Figure B4.3.1). Does this mean that the semi-erect stance is an adaptation for a combined aquatic and terrestrial existence? Clearly not, because a semi-erect stance is present in the large, fully terrestrial monitor lizards of Australia (goanna) and Indonesia (Komodo dragon). If adaptation is the only factor driving the evolution of features, why don't completely terrestrial lizards have a fully erect stance, and why don't aquatic crocodiles have a fully sprawling stance?

The issue is more complex and is best understood through adaptation to a particular environment or behavior, *as well as* through inheritance. If we consider stance simply in terms of ancestral and derived characters, the ancestral condition in tetrapods is sprawling. An erect stance represents the most highly derived state of this character, but are animals with sprawling stances not as well designed as those with erect stances?

In 1987, D. R. Carrier of Brown University, Rhode Island, USA, hypothesized that the adoption of an erect stance represents the commitment to an entirely different mode of respiration (breathing) as well as locomotion (see Chapter 14 on "warm-bloodedness" in dinosaurs). Those organisms that possess a semi-erect stance may reflect the modification of a primitive character (sprawling) for greater efficiency on land, but they may also retain the less-derived type of respiration.

Dinosaurs (see Figure 4.11) and mammals both have fully erect stances, which represent a full commitment to a terrestrial existence as well as to a more derived type of respiration. The designs of all these organisms are thus compromised among inheritance, habits, and mode of respiration. Who can say what other influences are controlling morphology?

Interestingly, the cladogram (see Figure 4.7) shows that the most recent common ancestor of dinosaurs and mammals – some ancestral amniote – was itself an organism with a sprawling stance. Because dinosaurs and mammals (or their precursors) have been evolving independently since their most recent common ancestor, an erect stance must have evolved at least twice in Amniota: once among the synapsids and once in dinosaurs' immediate precursors.

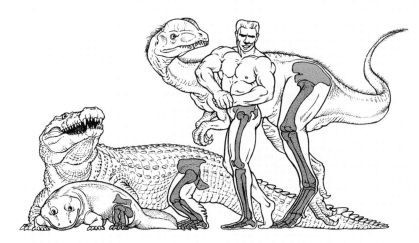

Figure B4.3.1. Stance in four vertebrates. To the left, the primitive amphibian and crocodile (behind) have sprawling and semi-erect stances, respectively. To the right, the human and the dinosaur (behind) both have fully erect stances.

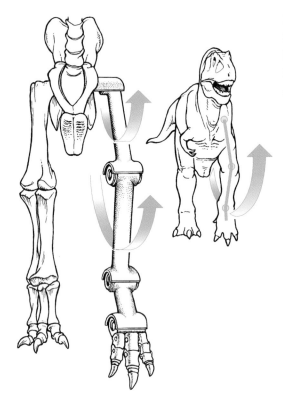

Figure 4.16. The fully erect posture in dinosaurs. Unlike in, for example, a human, the bones of the leg restrict movement to only forward and backward, effectively restricting movement to just one plane.

course, highlights another quality of dinosaurs: dinosaurs were – and are – terrestrial beasts through and through.

SUMMARY

In this chapter, cladograms are used to map the relationship of Dinosauria to the rest of the biota. Dinosaurs fall within Vertebrata, backbone-bearing animals within the larger group Chordata (bilateral creatures bearing a dorsal nerve cord and, at least embryonically, a notochord). Within vertebrates, dinosaurs are tetrapods, or animals with four limbs. Dinosaurs are amniotic tetrapods (or amniotes), which means that, like ourselves, they possess an amnion.

Amniotes are generally identified by the number and type of temporal fenestrae. Synapsida (the group that includes mammals) is united by the possession of lower temporal fenestrae, while Diapsida (the group that includes snakes, lizards, crocodilians, birds, and possibly turtles) all possess lower and upper temporal fenestrae. The split between these two major groups of reptiles took place about 320 Ma.

Dinosaurs (recall that in this book, we use the term to specify nonavian [nonbird] dinosaurs) are diapsids, most closely related to crocodiles and birds, both of which are archosaurs, a group of diapsid reptiles united by the presence of an antorbital opening. Calling birds reptiles is, of course, contrary to conventional Linnaean classification, which in this case fails to accurately depict their evolutionary relationships as revealed by cladistic analysis.

Finally, this chapter contains a brief summary of basic diapsid bone morphology. Elements of the postcranial skeleton are presented, including centra, neural arches, **hemal** arches, cervical vertebrae, caudal vertebrae, ribs, gastralia, pelvic and pectoral girdles. Limb skeletal structure is presented, including humerus, radius, and ulna; femur, tibia, and fibula; carpals, tarsals, metapodials, and phalanges. The elements of the skull and mandible are also presented.

SELECTED READINGS

Bakker, R. T. and Galton, P. M. 1974. Dinosaur monophyly and a new class of vertebrates. *Nature*, 248, 168–172.

Benton, M. J. 1997. Origin and early evolution of the dinosaurs. In Farlow, J. O. and Brett-Surman, M. K. (eds.) *The Complete Dinosaur*. Indiana University Press, Bloomington, IN, pp. 204–214.

Benton, M. J. 2004. Origin and relationships of Dinosauria. In Weishampel, D. B., Dodson, P., and Osmólska, H. (eds.) *The Dinosauria*, 2nd edn. University of California Press, Berkeley, pp. 7–20.

Brusatte, S. L. 2012. *Dinosaur Paleobiology*. Wiley-Blackwell, Oxford, 322 pages.

Carrier, D. R. 1987. The evolution of locomotor stamina in tetrapods: circumventing a mechanical constraint. *Paleobiology*, 13, 326–341.

Gauthier, J. A., Kluge, A. G., and Rowe, T. 1988. Amniote phylogeny and the importance of fossils. *Cladistics*, 4, 105–209.

Parrish, M. J. 1997. Evolution of the archosaurs. In Farlow, J. O. and Brett-Surman, M. K. (eds.) *The Complete Dinosaur*. Indiana University Press, Bloomington, IN, pp. 191–203.

Sereno, P. C. 1991. Basal archosaurs: phylogenetic relationships and functional implications. *Journal of Vertebrate Paleontology*, 11 (suppl.), 1–53.

TOPIC QUESTIONS

1. Define: Chordata, notochord, gnathomstome, Sarcopterygii, skull, mandible, girdles, neural arch, centrum, process, ilium, ischium, pubis, acetabulum, sternum, humerus, femur, radius, ulna, tibia, fibula, phalanx, ungual, metacarpals, metatarsals, metapodials, occipital condyle, foramen magnum, stapes, skull roof, tympanic membrane, nares, orbit, palate, amniote, anamniote, amnion, anapsid, synapsid, diapsid, upper temporal fenestra, lower temporal fenestra, archosaur, Archosauromorpha, lepidosaur, Crurotarsi, Ornithodira, Dinosauria, pterosaur, mesotarsal, Ornithischia, and Saurischia.
2. Explain the importance of the amnion in the evolution of terrestrial tetrapods.
3. Describe the basic structure of the vertebrate limb.
4. Draw a skull and lower jaws, indicating the skull roof, braincase, temporal region, orbit, nares, and snout.
5. For how long have the mammal and bird lineages been evolving separately?
6. Why is it that we said that a bird was more closely related to a crocodile than a crocodile is to a lizard?
7. Construct a cladogram with only Vertebrata, Diapsida, Dinosauria, Synapsida, and Lepidosauria marked on it.
8. Construct a cladogram with just Dinosauria, Ornithischia, Archosauria, and Crocodylia marked on it.
9. How can birds be reptiles?
10. Within Amniota, warm-bloodedness and flight occur in bats, in birds, and in pterosaurs. Use a cladogram to show how many separate evolutionary events this required.

In the Beginning...

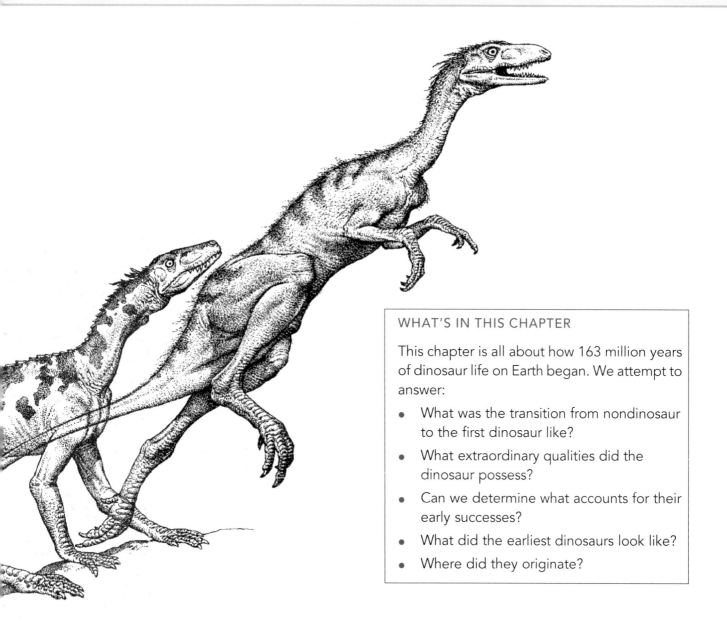

WHAT'S IN THIS CHAPTER

This chapter is all about how 163 million years of dinosaur life on Earth began. We attempt to answer:

- What was the transition from nondinosaur to the first dinosaur like?
- What extraordinary qualities did the dinosaur possess?
- Can we determine what accounts for their early successes?
- What did the earliest dinosaurs look like?
- Where did they originate?

Finding the First Dinosaur

How did it all begin? Was there something special, or different, or *better* about that first dinosaur, that led to all the rest? Did it show up, modest but fetching, on a half-shell? Do we even know who the first dinosaur was? Shouldn't the ancestor of all dinosaurs have been something *different* and *unique*?[1]

Twentieth-century British paleontologist Alan Charig certainly thought so, and proposed that it was their erect stance that conferred upon dinosaurs an irresistible selective advantage that ensured the success of the group. But as we shall see, dinosaurs didn't invent the fully erect stance, and many nondinosaurs of the time had it as well. Maybe the first true dinosaur was just a slightly modified Middle Triassic nondinosaur archosaur doing its Middle Triassic archosaur thing. If so, is it possible that the importance of that first dinosaur was just the subsequent proliferation of all its descendants?

A good guess for the age of the very first dinosaur could be, perhaps, towards the end of the Middle Triassic. Still, even if such rocks and fossils were extremely abundant, the likelihood of finding the actual *ur*-dinosaur – the very beast that was the great granddaddy (or mommy) of a lineage – is vanishingly small; even if you did find the remains of that animal, how would you know it was *the one* (and not, for example, its first cousin or next-door neighbor)?

So we'll resort again to the cladogram: the hierarchy of characters that specifies what features ought to be present in an ancestor. It is then a question of finding a creature that most closely matches the expected combinations of characters and character states. We'll search among primitive – but not *too* primitive (!) – Archosauria (sometimes termed basal archosaurs, for their positions near the base of the archosaur cladogram) for fossils that embody the unique features, specified by the cladogram, of the hypothetical dinosaur ancestor. In the case of dinosaurs, we have several small basal archosaurs that are very good candidates.

Archosauria

Crurotarsi vs. Ornithodira

Recall (Chapter 4) archosaurs: that group of reptiles bearing an antorbital fenestra (see Figures 4.11 and 4.12), whose living representatives are crocodiles and birds. Crocodiles and birds represent a fundamental split early in the evolution of archosaurs, with two groups emerging: Crurotarsi, the crocodile line, and Ornithodira (Figure 5.1), the bird line. These two are distinguished by their ankle bones (Figure 5.2). In crurotarsans, the two bones of the ankle, the calcaneum and the astragalus, are more or less equally sized; this is called, unsurprisingly, the crurotarsal ankle. In ornithodirans, however, the astragalus is somewhat larger, and the ankle allows movement in only one plane: this is the familiar (again from Chapter 4) mesotarsal ankle. These differences reflect different modes of walking; crurotarsans tend to have a sprawling, or semi-erect stance, while ornithodirans generally – but not always – have an erect stance (see Figure 4.16).

Crurotarsi is composed of a variety of ground-hugging, quadrupedal, armored behemoths, several of which are showcased in Figure 5.3. Most crurotarsans didn't make it into the Jurassic. The exceptions are crocodilians and their relatives (Crocodylomorpha), which still reside among the living.

---

[1] Answers: Not sure; No; No; Not necessarily!

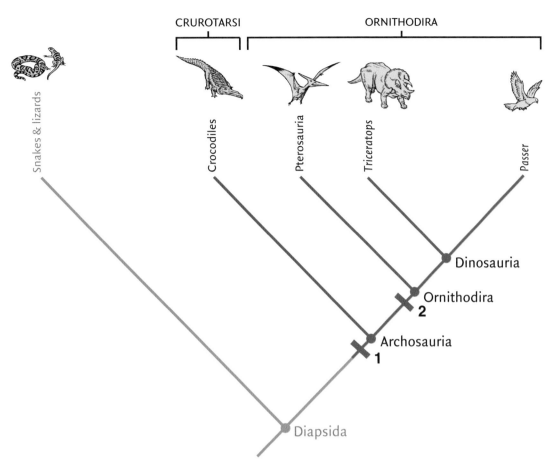

Figure 5.1. Cladogram of Archosauria, showing the fundamental split between Crurotarsi and Ornithodira. Derived characters include: at **1**, antorbital opening; at **2**, distal ends of neural spines not expanded; second phalanx of hand digit II longer than first phalanx; tibia longer than femur; compact metatarsus.

Ornithodira consists of two groups: Dinosauromorpha (*morph* – shape), which includes true dinosaurs and their near-relatives, and Pterosauria, the magnificent flying reptiles that aren't birds (Figures 4.13 and 14.7, respectively). Birds are theropod dinosaurs and sadly, are the last standing ornithodiran representatives (Figure 5.1; see Chapters 7 and 8).

Dinosauromorpha

Paleontology is notoriously terminology-rich, and as we get successively closer to Dinosauria, the names come thick and fast. Dinosauromorpha is a formal name encompassing dinosaurs *and* their very close relatives. To add to the confusion, dinosauromorphs exclusive of dinosaurs were once called "pseudosuchians." Some paleontologists recognize a group – not quite true dinosaurs – snuggled even closer to Dinosauria than dinosauromorphs, called Dinosauriformes (*formes* - in form of). Finally, a useful monophyletic group of dinosauromorphs exclusive of pterosaurs is the clade Avemetatarsalia ("bird feet"). Don't panic; we'll stick to the KISS Principle here and keep it simple, with just the clades Ornithodira and, within it, Dinosauromorpha (including Dinosauria; Figure 5.4).

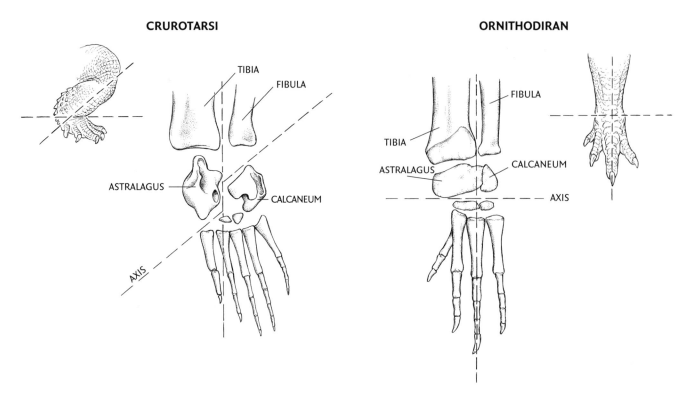

CRUROTARSI   **ORNITHODIRAN**

Figure 5.2. Crurotarsan and ornithodiran ankles compared. In crurotarsans, the ankle must accommodate a complicated range of motion between the moving leg – which is held at a constantly changing acute angle to the ground, and the foot, which is parallel with the ground. The mesotarsal ankle, on the other hand, transmits the single-plane movement of the legs to the same, single-plane movement of the foot.

Figure 5.3. Assorted primitive archosaurs. (a) A **phytosaur** *(Rutiodon)*; (b) an aetosaur *(Stagonolepis)*; (c) a rauisuchian *(Postosuchus)*; and (d) a primitive crocodile *(Protosuchus)*.

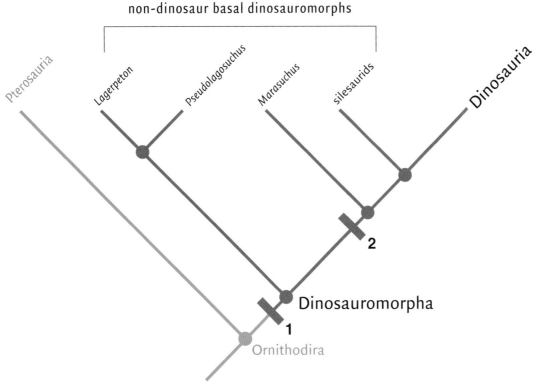

Figure 5.4. Cladogram of Dinosauromorpha. Derived characters include, at **1**, offset head of femur, straight cnemial crest, longest metatarsal > 50 percent of length of tibia, metatarsal V without phalanges and pointed; at **2**, elongate deltopectoral crest on humerus; tibia with transversely expanded subrectangular distal end.

So who are these dinosauromorphs? The earliest, most primitive forms, nondinosaur (basal) dinosauromorphs, were small (about 1 m or less, including tail), lightly built, likely insectivorous or carnivorous creatures, many of whom walked on all fours but ran bipedally, and could have starred in the opening scene of *Jurassic Park II*. From Argentina come *Marasuchus* (once known as *"Lagosuchus")* and *Lagerpeton*. First from Poland, and then Africa, Argentina, and North America, are the diminutive silesaurids, including *Silesaurus, Asilisaurus,* and *Pseudolagosuchus*, respectively. Also among the silesaurids is the poorly understood *Lewisuchus*[2] from Argentina. Only recently recognized, silesaurids apparently had a world-wide distribution. Silesaurids were likely herbivorous, with teeth that superficially resemble those of ornithischians. Several nondinosaur basal dinosauromorphs are shown in Figure 5.5; there are many others; here we've selected just a few to represent the group.

Dinosauria

Sometime around 245 Ma to 230 Ma, a true dinosaur likely first appeared within some population of nondinosaur dinosauromorphs. We can be pretty certain of this because the line between

[2] There is some evidence that *Lewisuchus* is the same animal as *Pseudolagosuchus*. If so, because *Pseudolagosuchus* is the younger (more recently named) name, all specimens of *Pseudolagosuchus* will come to be called *Lewisuchus*.

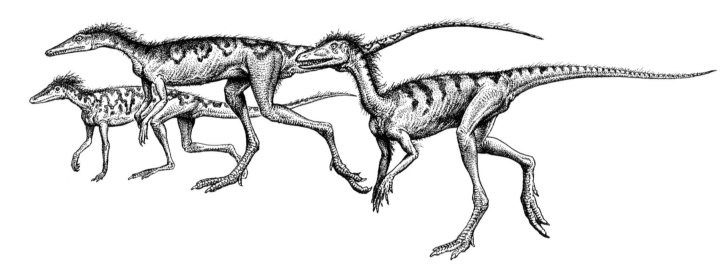

Figure 5.5. Basal dinosauromorphs reconstructed – *Silesaurus, Marasuchus, Pseudolagosuchus.*

nondinosaur and dinosaur is very finely shaded. S. Brusatte (see Chapter 15), in his book, *The Rise and Fall of the Dinosaurs* (2019) says it masterfully:

> At some point one of these primitive dinosauromorphs evolved into true dinosaurs. It was a radical change in name only. The boundary between nondinosaurs and dinosaurs is fuzzy, even artificial, a byproduct of scientific convention. The same way that nothing really changes as you cross the border from Illinois to Indiana, there was no profound leap as one of these dog-sized dinosauromorphs changed into another dog-sized dinosauromorph that was just over the dividing line [between nondinosaurs and true dinosaurs]. (pp. 33–34)

Here is where understanding the distinction between definition and the identification of diagnostic characters becomes essential. Recall the definition of Dinosauria from Chapter 4: dinosaurs are *the monophyletic group (e.g., the clade) that contains both* Passer *and* Triceratops *and all organisms descended from their most recent common ancestor* (Figure 4.14). This precisely limits who can and who can't be considered a dinosaur regardless of how closely it resembles a dinosaur.

Now, recall how a cladogram works: it identifies the monophyletic group (Dinosauria), and then successively more distant forms, based upon hierarchical character distributions. As more and more of these closely related nondinosaur dinosauromorphs are discovered, characters that were originally thought to pertain exclusively to dinosaurs now appear more primitively in the cladogram – outside of the dinosaur group that is circumscribed by the definition of Dinosauria. This leaves Dinosauria with fewer and fewer diagnostic characters.

As recently as 2010, a clear suite of characters was thought to diagnose Dinosauria: these are shown in Figure 5.6. Of these, several "unambiguous" characters are maybe a teeny bit more ambiguous than previously thought; not just the characters themselves, but their very sizes may differ between dinosaurs and nondinosaur dinosauromorphs, and – recalling that these are broken, fragmental fossils, and not beautifully preserved 100 percent complete skeletons – interpretations of what is and is not present – and what is and is not significant – on a particular fossil bone may vary. In short, the distinction between nondinosaurs and dinosaurs is such a knife edge that Sterling Nesbitt (Chapter 16), one of the world's most accomplished authorities on dinosaur origins, has

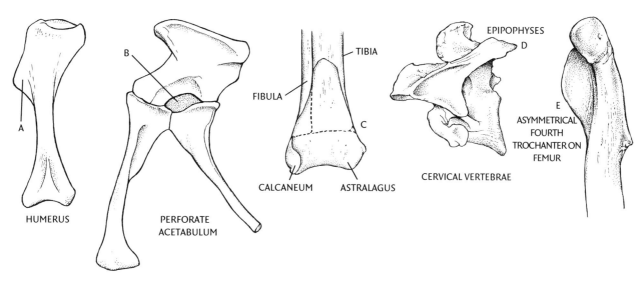

Figure 5.6. Some of the derived characters once thought to unite Dinosauria. (a) Elongate deltopectoral crest on humerus (increasing power of forearms); (b) perforate acetabulum; (c) fibula contacts ≤30 percent of astragalus (part of the mesotarsal ankle); (d) epipophyses on cervical vertebrae (decreasing rotation around neck); and (e) asymmetrical fourth trochanter on femur.

concluded that the only character that he can be confident actually diagnoses dinosaurs is the perforate acetabulum (Figures 4.14 and 5.6b).[3]

Early Dinosauria

The oldest record of dinosaurs *might* be some tantalizing footprints from Middle Triassic rocks in Europe and South America (in the range of 242 Ma to 237 Ma). These prints appear to match what is known of feet of the earliest dinosaurs, and could be a reasonable age for the first appearance of dinosaurs on Earth. On the other hand, can we be *absolutely* sure that they were not made by nondinosaur dinosauromorphs?

The oldest known confirmed dinosaur material, preserved as body fossils, comes from the Upper Triassic Ischigualasto Formation of Argentina. The Ischigualasto Formation has been definitively dated: its base could be as old as (or even slightly older than) 229 Ma; the top of the fossil-bearing section is about 2 million years younger.

The Ischigualasto produced several primitive bipedal dinosaurs, including *Eoraptor* and *Herrerasaurus* (Figure 5.7). Not only are these among the oldest known dinosaur body fossils, but they are by their characters the most primitive.

Their very primitiveness has made them difficult to classify definitively. They were once believed to be generic dinosaurs that belonged to neither of the two great groups of Dinosauria (Saurischia and Ornithischia; see below); then they came to be identified as primitive theropods (a type of saurischian); recently, it was suggested that *Eoraptor* is a very primitive sauropodomorph (yet

---

[3] In 2018, the molecular mechanisms that control the development of the perforate acetabulum were described by biologist S. Egawa and colleagues. In fact, the change was a very small one, mirroring the fine distinctions between true dinosaurs and their antecedents. Egawa, S., Saito, D., Abe, G., and Tamura, K. 2018. Morphogenetic mechanism of the acquisition of the dinosaur-type acetabulum. *Royal Society Open Science*, 5, 180604. http://dx.doi.org/10.1098/rsos.180604

(a)

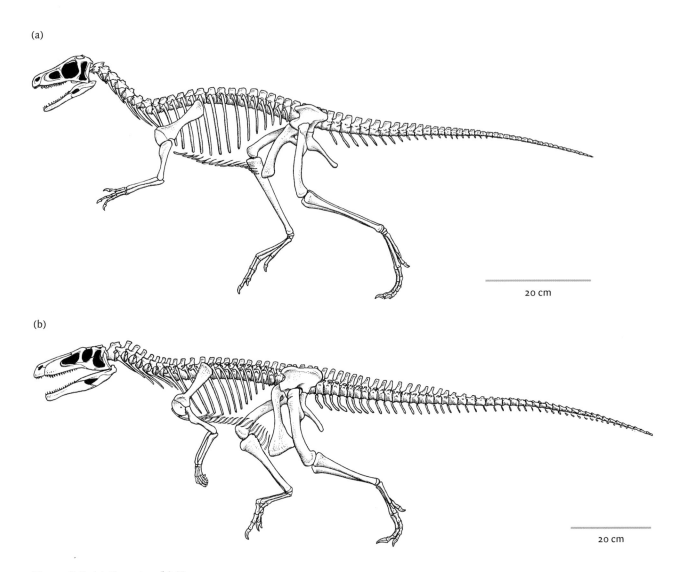

20 cm

(b)

20 cm

Figure 5.7. (a) *Eoraptor;* (b) *Herrerasaurus.*

another type of saurischian); while *Herrerasaurus* and a newly discovered saurischian, *Eodromaeus*, are both theropods!

From the same rocks also came a frustratingly fragmental likely oldest ornithischian: *Pisanosaurus*, as well as another sauropodomorph: *Panphagia,* and the theropod *Sanjuansaurus. Eocursor*, a fragmentary, but unquestioned ornithischian, was found in what are thought to be Upper (Late) Triassic rocks in South Africa.

Nearby rocks (in Brazil) of presumed equivalent age, have produced the basal saurischian *Staurikosaurus* and sauropodomorph *Saturnalia*. In rocks slightly younger than those that produced *Staurikosaurus*, also in Brazil, the incompletely known *Guaibasaurus* was discovered (Figure 5.8). Originally described as a basal saurischian, it was reinterpreted as a theropod, and then again interpreted as a saurischian too primitive to be properly considered either a theropod or sauropodomorph. *Staurikosaurus* comes from rocks dated at 233 Ma, and thus *could* be as old as that, making it potentially the oldest dinosaur known; its precise age, however, remains equivocal.

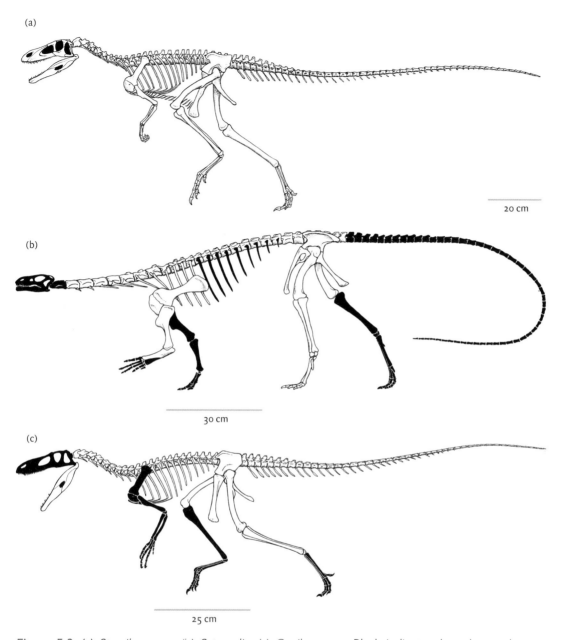

Figure 5.8. (a) *Staurikosaurus*; (b) *Saturnalia*; (c) *Guaibasaurus*. Black indicates those bones that are actually known from these animals.

Ornithischia and Saurischia

With the Late Triassic appearance of these dinosaurs, it's high time we formally met the two great clades of Dinosauria: Ornithsichia and Saurischia. In 1887, English paleontologist Harry Seeley first recognized a fundamental division among dinosaurs. **Saurischia** (*saur* – lizard; *ischia* – hip) were those that had a pelvis more like a lizard, in which the pubis is directed anteriorly, and slightly downward (Figures 5.9). **Ornithischia** (*orni* – bird) were all those dinosaurs that had a bird-like pelvis, in which at least a part of the pubis runs posteriorly, along the lower rim of the ischium (Figure 5.10). This type of pubis is called **opisthopubic**. This pelvic distinction has held sway ever since Seeley proposed it.

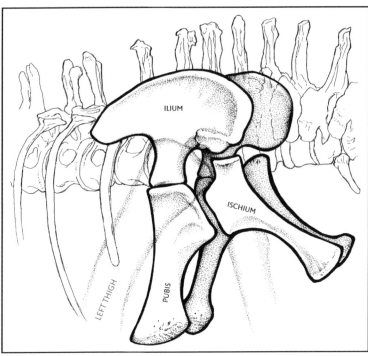

Figure 5.9. The pelvis of the saurischian *Apatosaurus*, viewed from the left side. The pubis is directed anterior only, exemplifying the saurischian condition.

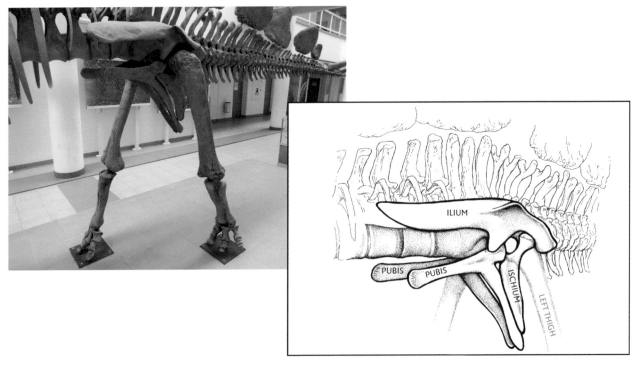

Figure 5.10. The pelvis of the ornithischian *Stegosaurus*, viewed from the left side. A piece of the pubis is pointing posterior, along the base of the ischium, exemplifying the ornithischian condition.

Seeley described the shape of the pelvises exactly as he (correctly) saw them: the saurischian pelvis looks rather similar to that found in a crocodile or lizard; while the ornithischian pelvis, with part of the pubis pointing backwards and running along the base of the ischium, is superficially reminiscent of a bird pelvis. But he wasn't thinking cladistically: the saurischian pelvis is in fact the primitive condition for all tetrapods. Moreover, ironically, birds are *saurischian* dinosaurs (Chapters 6 to 8), and thus, their pelvis is thought to have independently (separately) evolved the rearward-facing pubic branch of ornithischians. From our (living) vantage point, Ornithischia is, evolutionarily speaking, a dead end.

That dinosaurs had one or the other kind of pelvis was of great importance to understanding the evolution of these animals and, for Seeley, it implied that the ancestry of Ornithischia and Saurischia was to be found separately and deep within a heterogeneous group of primitive archosaurs once called Thecodontia.[4] Had he known the term, Dinosauria would not have been a monophyletic group for Seeley. In the end, however, cladistic analyses have seemed to support Seeley's Saurischia–Ornithischia split, but always within a monophyletic Dinosauria (Box 5.1).

Is Saurischia More Primitive than Ornithischia?

The distinction between Ornithischia and Saurischia is valid and both groups are diagnosed by suites of well-established characters (Figure 5.11). But is one more primitive? Saurischians, with their claws, and teeth, appear to be a lot like their basal dinosauromorph forebears – claws, teeth, etc. – and in many books, including this one, they are treated first, suggesting the intuitive notion that saurischians are more primitive and that somehow ornithischians must have evolved from a saurischian ancestor. Saurischians also appear – if not earlier, well, more abundant and diversified – than ornithischians.

No conclusive evidence for this, ornithischians evolving from saurischians, has ever surfaced. Figure 5.11 shows that, cladistically speaking, Saurischia is no less (or more) derived than Ornithischia; thus, we really don't know which came first. And the record of dinosaur body fossils shows that the two great clades of dinosaurs, Ornithischia and Saurischia, very likely go back to the Late Triassic.

The Evolution of Dinosauria

The earliest days of dinosaur evolution remain shrouded in mystery. As we have seen, the oldest dinosaurs known are from South America; Argentina and, possibly, Brazil. In North America, dinosaurs are also found in the southwest part of the United States, in rocks that range in time from nearly as old as, to up to 10 million years younger than, the Late Triassic dinosaurs of South America.

The dinosaurs from the American southwest are not primitive; animals such as the small carnivorous bipeds *Coelophysis bauri* and *Tawa hallae* are unambiguous examples of theropods, a type of saurischian that we shall formally meet in Part II of this book.

The relationships among dinosauromorphs, dinosaurs, and their near relatives, pterosaurs (see Chapter 4; Figures 4.11 and 4.13) have been somewhat confusing – mainly because the characters of basal dinosauromorphs and early dinosaurs are in the main, very primitive, and those of pterosaurs

---

[4] The group "Thecodontia" is based on the same characters that diagnose all archosaurs. If we're discussing archosaurs, we can't cherry-pick a few basal ones; we need to include *all* members of the group (including dinosaurs and pterosaurs) that bear the diagnostic characters. In short, the term "Thecodontia," though venerable, has not withstood cladistic scrutiny (see Chapters 8 and 15).

Box 5.1 But What if We've Had it TOTALLY Wrong for Over a Century?

Seeley's work was carried out long before cladistic methods were *de rigeur, au courant, à la mode,* in vogue, and all those good things. Nonetheless, his distinction between Ornithischia and Saurischia survived the cladistic revolution, since both groups seemed to be nice, well-mannered, and monophyletic – within a nice, well-mannered, monophyletic Dinosauria.

However, in 2017, English paleontologists M. Baron, D. B. Norman, and P. M. Barrett proposed a novel, and truly radical division of Dinosauria.[a] Analyzing 74 dinosauromorphs (including nondinosaurs as well as dinosaurs, some of whom we've met in this chapter) and 457 characters, they concluded that really Dinosauria was better divided into two groups: **Ornithoscelida**, a group that includes [Theropoda (here, in Chapters 6 to 8) + Ornithischia (here, in Chapters 10 to 12)], and a reworked **Saurischia**, which would include (as it does now) Sauropodomorpha (Chapter 10) and Herrerasauridae (currently thought to be a group of theropods as described in this chapter). Figure B5.1.1 compares Seeley's classification with this new one.

But Figure B5.1.1 reveals something else as well: in this brave new world, Sauropodomorphs – *Brontosaurus*-type animals – would no longer be dinosaurs by the conventional definition of dinosaurs (*Triceratops*, the sparrow, and all members of the clade descended from their most recent common ancestor) because those brackets (*Triceratops* and the sparrow) outline Ornithischia and Theropoda – both comprising Ornithoscelida, which does not include Sauropodomorpha. But how can you have a Dinosauria without *Brontosaurus* and its relatives?! So the new phylogeny, therefore, must also include in its definition a sauropodomorph – in this case, *Diplodocus* (Figure 9.1). Because of this, Baron and colleagues redefined Dinosauria as a clade that includes *Passer, Triceratops, Diplodocus,* and all members of the clade descended from their most recent common ancestor.

Critiques of this new phylogeny came in thick and fast; yet, all agreed that the work was careful, comprehensive, and impressive. Still, the conclusions were problematic; as strong as some of the shared characters of Ornithoscelida were, other characters, that worked well in the Seeley

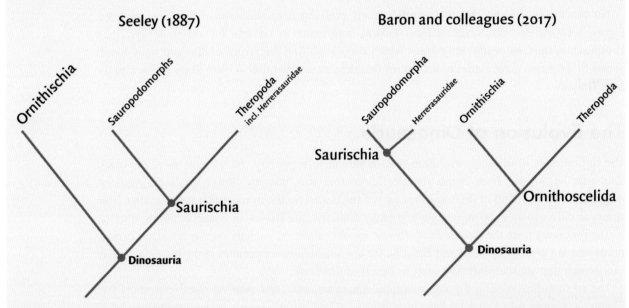

Figure B5.1. The 1887 (Seeley) and 2017 (Baron and colleagues) Dinosauria phylogenies compared. Cladograms simplified to highlight key features.

[a] Baron M. G., Norman, D. B., and Barrett, P. M. 2017. A new hypothesis of dinosaur relationships and early dinosaur evolution. *Nature,* **543,** 501–505. doi:10.1038/nature21700

Box 5.1 *cont'd*

phylogeny, were more problematical from a parsimony standpoint in the new phylogeny.

Are we looking at the future, here? The jury is out, but you may be certain that in the coming years this new phylogeny will be scrutinized with great care, and it is conceivable that in some future edition of this book, the Baron and colleagues analysis will become the norm, and Seeley's venerable, precladistic phylogeny will be relegated to a footnote – or a box like this one!

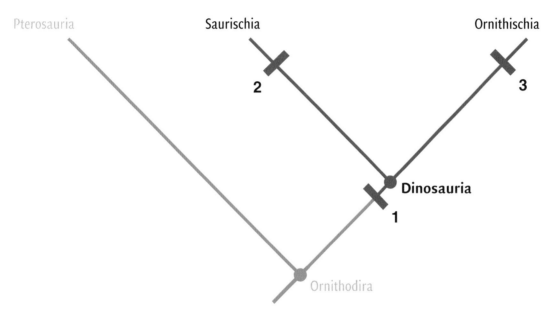

Figure 5.11. Cladogram of Dinosauria. Derived characters at **1**, perforate acetabulum (see also Figure 5.6); Derived characters at **2**, elongate cervical vertebrae, fossa expanded into the anterior corner of the external naris, lacrimal expanded over the rear part of antorbital fenestra, a concave facet on the axial intercentrum for the atlas, elongation of the centra of anterior cervical vertebrae, distinctly large hand, loss of distal carpal V, twisting of the first phalanx of manual digit I, metatarsals overlapping. Derived characters at **3**, opisthopubic pelvis, predentary bone, toothless and roughened tip of snout, reduced antorbital opening, **palpebral** bone, jaw joint set below level of the upper tooth row, cheek teeth with low subtriangular crowns, at least five sacral vertebrae, ossified tendons above the sacral region, small **prepubic process** along the pubis, long and thin preacetabular process on the ilium.

are too advanced! This leaves us, on the one hand, with a cloud of primitive organisms lacking the key diagnostic characters upon which to build the phylogeny, but, on the other hand, with a group (pterosaurs) with so many derived characters, that where they came from and how they evolved is equally difficult to figure out. Nonetheless, within the past few years, the outline of a consensus has been reached regarding the relationships of these two groups.

It had been thought, as recently as the early 1980s, that pterosaurs – otherwise highly modified for flight – might be the closest archosaurian relatives to dinosaurs, together sharing derived features as ornithodirans (see Figure 4.11). More recently, however, with the discovery of silesaurids, as well as a deepened understanding of creatures like *Marasuchus* and *Lagerpeton,* the nondinosaur basal dinosauromorphs are clearly closest to dinosaur ancestry, with pterosaurs distant within the clade Ornithodira.[5] Among nondinosaur basal dinosauromorphs, silesaurids are believed to be closest to dinosaur ancestry, with *Marasuchus* and *Lagerpeton* successively more distant (Figure 5.4).

The general outlines of this consensus didn't come easily; Argentine paleontologist M. Langer and colleagues have shown that, between the years 1986 and 2007, new hypotheses of the relationships of nondinosaur basal dinosauromorphs to dinosaurs were published at the rate of about one a year! Part of the reason for this is that, as we have seen, all of these early dinosauromorphs have many primitive characters. It is also due to the abundance of new material that was discovered during those years. For example, once silesaurids were identified, paleontologists began to correctly recognize silesaurid fossils all over the world. Then, too, many of these fossils were, and are still, lacking skull material (note that the diagnostic characters listed in Figure 5.4 are not cranial). Skull material has been tremendously useful in sorting out the relationships of fossil taxa, and we can hope that when complete skulls for some of these taxa become known, their phylogenetic relationships will be clearer. For now, we offer Figure 5.12 as a current (but not likely final!) summary of the relationships among dinosauromorphs and many of the early dinosaurs mentioned in this chapter.

There is an interesting and perhaps surprising consequence of this phylogeny. With archosaurs like *Marasuchus* and the silesaurids close to dinosaurian ancestry, apparently dinosaurs were primitively **obligate bipeds**. This means that the earliest dinosaurs were creatures that were completely bipedal. Because the primitive stance for archosaurs is quadrupedal, and because Dinosauria is monophyletic, it follows that creatures like *Triceratops, Ankylosaurus,* and *Stegosaurus,* all *quadrupedal* dinosaurs, must have **secondarily evolved** (or reevolved) their quadrupedal stance. That is, they must have (phylogenetically) gotten back down on four legs, as it were, after having been up on two. In fact, you can see the remnant of bipedal ancestry when you look at a stegosaur or a ceratopsian, in which the back legs are quite a bit longer than those at the front. Even silesaurids may have flirted with partial quadrupedality, a character left over from their ancestry.

Let the Games Begin!

Our information is limited about both the place and pace of the spread of dinosaurs. The oldest dinosaur body fossils known (above) are all from a chunk of time of the Late Triassic called the Carnian Stage (Figure 2.4; Box 5.2). Because the Carnian-aged dinosaurs described above are all known from Brazil and Argentina, paleontologists have suggested that the cradle of Dinosauria was South America. Yet, terrestrial Carnian rocks that preserve dinosaur remains are not so common globally. So we end up asking, are the South American dinosaurs truly the oldest dinosaurs, and thus South America was the place where dinosaurs originated? Or, alternatively, are the terrestrial Carnian rocks of South America simply the only game in town, and thus they are most of what we know of the terrestrial Carnian Stage globally? We have no satisfactory answer for those questions. At this point all we can really say is that the evidence is strong that dinosaurs first appeared either in the latter part of the mid Triassic (remember the footprints?) or, without doubt, based upon their body fossils, during the Carnian Stage.

[5] Or at least so it was thought until this book was well into the publication process. A recent (September 2020) study linked pterosaur origins with a clade of basal non-dinosauromorphs, including *Lagerpeton* and *Dromomeron* (Figure 5.12). The work is important because for the first time, phylogenetic methods were used to understand the origins of the enigmatic Pterosauria. Pterosaurs may have been closer to dinosaurs than anybody ever thought. See Ezcurra, M. D., Nesbitt, S. J., Bronzati, M., *et al.*, 2020, Enigmatic dinosaur precursors bridge the gap to the origin of Pterosauria: *Nature*, https://doi.org/10.1038/s41586-020-3011-4.

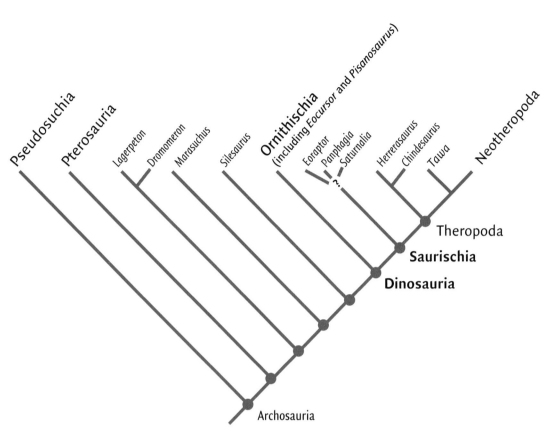

Figure 5.12. Summary of the phylogenetic relationships of the basal dinosauromorph and early dinosaur taxa discussed in this chapter.

What a Difference a Stage Makes

The Carnian Stage is thought to have ended at 228 Ma, which marks the beginning of the Norian Stage, the block of time that followed it (Figure 2.4; Box 5.2). By Norian time, dinosaurs clearly had acquired a startlingly pervasive global presence, and apparently populated all of the supercontinent of Pangaea (Figure 5.13).

What is striking, even among these very early Norian dinosaurs, is their diversity: examples of ornithischians, sauropodomorphs, and theropods are recognizable, suggesting that, if we understand the timing correctly (Box 5.2), the evolution and global dispersal of these groups took place relatively quickly after the first appearance of dinosaurs (wherever that was exactly). Ornithischians, later a pervasive group of dinosaurs, *may* be represented in the Carnian by the enigmatic *Pisanosaurus*, but are definitely present – if extremely rarely so – later in the Late Triassic, as represented by the meter-long *Eocursor parvus* (Figure 5.14).

Land Grabs: Competition, or Colonization?

This is not to say that the basal nondinosaurs dinosauromorphs politely stepped aside the first time that a true dinosaur showed up at the party. Both likely did similar kinds of things ecologically and,

Box 5.2 No Dates; No Rates

The late Samuel A. Bowring, a virtuoso geochronologist (a scientist who determines the numerical ages of rocks), used to say, "No dates; no rates!". By this he meant that if we don't know when, numerically, events occurred, we can't really determine how quickly or slowly they took place. In this chapter, we tell a story of rapid early dinosaur evolution; yet, without dates, our story could be a house of cards. Numerical dates from unstable isotopes, such as those discussed in Chapter 2, are rare for this time interval (one geochronologist called the Late Triassic a "black hole"), and those few that are known were not, until very recently, particularly precise.

Historically, in the absence of numerical ages, relative dating was carried out using a combination of lithostratigraphy and biostratigraphy (see Chapter 2; Figure 2.3). These techniques were used to divide the Triassic into three formal Epochs: Early, Middle, and Late. Each of these, in turn, was subdivided using biostratigraphy into smaller blocks of time called Stages. The key Stages for our story are the Carnian, the Norian, and the Rhaetian, the three Stages of the Late Triassic (Figure B5.2.1).

Numerical dates for these three Stages are very rare indeed; so while we know the sequence of time, exactly when that sequence occurred can be uncertain. For example, stratigraphers can still get a rude shock: in 2007, following a careful reanalysis of marine dates and faunas, the Norian gained almost 7 million years (at the expense of the Carnian!).

The situation is more of a problem in the terrestrial realm. Dinosaurs, as we have seen, are exclusively terrestrial beasts, and geographically dispersed dinosaur faunas around the globe have been correlated with each other. Yet, it has not been easy to securely tie these dinosaur faunas to the Norian and Carnian Stages, because these Stages are based upon marine organisms.

Still, there is progress: as we have seen, dates now exist for the Ischigualasto Formation of Argentina, the unit that produced *Eoraptor* and *Herrerasaurus*. The ages of *Guaibasaurus* and *Staurikosaurus*, from equivalent rocks in Brazil, remain less certain. Recently, a suite of extremely precise dates has been obtained for the Chinle Formation of the North American southwest, a rock unit that produced Late Triassic dinosaurs such as the theropod *Coelophysis* and the primitive saurischian *Chindesaurus*. For the time being, though, those dates stand alone, because their exact relationship to the Carnian and

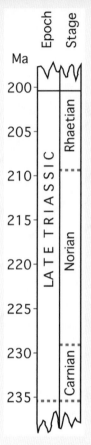

Figure B5.2.1. The Late Triassic time scale. Numerical dates from below the base of the Carnian approximate the base of the Carnian to slightly younger than 236 Ma. With the exception of a date from the upper Carnian of Italy, there are generally no reliable isotopic ages for the Late Triassic at this time. Therefore, ages for the Carnian–Norian boundary (227 Ma to 228 Ma) and the Norian–Rhaetian boundary (208 Ma to 209 Ma) have been interpolated using magnetostratigraphy, a technique based upon global magnetic polarity reversals detected in key stratigraphic sections. The Triassic–Jurassic boundary, on the other hand, has been precisely dated by isotopic methods at 201.3 Ma – but in the marine realm, ecologically and geographically far from dinosaurs.

Norian Stages is only beginning to be better understood – since the numerical ages of those Stages are still not well known. And this means that exactly how these North American faunas tie into other North American Late Triassic faunas, let alone global Late Triassic faunas, is not so clear, either. Stay tuned!

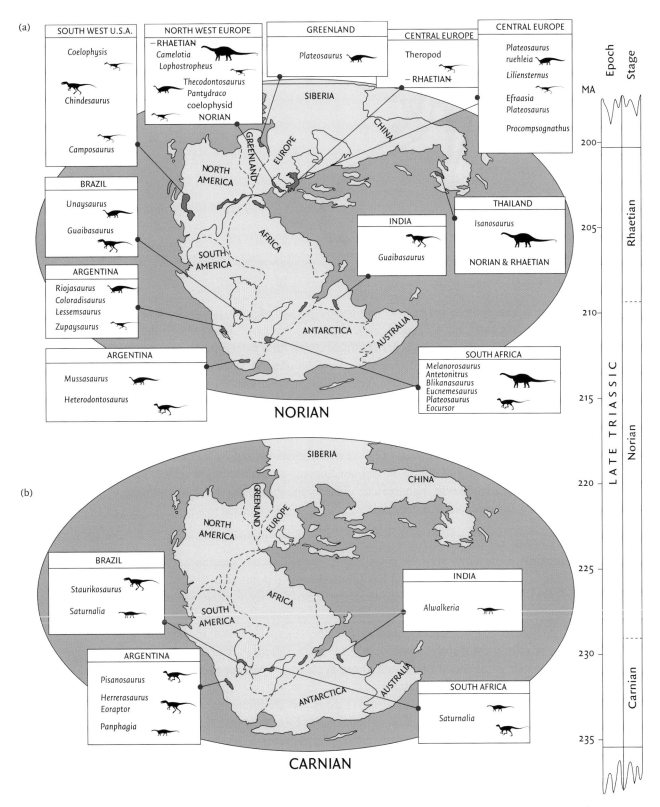

Figure 5.13. The distribution of dinosaurs around the Late Triassic supercontinent of Pangaea (a) Norian; (b) Carnian. Redrawn from Langer *et al.*, 2010.

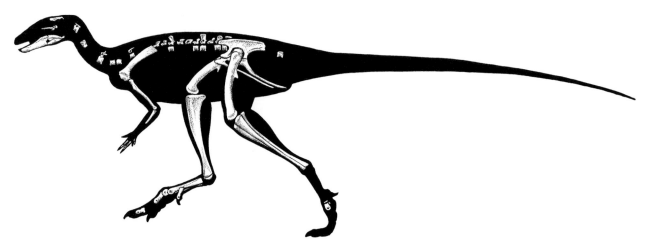

Figure 5.14. The primitive, early undoubted ornithischian, *Eocursor parvus*, from South Africa.

indeed, the earliest dinosaurs were rather primitive representatives of their group/clade, looking superficially (and in some cases not-so superficially) similar to the basal dinosauromorphs from which they are thought to have derived. Yet a strange signal comes from the record. The rich Carnian deposits of Argentina contain both early dinosaurs and nondinosaur basal dinosauromorphs, but rarely are the two found together; dinosaurs appear to supplant basal dinosauromorphs in South America. Yet, in the Norian of the North American southwest, by contrast, saurischian dinosaurs like *Coelophysis* and *Chindesaurus* lived for almost 12 million years with nondinosaur basal dinosauromorphs such as *Dromomeron*, an animal closely related to *Lagerpeton* (see Figure 5.12). Indeed, in North America at least, the Late Triassic was a time generally characterized by a relatively stable, long-lived global fauna consisting of dinosaurs and nondinosaur dinosauromorphs coexisting. By the end of the Late Triassic, however, nondinosaur dinosauromorphs phased out of the picture globally, and true dinosaurs were off and cursorial!

The Carnian Pluvial Episode

Exactly how dinosaurs managed to grab the brass ring from the host of dinosaur wannabes that inhabited the Carnian has been a question that has troubled many paleontologists – and the more similar to dinosaurs the nondinosaur dinosauromorphs look, the more enigmatic the whole process appears.

In 2018, M. J. Benton and colleagues proposed that the "Carnian pluvial episode," a well-documented time of significant global climate change from dry, to humid, and back to dry, may have been a causal factor in the rise of Dinosauria. Dated to about 232 Ma, the event seems to have involved extensive volcanism, in turn affecting global climates as well as oceanic and atmospheric circulation. These are familiar kinds of environmental changes that are connected with several mass extinctions historically, and indeed nondinosaur extinctions are coincident with this episode. In the case of the Carnian pluvial episode, Benton and colleagues have proposed that "ecological perturbation" caused by these climate changes initially caused extinctions, after which dinosaurs radiated dramatically – as we have seen – into what was then an Earth (ecospace) that was effectively depleted of vertebrates. The idea is supported by the extinction of many archaic land plants which accompanied the diversification of conifers after the "Carnian pluvial episode," and the simultaneous extinction of a variety of Triassic amniote herbivores, both synapsid and diapsid. Carnivore extinctions would have followed.

Most paleontologists would now agree that the explosive rise of Dinosauria that we have described in this chapter was likely not due to head-to-head competition with local thuggish nondinosaur amniotes, as had been once thought by paleontological illuminati, such as the Natural History Museum's Alan Charig, a couple of generations ago. The general consensus among paleontologists today is that the dinosaurs' success was likely a radiation that previously (re)filled abandoned and emptied ecospace. Less dramatic, perhaps, than a Late Triassic cage match between nondinosaur dinosauromorphs, and dinosaurs, but more the way things work in natural systems: we'll see mammals and avian dinosaurs do this 166 million years later, in Chapter 17 (although the plan is to get there is less than 166 million years).

Feathers

Before we begin our stroll through the various groups of dinosaurs, one last remarkable and unexpected subject intrudes here: feathers. Feathers? For previous generations of paleontologists and dinosaur lovers, life was simple: only birds have feathers, and dinosaurs, being reptiles, must have had scales. A few bumpy skin impressions (see Chapter 12) sufficed to make the point. Feathers were on *nobody's* radar. Even now, ask almost anybody what feathers are for, and they'll surely answer, "for flight." Hmmm.

Who Invented Feathers?

Thinking began to change when John Ostrom (see Chapter 16), in the early 1970s, showed that the oldest "bird" known – the winged wonder *Archaeopteryx* (Chapter 8), was also a dinosaur, a conclusion that provoked great confusion in those precladistic, Linnaeus-dominated days. If it is a bird, how could it *also* be a reptile? Linnaean taxonomy, as we have seen, does not play very nicely with evolutionary relationships.

So, most (but not all) paleontologists then concluded, OK, a few weirdo, highly derived primitive bird-dinosaur thingys must have had feathers; what's so amazing about that? But in the 1990s it became strikingly obvious that actually *nothing* was "amazing about that," since all kinds of nonflying, saurischian dinosaurs – even *T. rex* – evidently had feathers. Feathers appeared to be the invention of Theropoda… until feathers – or something like feathers – were found in Ornithischia! With feathers potentially on both saurischians and ornithischians, were feathers actually an invention of Dinosauria? Or could they go even farther back on the cladogram?

We'll meet these various feathered dinosaurs as we explore each group, but regardless, it was clear that after 150 years of smooth-skinned, polished, shiny, greeny-brown dinosaurs, it was time to change the paradigm to scruffy, feathery, dander-spewing, brightly-colored dinosaurs (Figure 5.15)! Because feathers are present in both Saurischia and Ornithischia, we'll introduce them here.

Feathers Without Flight

Feathers in modern birds do lots of things: they insulate (think down sleeping bags and vests), they are used for display (think peacock), they are used as sensory organs, and, most famously, they permit flight (think airfoil). One anatomical structure; multiple uses.

(a)

(b)

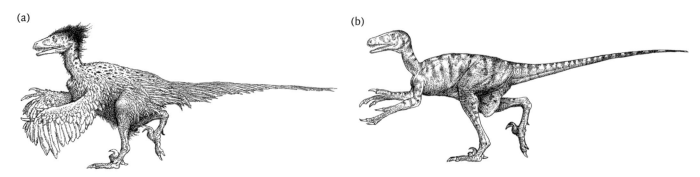

Figure 5.15. The carnivorous "raptor" (in *Jurassic Park* Speak) *Deinonychus*. (a) A vibrant, feathered "land eagle;" (b) the plucked image. The original would not have recognized itself in its featherless state.

Insulation

The insulator properties of down feathers are well known. Could feathers have evolved first for insulation, and then only later gotten coopted for flight? This idea is less far fetched when we remember that all living birds are warm-blooded; and most warm-blooded creatures, especially small ones, require insulation to keep from losing heat. It makes no sense for a cold-blooded animal, which needs an exterior source of heat to warm up, to be insulated.

If feathers were actually an adaptation for insulating a warm-blooded creature, then fossils of small, nonflying dinosaurs ought to be found with feathers. When evidence finally came that showed that feathers were used for insulation (and not for flight), it was spectacular (see Chapter 7). But did feathers *originate* for insulation?

Stylin'

Modern birds are brightly colored creatures, and it is no secret that, along with flight and insulation, birds use feathers for display. While shape is important (think about a lyrebird or a peacock), and excluding behavior and calls, color is probably the key criterion for reproductive success. So if feathers today have an important display function, did feathers also serve a display role in dinosaurs? Recently, paleontologists have begun to gain insights into the actual colors and patterns of ancient feathers, and blacks, browns, reds, iridescent colors, as well as color patterns, have been identified. We'll visit this in greater detail in Chapters 6 and 7 (see Chapter 6 "The wonderful world of color," and Figures 7.10 and 7.23). The murky crocodile-green of your parents' dinosaurs is a thing of the past: dinosaurs were colorful, brightly patterned animals. Did feathers therefore originate for display?

Senses

Anatomist W. S. Persons and paleontologist P. J. Currie recently pointed out that feathers in birds have yet another key function: sensing. Like the whiskers on mammals or the little sensory projections (vibrissae) on insects, single-filament feathers on birds – particularly on the snout – occur in low concentrations and serve as sensing organs. The authors note that, for feathers to have insulated, they needed to be densely concentrated. Perhaps, these authors suggested, simple, widely scattered feathers originally began as tactile organs, eventually coming to serve display, insulation, and finally flight functions.

Embryology

Feathers were long thought to be an outgrowth of "reptilian" scales. Somehow the scales grew longer and divided into barbs and barbules. Work in the past ten years, however, suggests that the development of feathers may occur by the interaction of specialized follicles and a series of genes

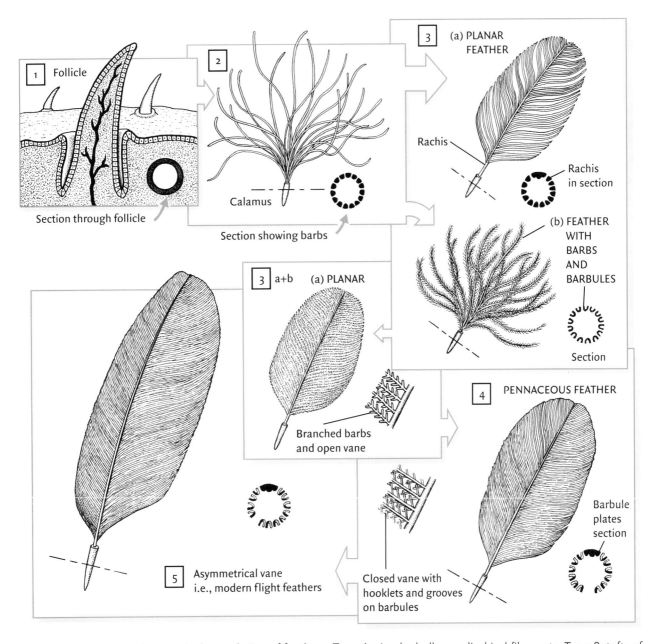

Figure 5.16. Sequential stages in the evolution of feathers. Type 1: simple, hollow, cylindrical filaments. Type 2: tufts of elongate, multiple filaments, attached at one end. Type 3: filament tufts align in a single plane (Type 3a) while also developing barbs and barbules (Type 3b). Eventually, a new planar (vaned) barbed form evolves (Type 3a+b). Type 4: vane becomes "closed"; that is, tiny hooks on the barbule attach to grooves on adjacent barbules, producing an integrated semi-rigid vane that does not allow much air to pass through. Type 5: vane becomes asymmetrical (for example, a flight feather).

that control the onset and termination of growth. Four sequential stages of feather evolution have been identified, each stage a developmental modification of the previous stage, and each found in living birds (Figure 5.16). These stages are:

(1) Keratin (protein)-based, single filament, constructed as a hollow cylinder (**monofilamentous**, hair-like feathers).
(2) Loosely associated, unconnected, unhooked barbs (**downy** feathers, sometimes called "plumulaceous").
(3) Hooked barbs on a symmetrical vane (**contour** feathers, such as wrap around the body, the shape is termed "pennaceous").
(4) Hooked barbs on an asymmetrical vane (**flight** feathers, also described as "pennaceous").

That there are multiple stages of feather development explains how it comes to be that different feather types are found in living birds: the *embryological* sequence shown in Figure 5.16 may reflect the *evolutionary* sequence of feather development. Thus, for example, single filamentous feathers (Stage 1) may represent a very early stage of feather development, whereas the hooked barbs and asymmetrical vanes found in flight feathers likely evolved later. Indeed, feathers in some nonflying dinosaurs were simply a hair-like coating (monofilamenous) and in smaller, active nonflying forms, down may have been used as an insulator; perhaps covered with contour feathers. Dinosaurs that flew, of course, developed specialized flight feathers for that purpose.

So feathers are associated with sensory, display, insulation, and flight in all these animals, and may very well be the quintessential diagnostic character for Dinosauria that we have sought throughout this chapter. Alternatively, it may be a bit more complicated. In 1971, a small pterosaur called *Sordes pilosus* demonstrated unequivocally that pterosaurs were covered with a soft (?) fur-like covering. Were feathers actually the invention of Ornithodira, thus going even farther back in the phylogeny than Dinosauria?

The answer came back as an unequivocal "yes" in 2019, when it was demonstrated in M. J. Benton's paleobiology lab in Bristol, England, that pterosaurs are coated with material that shows the diagnostic features of feathers, including a nonvaned filamenous morphology akin to that in dinosaurs. Either dinosaurs and pterosaurs independently evolved feather coatings separately or, more parsimoniously, feathers are a feature that characterizes Ornithodira as a group. We'll go out on a limb here and predict that the latter hypothesis is correct.

The very earliest dinosaurs were evidently carnivorous, bipedal, and small- to medium-sized, and were very likely covered with some kind of monofilamentous plumage. Unfortunately, to date, the fossil record – in the case of the Triassic forms – is mute on this point. With all that feathery integument, they might even have taken a tentative bipedal step down the road towards endothermy (warm-bloodedness; see Chapter 14).

SUMMARY

The earliest and morphologically most primitive dinosaurs were small, bipedal, carnivorous, and/or insectivorous creatures, known from deposits in Argentina and Brazil, likely appearing about 245 to 230 million years ago. These animals lived with, as well as were preceded by, several other groups of archosaurs – nondinosaur basal dinosauromorphs – of similar morphology and inferred behavior. The similarities of morphology suggest that dinosaur ancestry is to be found among the basal dinosauromorphs, likely, among the recently recognized silesaurid clade. The morphologies are so similar, that as we learn more about the nondinosaur dinosauromorphs, the number of diagnostic characters distinguishing dinosaurs becomes smaller and smaller.

The earliest known dinosaurs, although primitive, have been affiliated with somewhat derived dinosaurian groups, such as Theropoda and Sauropodomorpha. This suggests that some amount of evolution must have occurred preceding ~229 Ma and the appearance of the earliest known dinosaurs, *Herrerasaurus* and *Eoraptor*.

Two major groups of dinosaurs, Ornithischia and Saurischia, are currently recognized. These are identified by a variety of features among which only the orientation of the anterior position of the pubis is highlighted here. As the phylogeny is currently understood, ornithischians and saurischians are equally derived; thus it is not possible to say whether one is more primitive than the other. New phylogenies have been proposed, however, which may overturn the long-standing division of Dinosauria into Saurischia and Ornithischia.

Among the striking innovations of dinosaurs, among the most surprising may be the invention of feathers. Once thought to pertain only to birds, feathers are now known to have graced both saurischian and ornithsichian dinosaurs, with the possibility that feathers go as far back as basal Ornithodira on the cladogram. Early feather coats may have been monofilamentous and superficially fur-like; down likely evolved early as a means of insulation which, along with sensory functions and display, was the probable original function of feathers.

Once they made their Carnian (early Late Triassic) or even pre-Carnian appearance, dinosaurs radiated slowly at first, perhaps accelerating their diversification as a result of global climate change associated with the Carnian pluvial episode. After that event (middle Late Triassic), they had achieved a global distribution, perhaps resulting from their opportunistic ability to colonize abandoned ecospace.

SELECTED READINGS

Benton, M. J., Bernardi, M., and Kinsella, C. 2018. The Carnian Pluvial Episode and the origin of dinosaurs. *Journal of the Geological Society*, **175**, 1019–1026.

Brusatte, S. L. 2019. *The Rise and Fall of the Dinosaurs: A New History of a Lost World*. William Morrow, NY, 404 pages.

Brusatte S. L., Benton M. J., Ruta M., and Lloyd, G. T. 2008. Superiority, competition, and opportunism in the evolutionary radiation of dinosaurs. *Science*, **321**, 1485–1488. doi:10.1126/science.1161833

Brusatte, S. L., Nesbitt, S. J., Irmis, R. B., *et al.* 2010. The origin and early radiation of dinosaurs. *Earth-Science Reviews*, **101**, 68–100.

Desojo, J. B., Fiorelli, L. E., Ezcurra, M. D., *et al.* 2020, The Late Triassic Ischigualasto Formation at Cerro Las Lajas (La Rioja, Argentina): fossi tetrapods, high-resolution chronostratigraphy, and faunal correlations. *Nature Research*, **10**. https://doi.org/10.1038/s41598-020-67854-1.

Langer, M. C., Ezcurra, M. D., Bittencourt, J. S., and Novas, F. E. 2010. The origin and early evolution of dinosaurs. *Biological Reviews*, **85**, 55–110.

Nesbitt, S. J. 2011. The early evolution of archosaurs: relationships and the origin of major clades. *Bulletin of the American Museum of Natural History*, **352**, 1–292.

Nesbitt, S. J., Butler, R. J., Ezcurra, M. D., *et al.* 2017. The earliest bird-line archosaurs and the assembly of the dinosaur body plan. *Nature*, **544**, 484–487.

Padian, K. 2013. The problem of dinosaur origins: integrating three approaches to the rise of Dinosauria. *Earth and Environmental Science Transactions of the Royal Society of Edinburgh*, **103**, 423–442.

Person, W. S. IV and Currie, P. J. 2015. Bristles before down: A new perspective on the functional origin of feathers. *Evolution*, **69**, 857–862.

Prum, R. O. and Brush, A. H. 2003. Which came first, the feather or the bird? *Scientific American*, **288**, 84–93.

Xu, X. and Guo, Y. 2009. The origin and early evolution of feathers: insights from recent and neontological data. *Vertebrata PalAsiatica*, **47**, 311–329.

Yang, Z., Jiang B., McNamara, M., *et al.* 2019. Pterosaur integumentary structures with complex feather-like branching. *Nature Ecology & Evolution*, **3**, 24–30.

TOPIC QUESTIONS

1. What is Crurotarsi? Ornithodira? How would one tell the difference?
2. Give an example of a crurotarsan archosaur. When did the last crurotarsan live?
3. What do those names (Crurotarsi; Ornithodira) have to do with Dinosauria?
4. What is a dinosauromorph?
5. Please name two nondinosaurian dinosauromorphs.
6. What do those nondinosauromorphs tell us about the morphology of the first dinosaurs?
7. Describe the earliest known dinosaurs. Are these also the most primitive dinosaurs? What is the difference between "earliest known" and "most primitive"?
8. People have discussed whether or not the evolution of Dinosauria was "fast" or "slow." Which do you think it was? Please explain your answer: (a) what do you mean by "fast" (or "slow"), and (b) how do you measure fast or slow evolution?
9. Should "presence of feathers" be a diagnostic character for birds? For Dinosauria? For Ornithodira? Please explain your answer to each.
10. Construct and justify an evolutionary sequence for the various uses of feathers.

PART II
Saurischia: Meat, Might, Muscle, and Magnitude

Figure II.1 An *Allosaurus* meet-and-greet over a tasty rotting sauropod carcass.

Saurischians include the smallest of dinosaurs and the largest animals that ever lived on land; the most agile and ferocious of predatory dinosaurs and the most ponderous of plant-eaters; the brightest and, quite possibly, the dumbest of dinosaurs; the most Earth-bound and the most aerial. And those most familiar of saurischian dinosaurs, birds, remain with us today, very much alive and well.

Trouble in Paradise

Recall that the dinosaur world was traditionally divided into Saurischia and Ornithischia. Ornithischia – we will meet them in Part III – are an easy, obvious grouping; experienced eyes even observe a "family resemblance" among ornithishcians. Not so with saurischians; this is the group the lumps together dinosaurs like *T. rex* and *Brontosaurus*. Can that be right?

The short answer is, right now *we don't actually know*! Recall from Chapter 5 that, as early as 1887, H. G. Seeley recognized two great clades within Dinosauria: Ornithischia and Saurischia. He based this division upon the orientation of the pubis: in saurischians, they point forwards, towards the front of the animal, whereas in ornithischians, they point backwards, towards the tail of the animal (see Figures 5.9 and 5.10). But the pubes point forward, primitively, in all tetrapods. Because the character is primitive for tetrapods it cannot be used to unite monophyletic groups within Tetrapoda (see Chapter 3).

As we saw in the previous chapter, a dinosaur-shattering proposal was proposed recently (2017; see Box 5.1) in which Ornithischia and Theropoda are linked together within a group called Ornithiscelida, and Saurischia would consist of Sauropodomorpha and herrerasaurs. You may remember herrerasaurs as theropods from Chapter 5 (see Figure 5.7b), but among these very primitive dinosaurs, diagnostic characters can be quite difficult to identify, since these animals have not undergone much evolution from the primitive dinosaur condition. Whether this ends up being the preferable (more parsimonious) division of these groups is anybody's guess at this point; but the Ornithscelida–Saurischia division may yet prove to be the way of the future. As in boxing, however, the title must be taken from the champ; until that occurs definitively, we retain here Seeley's original Saurischia–Ornithischia division of Dinosauria. Figure II.2 highlights the monophyly of Saurischia as it is currently understood.

What Makes a Saurischian a Saurischian?

Saurischia is diagnosed by several derived features (Figure II.2), two of which are shown in Figure II.3. In general, these characters suggest a general tendency in the group for well-developed, grasping hands and powerful feet. In the case of the hands, the large size (almost half of the arm length), long fingers, and a distinctive thumb that, a bit like a human thumb, appears to fall across the palm[1] instead of simply pointing outwards from the hand; this suggests an organ designed for grasping, presumably prey. In the case of the feet, the tightly articulating, close-fitted metatarsals may have supported more efficient walking, but in carnivorous forms (including the earliest saurischians), they may have also aided in prey manipulation. While not all saurischians were carnivorous, the group from the very outset appears to have had some tendencies toward carnivory.

Keeping Seeley's Dinosauria concept for now, saurischians as presented here include the famous toothy carnivorous dinosaurs (Theropoda; *theros* – beast; Chapters 6 to 8) and their equally famous long-necked herbivorous cousins (Sauropodamorpha; *saurus* – lizard; *pod* – foot; *morph* – shape; Chapter 9).

---

[1] The distinctive position of the thumb is regarded as "semi-opposable" by some paleontologists.

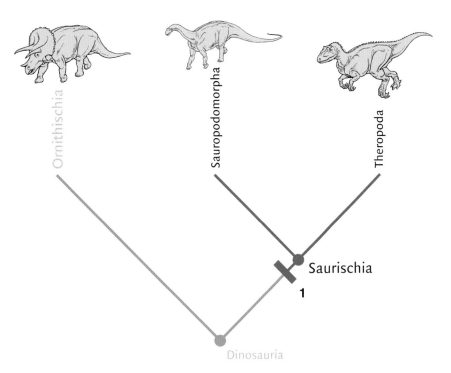

Figure II.2. Cladogram of Dinosauria, emphasizing the monophyly of Saurischia. Derived characters include: at **1**, elongate cervical vertebrae, fossa expanded into the anterior corner of the external naris, lacrimal expanded over the rear part of antorbital fenestra, a concave facet for the atlas on the axial intercentrum, elongation of the centra of anterior cervical vertebrae, distinctly large hand, loss of distal carpal V, twisting of the first phalanx of manual digit I, metatarsals overlapping. See Figure II.3.

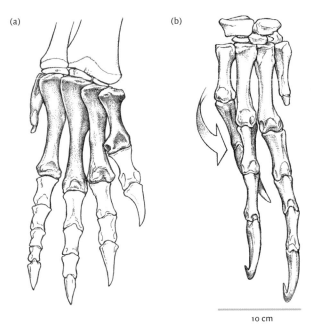

(a) (b)

10 cm

Figure II.3. (a) Overlapping metatarsals of the primitive saurischian foot (exemplified by the "prosauropod" *Ammosaurus*); (b) thumb (digit I of the hand) tends to angle across the palm.

SELECTED READINGS

Baron, M. G., Norman, D. B., and Barrett, P. M. 2017. A new hypothesis of dinosaur relationships and early dinosaur evolution. *Nature*, 543, 501–505. doi:10.1038/nature21700

Gauthier, J. A. 1986. Saurischian monophyly and the origin of birds. *Memoirs of the California Academy of Sciences*, 8, 1–55.

Langer, M. 2004. Basal Saurischia. In Weishampel, D. B., Dodson, P., and Osmólska, H. (eds.) *The Dinosauria*, 2nd edn. University of California Press, Berkeley, pp. 25–46.

Langer, M. and Benton, M. J. 2006. Early dinosaurs: a phylogenetic study. *Journal of Systematic Paleontology*, 4, 309–358.

Sereno, P. C. 1999. The evolution of dinosaurs. *Science*, 284, 2137–2147.

Sereno, P. C., Forster, C. A., Rogers, R. R., and Monetta, A. M. 1993. Primitive dinosaur skeleton from Argentina and the early evolution of Dinosauria. *Nature*, 361, 64–66.

Sereno, P. C. and Novas, F. E. 1992. The complete skull and skeleton of an early dinosaur. *Science*, 258, 1137–1140.

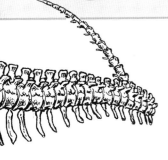

6

Theropoda I
Nature Red in Tooth and Claw

> **WHAT'S IN THIS CHAPTER**
>
> In this chapter we introduce theropods – more beautiful than your sweetest dream; more fierce than your worst nightmare – the group of dinosaurs that includes dinosaurs like *T. rex* and hummingbirds. Here you'll get a sense of:
>
> - Who theropods are.
> - Their basic relationships in Dinosauria.
> - How they lived... and died.

Figure 6.1. The tyrant king, *Tyrannosaurus rex.*

Theropoda

Eating Meat the Theropod Way

When dinosaurs did carnivory, they did it the theropod way: with steak-knife teeth, sinewy haunches, and grasping hands and feet tipped with scimitar claws (Figure 6.1). The combination was at once formidable and successful, and produced a rainbow palette of carnivores as well as, it turns out, a large pile of other-ivores. Carnivorous dinosaurs find their home among **Theropoda**, but why stick to meat when there there are so many other options out there? Reflecting this, Theropoda (*thero* – beast; *pod* – foot) is *really* diverse, and counts among its many members coelophysoids, neoceratosaurs, carnosaurs, therizinosauroids, ornithomimosaurs, oviraptorosaurs, troodontids, spinosaurids, dromaeosaurids, tyrannosauroids, and birds. This means that those "twitterpatin'" song birds that we hear on spring mornings, are theropod dinosaurs.

Theropods have had a long evolutionary history extending back from the Late Triassic right up until the end, 66 Ma. *Past* that "end," actually, since birds are still very much with us. But in this chapter, we'll concentrate on **nonavian** (that is, nonbird) theropods, holding off on the avian side of the story until Chapter 7.

Nonavian theropods (for simplicity, "theropods") have been found on every continent including Antarctica. They have been collected from virtually every kind of depositional environment. And, they have been found as isolated finds, but also in rich, monospecific (single-species), theropod bonebeds (see "Social behavior" below).

Theropods ranged in size from less than 1 m (3'; *Microraptor*) to animals growing to upward of 15 m (approximately 49') in length (*Tyrannosaurus, Carcharodontosaurus, Giganotosaurus, Spinosaurus*). For all the variety, though, nonavian theropod evolution seems to have been largely about tracking, attacking, and feeding.

Who are Theropods?

Theropoda is a monophyletic group within Saurischia that includes all saurischians that are not sauropodomorphs. Using a formal definition – that in effect says exactly the same thing – we would say that it incudes *Passer* (our old friend the sparrow – again! – see Figure 5.6). and all taxa sharing a more recent common ancestor with *Passer* than with any sauropodomorph (Figure 6.2). That definition includes birds within Theropoda.

Why not use the bracketing definition we've seen previously? That is, why not define theropods by a derived theropod on the one hand (*Passer*), a primitive theropod on the other, and call "theropods" all the descendants of their most recent common ancestor? Because in this case, as we've seen, it is rather difficult to be sure which, among the very earliest dinosaurs we've met, is a theropod and which is not. So this definition leaves the door open to new fossils and new interpretations, by including everything within Saurischia except Sauropodomorpha.

Viewed superficially, theropods are all clawed bipeds with sharp, flattened serrated teeth (Figure 6.3; but only when they haven't lost them, as has occurred in several groups), and distinctively hollowed vertebrae and limb bones. However, the serrated teeth, claws, hollow bones, and bipedal stance of theropods all appear to have been primitive carry-overs from earlier in dinosauromorph history. These characters give the *appearance* that theropods are somehow more primitive organisms than, say, ornithischians; the cladograms, however, show that they're not (see Figure 6.2).

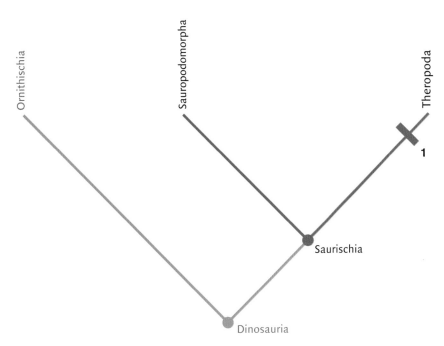

Figure 6.2. Cladogram of Dinosauria emphasizing the monophyly of Theropoda. With uncertainty surrounding the phylogenetic position(s) of the earliest dinosaurs such as *Eoraptor*, *Eodromaeus*, *Guaibasaurus*, and *Herrerasaurus*, appropriate diagnostic characters (e.g., those characterizing theropods as defined) can be tricky to find. Here are a few proposed in 2012 by theropod specialist Stephen Brusatte (University of Edinburgh): at **1**, knife-like teeth (thin, recurved, serrated; termed **ziphodont**), promaxillary fenestra on maxilla (an extra opening in front of the antorbital opening), large hands with advanced grasping ability.

Theropod Lives and Lifestyles

Theropod imagery is haunted by solitary, drooling brutes like *T. rex* mauling herds of herbivores. But is that really how it was? Did biggest = baddest? Evidence suggests that packs of highly social, sharp-eyed, large-brained, agile small theropods, armed with grasping hands and slicing claws were likely the real nightmares of Mesozoic landscapes (Figure 6.4). And there were other theropods as well – fleet-footed ornithomimids, hulking herbivorous therizinosaurs, and some predatory fish-eaters that likely would treat a modern crocodile like an *hors d'oeuvre*. Theropods tried everything – even flight.

Running for Your Life

We have seen how all dinosaurs had a fully erect stance built for running (cursorial) movement. In theropods, this is taken to an extreme. Theropods were (and are) obligate bipeds, like us, unable to walk or run on anything but their hind legs. The body was balanced directly over the solidly constructed pelvis, with the vertebral column held nearly horizontal (see Figure 6.5), balanced at the pelvis. Evidence from the skeleton and trackways reflects the fact that the hind legs were held close to the narrow body, feet so close to the midline that it appears that one foot was placed *ahead* of the other, rather than along its side (Figure 6.6).

Figure 6.4. Skeleton of the Late Cretaceous carnivore *Velociraptor*, in attack mode; seen at the Museum of Natural History, Los Angeles County, California, USA.

Figure 6.3. Typical large theropod tooth. Isolated tooth of *Allosaurus*, a Late Jurassic carnivorous theropod; note the distinctive serration, most visible on the right side (posteriorly) of the specimen.

Figure 6.5. *Allosaurus fragilis*, spotted at the Fukui (Japan) Prefectural Dinosaur Museum. The legs attach at the pivot point between the long tail, and the deep, but narrow body and skull. Note digitigrade foot position.

   In most theropods, the length of the leg was extended by the foot, which in effect becomes an extra joint (or segment) in the leg.[1] The foot was supported by three of its four toes (the fourth, like a dew claw, parked along side the foot), in what is termed a **digitigrade** stance (Figure 6.7). This stance

---

[1] A bit like being *en pointe* in ballet, only in theropods, the sharp claws at the very tips of the three weight-bearing toes did not bear the full weight of the body.

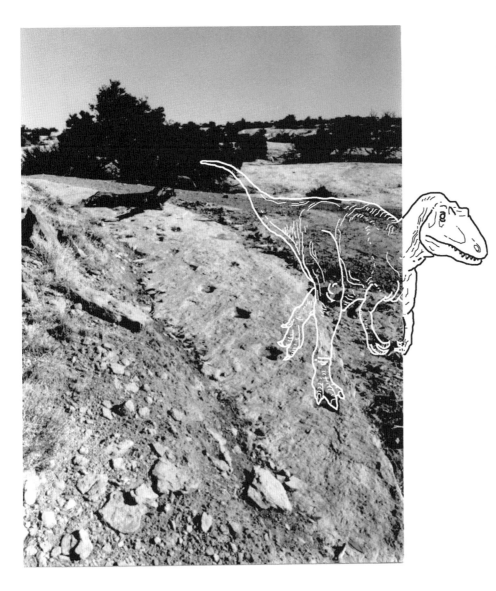

Figure 6.6. A theropod trackway from the Middle Jurassic Entrada Formation, Utah, USA. Note how closely the right and left prints are placed, with the steps very close to the animal's midline. When spectacular trackways like this are found, it is not hard to imagine the ghostly image of the trackmaker leaving a row of footprints in the soft mud.

Figure 6.7. Digitigrade stance in the right leg of the large, Late Jurassic, North American theropod *Allosaurus*, seen in the Museum of Natural History, Vienna.

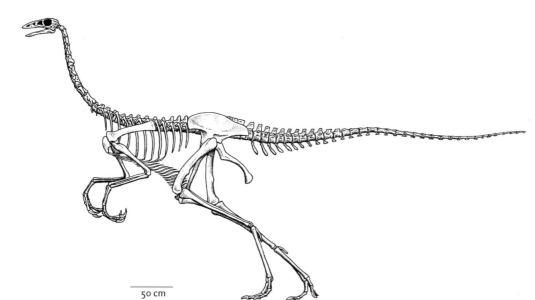

Figure 6.8. Many small- to medium-sized theropods, such as the ostrich-mimic (ornithomimosaur) *Struthiomimus* shown here, have a thigh (femur) that is significantly shorter than the shin (tibia), suggesting that these dinosaurs were capable of running fast.

50 cm

left a distinctive tridactyl (three-toed) footprint (see Figures 1.4 and 1.5). By contrast, our own foot rests not on its toes (except during ballet, of course!) but rather on the foot bones and heel. Our stance is termed **plantigrade**. There is an interesting size relationship in the proportionately long, slender hindlimbs of small- to medium-sized theropods: the thigh (femur) is short compared with the length of the shin (the tibia); this is a condition typical of fast-running bipeds (Figure 6.8).

Theropods were Surely Cursorial Animals!

But just how cursorial is cursorial? Calculations of running speeds on the basis of hindlimb proportions indicate that the fastest theropods probably clocked 40 to 60 kmh (up to 37 mph). Some footprint evidence bears these numbers out; for example, a Cretaceous trackway in Texas was made by a theropod that thundered away at between 30 and 40 kmh (18 to 25 mph; see Speed in Chapter 13). Of course most of the time, most theropods weren't running full tilt, and leisurely lakeside strolls of about 4 kmh (2.5 mph) have been recorded from Early Jurassic trackways in Connecticut.

In the case of large theropods, outrunning jeeps was probably not an option. No trackways are known from large, thundering theropods, but computer modeling suggests that, to achieve high speeds (in the 50 kmh [31 mph] range), so much leg musculature would have been required that the animal would have been grotesquely overmuscled. The balance of dispassionate evidence, therefore, suggests that the largest theropods likely could not exceed about 40 kmh (25 mph).

We've seen that theropods were spectacularly cursorial; we'll see in Chapters 7 and 8 how they mastered flight; we saw in Chapter 1 that they could burrow – even if it was not that common, and we also know, from a trackway in Spain showing the imprints of a medium-sized theropod (not to mention spinosaurs, which appear to have been amphibious), that some, at least, could swim. Theropods got around.

Paws and Claws

As in modern birds, the grasping, powerful, clawed feet were not only for locomotion, but were an important part of the theropod predatory arsenal (Figure 6.9). This character reached unparalleled

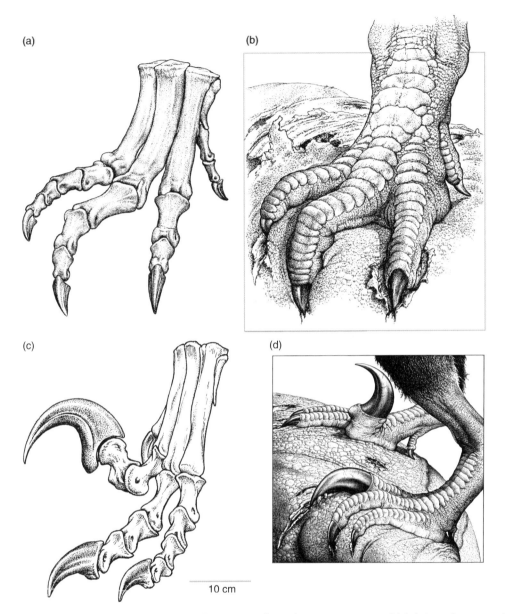

(a)

(b)

(c)

(d)

10 cm

Figure 6.9. Theropod feet. (a) Skeleton of *Allosaurus* foot; (b) reconstruction of (a) doing what nonavian theropod feet did best; (c) left foot of *Deinonychus* with its disemboweling second-toe claw. (d) Reconstruction of (c) in action.

sophistication in dromaeosaurids and troodontids, in which the claw on the second digit of the foot was enlarged, curved, and sharp, and capable of a very large arc of motion. During normal walking and running, it would have been held back or up, to protect it from abrasion or breakage. But, when needed, some paleontologists have suggested, it could be brought forward and, with the powerful kicking motion of the rest of the leg, used to eviscerate the bellies of hapless prey, lethally disemboweling them in one rapid stroke (Figure 6.9). Others find this scenario unconvincing; they argue instead that the claw was used to hold prey still. Either way, contact with that claw wasn't the beginning of a beautiful friendship.

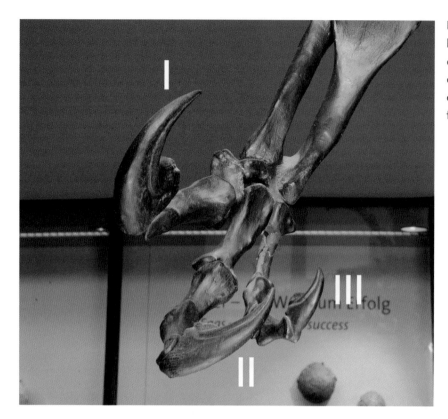

Figure 6.10. Three-fingered right hand of *Allosaurus*. Digit I was semi-opposable, conferring a grasping capability. In the highly evolved dromaeosaurs, this ability was even further developed.

Strong arms and dexterous, three-fingered hands characterized most theropods, although some primitive theropods had four digits (digit V is lost), and a few highly derived forms – most notoriously tyrannosauroids had only two (I and II, and II and III, respectively; see Chapter 7). In general, the three theropod digits, I (thumb), II, and III, were long and capable of extreme extension, and tipped with powerful claws. Digit I (the thumb) could fold across the palm in a semi-opposable fashion; that is, somewhat like a human thumb. There is no mistaking the function of this hand: it's all about grasping (Figure 6.10).

Even highly specialized large theropods such as the tyrannosauroids, with their notoriously short arms (the hands could not reach the mouth) had stout, yet powerful bones and fingers tipped with large claws, suggesting active use. By contrast, the tiny arms and hands of *Carnotaurus* and its relatives lost many of the bones in the hand, and were likely vestigial, that is, relict evolutionary features no longer used by the animal (Figure 6.11).

Teeth and Jaws

As with many other meat-eaters, carnivorous theropod heads tended to be proportionately large. In the case of the biggest, the heads could be upward of 1.75 m (5'8") in length. In general, theropod skulls are superficially reminiscent of those of many nondinosaurian ornithodirans. Yet there are differences among theropods: tyrannosauroids had robust, deep-jawed skulls, suggesting a powerful bite. Other theropods – even large ones like *Carcharodontosaurus* – had much more lightly built skulls (Figures 6.12 and 6.13). Noncarnivorous theropods, such as the toothless ornithomimosaurs and toothed, but noncarnivorous therizinosaurs, had proportionally smaller heads.

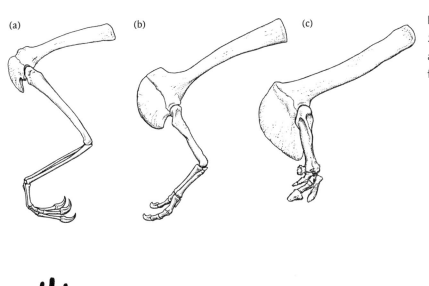

Figure 6.11. Left forelimb of (a) *Struthiomimus*, (b) *Tyrannosaurus*, and (c) *Carnotaurus*. Human hand for scale.

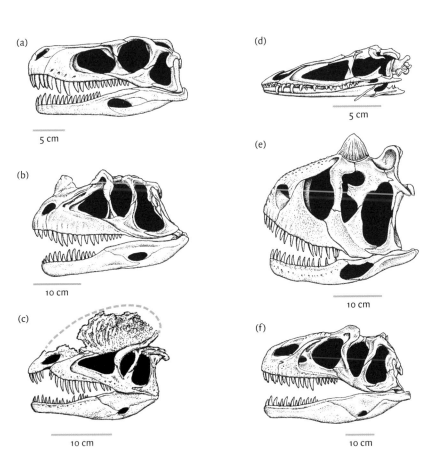

Figure 6.12. Left lateral view of the skull of (a) *Herrerasaurus*, (b) *Ceratosaurus*, (c) *Dilophosaurus*, (d) *Coelophysis*, (e) *Carnotaurus*, and (f) *Allosaurus*.

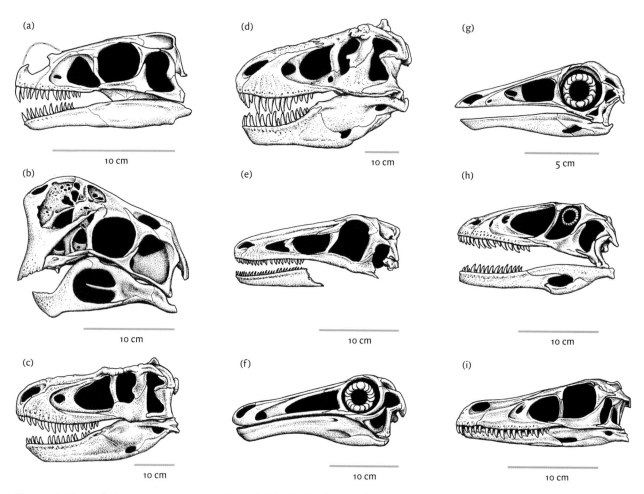

Figure 6.13. Left lateral view of the skull of (a) *Ornitholestes*, (b) *Oviraptor*, (c) *Albertosaurus*, (d) *Tyrannosaurus*, (e) *Saurornithoides*, (f) *Gallimimus*, (g) *Dromiceiomimus*, (h) *Deinonychus*, and (i) *Velociraptor*.

Theropod teeth tend to be flattened from side to side, **recurved** (curved backward), pointed, and serrated (see Figure 6.3). With the jaw joint at the level of the tooth row, the effect was like that of a pair of scissors, slicing from back to front (Figure 6.14). While these teeth are clearly about puncturing and slicing, they are equally clearly *not* about chewing. Chewing, as we shall see in Part III, is a behavior independently invented by mammals and ornithischian dinosaurs, and the fact is that most tetrapods (and most vertebrates, generally) don't really chew their food to any great degree. So in this sense, it is no surprise that theropods didn't bother with chewing and with all of the tooth and mouth modifications that this behavior entails.

It was the sharply pointed, recurved, and serrated teeth in the upper and lower jaws that handled the prey. The recurved shape kept prey from escaping from the mouth, but also accommodated the geometry of the theropod jaw: the teeth tended to be more curved toward the back, and straighter towards the front, which seems to have been about properly positioning them for the most efficient puncturing as the jaw closed. The teeth of

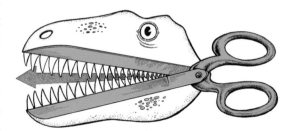

Figure 6.14. In its manner of slicing and in the placement of its hinge, the carnivorous theropod mouth was very similar to the design of a pair of scissors.

smaller theropods such as *Velociraptor*, with their prominently pointed serrations and narrow cross-sections, would have sliced like hacksaw blades. The mineral crystals that make up the enamel of these teeth are arranged in a simple pattern, suggesting greatest strength in only one dimension. This implies that the teeth were subjected to uniform stresses, perhaps associated with slicing, and were not subjected to complex stresses such as holding a struggling prey.

The teeth of tyrannosauroids, at the other end of the spectrum, with bulbous teeth and rounded serrations, had a weaker cutting ability, but greater bite power, suggesting that they could withstand complex, strong, and violent forces, such as might occur with a powerful, actively struggling prey (Figure 6.15). The enamel of tyrannosaur teeth has a complex pattern of crystal orientations, suggesting increased strength in multiple directions, such as might be useful when crushing bone (see below). More enigmatic, perhaps, the ambiguously carnivorous troodontids – to judge from their powerful, well-clawed hands and feet – had distinctively large denticles on their teeth instead of the usual serrations, a morphology that has been linked by some paleontologists with a partially herbivorous diet (Figure 6.15a).

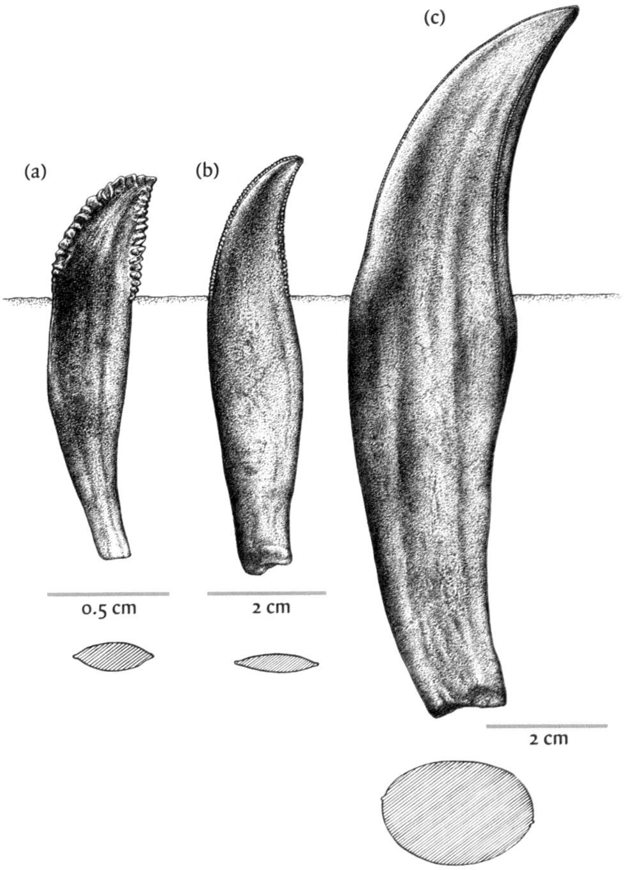

With such differences in skull and teeth, theropods likely bit in different ways. Recent studies have paired CT scans and computer-modeled stress analyses to the architecture of theropod skulls (Figure 6.16). We now know, for example, that *Allosaurus*, with its relatively lightly built skull, used a "slash-and-tear" attack on its prey, in which powerful neck muscles drove the skull downward rather

Figure 6.15. Extremes of theropod teeth compared. (a) *Troodon* tooth, with its large denticles in place of the serrations more commonly seen in theropods; (b) Blade-like meat-slicing tooth of *Dromaeosaurus*; (c) bulbous bone-crunching tooth of *Tyrannosaurus*.

than delivering a crushing bite with the jaw muscles alone. When the head was retracted, the teeth sliced and tore flesh. Such a wound might not kill prey immediately – but blood loss and possible bacterial infection would work their relentless damage. Tracking and waiting may have been part of the killing technique of dinosaurs with lightly built skulls and thin, blade-like teeth.

This contrasts with tyrannosauroids (see Figures 6.13d and 6.17) and perhaps abelisaurids such as *Carnotaurus* (see Figure 6.12e), whose more bulbous teeth and larger, more heavily built, deeper skulls likely delivered a bone-crushing, heart-stopping bite. They may also have suffocated their victim by seizing their snout or neck between their jaws and clamping down. This kind of attack is consistent with the large gape, powerful jaws, and stout teeth of these predators. Powerful neck muscles supported the large skull. In all cases, however, the skull must have had considerable mobility on the neck because of a well-rounded occipital condyle and its articulation with the first part of the cervical (neck) vertebrae (see Box 6.1).

Toothless

At least two times in their history (birds make a third time), theropods drastically reduced or lost all of their teeth. First (excepting one primitive genus) all ornithomimosaurs, a group of small-skulled,

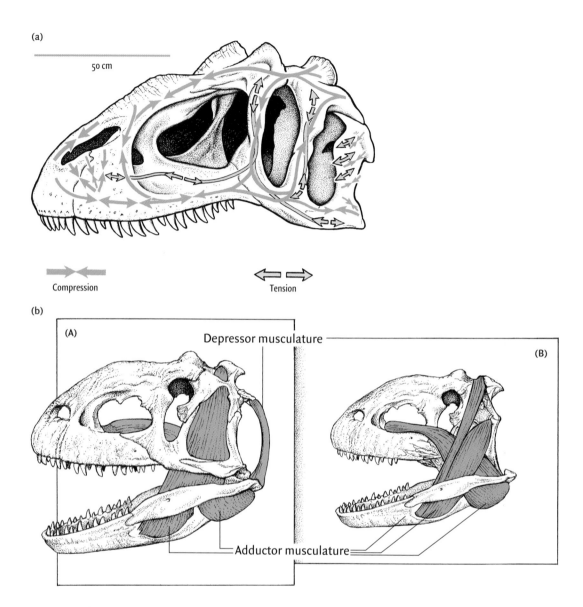

Figure 6.16. Reconstructing musculature. (a) The skull of *Allosaurus* marked with its finite element analysis model, indicating the regions of stress that pass through it as a function of biting. (b) Reconstructed jaw musculature in the skull of the closely related carnivorous theropod *Majungasaurus* showing (A) interior musculature blocked by the exterior muscles and (B) exterior musculature. Recalling that muscles can only retract, the jaw muscles are antagonistic: they work against each other, and one contracting will stretch the relaxed one. The adductor muscles close the jaw when they contract; the depressor muscles open the jaw when they contract. Not surprisingly, jaw-*closing* muscles are far larger than the jaw-*opening* muscles. Finite element analysis, in combination with CT scanning and anatomical comparison with the closest living relatives of these dinosaurs (birds and crocodiles; e.g., "extant phylogenetic bracketing;" see Box 6.1), allows the accurate positioning and reconstruction of muscle groups. From Holliday (2009) and Brusatte (2012).

long-legged theropods that look very much like ostriches with long tails (see Figure 6.10), lost all their teeth (see Figure 6.13f, g). Ornithomimosaurs had a beak, and in the case of *Gallimimus*, at least, the beak had a ridged feature that appeared almost sieve-like along its margin, provoking a controversial suggestion that it fed aquatically, much like a modern duck. Later interpretations of the beak edge suggest that it is less a sieve and more of a shearing feature, consistent with grinding fibrous plant matter. Either way, ornithomimosaurs were edentulous (toothless).

Ornithomimosaurs are also known to have had gastroliths, stomach stones that are stand-ins for teeth. Instead of chewing (e.g., processing food in the mouth), gastroliths were swallowed and kept in a muscular gizzard; then, swallowed but unchewed plant matter was ground down in the gizzard by the stones to make digestion more efficient (Figure 6.18). These, in combination with powerfully clawed hands and superb running capability, suggest a terrestrial existence rather than a duck-like lifestyle. Consistent with the beak, these animals used their gastrolith-lined gizzards to grind up fibrous plant matter, as do modern plant-eating birds.

The second edentulous group were oviraptorosaurs (see Figures 6.13b and 6.19; see also Chapter 7 and Figure 7.18). The skull was very short with apparent pneumaticity (see Chapter 8), and the jaw musculature was very well developed. Located between their shortened upper jaws is a pair of peg-like projections dead center in the middle of the palate. One analysis of the mechanics of the oviraptorosaur skull suggested that the jaws were designed to feed on hard objects that required crushing, such as clams, oysters, and mussels. Oviraptorosaurs, so the hypothesis goes, cracked them open by the brute force of their jaw muscles acting on the thick horny bill covering the margins of the mouth and the palate, and especially the stout pegs in the center. As we discuss in Chapter 7, however, what oviraptorsaurids did with their distinctive skulls is still an all-too-open question.

Figure 6.17. The teeth of *Tyrannosaurus* as seen from the left side of the skull.

Figure 6.18. Gastroliths.

Senses

To locate and track their prey, theropods of all kinds needed a keen awareness of their environment. Thanks to brain endocasts and, more recently, CT-scan reconstructions of the interiors of braincases (Figure 6.20; see also Box 11.1), we've begun to learn a lot more about how these animals operated. Because the brain represents the center for the senses, we look for similarities to the brains of related living creatures (see Box 6.1) such as birds and crocodiles, whose behavior is known, as well as to regions of enlargement, which suggested increased or heightened function of that part of the brain.

Clearly, however, sharp vision was key for theropods, so it is not surprising that their eyes were large. In deinonychosaurs generally, but especially in troodontids, the eyes have migrated to a more forward-looking position, indicating overlapping fields of vision. Overlapping fields of vision almost certainly means that these animals saw stereoscopically – that is, they merged the two separate independent images from each eye into a single image, much as humans and many modern carnivorous birds do today. Studies suggest that the narrow snouts of even tyrannosauroids allowed a 55° range of binocular vision, not nearly as much as a human or an owl, but far exceeding what one might find in a herbivorous dinosaur, such as hadrosaurid or ceratopsian (Figure 6.21).

Hearing, too, is important to predatory animals and so it is not surprising that many theropods likely had good sound perception. Indeed, the inner ear cavity of troodontids and ornithomimosaurs was greatly enlarged, suggesting that these theropods were especially able to hear low-frequency sounds. In troodontids, detailed anatomical study of the outer and middle ears suggests that the group was capable of identifying the direction from which sounds came; knowledge that would have been of extreme use to a predator.

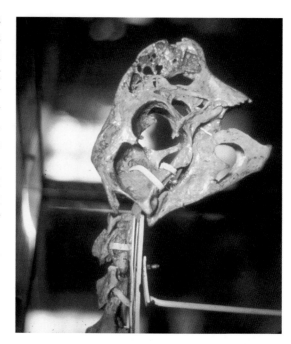

Figure 6.19. The box-like, toothless skull of the Late Cretaceous Mongolian theropod *Oviraptor*.

Making Sense(s) of Tyrannosaurs

With nothing close to a living analog, the behavior and lifestyle(s) of tyrannosaurs (there are many; see Chapter 7 for their place in the scheme of things) have always been enigmatic, with the mighty *Tyrannosaurus* being among the most perplexing – yet compelling – of all (see Box 7.1). Recent CT-scan-based work by L. M. Witmer, R. C. Ridgely, and associates at Ohio University has provided some exciting insights about *Tyrannosaurus* and, presumably, some of its near relatives (see Chapter 7). The brain has been fully reconstructed (Figure 6.20a), and with its relatively enlarged cerebral hemispheres, using the EQ metric (see Dino Brains in Chapter 13), Edinburgh paleontologist Stephen Brusatte has – perhaps optimistically – likened them to the thinking equivalent of a chimp; smarter than a dog or a cat. And what if he's only half right?!

The brain of *Tyrannosaurus* shows that it had a well-developed sense of smell. By comparison with other theropods, tyrannosaurs had relatively large olfactory bulbs, as well as a particularly large nasal cavity. The likelihood is that these features reflect behaviors that involve the ability to smell.

Like other theropods, tyrannosaurs were particularly adept at hearing low-frequency sounds. In the case of tyrannosaurs, however, the design and construction of the ears suggest that hearing was clearly an unusually important sense. Based upon tyrannosaur sensory capabilities, Witmer and colleagues have reconstructed tyrannosaurs as predators, with an ability to track prey rapidly by sight, smell, and sound.

Balance

The formidable armament of theropods, particularly the small- to medium-sized ones, must have been lethally coupled with exceptional balance. The nearly horizontal position of the vertebral column took advantage of the center of gravity being positioned near the hips.

Box 6.1 Putting Meat on the Bones... Through "Extant Phylogenetic Bracketing"

In this chapter and the ones that follow, there is a lot of glib talk about gnashing (teeth), slashing (claws), bashing (usually skulls); crunching (bone) and munching (plants). All of this takes muscles, of course; but how do we know where the muscles go – and how much muscle goes there?

We've all heard that skeletons provide support for the body; without them we would be quivering gelatinous puddles on the ground. While this is in a very broad way more or less true, it doesn't really explain why, for example, your back muscles get tired if you take a very long walk; isn't your skeleton supporting you?

But of course it isn't actually your skeleton that does the majority of the supporting – it is your musculature. The skeleton provides well-positioned support and attachment for the muscles, and the muscles do the heavy lifting. But how do we understand the musculature of extinct animals, whose actual muscles are long gone?

We start with homology – just as the bones show their phylogenetic history through homology, so do the muscles. That is, different muscles and muscle groups – sometimes unmodified and sometimes somewhat modified – are traceable via homology, just as the bones were.

That being the case, if you want to know about the muscles in nonavian dinosaurs, you begin with understanding the positions and functions of the muscles of the two closest living groups: birds and crocodiles, and groups successively distant from archosaurs (such as lizards and even mammals). This process, which is actually all about determining muscle homologies and potentially even function, is termed **extant phylogenetic bracketing**, because we are using the known muscle arrangements in closely related *living* forms – i.e., they are bracketed on the cladogram by their nearest living ("extant," not extinct) relatives – to reconstruct how the muscle groups would have appeared and perhaps functioned in extinct forms.

The skeleton, as we now know, is for muscle support, and therefore its shape is commonly (but not always) responsive to stresses placed upon it by those muscles: where the stresses are highest, the bones are largest and thickest; commonly prominent ridges develop; the shape of bone is, in large part, a reflection of the stresses it undergoes. Moreover, where the muscles attach to the skeleton, the bone shows marks called "muscle scars." The skeleton, therefore, in combination with phylogenetic bracketing, provides a kind of guide about which muscle groups ought to lie where. The keys to this process are well preserved, undistorted skeletal elements. Here, novel techniques such as CT scanning (see Figure 6.20) are very important, because they give us high-resolution, undistorted images of dinosaur skeletons.

Nobody has ever seen the full musculature of a large Mesozoic theropod; yet we may be reasonably certain that reconstructions, such as that shown in Figure B6.1.1, are accurate, thanks to phylogenetic bracketing.

This box has been about adding muscles to bone, but obviously the uses of phylogenetic bracketing are not restricted solely to muscles. In fact, phylogenetic bracketing is used for understanding the anatomy and function of many soft tissues, reflecting much about dinosaurs and other extinct creatures that rarely preserve. We'll use it again and again in this book, as we try to move past just the anatomy of the bones, and understand the lives of extinct dinosaurs.

Figure B6.1.1. Reconstruction of the musculature of *Tyrannosaurus*. Vanishingly little of the actual musculature of *Tyrannosaurus* has ever been preserved, yet, extant phylogenetic bracketing, in combination with a careful assessment of the animal's size and well-known skeletal structure, allows us to have confidence that its leg, displayed to shapely advantage in the Fukui (Japan) Prefectural Dinosaur Museum, must have looked very much like this in life.

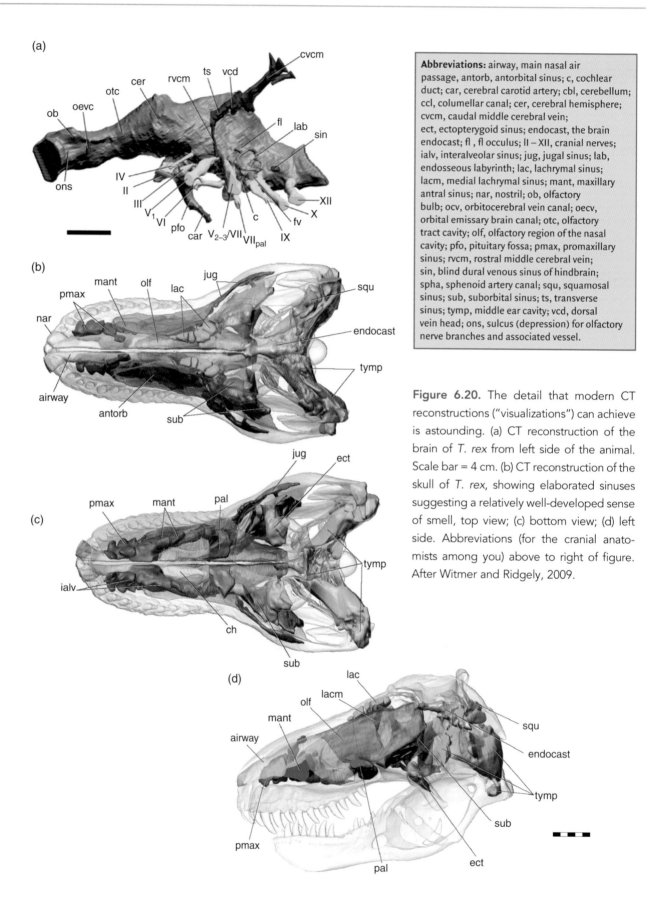

Abbreviations: airway, main nasal air passage, antorb, antorbital sinus; c, cochlear duct; car, cerebral carotid artery; cbl, cerebellum; ccl, columellar canal; cer, cerebral hemisphere; cvcm, caudal middle cerebral vein; ect, ectopterygoid sinus; endocast, the brain endocast; fl, fl occulus; II – XII, cranial nerves; ialv, interalveolar sinus; jug, jugal sinus; lab, endosseous labyrinth; lac, lachrymal sinus; lacm, medial lachrymal sinus; mant, maxillary antral sinus; nar, nostril; ob, olfactory bulb; ocv, orbitocerebral vein canal; oecv, orbital emissary brain canal; otc, olfactory tract cavity; olf, olfactory region of the nasal cavity; pfo, pituitary fossa; pmax, promaxillary sinus; rvcm, rostral middle cerebral vein; sin, blind dural venous sinus of hindbrain; spha, sphenoid artery canal; squ, squamosal sinus; sub, suborbital sinus; ts, transverse sinus; tymp, middle ear cavity; vcd, dorsal vein head; ons, sulcus (depression) for olfactory nerve branches and associated vessel.

Figure 6.20. The detail that modern CT reconstructions ("visualizations") can achieve is astounding. (a) CT reconstruction of the brain of *T. rex* from left side of the animal. Scale bar = 4 cm. (b) CT reconstruction of the skull of *T. rex*, showing elaborated sinuses suggesting a relatively well-developed sense of smell, top view; (c) bottom view; (d) left side. Abbreviations (for the cranial anatomists among you) above to right of figure. After Witmer and Ridgely, 2009.

Deinonychosaur balance was aided by a remarkable weapon in their arsenal: a tail stiffened by immensely elongated processes (zygopophyses) along the neural arches (Figure 6.22). The tail was flexible only at its base, just behind the pelvis; the rest was completely rigid. This stiffening allowed the rigid tail to move as a unit in any direction. It thus functioned as a dynamic counterbalance against the motions of the long, powerful arms, grasping hands, and large skull and jaws.

On the basis of their light, yet powerfully built skeletons, dromaeosaurids and troodontids must have had an extraordinary degree of agility. We imagine them flinging themselves at fleeing prey, kicking with great accuracy with one, or even both, of their dangerous feet , and then holding the struggling creatures down while they finished them off.

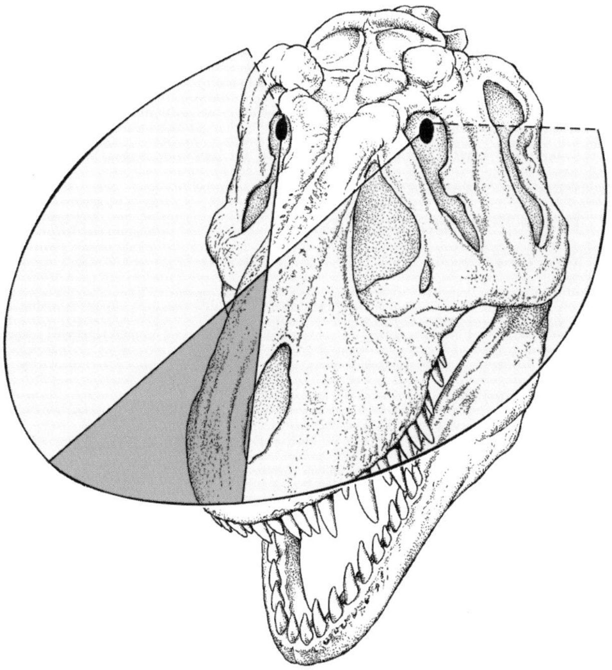

Thoughts of a Theropod

All theropods for which there is meaningful information about the shape and structures of the brain have surprising (to humans!) cerebral powers: their brains are comparable crocodilian or lizard "blown up" to the proper body size. Indeed, deinonychosaurs had the largest brains for their body size of any nonavian theropods, and were likely within the bird range of inferred intelligence (see Chapter 13). This suggests that these animals probably had more complex perceptual ability and more precise motor–sensory control than some of their smaller-brained brethren. We've seen that by some estimates, at least, even *T. rex* had some cranial capability. All of this implies sophisticated inter- and intraspecific behavior, attributes (a) for which there is independent evidence (as we shall see), and (b) that belie the image of irremediably stupid dinosaurs. Your parents' (and their parents') dumb dinos truly *are* extinct.

Figure 6.21. Three-quarter view of the skull of *T. rex* showing a 55° range of binocular vision (see text).

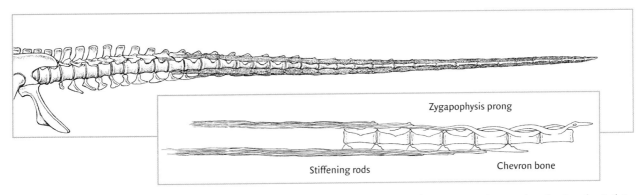

Figure 6.22. The rigid tail of *Deinonychus*. The elongate, intertwined zygapophyses on each neural arch give the tail its rigidity.

The Skinny on Skin

As we saw in Chapter 5, a modern conception of dinosaurs comes with feathers attached. Which dinosaur and which feathers is still something of an open question, but anyone who has been paying attention would agree that feathers are becoming more widespread in Dinosauria rather than less. Ultimately they may prove to be ubiquitous.

When birds and other exceptionally well-preserved fossils began showing up in China in the mid 1990s, paleontologists first recognized feathers and feather-like structures on nonavian theropods. Among the most famous of these are *Caudipteryx*, *Protarchaeopteryx*, *Sinornithosaurus*, and *Microraptor* (see Chapter 7). Two other theropods from the same region – *Sinosauropteryx* and *Dilong* – are extensively covered with filaments that have been interpreted as feather precursors. But feathered dinosaurs do not come exclusively from Asia; ornithomimisaurids from North America have also been reported with feathers. And, unsurprisingly, the feathers are all colored and patterned, suggesting that these animals interacted visually (see The Wonderful World of Color below).

We do not have skin impressions with feathers for most theropods. Still, some kind of insulatory covering (feathers) has been suggested for highly active theropods, such as dromaeosaurids and troodontids, and some paleontologists have speculated that, as juveniles, even large adult theropods may have been covered with a downy feather-like insulation.

Yet, the image of very large, powerful, heavy theropods, like *T. rex* or *Carcharodontosaurus*, covered with feathers is perhaps a little *outré*. Embrace it, because in 2011 Chinese paleontologist Xu Xing and colleagues announced the discovery of *Yutyrannus huali*, an 8 m (26') tyrannosaur from China. *Yutyrannus* was evidently covered, at least partially, with filamentous feathers 15 cm (6") in length (see Figure 7.8). With feathers clearly present in its near relative *Yutyrannus*, the "crazy" probability that *T. rex* was also feather-covered becomes very high indeed. We'll revisit feathers and theropods in greater detail in Chapter 7.

Eaters and Eatees

With all the different theropods (see Chapter 7), prey surely varied. The most dynamic and irrefutable evidence about the preferred prey of *Velociraptor* is the extraordinary "fighting dinosaurs" specimen, in which the fossil record captures a truly dramatic life-and-death moment: a theropod, *Velociraptor*, with its hindfeet half into the belly of a subadult *Protoceratops*, left hand grasping, right hand gripped held in the tightly clenched jaws of the evidently struggling prey (Figure 6.23).

A specimen of the Late Jurassic coelurosaur *Compsognathus* is known that contains most of the skeleton of a fast-running lizard. Not only did *Compsognathus* swallow nearly whole this delectable meal, but it must have captured its victim through its own speed and maneuverability. Other evidence of theropod stomach contents and diet comes from *Sinosauropteryx* (lizards and mammals), *Baryonyx* (fish remains), and *Daspletosaurus* (hadrosaurid bones).

Occasionally, the prey can be more easily identified than the predator. A 44 cm long, 13 cm high, and 16 cm wide (17" × 5" × 6") coprolite was found, containing between 30 percent and 50 percent of bone fragments, thought to be the remains of limb bones and parts of a ceratopsian (*Triceratops*?) frill. The source of this coprolite? Age (latest Cretaceous), geography (Hell Creek Formation, eastern Montana, USA; see Figure 17.12), and, uh, size, point to *Tyrannosaurus rex* as the perpetrator. The specimen, in combination with other information about theropod diets, provides physical evidence that tyrannosauroids, at least, crushed, consumed, and incompletely digested large quantities of bone.

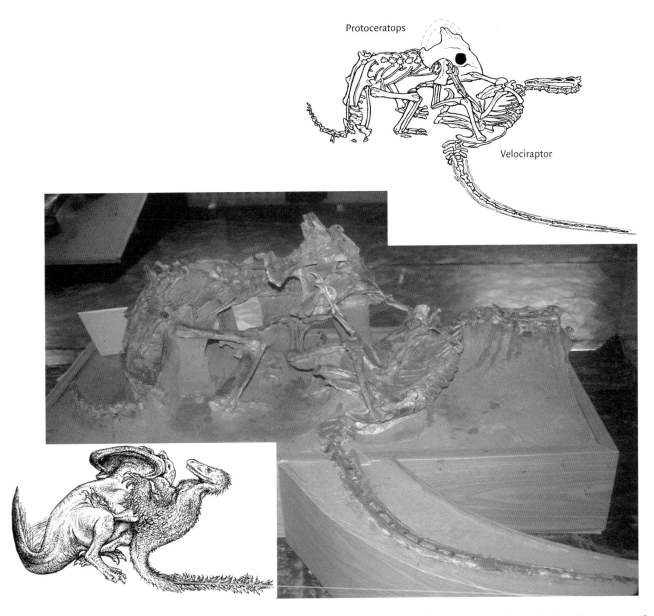

Figure 6.23. The famous fighting dinosaurs specimen; *Velociraptor* wrapped around *Protoceratops* from the Late Cretaceous of Mongolia.

Along those same lines, toothmarks attributed to *Tyrannosaurus* are known from marks in a *Triceratops* pelvis, and from a thoroughly crunched tail of the duck-billed dinosaur *Edmontosaurus*. Evenly spaced grooved toothmarks on the bones of the sauropods *Apatosaurus* and *Rapetosaurus* have been attributed to the local large theropods: *Allosaurus* and *Majungatholus* (from the USA and Madagascar, respectively; see below). And finally, the bite patterns of the large Late Jurassic theropod *Allosaurus* have been tied to marks on *Stegosaurus* neck plates and, conversely, *Stegosaurus* spikes have been implicated in a partially healed *Allosaurus* puncture-shaped wound. These dinosaurs lived together; were they also predator and prey?

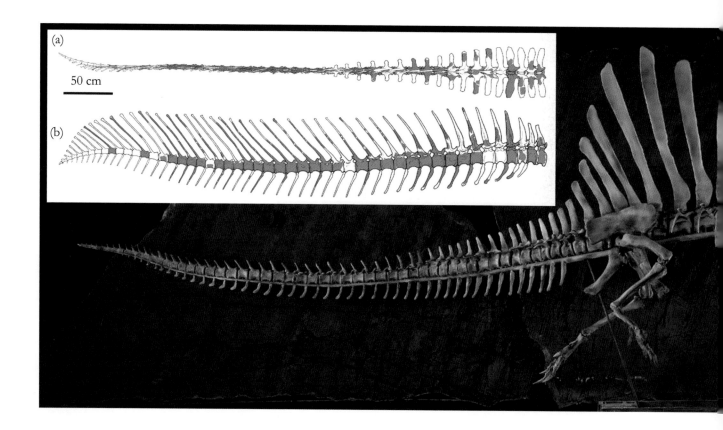

There had been hints for a long time that the large Cretaceous theropod *Spinosaurus* was fish-eating (piscivorous): equipped a crocodile-like snout, high nostrils, and tall conical teeth (hallmarks of fish-eaters; see Figure 6.24; see also Chapter 7), it was not a completely improbable idea, although swimming is not a common theropod adaptation.

In 2014, newly discovered material[2] suggested the interpretation that *Spinosaurus* spent a lot of time in the water, and that it might have been amphibious. The thing was huge – up to 15 m (49') – and showed a lot of unusual adaptations, including surprisingly long, powerful arms for a theropod that size (see Chapter 7). Moreover it had dense bone, a common features of many aquatic organisms, but a most unusual character in a theropod, animals that normally have hollow bones. It had flattened toe bones suggestive of a foot used for paddling (could it have been webbed?). Remarkably, it has a relatively weak, small pelvis (recall that the strong pelvis is the key balance point for this very terrestrially adapted group), and nostrils towards the middle (rather than the front) of the skull. All evidence suggested that the animal spent a lot of time in the water.[3]

If *Spinosaurus* was an aquatic predator, as all of these adaptations appear to suggest, it would likely have been piscivorous – but which fish? *Spinosaurus* teeth have been found in the same

---

[2] The original material was largely destroyed during the bombing of Munich by the Royal Air Force during World War II (see Box 16.7).

[3] The aquatic theropod deal was sealed, so to speak, in April 2020, when the full shape of the tail of *Spinosaurus* was first reported. As it turns out, *Spinosaurus* had a broad, flattened, finned tail, formed by the elongation of the neural arches and hæmal spines (chevrons). The tail would have provided aquatic propulsion with an undulatory motion. The existence of this tail was unfortunately not completely understood when the skeletal mount shown in Figure 6.24 was made.

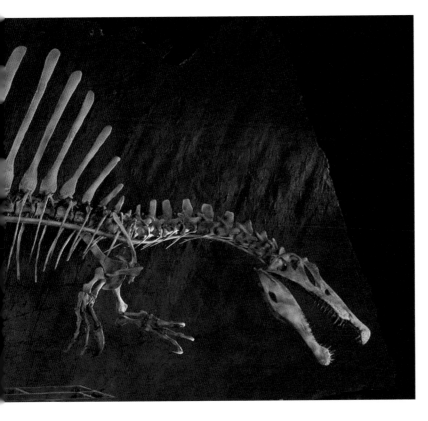

Figure 6.24. *Spinosaurus* rediscovered, and now re-realized! The reconstructed dinosaur, mixing elements form the original specimen and from newer specimens (see Box 16.7). With the tailfin now revealed, this mount will eventually need to be updated. Photograph courtesy of Paul Sereno. Inset: The newly discovered tail of *Spinosaurus*, as reconstructed in the pages of the prestigious journal *Nature*. (a) top view; (b) right-side view. From Ibrahim, N., *et al.*, 2020. Tail-propelled aquatic locomotion in a theropod dinosaur. *Nature*, **581**, 67–72.

deposits as those of the large sawfish *Onchopristis*; this and the fact that identifiable *Onchopristis* fragments were found embedded in the tooth sockets of *Spinosaurus* have led to media[4] speculation that *Spinosaurus* dined on *Onchopristis*. Could the sawfish fragments simply have lodged like pebbles in the empty sockets when the bones were washed together during burial? Of course; but the idea of *Spinosaurus* as a gigantic fish-eating amphibious predator is surely viable.

It took us about 100 years to begin to interpret the unusual morphology of *Spinosaurus* in a way that generates a coherent behavioral picture (admittedly, delayed by World War II). But what about its most striking feature: the row of 2 m high dorsal neural spines? Although it has been suggested that these were for display, you may have to check back in another 100 years for anything definitive about those!

Cannibals

Some dinosaurs didn't shy away from a bit of cannibalism: for example, *Majungatholus* apparently didn't avoid the odd conspecific snack: grooved toothmarks matching the spacing of its own teeth have been found on (other!) *Majungatholus* specimens (Figure 6.25). With only one carnivore known from Madagascar from that time with that tooth spacing, the evidence is circumstantial, but damning. Of course, we still don't know if such cannibalism occurred on the hoof, or whether it was the scavenging of sick or dead individuals (or both).

Other dinosaurs got undeserved reputations for cannibalism. It was once thought that the Late Triassic *Coelophysis*, based upon the supposed presence of its own juveniles within the rib cage, was cannibalistic; not a great recipe for species survival. However, recent work indicates that the

---

[4] The British television show *Planet Dinosaur* (2001).

Figure 6.25. Cannibalism in theropods. Above: the teeth of the large Malagasy theropod *Majungatholus*. Below: toothmarks on *Majungatholus* bone. Note that the spacing of the grooves in the bone matches that of the teeth; note also the scratches next to each groove, reflecting the serrations of the teeth.

juveniles in the rib cage were actually nondinosaurian archosaurs; a better species survival strategy, it seems to us.

Triceratops Spoils or Spoiled *Triceratops*

In 1917, L. M. Lambe of the Geological Survey of Canada suggested that the large carnivorous dinosaur *Gorgosaurus* was not so much an aggressive predator, but instead scavenged. The basis for his remarks was the apparent absence of heavy wear on the teeth of this theropod – these animals, he proposed, must therefore have fed primarily on the softened flesh of putrefying carcasses. This interpretation has appeared on and off again in discussions of theropod diet and hunting behavior, frequently enough to be something like a cottage industry in anecdotal "knowledge" about these animals.

The notion of large theropod scavenging rests on the assumption that tooth wear was usually absent and that carcasses were readily available. It is further bolstered by the lack of a convincing account of why the forelimbs of these animals are so small. We take each of these in turn. Firstly, it turns out that tooth wear is present on the teeth of nearly all large theropods. This doesn't "prove" that large theropods had to have been active predators; both modern scavengers and active predators alike can have a high degree of wear on their teeth. More to the point, tooth wear is present in all animals that have teeth; however, chewing, as we shall see, is a typically mammalian characteristic. Theropod dinosaurs never really chewed their food; rather, they used those marvelous serrated theropod teeth to slice off chunks, which they then sent down and back for processing.

The commonness of carcasses, putrefied or otherwise, was probably dependent on the season – dry, stressful seasons, as well as massive flooding events surely claimed their share of hadrosaurids, ceratopsians, sauropods, and other creatures (there is considerable evidence for flood fatalities in herbivorous herds; see Chapters 11 and 12). Interestingly, preserved specimens from river-associated deposits, the most common source of dinosaur bone, suggest carcasses of tough, dry flesh – dinosaur jerky, really – as the bodies dried out on the floodplains; we do not find preserved the kind of soft, ripe, malodorous predigested carrion imagined by Lambe. Perhaps it's just preservation; but here the fossil record duplicates what one observes in such settings today.

Finally, short forelimbs were likely used to slice and dismember prey, and are fully consistent with a mouth-first attack.

Tyrannosaurids in particular were singled out as carrion eaters, an idea that came from the observation that these theropods had surprisingly broad, bulbous teeth (Figures 6.15 and 6.17) and, due to their large size, could not have be fast enough to catch fleeter prey. Modern scavengers such as the hyena likewise have broad teeth, which are used to crush the bones of carcasses. Still, tyrannosaurids as carrion-eaters has been pooh-poohed by scientists who cannot imagine a dinosaur with the size and obvious carnivorous equipment of *T. rex* being a scavenger. Moreover, their argument goes, *T. rex* was a faster runner than the scavenger school suggests (see above).

What *is* clear is that the bulbousness of *Tyrannosaurus* teeth clearly exceeds the size increase that might be expected from the size of the animal. Here, the bone fragments in the large latest Cretaceous Montana coprolite tell the story: *T. rex* was clearly crunching bones as well as killing when called for, and getting at the protein in the marrow in the process. As we shall see (Chapter 13), growth rates in tyrannosaurids were such that these animals needed all the protein they could get.

In the end, we think it likely that theropods like *Tyrannosaurus* were capable, indeed terrifying, active predators, but not so picky that they would turn up their noses at a tasty carrion lunch. Indeed, it is probably better to view all theropods, from the exceptionally large tyrannosauroids down to much smaller troodontids, dromaeosaurids, and coelophysoids, as equal-opportunity consumers, taking whenever available large herbivores, smaller flesh eaters, and perhaps even the occasional carcass.

Jurassic Pack

Beyond these few direct observations of dietary preferences, we are left to speculate on who ate whom. Our best guesses are informed by the known theropods and potential prey in a particular place and time – we might pair *Tyrannosaurus* with *Triceratops* (the classic latest Cretaceous confrontation-of-the-titans; see Box 7.1), *Tarbosaurus* with *Saurolophus*, or *Troodon* with small ornithopods or juveniles of much larger coexisting dinosaurs (such as hadrosaurids).

Trackways referred to tyrannosaurids have been reported from the Late Cretaceous of northeastern British Columbia (Canada). These suggest that these animals were gregarious, and that the legendary *T. rex*, if not some of its nearest relatives, may have moved in groups of several dinosaurs (see below).

The invention of pack-hunting could have allowed relatively small animals to bring down much larger coexisting prey. Bones of different dinosaurs are often found together, leading to media-fueled speculation about diners and dinners. Teeth of *Deinonychus* (a dinosaur of around 3 m [nearly 10']) have been found in association with the large ornithopod *Tenontosaurus* (around 7 m; 23'), suggesting the former ate the latter. But the size difference between the two made this proposal a tall order, unless *Deinonychus* hunted in packs. Thus, the genesis of the idea of dromaeosaur pack-hunting, which was distorted almost beyond recognition in the *Jurassic Parks*. Yet, lest one think that pack-hunting in deinonychosaurs is all smoke and mirrors, a deinonychosaur trackway showing multiple individuals was reported from China in 2008.

Trackways provide potent suggestions of pack-hunting. In Figure 1.6, we saw a trackway from the Gobi Desert, Mongolia, that showed a pack (flock?) of small theropods likely tracking a larger theropod. The location, age, and morphology of the imprints point to *Velociraptor* for the origin of smaller theropod tracks; the larger imprints are likely those of single ornithomimid. The inference is that *Velociraptor* packs could and did take on much larger animals, such as ornithomimids.

The most successful attacks come from where they are least expected, and in 2009, paleontologists D. Varrachio and J. R. Horner, working with scientists studying modern avian infections, suggested that theropods, including the mighty *T. rex*, were under regular, and potentially crippling, attack by a lowly protozoan. These researchers noted that the jaws of many specimens of *Tyrannosaurus* show

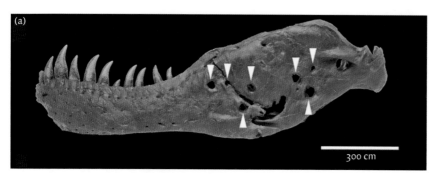

Figure 6.26. (a) Left lower jaw exhibiting multiple lesions (indicated by arrows). (b) Hypothesized reconstruction of the infection in a living tyrannosaur. Lesions within the lower jaw perforate the bone. Reconstruction based on photographs of living birds with similar pathologies.

300 cm

numerous lesions (Figure 6.26a). Once thought to be the work of other tyrannosaurs, these lesions are very similar in morphology to those found in the jaws of modern chickens, turkeys, and falcons. In the modern birds they are the handiwork of a parasitic protozoan, *Trichomonas gallinae*, an organism that infects jaws, eventually eating out chunks of the jaw bone, deforming and/or damaging the jaw. Similarities in the morphology of the lesions strongly suggest that this same organism, or a very close relative, was at work on Cretaceous nonavian theropods as well. The handiwork of other bacteria is also evident in theropods. The redoubtable *Tyrannosaurus* "Sue" (Box 1.1) shows the results of a chronic bacterial bone infection, as does "Big Al," an *Allosaurus* at the Field Museum in Chicago.

Social Behavior: Sex and the Single *rex*

Single-species bonebeds provide strong suggestions that theropods lived – and not just hunted – in packs. Many of these mass graveyards – which include both juveniles and adults – pertain to the coelophysoid group of theropods (Coelophysoidea; see Chapter 7) and include such luminaries as *Syntarsus* (Zimbabwe, South Africa, and Arizona) and *Coelophysis* (New Mexico).

Other theropods, though, are also known in bonebeds: *Allosaurus* (Utah), as well as *Giganotosaurus* and *Mapusaurus* (Patagonia). Could it have been that some theropods were gregarious, living in large family groups that perished together? Or perhaps each accumulation represents a communal feeding site?

Over the last 20 or so years, paleontologists keep noticing that even those most supposedly solitary of killers – the *über*-tyrannosaurs *Daspletosaurus* and *Tyrannosaurus* – are commonly found in pairs, at a minimum, and are therefore potential candidates for the comforts of at least family. Nonavian theropods as gregarious beasts – even the largest theropods – is becoming more and more the expected, and less and less the anomalous, particularly in the face of flocks of avian theropods.

Sexual Dimorphism and its Role in Sexual Selection

Many vertebrates show strong sexual dimorphism; that is, males and females are not identical, but instead have different attributes. Commonly, one gender is larger than the other, or they differ by other features. In diapsids generally, and modern birds in particular, coloration between males and females can be strikingly different, with the males commonly assuming the brightest hues. These traits are generally related to sexual selection, which can be described as selection not occurring equally on all members of a particular population or species (such as one sees in natural selection – see Appendix 3.1), but rather a kind of selection based upon gender. Male deer clashing for reproductive rights with an insouciant female would be a familiar example of sexual selection. In birds, sexual dimorphism is deeply tied into sexual selection, and commonly goes hand in hand with fancy, bright plumage, virtuosic vocalizations, complicated feather dances, and all types of display. How different is this from humans?!

What can the skeletons of theropods tell us about how these animals related to each other socially? Like frills and horns in neoceratopsians (Chapter 11), or crests in hadrosaurids (Chapter 12), quite a number of predatory dinosaurs – including *Syntarsus*, *Dilophosaurus*, *Proceratosaurus*, and possibly *Ornitholestes*, to *Ceratosaurus*, *Cryolophosaurus*, *Alioramus*, and *Oviraptor* – flaunted highly visible cranial crests (see Figures 6.12c and 6.27). Some are made of thin sheets of bone, while others are hollow, presumably part of the cranial air–sinus system. Beyond these, theropods such as *Yangchuanosaurus*, *Allosaurus*, *Acrocanthosaurus*, and the tyrannosauroids bore slightly elevated upper margins on the snout and raised and roughened bumps over the eyes. These structures are believed to have been cores for hornlets (small horns) made of keratin, which must have given the face a spiky punk-rock look (Figures 6.12b, e, f, and 6.13 a, c).

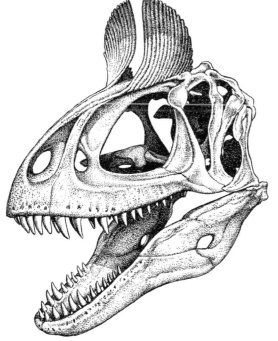

Figure 6.27. The skull of the Late Jurassic southern hemisphere polar theropod *Cryolophosaurus*, with its distinctive crest.

The crests and hornlets must have functioned in display and – at least for the latter – may have been used occasionally in squabbles over territories and mates. If crests and hornlets functioned in visual display, particularly in those theropods that lived in large groups (see above), we might expect them to be species-specific and probably sexually dimorphic so as to signal a given animal's identity and sex. And likewise we might expect crests to show their greatest development in reproductively mature individuals; youngsters should have small, poorly developed crests and hornlets.

Are these expectations met in any theropods? To a degree; sexual dimorphism is found in *Syntarsus* and *Coelophysis*. In these two theropods, one of the two morphs had a relatively long skull and neck, thick limbs, and powerfully developed muscles around the elbow and hip, while the other form had a shorter skull and neck, and slender limbs. In *Tyrannosaurus*, the case has been made for sexually dimorphic ornamentation. The larger, more robust morph was hypothesized to be the female. In the cases of other theropods, we just don't yet know; what we would give to see one (or two, or a pack) alive!

The Wonderful World of Color

With the evidence (above) for display and its role in sexual selection so strongly imprinted in the skeletal remains of extinct theropods (as well as in the behavior of living theropods), could color have played a role in nonavian theropod sexual selection, as it does today in *avian* sexual selection, in which – generally – dolled-up males strut and preen for the favors of more reasonably turned-out females?

Incredibly, some feather patterns have actually survived the process of fossilization. For example, a striped pattern in the tail of *Sinosauropteryx* is clearly visible with the naked eye (Figure 6.28). Beyond rudimentary patterns, however, color in any dinosaur – including those covered with feathers – has been, until recently, very elusive.

In modern birds, feather color is based in part upon microscopic capsule-shaped organelles called **melanosomes**. These organelles have different shapes, sizes,

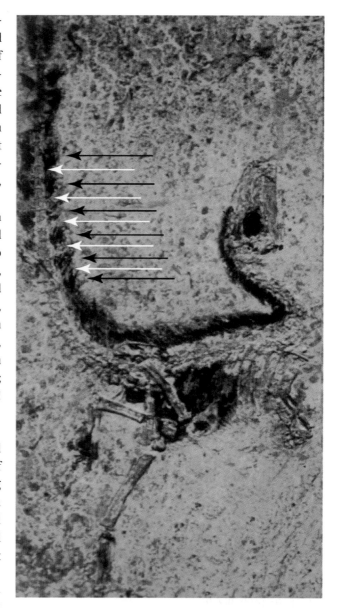

Figure 6.28. Patterns of color of the feathered, nonavian theropod *Sinosauropteryx*. Filamentous features – interpreted to be early, nonflight feathers – are preserved down the back of the animal, extending the length of the tail. Arrows indicate color banding, particularly well preserved on the tail.

and distributions, which correlate in modern birds with colors such as blacks, reds, browns, and yellows (Figure 6.29a). In 2007, using a high-powered electron microscope, melanosomes identical to those found in modern birds were identified by then Yale graduate student Jakob Vinther in extinct bird feathers. Vinther's team, working in parallel with a team lead by English paleontologist Michael Benton, and using the shapes and distributions in modern birds as a guide, for the first time reconstructed the actual colors present in long-extinct dinosaurs.

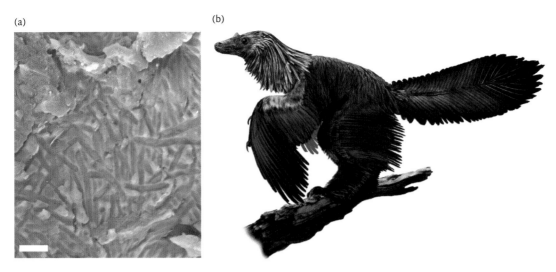

(a)

(b)

Figure 6.29. Melanosomes and dinosaur coloration. (a) Melanosomes, cells that produce color in feathers, seen here under an electron microscope. Their shapes and distributions give an indication of some of the colors that graced these nonavian theropods; these elongate shapes are indicative of blacks, browns, and grays. (b) A 2019 reconstruction of *Caihong*, an iridescent feathered theropod from Jurassic rocks in China.

Ultimately colors were reconstructed in a primitive bird, *Confuciusornis* (see Figure 8.7), in the primitive bird *Archaeopteryx* (Chapter 8), as well as in the primitive feathers of non-flying theropods *Sinosauropteryx* (Figure 6.28) and *Sinornithosaurus* (see Chapter 7; Figure 7.21), and the troodontid *Anchiornis* (Figures 7.23 and 7.24). As Benton tells it in his fascinating 2019 book *The Dinosaurs Rediscovered*, his team managed to reconstruct the colors of *Sinosauropteryx* just ahead of Vinther's team, which ultimately reconstructed colors in many dinosaurs, some of which grace this text. What Benton's team determined was that banding in the tail of *Sinosauropteryx*, as we have seen, still visible, was once white and ginger striped; a crest along its back may have had tan–reddish-brown coloration. Now, other colors have been recognized in fossil feathers of nonavian theropods; even iridescence, such as in the spectacular colors of the newly identified Jurassic primitive bird *Caihong* (Figure 6.29b).

The colors and patterns present in these nonavian theropods reinforce the idea that nonavian theropods probably used color and pattern for a variety of complex social behaviors, including display and sexual selection, just as we see in living (avian) theropods. We'll see other examples of nonavian dinosaur color in Chapter 7.

Mama's (?) Little Theropod

Regardless of sexual dimorphism in nonavian theropods, what we know about their reproductive biology has been greatly enhanced by the discovery of brooding oviraptorosaurs. Eggs were first found in Mongolia by Roy Chapman Andrews' 1920s expedition (Figure 1.9; Box 16.1) and attributed to the small Asian ceratopsian *Protoceratops* (Chapter 11). Because the eggs were associated with theropod skeletons, it was assumed that the theropods were stealing the "*Protoceratops* eggs" and were given the name *Oviraptor* (= egg stealer). It was a bit shocking, therefore, when, in the mid 1990s, the eggs were found to contain not *Protoceratops* but *Oviraptor* embryos; poor *Oviraptor* had languished, falsely accused, for some 70 years.

(a)

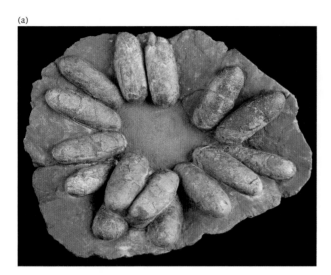

Figure 6.30. *Oviraptor* eggs and babies. (a) *Oviraptor* eggs. Note the paired eggs organized in a circle. These eggs appear to be pointing down; that may be an artifact of compaction after burial. In life, they were likely angled upwards. (b) Specimen of an oviraptorosaur adult on its haunches, nesting its eggs. Most of the body of the adult is gone, but the legs and arms indicate where it once was, arms cradling the eggs.

(b)

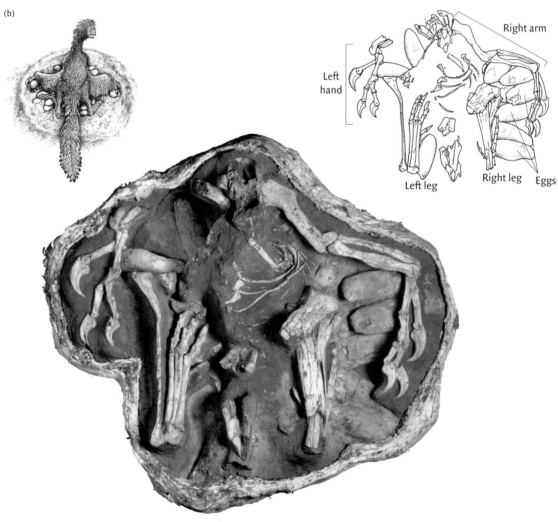

Since the discovery in the 1990s, as many as eight articulated oviraptorosaur skeletons have been found preserved overlying nests of eggs. The embryos are the same species as the adult skeleton overlying them. The oviraptorosaur skeleton (Mom? Dad?) is positioned directly above the center of the nest, with its limbs arranged symmetrically on either side and its arms spread out around the perimeter as if protecting the nest (Figure 6.30). Were these specimens incubating eggs on open nests?

The nests consist of paired eggs, deposited in concentric circles (Figure 6.30a). Interestingly, the innermost part of the nest was a dirt mound, around which the first circle of paired eggs appear to have been laid, so each egg was inclined at about 30° to 40°, blunt end upwards. More dirt seems to have been laid down by the parent (Mom? Dad?), and then a second circle of paired eggs was deposited (Mom!). How many eggs were laid by a single *Oviraptor* remains a question; estimates run from 15 (almost certainly low) to 30. The process was then repeated and a nest could contain as many as three concentric circles of egg pairs.

Knowing certain compounds responsible for the coloration in eggs, a team of scientists from the University of Bonn were able to isolate pigments in the Late Cretaceous eggshell genus *Macroolithus* – attributed to the Chinese oviraptorosaur *Heyunannia*. They reconstructed the color of these eggs as ranging from blue to green, presumably matching the background of the substrate on which they were laid. This work reinforced conclusions reached from the fossil shown in Figure 6.30: the eggs were likely laid in an open nest.

And what exactly was that parent (Mom? Dad? Mom and Dad?) doing there? At first one might suppose that (s)he was (a) incubating the eggs, but other roles can be conceived as well, including (b) (s)he provided protection and, (c) the fossil caught the adult oviraptorosaur *in flagrante delicto*, as it were, in the very act of egg-laying. Likely as (a) seems, and unlikely as (c) seems, (a) has some problems. First, the eggs, angled upwards as they are, and separated by layers of nest, are not ideally positioned for being warmed by a dinosaur (or anything else) sitting on top of them. That also applies to the center of the next layer, which, it will be recalled, is not eggs at all, but a rather small egg-free dirt mound. Was the parent incubating the mound?! Second, clumped isotopic studies from 2015 (see Chapter 14 for a discussion of clumped isotopes) suggest that the parents' body temperatures were a bit too cool. This result, however, was contradicted by an isotopic study in 2017 showing higher temperatures during oviraptorosaur egg incubation. The upshot of this is that no definitive conclusion has yet been reached about body temperatures in these dinosaurs and thus, other than being the parent, the role(s) of the adult oviraptorosaurs in this drama is still uncertain.

What we know of the posthatching growth of nonavian theropods mostly comes from bonebeds (for example, Ghost Ranch (New Mexico) and Cleveland-Lloyd (Utah)), as well as from some of the Upper Cretaceous localities of the Gobi Desert in Mongolia. For *Coelophysis* and *Syntarsus*, apparently there was a 10- to 15-fold increase in body size from hatchling to adulthood, and this growth is thought to have been quite rapid. We'll revisit the question of growth rates in dinosaurs in Chapter 13.

For now, we leave the details of theropod lives and, in the next chapter, we explore the extraordinary diversity of theropods.

SUMMARY

Theropods are among the most iconic of dinosaurs, including beasts such as *Tyrannosaurus rex*. Clawed bipeds all, with distinctive hollow bones, the earliest known dinosaurs were theropods, and it is theropods that are still living (as birds). Most of the Mesozoic forms had claws with a semi-opposable thumb on a grasping three-fingered hand; possessed recurved, serrated, laterally compressed teeth; and were carnivorous. As a group, they were cursorial through and through.

Theropods were the most intelligent of dinosaurs, with highly developed sensory and motor skills for tracking and killing their prey. All had very good vision – in many cases, stereoscopic – and *T. rex*, at least, enjoyed a highly developed sense of smell. A strong impression of high agility in small- to medium-sized theropods is conveyed by the skeletons.

Theropods were all likely feathered to a greater or lesser extent; and almost certainly rather colorful animals, a characteristic that was probably associated with display and sexual selection. With modern birds as living (if highly derived) examples, social behavior can be inferred in theropods. A variety of facial features such as hornlets adorned theropods, and some evidence suggests that many of them hunted in packs. Particularly bird-like is theropod maternal behavior: in those forms in which it is known, nonavian theropod mothers (?) incubated clutches of eggs, very much like avian theropods.

SELECTED READINGS

Abler, W. L. 1992. The serrated teeth of tyrannosaurid dinosaurs, and biting structures in other animals. *Paleobiology*, **18**, 161–183.

Benton, M. J. 2019. *The Dinosaurs Rediscovered: How a Scientific Revolution is Rewriting History*. Thames & Hudson, London, 320 pages.

Brusatte, S. L. 2012. *Dinosaur Paleobiology*. Wiley-Blackwell, New Jersey, 322 pages.

Carpenter, K., Sanders, F., McWhinney, L. A., and Wood, L. 2005. Evidence for predator-prey relationships: examples for *Allosaurus* and *Stegosaurus*. In Carpenter, K. (ed.) *Carnivorous Dinosaurs*, Indiana University Press, Bloomington, IN, pp. 325–350.

Clark, J. M., Maryańska, T., and Barsbold, R. 2004. Therizinosauroidea. In Weishampel, D. B., Dodson, P., and Osmólska, H. (eds.) *The Dinosauria*, 2nd edn. University of California Press, Berkeley, pp. 151–164.

Currie, P. J. 1990. Elmisauridae. In Weishampel, D. B., Dodson, P., and Osmólska, H. (eds.) *The Dinosauria*, 1st edn. University of California Press, Berkeley, pp. 245–248.

Currie, P. J., Trexler, D., Koppelhus, E. B., Wicks, K., and Murphy, N. 2005. An unusual multi-individual tyrannosaurid bonebed in the Two Medicine Formation (Late Cretaceous, Campanian) of Montana (USA). In Carpenter, K. (ed.) *Carnivorous Dinosaurs*. Indiana University Press, Bloomington, IN, pp. 313–324.

D'Amore, D., 2009. A functional explanation for denticulation in theropod dinosaur teeth. *Anatomical Record*, **292**, 1297–1314.

Holliday, C. M. 2009. New insights into dinosaur jaw muscle anatomy. *Anatomical Record*, **292**, 1246–1265.

Holtz, T. R. 2004. Tyrannosauroidea. In Weishampel, D. B., Dodson, P., and Osmólska, H. (eds.) *The Dinosauria*, 2nd edn. University of California Press, Berkeley, pp. 111–136.

Holtz, T. R. Jr. 2012. Theropods. In Brett-Surman, M. K., Holtz, T. R. Jr., Farlow, J. O., and Walters, R. (eds.) *The Complete Dinosaur*, 2nd edn. Indiana University Press, Bloomington, pp. 346 – 378.

Holtz, T. R., Molnar, R. E., and Currie, P. J. 2004. Basal Tetanurae. In Weishampel, D. B., Dodson, P., and Osmólska, H. (eds.) *The Dinosauria*, 2nd edn. University of California Press, Berkeley, pp. 71–110.

Horner, J. and Gorman, J. 2009. *How to Build a Dinosaur*. Dutton, NY, 246 pages.

Makovicky, P. J. and Norell, M. A. 2004. Troodontidae. In Weishampel, D. B., Dodson, P., and Osmólska, H. (eds.) *The Dinosauria*, 2nd edn. University of California Press, Berkeley, pp. 184–195.

Makovicky, P. J., Kobayashi, Y., and Currie, P. J. 2004. Ornithomimosauria. In Weishampel, D. B., Dodson, P., and Osmólska, H. (eds.) *The Dinosauria*, 2nd edn. University of California Press, Berkeley, pp. 137–150.

Norell, M. A. and Makovicky, P. J. 2004. Dromaeosauridae. In Weishampel, D. B., Dodson, P., and Osmólska, H. (eds.) *The Dinosauria*, 2nd edn. University of California Press, Berkeley, pp. 196–209.

Osmólska, H. and Barsbold, R. 1990. Troodontidae. In Weishampel, D. B., Dodson, P., and Osmólska, H. (eds.) *The Dinosauria*, 2nd edn. University of California Press, Berkeley, pp. 259–268.

Osmólska, H., Currie, P. J., and Barsbold, R. 2004. Oviraptorosauria. In Weishampel, D. B., Dodson, P., and Osmólska, H. (eds.) *The Dinosauria*, 2nd edn. University of California Press, Berkeley, pp. 165–183.

Rega, E. 2012. Disease in dinosaurs. In Brett-Surman, M. K., Holtz, T. R. Jr., Farlow, J. O., and Walters, R. (eds.) *The Complete Dinosaur*, 2nd edn. Indiana University Press, Bloomington, pp. 666 – 712.

Sereno, P. C., Martinez, R. N., Wilson, J. A., *et al.* 2008. Evidence for avian intrathoracic air sacs in a new predatory dinosaur from Argentina. *PLoS ONE*, **3**, e3303. doi:10.1371/journal.pone.0003303.

Schweitzer, M. H. 2011. Soft tissue preservation in terrestrial Mesozoic vertebrates. *Annual Reviews of Earth and Planetary Sciences*, **39**, 187–216.

Schweitzer, M. H., Moyer, A. E., and Zheng, W. 2016. Testing the hypothesis of biofilm as a source of soft tissue and cell-like structures preserved in dinosaur bone. *PLoS ONE*, **11**, e050238. doi:10.1371/journal.pone.050238.

Stokosa, K. 2005. Enamel microstructure variation within the Theropoda. In Carpenter, K. (ed.) *Carnivorous Dinosaurs*. Indiana University Press, Bloomington, IN, pp. 163–178.

Sues, H. D. 1990. *Staurikosaurus* and Herrerasauridae. In Weishampel, D. B., Dodson, P., and Osmólska, H. (eds.) *The Dinosauria*, 2nd edn. University of California Press, Berkeley, pp. 143–147.

Therrien, F., Henderson, D. M., and Ruff, C. B. 2005. Bite me: biomechanical models of theropod mandibles and implications for feeding behaviour. In Carpenter, K. (ed.) *Carnivorous Dinosaurs*. Indiana University Press, Bloomington, IN, pp. 179–237.

Tykoski, R. S. and Rowe, T. 2004. Ceratosauria. In Weishampel, D. B., Dodson, P., and Osmólska, H. (eds.) *The Dinosauria*, 2nd edn. University of California Press, Berkeley, pp. 47–70.

Witmer, L. M. and Ridgely, R. C. 2009. New insights into the brain, braincase, and ear region of tyrannosaurs (Dinosauria, Theropoda), with implications for sensory organization and behavior. *The Anatomical Record*, **292**, 1266–1296.

Yang, T.-R, Wiemann, J., Xu, L., *et al.* 2019. Reconstruction of oviraptorid clutches illuminates their unique nesting biology. *Acta Palaeontologica Polonica*, **64**, 581–596.

TOPIC QUESTIONS

1. Describe the general features of theropods.
2. Describe how we know that running – cursoriality – was a key feature of theropod behavior.
3. What features of the theropod hand strongly suggest dexterity and grasping ability?
4. What kinds of evidence exist for what theropods ate?
5. Can a toothless animal be carnivorous?
6. Describe the range of skull and tooth design in theropods. How do those relate to our understanding of how theropods bit and killed?
7. What are the key clues that suggest that, as a group, theropods were carnivorous?
8. Highlight the evidence for pack-hunting by theropods.
9. Consider the design of ornithomimosaurs. Why is it difficult to explain the lifestyle of these animals?
10. We said that the most unambiguous features of a carnivorous lifestyle are seen in small- to medium-sized theropods. What features caused us to say this? Speculate on why this might be the case.
11. Theropod evolution was complex, so here is an exercise to help you better understand their relationships. Construct a single cladogram with the groups Theropoda, Coelophysoidea, Neoceratosauria, Tetanurae, Avetheropoda, Coelurosauria, Carnosauria, Maniraptora, Eumaniraptora, and Aves on it.

Theropoda II
Meet the Theropods

WHAT'S IN THIS CHAPTER

Theropoda encompasses a *lot* of dinosaurs, some familiar and some not-so-familiar. In this chapter we'll sort out theropods, by highlighting important groups and their relationships to each other. We begin with a simple framework onto which we can "hang" a bit of theropod diversity. As the complexity mounts, the cladogram is our friend; it helps us distill the welter of dinosaur names into a few large, simple, and connected groups. The basic relationships of these large groups are key, because these allow us to identify the dominant themes in theropod evolution.

Major Events in Theropod Evolution

Figure 7.1 shows the core relationships of the major groups of theropods. We start with Neotheropoda (*neo* – new) – all theropods excepting some basal (primitive) ones we met in Chapters 5 and 6.

Within neotheropods, there is a fundamental split between the Late Triassic *Coelophysis* and some near relatives on the one hand, and on the other Tetanurae (*tetanus* – stiff; *uro* – tail). And within Tetanurae, we'll see yet another division between the megalosaur group (of whom we met *Spinosaurus* in Chapter 6), and Avetheropoda (*aves* – birds, denoting the fact that this group is on the evolutionary road to birds). Finally within Avetheropoda, we meet Coelurosauria, a group that includes all kinds of dinosaurs, from *Tyrannosaurus* to birds. Here the cladogram maps out these basic relationships (Figure 7.1).

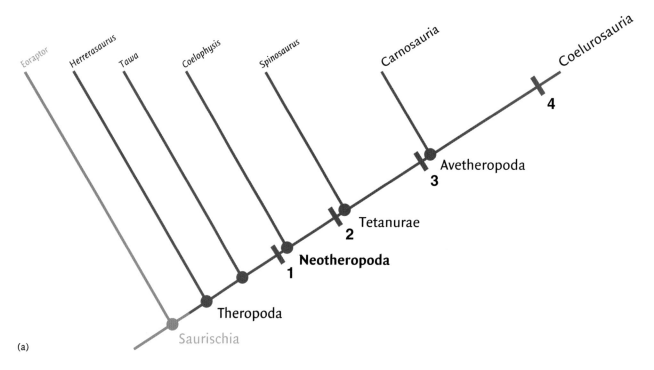

(a)

Figure 7.1. (a) The basic relationships of the major groups of theropods. Again, these characters are taken largely from the work of theropod specialists Tom Holtz (2012) and Stephen Brusatte (2012) of University of Maryland and University of Edinburgh, respectively: at **1**, furcula present; digit V on hand lost; five or more sacral (pelvic) vertebrae; foot functionally three-toed (tridactyl); at **2**, teeth restricted to the front of the jaws; large hands (proportionately); interlocking zygapophyses on tail vertebrae; maxillary fenestra (a second antorbital opening, anterior to the first); at **3**, three-fingered hand (loss of digit IV); complex chambers in vertebrae suggesting elaborate air sac system; at **4**, enlarged brain; long, narrow foot.

(b) The cladogram seems to suggest that theropod evolution was aimed, from the beginning, towards the evolution of birds (or at least towards Coelurosauria). But of course it was not; that is just an artifact of how the cladogram appears. Evolution is never *aimed*, nor is it *trying* to get to a particular accomplishment (say, transitioning from legs to fins) or behavior (for example, flight). To underline this point, we show the same cladogram, with the appearance that theropod evolution was "aimed" towards the evolution of *Coelophysis*. All we have done is rotate the cladogram around the Neotheropoda node, which as we saw in Figure 3.5, changes nothing about relationships. The illusion of directionality of evolution is only possible to observe after the fact.

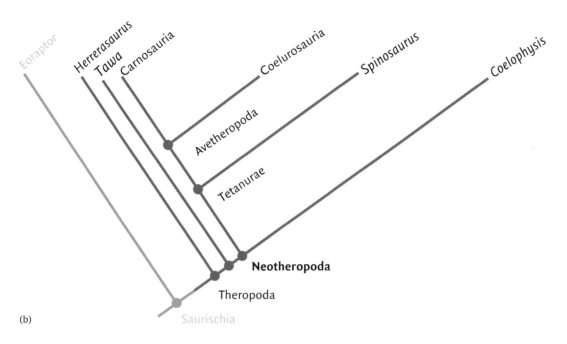

Figure 7.1. (*cont.*)

Neotheropoda

Neotheropoda (*neo*, new) is a well-diagnosed clade, among whose many characters are the development of the familiar "three-toed" theropod foot (recall that three are weight-bearing, the fourth, actually digit I, is very reduced and was evidently held above the ground, to the side or back (see Figures 6.7 and 6.9). Another distinctive neotheropod character is the furcula, a single bone composed of the two clavicles (collar bones) fused together. The furcula is the familiar Thanksgiving "wishbone" of chickens and turkeys. In *living* tetrapods, a furcula is known only in birds: and all living birds have a furcula. But the cladogram tells the more accurate story: rather than being invented by birds, the furcula is a carryover from the neotheropod ancestors of birds. Figure 7.1 also lists other neotheropod characters.

Famous neotheropods include the Late Triassic small biped *Coelophysis* (Figures 7.2a and 6.12d), the larger Early Jurassic *Dilophosaurus* (Figures 7.2b, 6.12c), and the large, stubby, carnivorous, south American abelisauroids such as *Carnotaurus* (Figures 7.2c and 6.12e), and their near relatives, the ceratosaurids (including most famously the Late Jurassic *Ceratosaurus*; Figure 7.2d), and of course tetanurans (see below).

Tetanurae

Linking the members of the great dinosaur clade Tetanurae – and the source of its name – is the possession of an inflexible tail: beyond its base, the tail simply doesn't bend as it does in other tetrapods. This is because the back (distal) half of the tail, towards the tip, is stiffened by interlocking zygapophyses, fore-and-aft projections from the neural arches (Figure 7.3). Recall that we saw an extreme example of this tendency in *Deinonychus* (Figure 6.22). The tail is an important counterbalance in theropod architecture (Chapter 6), and through the evolution of tetanuran functional design, there is a marked tendency to *decrease* its flexibility (except at its base).

Figure 7.2. Some representative (nontetanuran) neotheropods. (a) *Coelophysis* (Late Triassic, Arizona, USA); (b) *Dilophosaurus* (Early Jurassic, Arizona, USA); (c) *Carnotaurus* (Early Cretaceous, Argentina); (d) *Ceratosaurus* (Late Jurassic, western USA). Human for scale.

It was in Tetanurae that some of the most remarkable and important theropod evolution took place. For starters, tetanurans evidently took breathing to new places. All saurischians (including birds) appear to have had an internal system of cavities in their skeletons, called **pleurocoels**, as well as small openings called **pneumatic foramina**. The pleurocoels augment the volume of air in their lungs by a series of interconnected auxiliary air sacs (see Chapter 13), which reach (and expand into) the internal cavities via the pneumatic foramina. A basal tetanuran, *Aerosteon*, shows the openings and internal cavities indicative of a branching suite of air sacs like those found in modern birds. These

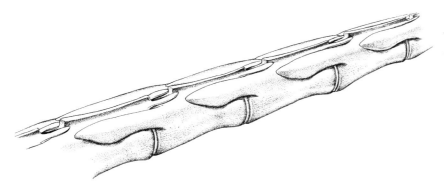

Figure 7.3. Zygapophyses in tetanurans. Note how these processes extend across the adjacent vertebrae both anteriorly and posteriorly, hindering flexibility.

adaptations only increased in complexity as theropods evolved, so that highly derived theropods had (and have) some of the most efficient breathing ever invented. This doubtless enabled them to sustain high levels of activity, and led to a respiratory system capable of supporting sustained flight. We take up the subject of dinosaur breathing and metabolism more fully in Chapters 13 and 14, respectively.

Tetanurans share a large number of other derived features (Figure 7.1), and the group includes a host of large theropods – including megalosaurids such as *Megalosaurus* and *Torvosaurus* (Figures 7.4a and b, respectively) and spinosaurids such as *Baryonyx* and *Spinosaurus* (see Figures 7.4c and d, respectively; also Chapter 6). Finally, Avetheropoda, our next major group, is also a member of Tetanurae.

Avetheropoda

Avetheropoda is that group of theropods more closely related to birds than to the preceding genera and their near relatives. Avetheropods share many derived features, and consist of two clades, Carnosauria (*carn* – meat; no explanation necessary!) and Coelurosauria (*coel* – hollow; a reference to their delicately built, hollowed-out bones; Figure 7.1).

Carnosauria and Coelurosauria

There was a time when people thought of all carnosaurs as large, sluggish, thuggish meat eaters, and coelurosaurs as small, delicate, nonflying lizard- or insect-eaters. While cladograms show that the dichotomy between the two is real (see Figure 7.1), the terms carnosaur and coelurosaur now have rather different meanings.

Carnosaurs include some really large dinosaurs, such as the allosaur and carcharodontosaur groups. The well-known *Allosaurus* (Figure 7.5a) and *Sinraptor* (Figure 7.5b) are typical allosaurs; carcharodontosaurs are here represented by the truly gargantuan *Carcharodontosaurus* and *Giganotosaurus* (Figures 7.5c and 7.5d, respectively). At about 15 m, these last two are contenders – but not the only ones! – for the title of "World's Biggest Land Carnivore."

Coelurosauria includes both small and large forms, most famously the tyrannosauroids. It also includes, as we shall see, some of the most interesting dinosaurs that ever lived. But coelurosaurs are sufficiently compelling that we'll seat them separately (with their own private cladogram) a few paragraphs down the road. For now, we pay our respects to the Big Boys.[1]

---

[1] In a somewhat transparent quirk of human psyche, even the most socially enlightened nonprofessionals commonly use the pronoun "he" when referring to large theropods. But obviously this gender designation can only be right about 50 percent of the time.

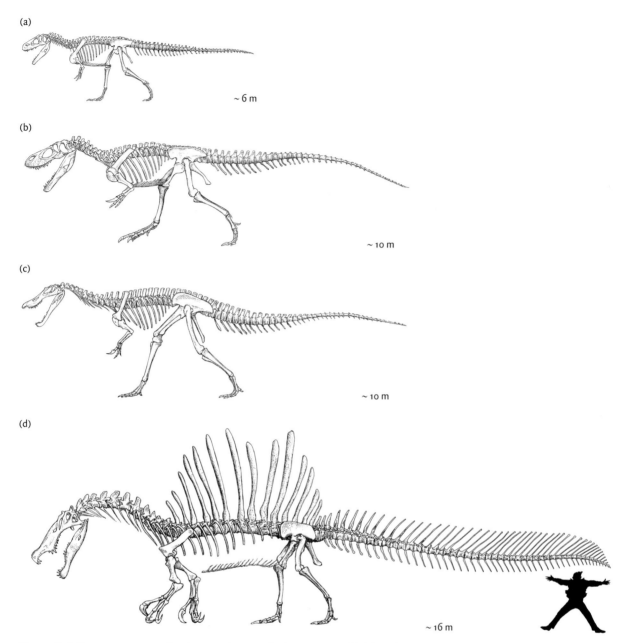

Figure 7.4. Some representative (nonavetheropodan) tetanurans. (a) *Megalosaurus* (Late Jurassic, UK); (b) *Torvosaurus* (Late Jurassic, western USA; Portugal); (c) *Baryonyx* (Early Cretaceous, western Europe); (d) *Spinosaurus* (mid Cretaceous, northern Africa). Human for scale.

Convergent Evolution in Large Theropods

Looking within Tetanurae, we see a striking quality of theropod evolution. Put most simply, superficially, big theropods all resemble each other (hence, the old "Carnosauria" designation for them). Clearly, however, as theropods evolved to large sizes, lineages *independently* developed some of the same features. Such similar, although independent, evolution is called convergent. In the case of large theropods, features such as proportionally large heads and a tendency toward shorter arms occurred convergently. The same features occur *independently* in neoceratosaurs (for example,

Figure 7.5. Avetheropod carnosaurs. (a) *Allosaurus* (Late Jurassic, western USA); (b) *Sinraptor* (Late Jurassic, northwestern China); (c) *Carcharodontosaurus* (mid Cretaceous, North Africa); (d) *Giganotosaurus* (early Late Cretaceous, Argentina). Human for scale.

Carnotaurus), in primitive tetanurans (*Szechuanosaurus*, *Megalosaurus*, and *Afrovenator*), and in avetheropods (*Giganotosaurus*, *Allosaurus* and, again independently, in some tyrannosaurs). Here, then, is why the gigantic – and *amphibious* – *Spinosaurus*, with its longish arms (by large theropod standards), appeared so aberrant to paleontologists!

Despite their superficially convergent morphology, all of these big-headed, short-armed dinosaurs likely behaved very differently from each other; as we have seen, some are lightly built – perhaps more cursorial – and others are ponderous and tank-like. To judge from the teeth and jaws, they likely attacked and killed their prey differently too. Why, then, did they independently develop similar gross morphologies?

The answer *may* reside in the logistics of growing really BIG. It is possible that to increase in size yet maintain reasonable agility as bipeds, compromises had to be made in the size and power of various body parts, and it has been assumed that shortened arms, reduced numbers of fingers, and large heads were part of such compromises.

Continuing through Theropoda, we'll see that smaller theropods tended to have much longer arms and long-fingered, grasping, dexterous hands, to which we assign active roles in prey capture. The fact that large, carnivorous theropods tended to reduce the size of the arms suggest that prey were obtained in a different manner but, we can safely assume, no less successfully. The manner of capture may have been different, as indeed might have been the nature – particularly the size – of the prey.

Coelurosauria

By almost any measure, coelurosaurs are among the most remarkable group of animals that ever lived. They are startlingly diverse: there were all kinds (hunters and herbivores); they came in all sizes (tyrannosaurs to hummingbirds); many were extraordinarily intelligent; some were hypercarnivorous; and some... well, some still defy easy explanation. And of course out of this wellspring of genetic wealth, diversity, and evolutionary ingenuity, come today's (and yesterday's) birds.

It was within Coelurosauria that the first obvious nonflying feathered dinosaurs were discovered, and by far the preponderance of feathered dinosaur specimens known to date are coelurosaurs. It would not be brash to state that the evidence suggests that *all* coelurosaurs, at a minimum, were feathered.

Because coelurosaurs contain so much diversity, it has been tricky to diagnose them precisely. Paleontologist Steve Brusatte (see Figure 16.20), whose expertise is, among other subjects, in coelurosaur phylogeny, has proposed that the hallmark of all coelurosaurs in an enlarged brain.[2] An important character, and one that as we shall see is critical in this group as evolution proceeded. The coelurosaurs we'll meet here include tyrannosaurs, diminutive compsognaths, alvarezsaurs; sloth-like therizinosaurs, oviraptors, deinonychosaurs, and avialans (Figure 7.6). Many of these we met in the preceding chapter as we discussed some of the things theropods do; now, we'll put all of them into their phylogenetic context.

---

[2] Other characters that he suggests include "an L-shaped quadratojugal without a prominent posterior process, an astragalus whose ascending process is separated from the condyles by a transverse groove or fossa, a jugal with a weakly concave orbital margin that is not deeply U-shaped, and a femoral fourth trochanter that is positioned near the center of the posterior surface of the shaft distally and extends proximomedially to become confluent with the posteromedial corner of the shaft proximally." Wow! These are a bit less accessible, so we'll stick with the enlarged brain. Brusatte, S. L., Lloyd, G. T., Wang, S. C., and Norell, M. A. 2014. Gradual assembly of avian body plan culminated in rapid rates of evolution across the dinosaur–bird transition. *Current Biology*, **24**, 2386–2392.

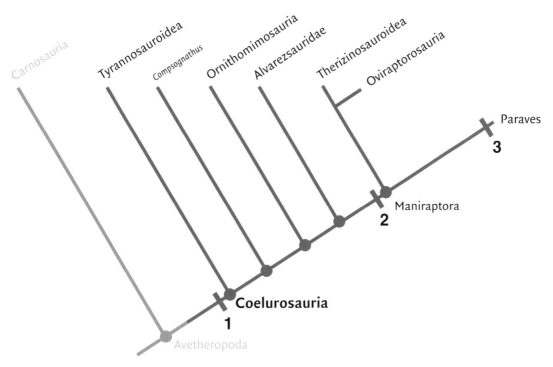

Figure 7.6. Cladogram of Coelurosauria, showing basic relationships of the major groups. At **1**, enlarged brain; narrow foot; at **2**, proportionately elongated forelimbs, bony sternum; at **3**, extremely long arms and hands, wings made up of multiple layers of quill-like (pennaceous) feathers, downwards- and/or backwards-facing pubis (**retroverted** pelvis), large retractable sickle claw on the second digit of the foot. Cladogram after Brusatte (2012).

Tyrannosauroidea

Tyrannosaurs are a rich clade, comprising some 20 known types. They first appeared in the Middle Jurassic, but stayed more or less medium-sized, under the semi-opposable thumbs, perhaps, of carnosaurs. In their earliest incarnation, they were light-bodied, small, and highly cursorial; characterized by North Carolina paleontologist Lindsay Zanno (see Figure 16.28) and her colleagues as "marginal predators." With the disappearance of carnosaurs, tyrannosaurs flourished and by mid Cretaceous time took a dramatic step towards huge sizes (termed gigantism), and closed the Mesozoic down at the dinosaur extinction, 66 Ma (Chapter 16).

Like so many dinosaurs, tyrannosauroids had an Asian and North American distribution. The earliest (and most primitive) forms are Asian, and throughout much of their time on Earth, they were largely Asian. At some point during the mid Cretaceous, however, they migrated (with many other kinds of dinosaurs (see Chapter 11), presumably across the Bering land bridge, to North America. There, life was evidently very congenial, and the classic huge North American tyrannosaurids evolved: *Albertosaurus* (Figure 7.7), *Dryptosaurus*, *Gorgosaurus*, and *Daspletosaurus*, sizeable (in the range of around 10 m) creatures all. The largest tyrannosauroid was the very last: the redoubtable end-Cretaceous, North American *T. rex* (Box 7.1). In general, tyrannosaur evolution seems to have been characterized by an overall increase in size through time, but concomitantly a gradual decrease in the size of the arms, perhaps compensated for by a gradual increase in the size and power of the skull.

Figure 7.7. Albertosaurus skeleton

The 2012 discovery of *Yutyrannus huali*, an approximately 9 m relatively primitive Early Cretaceous tyrannosauroid from China, raised the possibility that all tyrannosauroids were feathered. *Yutyrannus* is coated in monofilamentous feathers (Figure 7.8), but the issue may end up being about size. We all know that small mammals have full coats of fur; very large ones (elephants, hippos, rhinoceroses), do not. The fur coating in mammals is largely about insulation; and because small mammals have a lot of surface area relative to their total volume, they can lose heat easily through their skin. This is the exact opposite with the very largest mammals, which have proportionately less surface area than volume; in turn, allowing them to retain heat more easily (this question, surface-to-volume ratios in insulating animals, is treated in greater detail in Chapter 14). So the question arises, did the dramatic jump to gigantism in tyrannosauroids in the mid Cretaceous preclude their retaining their feathers, since feathers were no longer needed to keep the animals warm? As we note in Box 7.1, the issue of whether very large tyrannosaurids of the latest Cretaceous were feathered, or not, remains very much an open question.

Compsognathidae and Ornithomimosauria

Compsognathids were small (1 to 2 m), lightly built highly cursorial animals, and also good candidates for the opening scene in *Jurassic Park II*. The bladed, recurved teeth suggest carnivory; small vertebrates or perhaps insects.

Here, we'll highlight two compsognathids: the diminutive Late Jurassic *Compsognathus*, and *Sinosauropteryx*, from the Early Cretaceous of China. *Compsognathus* was found in Solnhofen, Bavaria (Germany), a very famous Lagerstätte, or locality of extraordinary preservation, which had been quarried for building stones since before the 1700s (Figure 7.9). *Sinosauropteryx* (Figure 7.10) was found in another Lagerstätte, in Liaoning Province, in China. The fossil-rich rocks of the Early Cretaceous Jehol Group of Liaoning Province look superficially like those of Solnhofen. In both, preservation is spectacular; the specimens are generally complete *and* completely articulated, and the fine mudstones in which they are preserved show not only the impressions of the animals' coverings but also some darkened staining, possibly representing the original soft tissues.

Box 7.1 Will the Real *Tyrannosaurus rex* Please Stand Up?

Like sirens in *The Odyssey*, *Tyrannosaurus rex* – icon, the biggest and baddest carnivore ever to walk the planet – compels attention. It seems as if hardly a month passes without a new study – with new "insights" – into this creature. But *T. rex* has proven enigmatic, and historically there have been deep disagreements about its (paleo) biology. Here, however, like the venerable TV show *To Tell The Truth*, we eschew speculation and the tyrannosaur imposters it brings. We want just the real *Tyrannosaurus rex* to please stand up!

Shorn of all the mythology, the problem is really quite simple: *T. rex* simply has no modern analog from which we can drawn conclusions. A 45′, multiton, carnivorous bipedal, short-armed, two-fingered dinosaur is *sui generis*, and this leaves a lot of latitude for speculation. But with upwards of 50 specimens known (most found in the past 30 years, although few anywhere near complete), and with all that paleontological brain-cell energy expended on them, we ought to know something reliable about these beasts. And, in fact, we're getting there.

T. rex was first discovered in the ranchlands of eastern Montana in 1902 by the legendary fossil collector Barnum Brown. Two individuals were discovered together, and, after preparation, H. F. Osborne, then director of the American Museum of Natural History, interpreted them to be different genera (and thus species), and gave each a different name: *Tyrannosaurus rex* ("tyrant king") and *Dynamosaurus imperiosus* ("imperial power"). These names suggest that Osborne allowed the siren song that is *Tyrannosaurus* to weave its way into his psyche; having named it, he was its first victim.[a] Ultimately the two specimens were found the be the same creature, and so it came to pass that Marc Bolan's 1970s rock band was not named "*D. imperiosus.*"

T. rex was Big

Bigness can be measured in length or in weight. While a quick visit to the internet and many dino books will give you a dinosaur's weight (*T. rex*'s is estimated in tonnage), weight measurements in extinct animals – particularly ones

that have no modern analogs – vary dramatically depending upon a number of factors (see Tonnage, Chapter 13). Length would seem to be the better metric, because it can be known with precision: and with all the specimens in which length can be determined, growth seems not to have gotten much past 14 m (45′) (but see Life was Tough, below).

T. rex was Robust

By comparison to some of its contemporaries (for example, *Albertosaurus*), *T. rex* was a tad, er, stocky. Most paleontologists would agree that this limited its running speed to not much more than 25 kmh (15 mph), although some still hold out for its being more fleet of foot. The heavy build is a source of some confusion: whereas *Giganotosaurus* and *Carcharodontosaurus* approached the length of *T. rex*, they were much more lightly built animals. *Spinosaurus* was slightly longer, but then it was very likely aquatic! Compared to all these, *T. rex* had *gravitas* (literally).

T. rex had Tiny Arms and Each Hand had Two Fingers (Digits I and II)

The arms in *T. rex* are about 20 percent the length of the legs (they are about 70 percent in humans) and cannot reach the mouth. Yet they were not likely vestigial; they are powerfully built, and reconstructions suggest that they could manage 140 kg (about 300 lb). Finally, they were tipped with impressive claws. They clearly had a function, whether or not they were used to put food into the mouth. An old idea – proposed in complete seriousness – was that the arms and hands were used like door jams to allow the animal to get up when lying down; since the front end was locked in place by the arms against the ground, the back end would rise up and then the animal could stand. And yet, as recently as January 2021, a footprint was reported from New Mexico that was interpreted to show that very behavior. Over-interpretation or a great insight?

My, What a Big Head and Big Teeth You Have!

The head was large (approximately 2 m (6′)); the jaws were deep, to accommodate the massive adductor muscles; the bones of the skull were tightly sutured together; the teeth

[a] But not its last. In 2018, a new tyrannosaur was discovered and dubbed "*Dynamoterror*" in conscious homage to Osborn and the now-abandoned name "*Dynamosaurus.*"

Box 7.1 *cont'd*

were, as we have seen, bulbous, banana-sized and -shaped, and were used for bone crushing (see Figure 6.15). What are commonly not remarked on by people describing this dinosaur are the long, stout neural arches on the neck, the sites of powerful muscle attachments to the back of the head.

University of Florida paleontologist Greg Erickson modeled the power of the bite in *T. rex* using fabricated teeth and hunks of modern bone, analogous to fossil bones from *T. rex* contemporaries, in which spectacular carnivorous punctures have been found. To reproduce those puncture marks, Erickson required as much as around 7000 lb (31 138 N) of force – for a single tooth! *T. rex* had a mouthful of these teeth and it arguably had the most powerful bite of any terrestrial animal that ever lived. The coprolites (Chapter 6) are definitive evidence that *T. rex's* bite was truly bone-crushing.

T. rex had Uncommon Good Sense(s)

We've seen the sensory apparatus of *T. rex* (see Figures 6.20 and 6.21) suggesting a well-developed sense of smell, and stereoscopic vision. We must confess that the jury is out on its intellect; the EQ of around 2.5 leaves it somewhere less than a modern bird (see Figure 13.3) below the smartest of the other coelurosaurs, although, as we have seen, its intellectual powers have also been favorably compared, by otherwise reasonable people, to those of a chimp. This may not do the chimp justice.

T. rex was Feather-Covered

No matter how much you like your dinosaurs shiny, scaly, clean, mean, and green, you may be out of luck. In 2012, Chinese paleontologist Xu Xing (see Chapter 16) and colleagues, described *Yutyrannus*, a large, feathered, primitive tyrannosaur (see text) The feathers were monofilamentous (Chapter 5), and relatively densely distributed on the body (Figure 7.8).

Yutyrannus is a compelling argument for feathering all tyrannosaurs, but is the evidence definitive? As recently as 2017, P. R. Bell and colleagues argued that, because scaly, unfeathered, skin impressions are known from various parts of various large tyrannosaurids from the Late

Cretaceous (e.g., *Albertosaurus*, *Daspletosaurus*, *Gorgosaurus*, and *Tarbosaurus*), in these large late-evolving theropods, extensive feather coatings, although present in more primitive forms, were lost in the Late Cretaceous forms. These authors see this as related to the move to gigantism, and regard the possibility of feathers in these big latest Cretaceous theropods, if they were present at all, as limited to a vestigial strip down the back. The jury is still out on this question.

Life was Tough

Many specimens of *T. rex* show bone damage, and fossils tend to be young (see Chapter 13), few exceeding 20 years in age. If this is a reliable sample of *T. rex* populations, then it might be said that these animals grew fast and lived hard. Many recorded bone pathologies, including healed breaks, sustain this interpretation. Recently, however, a particularly large, old specimen of *T. rex* was reported, suggesting that there would have been larger ones around had they lived longer. Maybe being the baddest carnivore of all time wasn't all that it is cracked up to be, because cracked up seems to have been part of its life.

T. rex as a Living Carnivore

All in all, a somewhat coherent picture emerges. *T. rex* was an active carnivore, likely seeking large, strong, and definitely unwilling, prey (ceratopsians and hadrosaurs, Chapters 11 and 12, respectively). It was brilliantly balanced at its pelvis, a long tail counterbalancing the powerful body forward of the pelvis, massive head, and compensating, reduced forearms. Ambushes would have been the preferred mode of attack; *T.rex* likely did not run prey down. The attack would have been head first, the gigantic, tooth-studded mouth substituting for the long powerful arms and grasping hands of other theropods (S. Brusatte, whom we have met earlier in these pages, characterized the attack as a "giant land shark"). The bite would have been crippling, to understate the process, enhanced by the neck muscles powerful enough to throw even large, powerful animals to the ground – conveniently positioning them within reach of the claw-studded *T. rex* feet. The hands might have helped immobilize prey, but

Box 7.1 *cont'd*

the head, neck, powerful legs, and clawed feet were surely the most important weapons in the arsenal.

The classic confrontation of the latest Cretaceous pairs the largest baddest latest Cretaceous carnivore (*T. rex*, of course) with the largest baddest herbivore of the time (the formidable, three-horned ceratopsian *Triceratops*). This isn't stricty whimsy: these two behemoths shared latest Cretaceous landscapes in what is the upper Great Plains of the United States: eastern Montana and western North and South Dakota. Too much ink has bled upon the outcome of such a contest – the ultimate dinosaur Main Event – and we could hardly add insights that haven't already been bestowed upon a voracious public. Instead, we offer for the evocative new mount in the National Museum of Natural History (Smithsonian Institution) that memorializes this real Clash of the Titans (Figure B7.1).

The fact that *T. rex* has commonly been found in pairs suggests some social structure. "Dinosaurologist" Gregory Paul in his inventive book, *Predatory Dinosaurs of the World* (1988), imagined, perhaps drawing inspiration from a pride of lions, a *Tyrannosaurus* family with the killing carried out by the female. The idea that hunting was a social event is reinforced by the fact that very close tyrannosaurid relatives – *Tarbosaurus* and *Albertosaurus* – are now known to have functioned in packs.

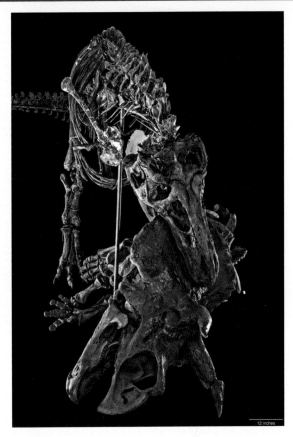

Figure B7.1. *Tyrannosaurus* meets *Triceratops* as newly (2019) displayed in the Smithsonian Institution, Washington DC, USA.

The 1997 discovery of the diminutive *Sinosauropteryx* revealed a small compsognathid that obviously didn't fly. Yet it was covered with monofilamentous feathers, a thick coat covering a clearly nonflying theropod. The point was driven home: feathers were not originally about flight (Figure 7.10; Figure 6.28 shows another specimen of this dinosaur, highlighting color patterns in the plumage).

The highly cursorial ornithomimosaurs (*mimus* - mimics) were a group of ostrich-like dinosaurs that must have had a similar lifestyle. They include medium-sized runners like *Gallimimus*, *Struthiomimus*, and *Dromiceiomimus*, all of whom had long legs, small heads and large eyes, and were toothless (Figure 7.11). As a group, they have been reasonably well understood for many years, although their precise diet had been uncertain, because the large, powerful hands and claws (although these are not strongly curved, as one might expect in a carnivore) seem at odds with the small heads and toothless jaws. Most paleontologists now suspect a more or less herbivorous diet for these dinosaurs. Like the tyrannosaurids, they too have a Late Cretaceous Asian and North American distribution.

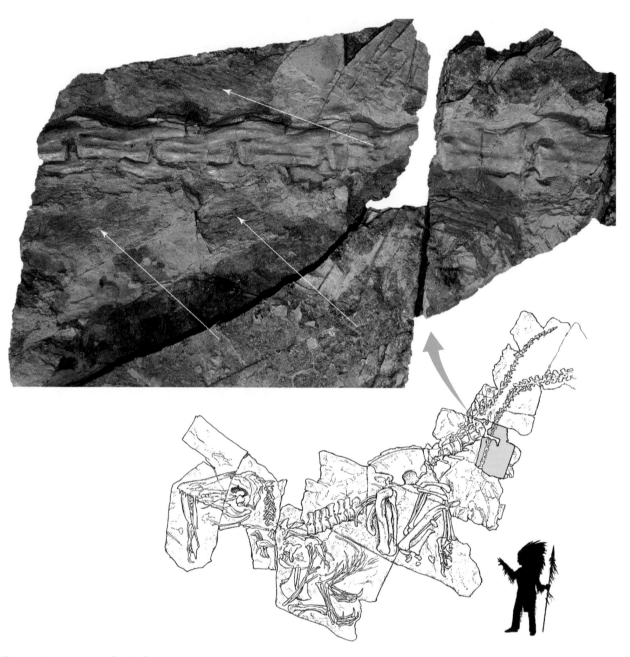

Figure 7.8. Section of tail of *Yutyrannus*. Arrows indicate filamentous features interpreted to be (nonflight) feathers. From Liaoning Province, China.

However, what we thought we knew about ornithomimds got upended in 1967 with *Deinocheirus* from the Gobi Desert of Mongolia. The late, legendary Polish paleontologist Zofia Kielen-Jawarowska describes her discovery of the first specimen (in a chapter fittingly titled "There ain't no such animal"), which consisted of only a pair of *huge* ornithomimid arms (and nothing else):

> While I was walking in the [Gobi] along the southern part of the outcropping, I suddenly noticed some fairly large bones showing on a small hill. There were more than a dozen of them...I began removing the sand from around the bones. Suddenly I saw there...an

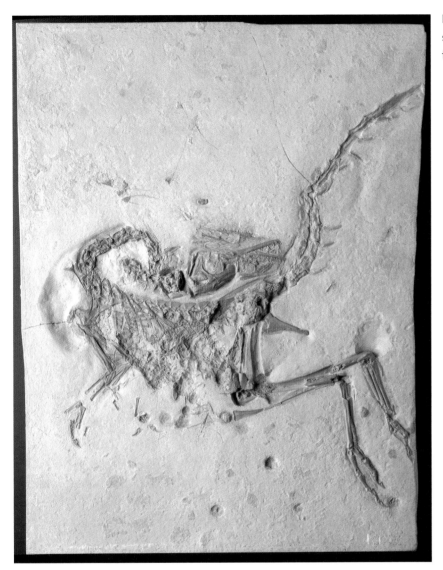

Figure 7.9. *Compsognathus*, the small Early Jurassic compsognathid from Solnhofen, Bavaria, Germany.

excellently preserved, powerful, strongly arched claw twelve inches long. It was undoubtedly a forelimb claw, but larger than any ever found before...the partly bare protruding bone looked like an arm bone; if this was actually the case it meant that I had come across a long forelimb bearing claws bigger than any ever seen before. Kielen-Jaworowska, Z. 1969. *Hunting For Dinosaurs*. MIT Press, Cambridge, MA, p. 141.

The arms, at longer than 2.5 m each, had massive hands with three elongate fingers, tipped with huge, powerful claws. If you blew up a conventionally sized ornithomimid to this size, the animal would be inexplicable. For the next 50 years, the arms hung on the walls of the Paleontological Museum in Ulan Baatar, Mongolia, a wake-up call for anybody who thought that we understand ornithomimid dinosaurs.

In 2009, tantalizing rumors began to circulate that much more of *Deinocheirus* had been found. But how much? And what did it look like? All rumors were put to rest in 2014, when a complete description was finally published. Instead of being a gargantuan fleetly cursorial monster ostrich mimic, *Deinocheirus* is hunch-backed, long-snouted, deep-jawed, and short-legged: absolutely *not* fleetly cursorial! Moreover, more than 1000 gastroliths for grinding and fish remains were preserved

Figure 7.10. The feathered, nonflying compsognathid *Sinosauropteryx*. Monofilamentous feathers (see arrows) – interpreted to be unrelated to flight – are preserved down the back of the animal, extending the length of the tail. From Liaoning Province, China.

in the stomach region – the thing ate fish?! The authors proposed that *Deinocheirus*, with its blunt-tipped feet and deep jaws, mucked around the rivers in which it lived, catching fish, and using those massive claws on the arms for digging plants, and its size for protection. They characterized it as a "megaomnivore." Figure 7.12 shows that you just can't make this stuff up.

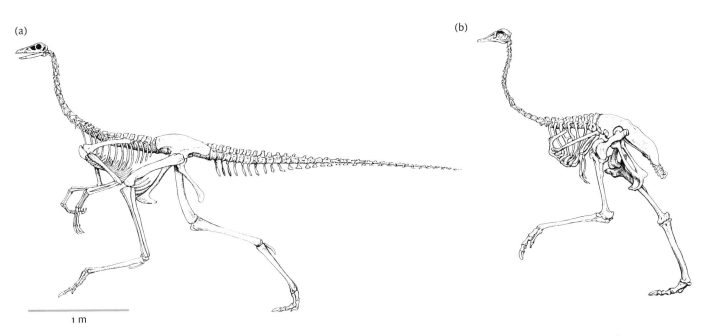

Figure 7.11. Skeletons of an ornithomimid and an ostrich compared. (a) *Struthiomimus*; (b) ostrich. The similarity in design suggests similarities in lifestyle.

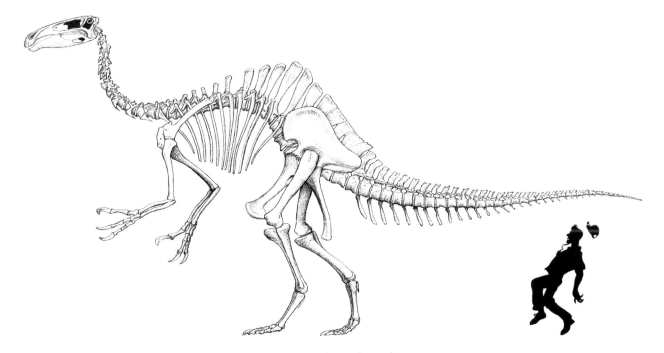

Figure 7.12. The skeleton of *Deinocheirus*, with a human drawn for scale.

Maniraptora

Maniraptora (*manus* – hand; *raptor* – stealer) includes everything else within Coelurosauria (see Figure 7.6). This is a grab bag of dinosaurs, from rare aberrant forms like therizinosaurs and alvarezsaurs, to oviraptorosaurs (we met these; see Figures 6.19 and 6.30) to the now-(in)famous and highly predaceous forms such as *Deinonychus* and *Velociraptor*, within Deinonychosauria. Chickens, penguins, sparrows, ostriches, and eagles (e.g., birds) as members of Avialae (see below and Chapter 8), are also found here. We've seen something of oviraptorosaurs and their nesting habits, as well as deinonychosaurs and their carnivorous habits in Chapter 6, so we'll devote more attention to the other groups, just touching on the origin of birds (which we'll address in greater detail in Chapter 8).

Maniraptorans as a group are identified by a number of characters (see Figure. 7.6), but a character that many share is the semi-lunate carpal, a wrist bone modification that dramatically increased the flexibility of the wrist joint (Figure 7.13).

We begin with alvarezsaurs and therizinosaurs. Alvarezsaurs are a strange, uncommon group of dinosaurs, known from South and North America, and Asia. Epitomized by the *Mononykus* (Figure 7.14), they are not large (up to 2 m), but have strangely fused hands (convergent on those of a bird; see below), short arms, small teeth restricted to the front of the jaw, and were evidently feather-covered. What they did – and how they did it – remain enigmatic. The distinctive fused, short hands, however, were clearly for a particular function, and strategies from herbivory to digging and termite-eating have been proposed.

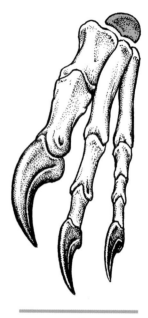

5 cm

Figure 7.13. Semi-lunate carpal (colored) in the left hand of *Ingenia*, an oviraptorosaurian maniraptoran.

Therizinosaurs are only slightly less enigmatic. Known from North America and Aisa, therizinosauroids represent yet another Cretaceous[3] theropod venture into – most likely – herbivory. Quite large and ponderous, these highly evolved and distinctive maniraptoran theropods have been compared to giant sloths, proposed as piscivorous, considered as tree climbers for wasp nests, and regarded as herbivores (Figure 7.15). Their behavior has thus been a matter of considerable confusion (as was, initially, their phylogenetic position), and the consensus was initially that they were an aberrant group of theropods; new finds, however, suggest that they were neither aberrant, nor sloth-like, nor wasp-eating tree climbers, but were likely – because stomach contents are to date unknown – herbivorous.

Several features of therizinosaurs are quite distinctive. The sloth analogy was originally suggested because of their extraordinarily large claws (Figure 7.16) which are quite distinct in shape from those one finds gracing the hands of carnivorous theropods. These are clearly not tightly recurved, sharp, meat-grabbing implements; exactly what they are, however, is less obvious. One thing that we do know about therizinosaurs, however, is that they were feather-covered – the kind of monofilamentous primitive feathers we saw in *Sinosauropteryx* (Figure 7.17).

Defying our old bugaboo "overall similarity" (see Chapter 3), the phylogenetically closest creatures to therizinosaurs were something that looked nothing like them: the very successful oviraptorisaurids (Figure 7.18; see also Figures 6.13b, 6.19, and 6.30). Oviraptorisaurids lived throughout the Cretaceous, started with teeth, but by the Late Cretaceous, were edentulous (without teeth). They ranged in size from around 1 m to around 8.5 m, and in estimated weight from around 5 kg to around 3000 kg (*Caenagnathasia* and *Gigantoraptor*, respectively). *Oviraptor*, who we met in Chapter 6, fits somewhere between these two extremes. These enigmatic (generally) toothless dinosaurs with their hollowed-out skulls (Figure 6.19) and strong beaks were well equipped for running, as well as for doing something with powerful, grasping clawed hands.

---

[3] One form, *Eshanosaurus*, is from the Early Jurassic. The specimen consists of a left mandible only and some authors have referred it to basal Sauropodomorpha. Restudy by Paul Barrett in 2009 confirms it as a therizinosaur, albeit a very early one.

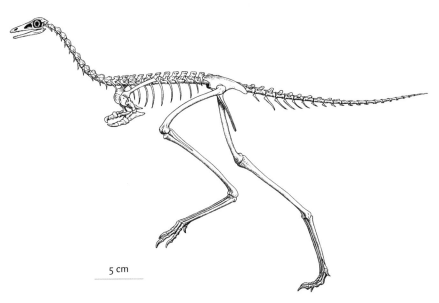

Figure 7.14. The skeleton of *Mononykus*, showing the unusual **alvarezsaurid** fused hands and short arms.

5 cm

Something… but what? There has been a lot of loose talk, some of it mutually exclusive, about invertebrate feeding in shallow waters, herbivory, scraping in crevices with the large claws, tree-climbing, prey-catching, omnivory, some of the above, all of the above, none of the above. Specialized functions for particular parts of the anatomy seem to be ruled out, given the size range of oviraptorisaurids – unless (obviously) different oviraptorsaurids did different things. Yet the unique skull implies a similar range of behaviors. Regardless, the slim gracile bones suggest a generally active, cursorial lifestyle. And, thanks to the basal oviraptorid *Caudipteryx*, from the Liaoning Lagerstätte, we can now be sure that they were covered with a colorful (?) monofilamentous feather coating (Figure 7.19).

Paraves

The last two groups of maniraptorans are Deinonychosauria, and a group at once more and less familiar: Avialae. By any measure, these constitute a startling clade of creatures – highly predaceous and crazily intelligent Paraves (Figure 7.20).

Deinonychosaurs rightly ought to evoke more fear than any large theropod, for they include the sickle-clawed troodontids and dromaeosaurids. Dromaeosaurs, as we saw in Chapter 6, were remarkably intelligent (see Figure 13.3), hypercarnivorous, probably social animals. Gracile bones, large claws, powerful hands, and a locked tail bespeak an active lifestyle (see Figure 6.4) and, potentially, "warm-bloodedness" (see Chapter 14). And with that kind of metabolism, it wouldn't be surprising to find some kind of insulation keeping the animals warm. And here, once again, the Early Cretaceous Liaoning Lagerstätte served up a feast of feisty, feathered carnivores. These include *Sinornithosaurus* (Figure 7.20), and *Microraptor* (Figure 7.21).

Microraptor is an extraordinary creature, with *flight* feathers (see Chapter 5) on the legs as well as the wings – not just the monofilamentous types of feathers we've seen previously – leading to considerable speculation about exactly how this "four-winged" dinosaur flew. Gliding was surely involved; and likely some kind of flapping flight. Regardless, of how it flew, however, it must have shimmered when it did: microscopic melanosomes preserved in the feathers suggest that it was iridescent.

1 m

Figure 7.15. The therizinosaur *Nothorhynchus* meets its skeleton. See also Figure 7.16.

Figure 7.16. A cast of the hand of a therizinosaur. Each claw pictured is about 40 cm (~16″) in length. On display at the Fukui (Japan) Prefectural Dinosaur Museum.

Microraptor wasn't very large; it was less than 1 m long, and more than half of that was tail. By contrast, at least five other "four-winged" dromaeosaurids are known, including the very large *Changyuraptor*, whose tail feathers alone were up to 30 cm! These animals appear bird-like, but were not actually on the road to the evolution of birds. Rather, they seem to have been a unique radiation of gliding and/or flying dromaeosaurids (Figure 7.22).

Troodontids were another enigmatic group of animals, whose distinctive teeth (Figure 6.15a) caused them to be interpreted as ornithischians, as herbivores, and as *Deinonychus*-like carnivores. They too have the sickle-shaped claw seen in other dromaeosaurs; however, theirs is not as wickedly curved as those of deinonychosaurids, and may not have been held up off the ground as fully. Troodontids were nonetheless very large-brained, with stereoscopic vision, a rigid tail, and grasping hands. Troodontids, too, were feather covered, as clearly indicated by the Liaoning troodontid *Anchiornis* (Figure 7.23). But in this case, careful mapping of the distribution of color-indicating melanosomes preserved on the feathers offers us a startling glimpse of the original colors of the very extinct, very nonflying, feathered troodontid *Anchiornis* (Figure 7.24).

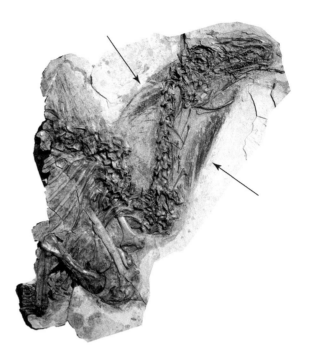

Figure 7.17. Monofilamentous feathers (see arrows) preserved with the skeleton of the Chinese therizinosaur *Beipiaosaurus* – skull, neck, and anterior portion of rib cage (including scapula and humerus). From Liaoning Province, China.

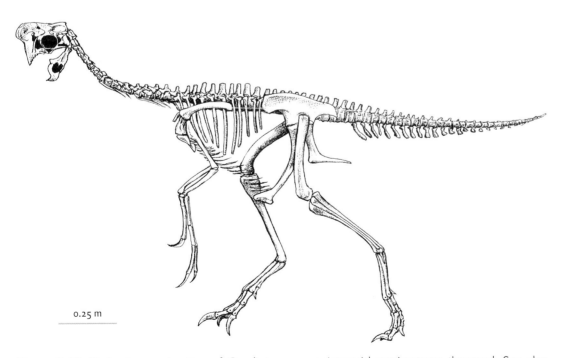

0.25 m

Figure 7.18. Skeletal reconstruction of *Caudipteryx*, an oviraptorid maniraptoran theropod. See also Figure 7.19.

Figure 7.19. Feathers preserved in a specimen of the oviraptorid *Caudipteryx*. Arrows point to preserved feather fragments. From Liaoning Province, China.

Along with those audacious, rapacious, hellaceous deinonychiosaurs, Paraves also contains Avialae (Figure 7.20). Avialae includes modern birds, as well as a pile of toothy ancient birds, whose acquaintance(s) we will make in Chapter 8.

Scansoriopterygidae

Our quick tour through Paraves would not be complete without the most unexpected appearance of a group of membrane-winged, superficially bat-like theropod dinosaurs, the scansoriopterygids (Figure 7.20). Two forms – *Epidendrosaurus* (= *Scansoriopteryx*?) and *Epidexipteryx* were initially found; these had elongated digit III (recall that in theropods digit II is amost always the long one). Other than this technical oddity, these tiny (sparrow-sized) theropods were not considered too noteworthy. But – and this is a big but – the full aberrancy of this group was not fully understood until the 2015 and 2019 discoveries of *Yi qi* and *Ambopteryx longibrachium* (respectively). These two paravians show that learning to fly involved unexpected experimentation (Figure 7.25a, b) – as it surely has for humans as well. *Yi* and *Ambopteryx* are theropods that have all the usual derived

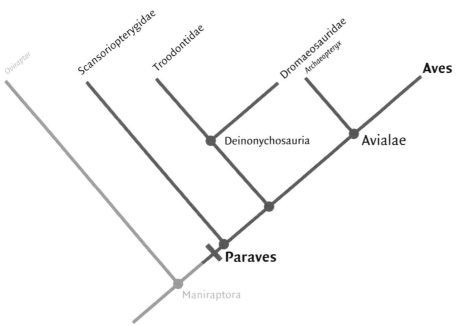

Figure 7.20. Cladogram of Paraves. At **1**, extremely long arms and hands, wings made up of multiple layers of quill-like (pennaceous) feathers, backwards-facing pubis (retroverted pelvis), large retractable sickle claw on the second digit of the foot. Cladogram after Brusatte, 2012; Wang et al., 2019.

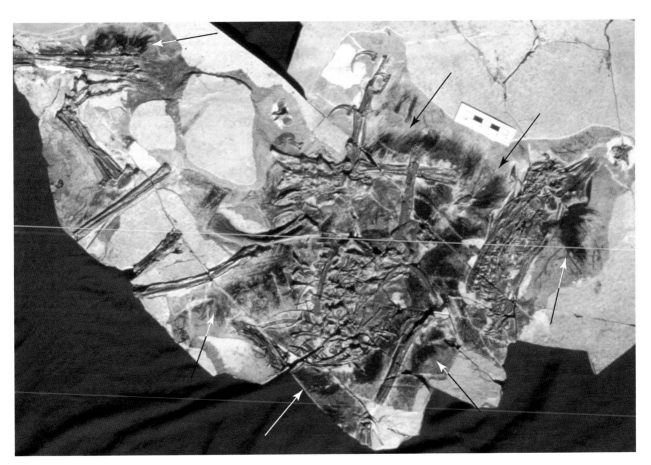

Figure 7.21. Complete skeleton of the feathered dromaeosaur *Sinornithosaurus* from Liaoning Province, China. Arrows indicate monofilamentous feathers.

Figure 7.22. *Microraptor gui*, a "four-winged" dromaeosaur from Liaoning Province, China, with well-developed flight feathers present on all four limbs.

maniraptoran features that we've come to know: the large hands, the three fingers, and lots of feathers. But they *also* have something else: a rod-like "styliform element" which appears to be a long, thin extra bone growing out from the wrist in each hand which, the evidence suggests, supported a wing membrane – decidedly un-bird-like and more bat-like. The remaining question is the extent of the membrane (Figure 7.25c), but *Yi* and *Ambopteryx* afford a glimpse into yet another unknown and utterly unanticipated *cul de sac* of theropod evolution.

Now Hold Yer Pegasi There, Cowboy!

All of these fun critters in the context of the paravian cladogram (Figure 7.20), nonetheless leave us slightly uncomfortable. Troodontids (*Anchiornis*)and dromaeosaurs (*Microraptor*) had long pennate feathers (legs and arms), and many likely glided in some kind of four-winged way. And, as we can see by looking outdoors (or in Chapter 8 – your choice!), birds – modern and ancient – fly and flew. Yet, according to the cladogram – and that is our best indicator of what happened evolutionarily – dromaeosaurs and troodontids were not the ancestors of birds. However, the common ancestor of those three groups must have been a primitive paravian. But did that ancestor fly?

It would be a hard sell to argue that primitive deinonychosaurs flew, so our best call is that, within Deinonychosauria, flight appears to have been at least attempted twice: within Troodontidae (represented by *Anchiornis*) and within Dromaeosauria (represented by microraptorids).

That still leaves us with flight, or attempted flight, occurring in three lineages independently: Avialae, Troodontidae, and Dromaeosauria. And that's just a little weird, since flight is not perhaps the easiest achievement to master evolutionarily (or any other way)!

(a)   (b)

Figure 7.23. A beautifully preserved, complete specimen of the troodontid *Anchiornis* preserved between two slabs of claystone from Liaoning Province, China. Feathers are clearly visible. (a) Slab; feathers marked by white arrows; (b) counterslab.

Figure 7.24. A reconstruction of the feathers and coloration of the troodontid *Anchiornis,* based upon preserved melanosome size, shape, density, and distribution.

But just when you thought you understood all that, also belonging to Paraves is the unexpected, uninvited, and unusual bat-winged Scansoriopterygia (Figure 7.25), of which *at least* two (*Yi* and *Ambopteryx*) and almost for-sure four (plus *Epidendrosaurus* and *Epidexipteryx*) very likely flew. Bat-winged, flying theropods are *definitely* an evolutionary dead end; nothing evolved from them, and neither are they some kind of derived troodontid, dromaeosaur, or avialan. They're out there on their own limb of the theropod evolutionary tree, which means that within just Paraves – if our cladograms are right – flight was attempted at least four times *independently*: in troodontids, dromaeosaurs, avialans, and scansoriopterygians. It is as if there was something about light, small bodies, long arms, pennate feathers for insulation and the warm-blooded metabolism that comes with it (see Chapter 14) that cried out, evolutionarily speaking, "you should try flight!". And so they did.

(a)

2 cm

(c)

(b)

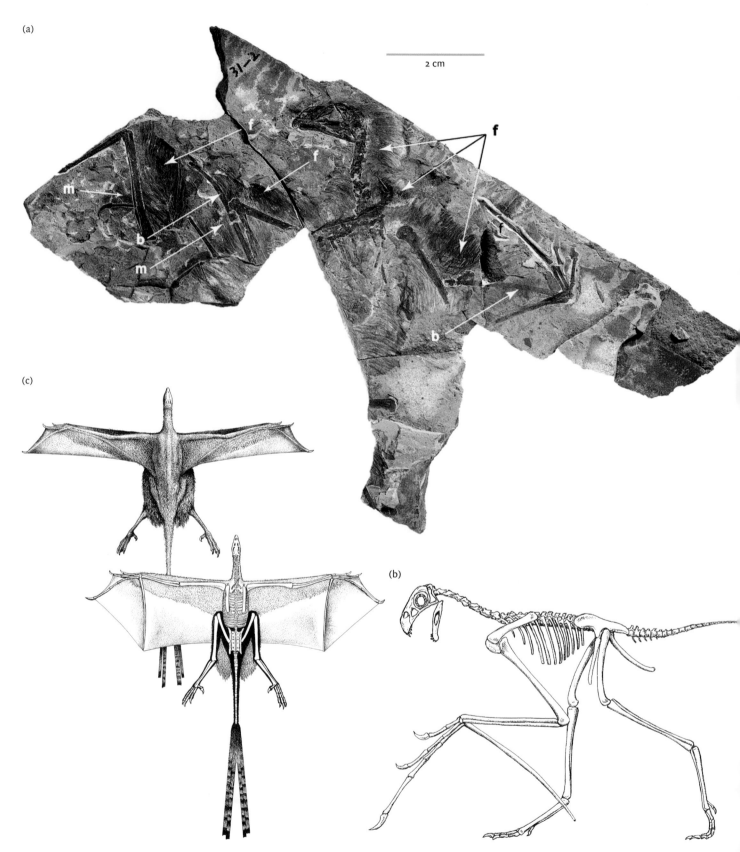

Figure 7.25. The bat-winged theropod *Yi qi*. (a) Specimen; arrows labeled "f" point to feathery coat; arrow labeled "m" points to preserve membrane; arrow labeled "b" points to elongate bony support for wing membrane. (b) Reconstruction of *Yi qi*; and (c) two possible membran reconstructions of *Yi qi*. Redrawn from *Nature*, doi:10.1038/nature14423.

Avialae

Now we're very close to Avialae (*aves* – birds), animals that you might call "birds," but surely not all avialans are modern birds. We've all seen birds, and so, with Avialae, things may get a little bit more familiar. Avialans are rightly deserving of their own chapter, so that's what we'll give them (Chapter 8).

SUMMARY

Theropods are a complex group, with a remarkable evolutionary history. The first major step in their evolution was the appearance of neotheropods; within these, *Coelophysis* and some related animals can be distinguished from tetanurans, whose zygapophysis-stiffened tails were used as dynamic counterbalances to grasping claws; it is among these that some of the most predatory dinosaurs reside. Tetanurans include a host of large theropods including *Megalosaurus*, *Spinosaurus*, and Avertheropoda, in which we see a modern realization of the venerable carnosaur–coelurosaur dichotomy. Carnosaurs include the North American *Allosaurus*, some even larger South American and African forms such as *Giganotosaurus* and *Carcharodontosaurus* (respectively); coelurosaurs include an astounding diversity of hypercarnivores, leggy herbivores, and possible omnivores. Coelurosaurs all bore feathers, and colors and patterns are now known from some of them.

The many discoveries of feathered, nonflying theropods from the Early Cretaceous Liaoning Province of China blur the line between bird (such as we know them today) and nonbird. Clearly, although feathers are diagnostic for birds among living organisms, the Liaoning fossils demonstrate that presence of feathers does not guarantee that one is dealing with a bird. These discoveries reinforce the viewpoint that feathers had multiple uses, and were evidently invented as a form of insulation or perhaps display, only later to be coopted for flight purposes. The feathered nonavian maniraptorans demonstrate that the road to flight was divergent and involved some dead ends.

SELECTED READINGS

Bell P. R., Campione N. E., Persons IV W. S., *et al.* 2017. Tyrannosauroid integument reveals conflicting patterns of gigantism and feather evolution. *Biological Letters*, **13**, 20170092. http://dx.doi.org/10.1098/rsbl.2017.0092

Brusatte, S. L. 2012. *Dinosaur Paleobiology.* Wiley-Blackwell, New Jersey, 322 pages.

Brusatte, S. L., Norrell, M. A., Carr, T. D., *et al.* 2010. Tyrannosaur paleobiology: New research on ancient exemplar organisms. *Science*, **239**, 1482–1485.

Holtz, T. 2012. Theropods. In Brett-Surman, M. K., Holtz, T., Jr., and Farlow, J. O. (eds.) *The Complete Dinosaur*. Indiana University Press, Bloomington, IN, pp. 347–378.

Hou, D., Hou, L., Zhang, L., and Xu, X. 2009. A pre-*Archaeopteryx* troodontid theropod with long feathers on the metatarsus. *Nature*, **461**, 640–643. doi:10.1038/nature08322

Lamanna, M. C., Sues, H.-D., Schachner, E. R., and Lyson, T. R. 2014. A new large-bodied oviraptorosaurian theropod dinosaur from the latest Cretaceous of western North America. *PLoS ONE*, **9**(3), e92022. doi:10.1371/journal.pone.0092022

Li, Q., Gao, K-Q., Vinther, J., *et al.* 2010. Plumage color patterns of an extinct dinosaur. *Science Xpress*, www.sciencexpress.org / 5 February 2010 / Page 4 / 10.1126/science.1186290

Wang, M., O'Connor, J.K., Xu, X., and Zhou, Z. 2019. A new Jurassic scansoriopterygid and the loss of membranous wings in theropod dinosaurs. *Nature*, **569**, 256–260. doi.org/10.1038/s41586-019-1137-z

Witmer, L. M. and Ridgely, R. C. 2009. New insights into the brain, braincase, and ear region of Tyrannosaurs (Dinosauria, Theropoda), with implications for sensory organization and behavior. *The Anatomical Record*, **292**, 1266–1296.

Xu, X., Tang, Z., and Wang, X. 1999. A therizinosaurid dinosaur with integumentary structures from China. *Nature*, 399, 350–354.

Xu, X., Wang, X., and Wu, X. 2000. A dromaeosaurid dinosaur with a filamentous integument from the Yixian Formation of China. *Nature*, 401, 262–266.

Xu, X., Wang, K., Zhang, K., *et al.* 2011. A gigantic feathered dinosaur from the Lower Cretaceous of China. *Nature*, 484, 92–95. doi:10.1038/nature10906

Xu, X., Zhou, Z., Wang, X., *et al.* 2003. Four-winged dinosaur from China. *Nature*, 421, 335–340.

Zanno, L. E., Tucker, R. T., Canoville, A., *et al.* 2019. Diminutive fleet-footed tyrannosauroid narrows the 70-million-year gap in the North American fossil record. *Nature Communications Biology.* doi.org/10.1038/s42003-019-0308-7, 12 p

Zhou, Z.-H., Wang, X.-L., Zhang, F.-C., and Xu, X. 2000. Important features of *Caudipteryx* – evidence from two nearly complete new specimens. *Vertebrata Palasiatica*, 38, 241–254.

TOPIC QUESTIONS

1. Describe some of the diversity of theropods. What are the different types, and what features characterize them?
2. What are the major formal groups of theropods; how are these distinct?
3. Given those groups, characterize the major events in theropod evolution.
4. In which groups do we have good evidence for feathers? How about the furcula?
5. Why is it that we no longer think feathers evolved for flight? Does this mean that they are unrelated to flight? Why?
6. The furcula and feathers are, in living animals, known only in birds. Does this mean that they evolved only once, in birds?
7. Which is more parsimonious – to say that feathers and the furcula evolved twice, once in modern birds and once in theropods; or to say that they evolved once, and thus birds must have a close relationship with theropods?
8. What are the possible original functions that feathers must have had? In what order do you think those functions accumulated through evolution? Why?
9. If *Microraptor* and its close "four-winged" relatives are not directly on the main road to bird evolution – as we said – and if they evolved at least gliding flight – as we said – *and* if birds are avialan theropods – as we said (!) – and if the scansoriopterygids were flying animals – as we said – then how many times must flight have evolved in maniraptoran theropods? How about powered flight? Why? Please explain your answer.

Chapter 8

Theropoda III
The Origin and Early Evolution of Birds

Q: Which came first, the chicken or the egg?
A: The egg. But it wasn't from a chicken.

WHAT'S IN THIS CHAPTER

In this chapter, we:

- Introduce the evolutionary transition from maniraptoran dinosaurs to modern birds.
- Understand birds for what they are: theropod dinosaurs.
- Learn how birds mastered flight.
- Meet some early birds (which "got" much more than worms).

Avialae

In Chapter 7 we tracked the evolution of theropods through Maniraptora, the theropod group that includes birds. As we moved though the evolutionary sequence towards Avialae, we picked up more and more diagnostic characters – the four-toed foot leaving tridactyl prints; the furcula; feathers; among others – that today are exclusively associated with birds, but back in the Mesozoic were surely not. We went as far as Avialae, a monophyletic group that includes living birds and as well as some extinct birds and bird relatives. Avialae, therefore, must contain the evolutionary steps to a modern bird.

Avialae is defined as including *Passer* (again, representing living birds), *Archaeopteryx* (a paravian theropod whom we shall meet next), and all the descendants of their most recent common ancestor. Figure 8.1 gives some diagnostic characters for Avialae.

Now, before you go too far in this chapter, you'll want to be sure that you know what a bird is. Birds are distinctive – and thus recognizable – *because* they contain many diagnostic characters of soft anatomy and skeleton. We focus on the diagnostic characters of the skeleton because they will be the ones that are preserved after millions of years. So, if you are unsure of the diagnostic characters of birds (e.g., what makes a bird a bird), be sure to take a cool, relaxing dip in Appendix 8.1 for a refresher on which diagnostic characters pertain to Aves (birds).

Meet *Archaeopteryx lithographica*

Archaeopteryx has been called the most important fossil ever found, and the road to the origin of birds still runs through this kingfisher-sized, feathered, paravian theropod. It was first discovered in

---

Box 8.1 **The Bird is a Word**

Oh well uh-everybody's heard, about the bird –
Bird bird bird, bird is a word!
"Surfin Bird," The Trashmen (1963)

It all goes back to the pre-evolutionary Carolus Linnaeus (we met *him* in Chapter 3), who distinguished between "Reptilia" and "Aves." Reptiles were crawling scaly things; avians were flying, feathered things. Yet, reasonable people, looking at the feathered *Archaeopteryx*, would (and did) call it a bird. Thinking evolutionarily, that ought to mean that Avialae by definition must all be birds.

So potentially, Aves = Avialae (because if *Archaeopteryx* is a bird, it must belong to Aves, and the evolutionary definition of Aves would therefore be the most recent common ancestor of *Archaeopteryx* and

Passer, and all its descendants); and within Aves would be **Neornithes** (the most recent common ancestor of all living birds and its descendants). This would of course end up removing the term Avialae, an outcome that might not be objectionable (at least to language-addled students). But following Linnaeus (who originally invented the word, after all!), Aves = Neornithes. So what to do?

In this book, we arbitrarily choose a hybrid terminology: we recognize the formal terms Avialae and Neornithes (and a bunch of others we'll meet as the chapter progresses), but we'll use the informal term "bird" to describe all obvious birds – and for us these will be feathered, maniraptoran theropods that likely flew to a lesser or greater degree – and we'll say that "modern birds" = Aves = Neornithes. Linnaeus would be pleased.

the Upper Jurassic, 150 myr-old, Solnhofen Lagerstätte[1] (Bavaria, southern Germany) as a feather in 1861 (Figure 8.2) and as bones shortly thereafter. The half-meter dinosaur seemed chimeric, because it had "bird" features like feathers and a furcula, which coexisted with "reptilian" features, such as a long tail and hands with claws (Figures 8.3 and 8.4; Table 8.1).[2] Nothing else like it had even been encountered, and people were confused: was it a bird or a reptile?

Archaeopteryx – most distinctively – has well-preserved, unambiguous feather impressions (Figures 8.3), now known, thanks to melanosomes (Chapter 7), to have once been black. The wings show flight feathers that are effectively indistinguishable from those of modern birds (see Figure 8.2), and the body was covered with symmetrical (nonflight) body feathers such as one might find on a bird today. There were long feathers on the hind legs; these, however, shortened dramatically down the leg, and were not likely as much a key part of the flight apparatus as they perhaps were in the *Microraptor* group of "four-winged" feathered theropods (Figure 7.22).

But going back to the late 1800s and the years immediately following the discovery of *Archaeopteryx*, people knew that birds have feathers, and this thing had feathers; thus, most scientists regarded it as the world's oldest bird (see Box 8.1). Some natural historians, however, cried foul; how could a thing with teeth and a tail, and scads of other reptilian features (Table 8.1) ever be a bird? It had to be a reptile! Besides, if it was a dinosaur (see Box 8.2), aren't all dinosaur reptiles?

Neither Fish Nor Foul

Fast-forwarding to the twenty-first century, after our sashay through Theropoda (Chapter 7), *Archaeopteryx* looks neither aberrant nor inexplicable; in fact, it looks like more of the same (Figure 8.4). With its additional diagnostic avialan characters, it comfortably takes its place among the many feathered maniraptoran theropods that we've already seen. So much so, of the 12 specimens of *Archaeopteryx* currently known, at least one was for many years believed to be a small theropod (which it is!) because the feather impressions were not well preserved on it and nobody *thought* to call it a bird!

The relationship between theropods, *Archaeopteryx*, and modern birds was cemented in 1976 by J. H. Ostrom (see Chapter 16), who in so doing inaugurated a quiet (at first), but utterly far-reaching revolution in vertebrate paleontology and evolutionary biology. At that point, feathered theropods had not yet begun to be identified, and many of finer shavings of the evolutionary sequence that we recognize today (and will describe below) were largely unknown. Cladograms were not widely understood, the very word "Maniraptora" didn't even exist and, as we said in Chapter 7, carnosaurs

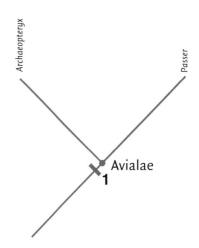

Figure 8.1. Cladogram of Avialae (the common ancestor of *Archaeopteryx* and *Passer*, and all its descendants), based upon the work of T. Holtz. Derived characters include: at **1**, fused scapula and coracoid; humerus longer than scapula; ulna longer than femur (again, a character indicating extremely long arms); $\leq$ caudal vertebrae. Multiple layers of pennaceous (quill-like) feathers *had been* a diagnostic character of avialans since the establishment of the group; however, it has become clear in the past seven years that these diagnose a group of theropods significantly more inclusive than Avialae; we put them at Paraves, but like so much else theropodan, time may show them to be significantly more primitive than even this.

---

[1] The species name, *lithographica*, is a nod to the fine-grained limestones in which it was found; limestones of such extremely fine clay-sized material that for hundreds of years they were engraved for making detailed lithographic prints. *A. lithographica* is a relatively rare animal; after 150 years of extreme vigilance, only 12 specimens (and the feather) are known.

[2] Darwin had just published *On the Origin of Species* in 1859, proposing that species evolved into other species. Here, a mere two years later, was discovered an apparent "missing link" that mixed "reptilian" and avian features.

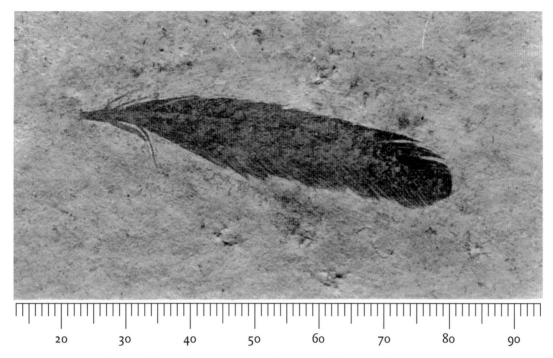

20   30   40   50   60   70   80   90

Figure 8.2. The first evidence for Jurassic-aged birds: the feather of *Archaeopteryx lithographica*, described in 1861, from the Solnhofen quarry in Bavaria (scale in centimeters). Photograph courtesy of J. H. Ostrom.

and coelurosaurs were thought to be rather different from what we now think them to be. Nonetheless, by the time Ostrom was done describing the dromaeosaur *Deinonychus* (1969), and re-studying *Archaeopteryx* (1970), the world of reptiles, birds, and dinosaurs would never be the same. This much Ostrom showed: *Archaeopteryx* seemed to have many dromaeosaur-like features and many bird-like features. Forty-six years later, Chiappe and Meng characterized *Archaeopteryx* in words of which Ostrom would surely have approved: "Like an avian version of the Roman god Janus, one half of this renowned fossil bird looks backward, while the other half looks forward." And yet, 160 years after *Archaeopteryx* was found, people are still surprised – and think that there's some cute stupid trick involved – when you tell them that those lovely songbirds they so enjoy on a spring morning are dinosaurs (see Box 8.2)!

The intermediate anatomical position occupied by *Archaeopteryx* between dinosaurs and birds is summarized in Table 8.1. There was still a wide evolutionary gulf between modern birds and *Archaeopteryx*, a gulf all the more difficult to fill in because of the extreme paucity of bird fossils; their size and delicacy make them among the rarest and hardest vertebrates to study.

Enter the Jehol Biota

Liaoning Province (along with slivers of some neighboring provinces), in the northeast part of the People's Republic of China, contains an extraordinary series of deposits that constitute one of the most fabulous Lagerstätten known on Earth. We know that an Early Cretaceous temperate forest there, associated with a network of interconnected lakes, was home to thousands upon thousands of animals, insects, plants. This idyllic setting was inconveniently and dramatically disrupted by repeated nearby volcanic eruptions wreaking havoc upon the landscape and biota.

Figure 8.3. The beautifully preserved, complete Berlin specimen of *Archaeopteryx*: No longer the "missing link" between two pre-evolutionary Linnaean "Classes" of vertebrates, Reptilia and Aves!

Figure 8.4. A perching reconstruction of *Archaeopteryx* in the Fukui Prefectural Dinosaur Museum. It looks like a typical maniraptoran theropod, posed with long arms angled back and large hands bent backwards at the flexible wrist (thanks to the semi-lunate carpel), as if preparing to launch itself into history.

Associated with these eruptions were mass deaths, with the bodies of the hapless vertebrates and invertebrates, not to mention plants, being carried in volcanic-ash-clogged streams into the lakes, where they sank to the bottom and were gently buried, as the fine ash and lime mud settled in the now quiescent waters. Today we find their remains exquisitely preserved in virtually undisturbed thinly laminated layers of fine claystones and volcanic ash. The result is thousands of fossils, bones completely articulated, with hair, feathers, soft tissues, skin impressions, scales, gizzard stones, and last meals, all in place (and waiting to be described by paleontologists, of course). Failing a trip to China, we recommend the breathtaking images of these fossils in Chiappe and Meng's *Birds of Stone* (2016).

The Jehol biota constitutes three formations, dating from 131 Ma to 120 Ma, preserving, for once, just about everything! Even the mighty Solnhofen, which coughed up *Archaeopteryx* (among so many other things), cannot boast this kind of fossil wealth.

Birds of the Jehol: Filling in the Gap Between *Archaeopteryx* and Modern Birds

The timing turns out to be spectacular, too, because it comes a mere 20 myr after Solnhofen, providing key data to fill that gap between *Archaeopteryx* and modern birds. The approximately 10 myr represented by the Jehol Biota is broad enough to serve up a rich slice of Early Cretaceous vertebrate life. Strikingly, Jehol has an avialan fauna whose most primitive members seem very close to the *Archaeopteryx* stage of anatomy, but these lived with other birds whose more derived characters can be laid out on the cladogram so that we can begin to reconstruct the sequence of character development bringing these avialans closer to modern birds.

The Mesozoic Avialary

Primitive Avialians

Within Avialae, very close to *Archaeopteryx* is *Rahonavis* from the Late Cretaceous of Madagascar. We've opted to emphasize its avian features in tentatively considering it more derived than

Box 8.2 **Plus ça Change...**

The relationship of birds to dinosaurs as outlined here is not new. The famous early Darwinian advocate T. H. Huxley, as well as a variety of European natural scientists from the middle and late 1800s, recognized the connection between the two groups. Indeed, one did not have to be a Darwinian to recognize the important shared similarities, and Huxley's opinions were widely accepted at the time. As noted in 1986 by Yale's J. A. Gauthier (Figure 16.12), Huxley outlined 35 characters that he considered "evidence of the affinity between dinosaurian reptiles and birds," of which 17 are still considered valid today.

So what happened? Why is it news now that birds are dinosaurs? During the very early part of the twentieth century, Huxley's ideas fell into disfavor, because many of the features shared between birds and dinosaurs were thought to be due to convergent evolution.

What evidence was there to argue for convergence in the case of dinosaurs and birds? Really, not terribly much. But in light of the limited knowledge of dinosaurs at the time, the group just seemed too specialized to have given rise to birds. Moreover, clavicles were not known from theropods (then, as now, the leading contender as the most likely dinosaurian ancestor of birds). Thus, fused clavicles (furcula) in birds had to have originated outside Dinosauria. What was needed was a more primitive group of archosaurs that did not seem to be as specialized as the dinosaurs.

In the early part of the twentieth century, such a group of archosaurs, the ill-defined "Thecodontia" (see also Chapter 5), was established by Danish anatomist G. Heilmann as the group from which all other archosaurs evolved. Since this was, by definition, the group that gave rise to all archosaurs, and since birds are clearly archosaurs, it was concluded that birds must have come from "thecodonts." Heilmann had in mind an ancestor such as *Ornithosuchus* (note the name: *ornitho* – bird; *suchus* – crocodile), a 1.5 m long carnivorous bipedal archosaur that, among living archosaurs, looks a bit like a long-legged crocodile. For over 50 years, Heilmann's detailed and well-argued analysis held sway over ideas about the origin of birds.

Several events caused the thecodont ancestry hypothesis to fall into general disfavor. The first was that clavicles were found in coelurosaurians among theropods. Moreover, it came to be recognized that "Thecodontia" is not monophyletic; that is, it is diagnosed by no unique, diagnostic characters pertaining to all its members and no others. How could one derive birds (or anything else) from a group that had no diagnostic characters?

The renaissance of the dinosaur–bird connection must be credited, as we have seen, to J. H. Ostrom of Yale University (Figure 16.9). Ostrom's ideas inspired his student R. T. Bakker and P. Galton (Figure 16.10), who in 1974 published a remarkable paper suggesting that birds should be included within a new vertebrate class: Dinosauria. The idea didn't catch on, in part because it involved controversial assumptions about dinosaur physiology and because the anatomical arguments on which it was constructed were not completely convincing. It may also have been just a bit too radical for its time.

In 1986, however, J. A. Gauthier applied cladistic methods to the origin of birds, and with well over a hundred characters demonstrated that *Archaeopteryx* (and hence, birds) is indeed a coelurosaurian dinosaur. The pendulum – finally – had begun to swing back.

Archaeopteryx. Slightly larger than *Archaeopteryx* (the size of a crow; see Figure 8.6), recent work places *Rahonavis* as a dromaeosaurid theropod, but its position on the cladogram above or below *Archaeopteryx* remains uncertain. It possessed an enlarged sickle-shaped claw on its feet (similar to that of dromaeosaurids and troodontids), and a long, *Archaeopteryx*-like tail. Younger than *Archaeopteryx* by 25 million years, it had forward-looking features such as pneumatic foramina leading into pleurocoels in its chest vertebrae, which as we will see in Chapter 13 implies efficient unidirectional breathing, among other bird-like characters (Figure 8.5). Time and further analyses will eventually consolidate its position. Another creature with a mix of features similar to those of *Archaeopteryx* is *Jeholornis* (unsurprisingly from Liaoning Province), whose characters are regarded

Table 8.1 Distribution of selected characters among nonavialan maniraptoran theropods, *Archaeopteryx*, and modern birds.

Nonavialan Maniraptora	*Archaeopteryx*	Neornithes
Teeth (+)	Teeth (+)	Teeth (−)
Braincase slightly enlarged	Braincase slightly enlarged	Swollen braincase
Rigid tail long, well developed	Rigid tail long, well developed	Pygostyle (+)
Hand three-fingered; I, II, and III	Hand three-fingered; I, II, and III	Carpometacarpus (+); fused digits I, II, and III
Legs:	Legs:	Legs:
Bipedal	Bipedal	Bipedal
Foot:	Foot:	Foot
Three toes in front; one in back	Three toes in front; one in back	Three toes in front; one in back
Unfused foot bones (tarsals)	Unfused foot bones (tarsals)	Fused foot bones (tarsometatarsus)
Digit V (−)	Digit V (−)	Digit V (−)
Claws	Claws	Claws
Pneumatic bones	Pneumatic bones	Pneumatic bones
Furcula (wishbone)	Furcula (wishbone)	Furcula (wishbone)
Flexible trunk	Flexible trunk	Rigid trunk
Sternum small; flat; cartilagenous	Sternum small; flat; cartilagenous	Keeled, bony sternum
Pelvic bones unfused	Pelvic bones unfused	Pelvic bones fused (synsacrum)
All vertebrae (+); primitive trunk and flexible spine	All vertebrae (+); primitive trunk and flexible spine	Some vertebrae (−); trunk region rigid
Flight adaptions (−); primitive theropodan shoulder, upper arm, and musculature	Flight adaptions (−); primitive theropodan shoulder, upper arm, and likely associated musculature	Dramatic rigid flight adaptations in trunk, wing (upper arm) musculature, and rigid trunk/bony sternum/synsacrum (see Appendix 8.1)
Feathers (+)	Feathers (+)	Feathers (+)
Feathers monofilamentous	Feathers vaned (pennaceous)	Feathers vaned (pennaceous)
Flight capability (−)	Flight capability (+?) Likely restricted to gliding and weak powered flight	Flight capability (+) Advanced, powered flight

The plus sign (+) indicates character present; the minus sign (−) indicates character absent.

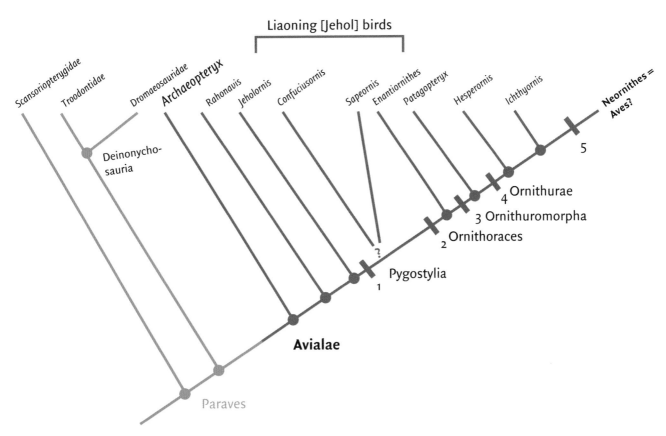

Figure 8.5. Cladogram showing proposed relationships of some of the avialans discussed in this chapter. Redrawn and modified from: Wang, M. and Zhou, Z. 2017. The evolution of birds with implicationsfrom new fossil evidences, in Maina, J. N. (ed.), *The Biology of the Avian Respiratory System*, Springer International Publishing, NY, pp. 1–26, and Chiappe, L. M. and Meng, Q. 2016. *Birds of Stone*, Johns Hopkins University Press, Baltimore, MD, 294 pages. Derived characters include: at 1, pygostyle; at 2, reduction in number of trunk vertebrae, flexible furcula, strut-like coracoid, alula, carpometacarpus, fully folding wings; at 3, further reduction in number of trunk vertebrae, loss of gastralia; at 4, reorientation of pubis to lie parallel to ilium and ischium, reduction of number of trunk vertebrae, decrease in size of acetabulum, patellar groove on femur; at 5, loss of teeth.

as slightly closer to modern birds than to *Archaeopteryx*, but which nevertheless sported the long, bony maniraptoran tail of its antecedents.

Confuciusornithidae, a group containing *Confuciusornis* (Figure 8.7) and *Changchengornis* from the Early Cretaceous of China, is characterized by the appearance of a pygostyle, the shortened, fused tailbones seen on modern birds (see also Appendix 8.1). *Confuciusornis* is known from many specimens; so many, in fact, that two morphs are known: the long-tail-feathered form (on display here in Figure 8.7) and one without decorative tail feathers at all. Otherwise largely indistinguishable, these features suggest sexual dimorphism and accompanying sexual selection (see Chapter 7), behaviors that are surely well known in modern birds. *Confuciusornis* has several relatively modern attributes, including a complete loss of teeth (and the corresponding development of a beak), and a bony sternum, the hallmark of powerful flight muscles. These, however, cooccurred with primitive features such as an unfused, three-fingered hand, relatively primitive forearms and torso (and associated musculature), and unfused foot bones. Both of these characters (loss of teeth and a bony sternum) likely evolved independently in this lineage of Mesozoic birds, because the primitive condition (teeth and a cartilaginous sternum) persists in otherwise more derived Mesozoic birds.

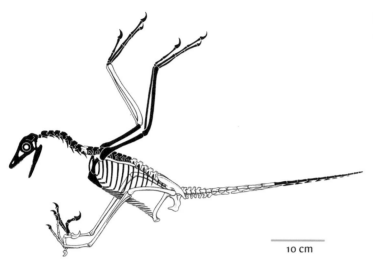

Figure 8.6. *Rahonavis*, either a primitive bird, or its near relative, a non-flying dromaeosaur, from the Late Cretaceous of Madagascar.

10 cm

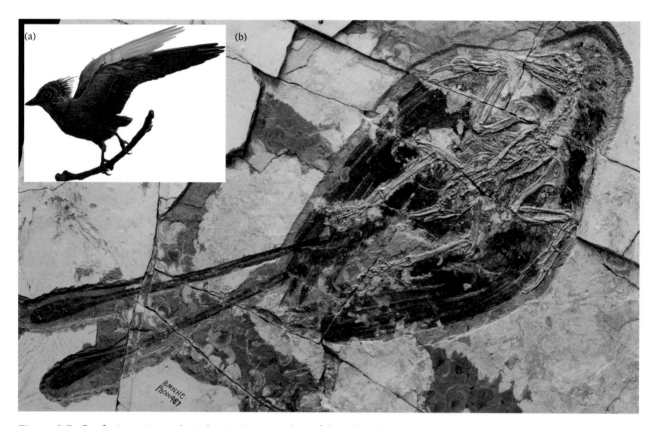

Figure 8.7. *Confuciusornis*, a relatively primitive member of the Jehol Biota, Early Cretaceous Liaoning Province, People's Republic of China. Not the paired elongate feathers, very likely used for display. Inset: A reconstruction showing striking coloration patterns in the feathers. The colors themselves were not recoverable; however, the patterns were obtainable using a variety of microscopic and spectroscopic techniques, and for this reason the image is shown in black and white. Photograph courtesy of S. Abramowicz, Dinosaur Institute, Natural History Museum of Los Angeles County.

By comparison, the closely related *Sapeornis* has no such tail feathers, but also has the well-developed pygostyle (Figure 8.8). Unlike *Confuciusornis*, *Sapeornis* did not have a bony sternum, suggesting that it would not have been as strong a flyer as more derived birds (or even, perhaps, as *Confuciusornis*).

Figure 8.8. A spectacularly preserved specimen of *Sapeornis chaoyangensis* from the Jehol Biota, Early Cretaceous, Liaoning Province, People's Republic of China. Note the preserved details in the wing feathers and skull. Photograph courtesy of S. Abramowicz, Dinosaur Institute, Natural History Museum of Los Angeles County.

Ornithoraces

Thereafter, birds reduced the number of trunk vertebrae, altered the shoulder joint, at least partially fused the digits of the hand into a carpometacarpus, and developed an **alula** (two to six symmetrical feathers and a vestigial digit on the leading edge of the wing, allowing a bird to resist stalling a very slow flight speeds), among other advanced flight-related features. All birds with these transformations are united within **Ornithothoraces** (*ornith* – bird; *thora* – thorax, or chest). Ornithothoracans have two main divisions: on the one hand, **Enantiornithes** (*enantios* – opposite),[3] and on the other **Ornithuromorpha** (*ornith* – bird; *uro* – tail; *morpha* – form), the lineage leading to Aves (see Figure 8.5).

Enantiornithes

If you were a Mesozoic birder, you would know some enantiornithine birds. They would have been common (to judge from their fossil record – they are known globally, except from Antarctica), and were very diverse: more than 70 species have been named. Enantiornithes were generally – but not always – sparrow-sized, arboreal birds that had variably, but for the most part well-developed flight capabilities. They were the most diverse clade of Mesozoic birds (Figure 8.9).

Enantiornithes modified the wrist joint to allow, for the first time, the wing to fold tightly against the body, and developed adaptations indicative of perching: the first toe is positioned fully opposite to the others. The perching foot is a clue that these Mesozoic birds lived in trees, reinforcing other evidence that flight was an integral part of their life habits.

The teeth and strong skulls of enantiornithine birds suggest a relatively powerful bite. Seeds, perhaps? In those few specimens in which stomach contents have been recovered (Jehol, naturally), invertebrates and sap have been found, suggesting an insectivorous and/or herbivorous diet for those enantiornithines, at least.

Still, Enantiornithes are not exactly modern birds (as our birders would surely have noticed); they (the birds not the birders) likely retained gastralia – the belly ribs from the primitive archosaurian condition; relatively numerous trunk vertebrae (a number intermediate between the 13 or 14 found in *Archaeopteryx* and the four to six found in living birds); an unfused tarsometatarsus, and an unfused pelvis.

Ornithuromorpha

Still relentlessly aiming towards Aves (but recalling that humans aim; evolution cannot), Ornithuromorpha is represented by here by *Patagopteryx* (Figure 8.10) from the Late Cretaceous of Argentina, and all remaining birds, in the clade known as **Ornithurae** (*ornith* – bird; *uro* - tail). *Patagopteryx* seems to have been a flightless bird (flightlessness having occurred convergently in several lineages of birds), about the size of a rooster, with powerful legs, and reduced wings. A feature pertaining to all ornithuromorphs is the widened pelvis, an adaptation presumed to allow these birds to lay larger eggs. Larger eggs allow for larger, more developed embryos.

Ornithurae is one of the most robust of all avialan clades, united by at least 15 unambiguous characters (for a few, see Figure 8.5). Not surprisingly, ornithurans include not only the closest relatives to living birds (Hesperorniformes and Ichthyorniformes), but also Neornithes (the group that includes the most recent common ancestor of all modern birds and its descendants) as well.

Hesperornithiform (*hesper* – western[4]) birds were a monophyletic clade of large, long-necked, flightless, diving birds that used their feet to propel themselves, much like modern loons or cormorants (Figure 8.11). They had highly reduced arms and developed powerful hindlimbs for propulsion. The hindlimbs were oriented to the side of the creature, and could not be brought under the body. For this reason, locomotion on land must have been a kind of seal-like waddling (at best).

---

[3] The name is surely obscure, referring the articulation between the scapula and coracoid bones, which was said by C. A. Walker, who named the group, to be "opposite" that found in other birds.

[4] When this bird was discovered in Kansas, USA, in the nineteenth century, Kansas was still considered part of the American "west."

(a)

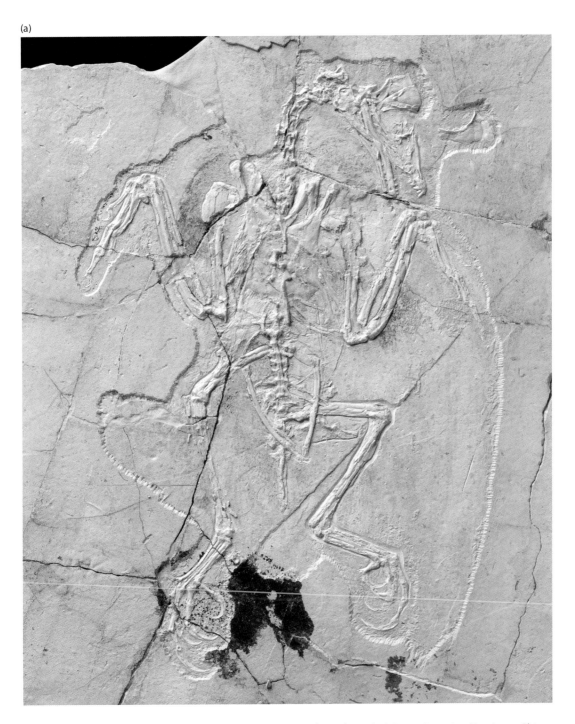

Figure 8.9. Two 125 Ma to 120 Ma enantiornithine birds from the Jehol Biota, Liaoning Province, China. (a) The pigeon-sized *Zhouornis hani*, slightly larger than most other enantiornithines, whose powerful jaws and claws suggest that it might have been predatory – a kind of small, middle Cretaceous raptor; and (b) *Sulcavis geeorum*, a sparrow-sized bird with powerful jaws, perhaps used to crunch hard food, like seeds. Photograph courtesy of S. Abramowicz, Dinosaur Institute, Natural History Museum of Los Angeles County.

(b)

Figure 8.9. (*cont.*)

In the water, on the other hand, hesperornithiform birds appear to have been supremely adapted for marine life. The long, flexible neck must have been useful in catching fish, a behavior indicated from coprolites preserved with their skeletons. In many respects, the group is quite close to modern birds, yet they retain teeth in their jaws. Like modern diving birds, some of the pneumaticity in the bones was lost. Presumably because strength in the chest and shoulder regions for flight was no longer needed, the furcula, coracoid, and forelimbs were highly reduced.

Figure 8.10. A rare mount of the Late Cretaceous Argentine ornithuromorph flightless bird *Patagopteryx*, seen in the Museo Histórico de Ciencias Naturales in Buenos Aires. Photo courtesy of Professor J. R. Hutchinson, Royal Veterinary College, Kings College, London.

All of these swimming adaptions indicate that this group had a long evolutionary history prior to their appearance in the Late Cretaceous. Hesperornithiformes include *Enaliornis* from the Early Cretaceous of England, *Hesperornis* (Figure 8.11) and its smaller relative *Baptornis*, and *Parahesperornis*, all from North America.

Closer related yet to Aves were the toothed Ichthyornithiformes (*ichthys* – fish; Figure 8.12; see Figure 8.5). Unlike Hesperornithiformes, Ichthyornithiformes were excellent flyers. *Ichthyornis*, from the Late Cretaceous of North America, had a massive keeled sternum and an extremely large process (deltoid crest) on the humerus that was probably an adaptation for powerful flight musculature. In other respects, it shared many of the adaptations of modern birds, including a shortened, fused trunk, a carpometacarpus, a pygostyle, a completely fused tarsometatarsus, and a synsacrum formed of ten or more fused vertebrae. Found exclusively in marine deposits, Ichthyornithiformes must have been rather like Mesozoic seagulls – but with teeth.

Neornithes (Aves)

Neornithes (Aves) is a well-supported clade, involving as many as 11 characters of the skull, pelvis, and ankle. Birds clearly suffered a major extinction when the nonavian dinosaurs went extinct at the end of the Cretaceous (see Chapter 17). Certainly the toothed birds of the Mesozoic appear not to have made it into the new Era. But the data are beginning to suggest that modern birds – such as we describe here, in Appendix 8.1 – actually first appeared before the nonavian dinosaurs went extinct; that is, some familiar modern families of birds coexisted with nonavian dinosaurs. Screamers and water-fowl (Anceriformes), loons (Gaviiformes), and possibly shorebirds such as sandpipers, gulls, and auks (Charadriiformes), landfowl (Galliformes), wing-propelled divers such as modern petrels (Procellariiformes), and parrots (Psittaciformes) all may have had Mesozoic origins.

The discovery of the bones of a small anseriform (water fowl), *Vegavis*, from the Cretaceous of Antarctica is physical evidence that modern groups of birds were alive and perhaps thriving along with nonavian dinosaurs (Figure 8.13). University of Texas avian paleontologist Julia Clarke

Figure 8.11. *Hesperornis*, the diving bird from the Late Cretaceous of the USA. Print courtesy of J. H. Ostrom.

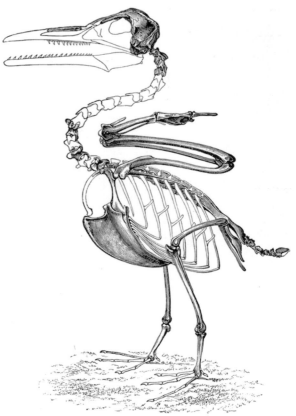

Figure 8.12. *Ichthyornis*, a relatively small, gull-like bird from the Late Cretaceous of the USA.

(see Figure 16.25), and her colleagues, have suggested that ducks, chickens, and ratites (including emus and ostriches) were likely present during the latter days of nonavian dinosaurs. Clearly, these early records of modern birds speak, however incompletely, to the origin, initial radiation, and establishment of Aves in the closing moments of the Cretaceous.

Molecular Clocks

Data on exactly when Neornithes first appeared can come from a source other than the fossil record. We've emphasized the fossil record mainly because this book deals with extinct organisms, the only record of which has been the fossil record. But in those cases in which we are dealing with living organisms (like birds!), a whole different type of technique is available for study: molecular clocks.

Molecular clocks involve measuring the timing of molecular changes. So, for example, take two somewhat closely related living organisms, A and B. Now, choose a particular protein that they share, say, serum albumin (a protein in their blood). Their serum albumins might be quite closely related, but if some time has elapsed since A and B last shared a common ancestor, the exact molecules will now differ slightly (the molecules are said to have evolved), in either form, composition, or both. If we knew the *rate* at which the molecules diverged, we would know how distantly in the past A and B shared a single common ancestor (whose serum albumin composition and form were the ancestral ones). This very technique (and indeed this very molecule!) was used in the case of humans and

Figure 8.13. Bones of the anseriform (modern bird clade including ducks, geese, and swans) *Vegavis*, from the Cretaceous of Antarctica. This fossil was the first body fossil evidence of a modern bird group coexisting with nonavian dinosaurs.

chimpanzees to show that the two shared a most recent common ancestor only around 5 million years ago, instead of the 15 million that had been inferred from the geological record.

More recently, molecular biologists have been using a technique called DNA hybridization. This technique works similarly to the one described above for proteins, except that it compares two strands of DNA (instead of proteins). In the same species, the strands of DNA should be virtually identical. DNA hybridization allows molecular biologists to measure the differences between the two strands. Knowing what rate substitutions (or changes) occur in the DNA allows us to calculate how long ago two different creatures shared identical DNA. That number should equal the time of divergence from a common ancestor.

And what of birds? Molecular clock estimates of the earliest Aves have been somewhat earlier than has been inferred from the fossil record. Until very recently, the fossil record seemed to suggest that the major radiation of birds took place *after* the extinction of nonavian dinosaurs. Yet, the fossil record of birds is, as we have seen, rather spotty, and perhaps most trustworthy only in its broadest outlines.

Estimates of the radiation of Aves, based on molecular data – primarily DNA hybridization – have put the time of the radiation well within the Cretaceous, *before* the boundary. How to resolve this contradiction?

Recently, the fossil record has begun to support the molecular record... a little bit. As we have seen, the fossil record of modern bird groups in the Cretaceous now includes the ancestral relatives of ducks, chickens, and large, flightless birds such as ostriches and emus. Paleornithologist Daniel Field and colleagues, addressing this very question in 2018, concluded that while some of the earliest members of these neornithine groups likely appeared in the Cretaceous, the major radiation that produced most of today's bird groups occurred after the field was abandoned, as it were, by Mesozoic birds. As in the case of the rise of Dinosauria, it was not the head-to-head competition and innate superiority of the modern bird group; it appears that the others were simply gone.

In short, then, the evidence for birds as dinosaurs, and thus the details of the transition from non-avian to avian theropods is overwhelming. But it wasn't always so; the battles began with Ostrom and continued throughout the following 30+ years (see Box 8.3). Pockets of resistance, ideas held by scientists who grew up in pre-cladistic time, remain, but the overwhelming consensus in ornithology today is that birds are living dinosaurs.

The End of an Era

No tooth-bearing bird survived past the end of the Mesozoic Era (see Chapter 16) and, not to put too fine a point upon it, nobody really knows why. Nonetheless, the 2018 study Daniel Field and colleagues suggested that ground-dwelling life strategies – and by this they did *not* mean nonflying birds; simply birds that were not arboreal, or tree-dwelling in their life habits – seemed to favor survival into the Cenozoic. What this is interpreted to mean, is that a few edentulous, ground-dwelling early neornithine birds appeared before the end of the Cretaceous, somehow survived the extinction and then, in the Brave New Cenozoic World that followed, they thrived, evolved, and radiated into the diversity of birds that we see today.

One last evocative thought. Living birds have no teeth (e.g., the enamel and dentyne features that develop in the jaws of vertebrates); in fact, no post-Mesozoic (Cenozoic) bird ever had any (although tooth-like bumps on beaks is not unknown). With toothlessness having come and gone several times in theropods, is it possible that the genetic machinery is still buried somewhere deep in the genome of modern birds? The answer is intuitively "most likely"; but an experiment carried out several times in laboratories in the 1980s and 1990s demonstrated that modern birds still possess the ancient genetic machinery for making teeth. The only thing that has changed is the genetic program for *activating* their biological manufacture. The same code for producing teeth exists deep in the genomes of all amniotes, and thus even today it is possible to induce tooth formation in Neornithes (in this case, a chicken). Inside of every modern bird resides it inner maniraptoran theropod.

Origin of Avian Flight

A look at the cladogram shown in Figure 8.5 shows that flight proficiency seems to have been a driving force in avialan evolution. The question of how this occurred was originally – and for a long

Box 8.3 Birds as Dinosaurs: at First, not Exactly a Love(bird)fest

This scientific hypothesis – birds as living theropod dinosaurs – is an impressive bit of evolutionary detective work and represents one of the best-known and most spectacular evolutionary transitions known. Like all good science, however, it wasn't presented as a *fait accompli* one day at some ornithology meeting, and everybody smiled harmoniously and loved it from the moment they heard the wonderful news. Rather, the story unfolded over many years and, along the way, unconvinced scientists raised many important challenges to the theory. The survival of the theory after all these challenges gives it its strength. Here we discuss a few of the objections to the idea of birds as living dinosaurs.

Counting on the Fingers

The most significant impediment to deriving birds from maniraptoran theropod dinosaurs was a dispute over the fingers involved in the carpometacarpus. Recall that the carpometacarpus of living birds is a fused structure composed of three fingers. Paleontology clearly tells us which three fingers these are: if Avialae is real, then the fingers must be I, II, and III, since the fingers in avialan theropods, including *Archaeopteryx*, are I, II, and III.

Embryologists since the 1870s have studied the hand of modern birds as the carpometacarpus develops. They repeatedly concluded that the fingers of the bird hand appear to be II, III, and IV. How could paleontology unambiguously identify the fingers as I, II, and III, when embryology identifies them as II, III, and IV? And, if the fingers are II, III, and IV as suggested by embryology, how could a bird come from a dinosaur that only had digits I, II, and III?

A possible solution to this conundrum was proposed by diapsid specialist J. A. Gauthier (see Figure 16.12) and developmental biologist G. P. Wagner. Embryologists identify the sequence of "condensations" on the hands and feet; that is, early developed buds of material that later become digits. As early growth occurred in bird hands, embryologists thought they saw condensation I become digit I, condensation II become digit II, and so on. Gauthier and Wagner thought they saw something else, however, something that they called a "frameshift in the developmental identities" of the fingers. According to them, the bud considered to be embryological

condensation II actually becomes adult digit I, condensation III actually becomes digit II, and condensation IV actually becomes digit III. More recent work now suggests that genes coding for the development of the original digit I cause its development in position II; and genes coding for the development of the original digit II cause its development in position III. The "frameshift" appears to be from the ancestral position to a new one.

With these new observations, the fingers identified in living birds and those in extinct theropods are the same. Wagner and Gauthier's work seemed to resolve the apparent discrepancy between embryology and paleontology, and reaffirms the avialan origin of birds.

However, a 2019 paper by evolutionary biologist Thomas Steward and colleagues (including Günther Wagner, who, with J. A. Gauthier, proposed the original frameshift hypothesis) has cast some doubt our ability to recognize digit identities (is it I, II, III, or even IV?) in amniotes, generally. According to these authors, the problem is that the genetic basis of all digits except number I is variable across Amniota; that is, for example, the genetic basis of digit III in one amniote may differ from that in another. The exception to this is digit I, which evidently is expressed consistently across Amniota. These authors suggested that the digital formula for modern birds is I, III, and IV. If this is all correct, it would suggest that the digital formulae for theropods also may not have been properly determined – but with no living nonavian theropods, it would be hard to know. And so the question remains unresolved; stay tuned.

Do Feathers of a Bird Stay Together?

Another potential hiccup had to do with the feathers in theropods. Initially, it was uncertain whether or not the monofilmentous coverings seen, for example, in *Sinosauropteryx* and *Sinornithosaurus*, are anatomically the same thing as the vaned (pennaceous) feathers seen in fossils like *Microraptor*, *Anchiornis*, and *Archaeopteryx*. If they are some other anatomical structure, then their presence can hardly be used to evolutionarily link these nonflying theropods to birds. Now, however, the distinctive shared compound structure of both monofilamentous and vaned feathers and the recognition of identical

Box 8.3 *cont'd*

melanosomes in both types of feather, as well as an improved understanding of the embryological of feather development, lead to the inescapable conclusion that both monofilamentous and vaned feathers are homologous.

"But They're Too *Young*!"

Perhaps the *least* scientifically meaningful – but most frequently raised – concern about the theropod–bird hypothesis is one of timing. Many of the known feathered nonavian theropods (some of which grace the pages of this book) come from the Jehol Biota in Liaoning Province, and are Early Cretaceous in age. *Archaeopteryx*, on the other hand, is a Late Jurassic form. Because *Archaeopteryx* is 20 myr to 30 myr older than the feathered Chinese dinosaurs, so the argument goes, the younger (Early Cretaceous) feathered fossils cannot have any bearing upon understanding the evolution that led to *Archaeopteryx*.

This reflects a fundamental misunderstanding of cladograms. Because of the inconsistencies of preservation, cladograms do not incorporate time into the analysis in a way that is testable; there is no way to know whether an organism is not present in the fossil record because it didn't exist at that time, or because it simply was not preserved. As the old adage goes,

"Absence of evidence is not evidence of absence!"

The preservation of feathers is rare and requires most remarkable circumstances; it is a highly unlikely proposition – and certainly not testable – that the few times (and places) in the fossil record in which feathers are preserved are the first occurrence of such features. Recalling how we reconstruct phylogeny (again, Chapter 3), we look for the distributions of primitive and derived characters, and develop our phylogenetic hypotheses based upon these data. The Early Cretaceous age of the Liaoning feathered dinosaurs is counterintuitively all but irrelevant; the mixes of primitive and derived characters that these dinosaurs possess is the key to understanding their relationships and reconstructing their evolutionary histories.

time, somewhat acrimoniously – framed in the form of two opposing hypothetical endpoints exist as regards the origin of bird flight (Figure 8.14). The first is the so-called **arboreal** (or "trees down") hypothesis: that bird flight originated by birds gliding down from trees (Figure 8.14a). In this hypothesis, gliding is a precursor to flapping (powered) flight; as birds became more and more skillful gliders, they extended their range and capability by developing powered flight. Perhaps flapping developed as a modification of the motions used in controlling flight paths. The second, antithetical to the arboreal hypothesis was the **cursorial** (or "ground up") hypothesis for the origin of flight (Figure 8.14b). The cursorial hypothesis states that bird flight originated by an ancestral bird running along the ground. In this scenario, perhaps as obstacles were avoided, the animal became briefly airborne.

In the face of what is known about the evolution of birds, both of these ideas seem somewhat restrictive and perhaps a bit naïve: either/or propositions when the fossil record tells us both had to be involved. Maniraptorans did not run faster and faster until one finally flew (even though they were cursorial). Maniraptorans did not leap from trees until one finally failed to crash.

Towards a Model for the Origin of Flight in Birds

The arboreal hypothesis is intuitively appealing, and getting airborne is easy. On the other hand, the cursorial hypothesis is strongly supported because ultimately the ancestor of birds had to have been a cursorial creature. A problem with the cursorial hypothesis is that it has so far proven nearly

Figure 8.14. Endpoints: The arboreal and cursorial hypotheses for the origin of bird flight. (a) The arboreal hypothesis, which suggests that bird flight evolved by "birds" gliding down from trees. (b) The cursorial hypothesis, which suggests that bird flight evolved by "birds" running along the ground until the animals became airborne.

insurmountable to model a cursorial theropod that developed flight by running along the ground. For this reason, an arboreal stage, intermediate in the development of flight, has been attractive to many scientists. Yet indications of a cursorial heritage are present in all living birds as, indeed, their limbs are little changed from the nonflying coelurosaurian condition. However, what would stop a group of small, cursorial maniraptorans from climbing into trees and developing an arboreal lifestyle?

Recently an interesting compromise position has been proposed. Perhaps flapping wings helped early cursorial theropods to get a purchase on steep slopes, overhangs, or even tree trunks. From this it would not have been a big leap, as it were, to flapping flight.

Paleontologists Q. Li, M. J. Benton (Figure 16.21), M. S. Y. Lee, and a variety of colleagues have proposed that the process of becoming a bird must have involved several steps: the decoupling of the forelimbs from the hindlimbs (which would have been already in effect accomplished since these animals were bipedal), the appearance of flight feathers (also in place, since pennaceous feathers were present – particularly on the arms – of nonflying theropods as far back as ornithomimids [Figure 7.11]), a startling plunge in overall size, and the assumption of an arboreal lifestyle (likely made possible by the diminution in size).

The drop in size was certainly precipitous: paravans shrank dramatically before the advent of true birds, the size decrease picking up dramatically around 178 Ma (Figure 8.15).

Within the basal avialan lineage, successful powered flight was clearly possible; however, because most members lacked well-developed, bony sternums for powerful flight-muscle attachment (enantiornithines being an exception), as well as significant modifications of the upper arm, it likely did not take place at the level seen in birds today. Our vote, therefore, is for brief stretches of weakly

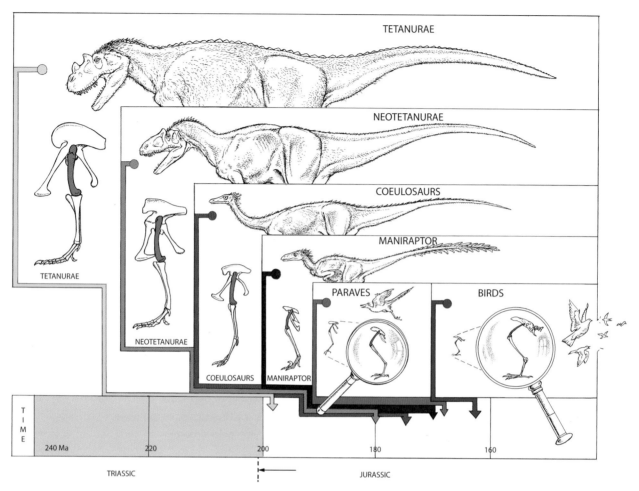

Figure 8.15. Dramatic decrease in size and change in limb proportions over 50 million years, in the theropod lineage leading to Aves. Here, the measured length of the femur (thigh) represents body size (here also represented by inferred weight). These are plotted against time, to show the dramatic decrease in body size and proportion that took place during in the theropod lineages leading to birds, during the first part of the Jurassic. From Lee, M. S. Y., Cau, A., Naish, D., and Dyke, G. J. 2014. Sustained miniaturization and anatomical innovation in the dinosaurian ancestors of birds. *Science*, **345**, 562–566.

powered flapping flight, stitched together by a large proportion of gliding. The habitats of these birds were very likely arboreal, an inference strengthened by the perching anatomy of the feet, by the fact that the Jehol Biota lived in a forested environment, and by the ease with which this type of flight behavior could be maintained in trees.

At and beyond the Ornithoraces stage of evolution, powered flapping flight must have been common and practicable for most birds. By this point, the muscles and soft tissue necessary to control the tail feathers (a key feature of modern bird maneuverability) would have evolved in association with the pygostyle, and we can safely assume that trees were no longer necessary as a crutch to support a volant lifestyle. We think it likely that an ornithuran like *Ichthyornis* flew as well as a neornith like *Larus* (seagull)! Figure 8.16 is a summary diagram highlighting the many changes that took place in paravian theropods through the latter part of the Mesozoic, to reach the grade, that is, the morphological stage that neornithine birds have achieved.

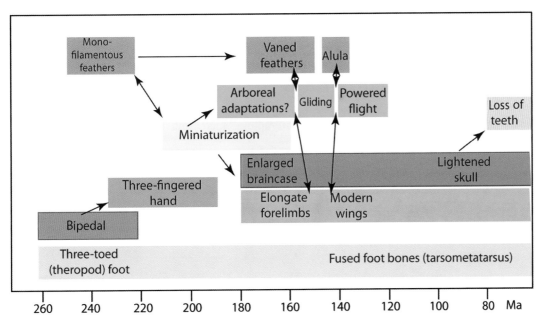

Figure 8.16. Diagram highlighting the key steps in the evolution of modern birds. Single-headed arrows suggest the direction in which one feature may have modified another through evolution; double-headed arrows suggest potential coevolution. Modified from Cau, A. 2018. Assembly of the avian body plan: a 160-million year long process. *Bollettino della Società Paleontologica Italiana*, **57**, 1–25.

Forgetting History

In our modern world, many features are unique to birds; this gives the illusion that the evolution of Aves required a simultaneous suite of radical changes to go from "reptile" to "bird." However, it turns out that these apparently unique bird features are distributed throughout Theropoda – some even more primitive than this (see Chapter 6 and 7). Birds inherited a lot of features shared by many extinct theropods, and the cladograms (Chapters 7 and 8) show at which points during theropod evolution apparently "unique" features of modern birds actually appeared. As we hope was clear from this chapter (at least), the evolutionary assembly of modern birds was gradational. Here, we highlight just how far back the supposedly "unique" features actually go; next to each feature, we've noted – in blue – the level in the hierarchy in which the character is diagnostic:

- Gastralia (belly ribs); found in *Archaeopteryx*, but lost in modern birds (Amniota).
- Bipedality (Dinosauromorpha).
- Mesotarsal ankle (Dinosauromorpha – Dinosauria).
- Feathers: monofilamentous feathers may occur as early as Ornithodira; they certainly occur as early as Dinosauria, being present in Ornithischia (as we shall see in Part III) as well as Saurischia. Pennaceous (quill-shaped) feathers occur in maniraptoran theropods, but long before flight was possible.
- Thin-walled ("hollow") long bones (Saurischia).
- Blade-like, unserrated,[5] recurved teeth (Theropoda).
- Foot with three toes in front, and a fourth to the side with well-developed claws (Neotheropoda).

---

[5] The loss of serrations seems to have been an avialan character.

- Furcula (Neotheropoda).
- Semi-opposable thumb (Tetanurae; developing through Paraves).
- Hand with three, fully moveable, separate fingers, tipped with well-developed, recurved claws (Tetanurae).
- Lack of flexibility in the tail through extended zygopophyses (Tetanurae).
- Semi-lunate carpal (Maniraptora).
- Large tibia; small fibula thinning toward the ankle (Maniraptora. increasing dramatically through Paraves).
- Large brain (Coelurosauria).
- Retroverted pubis (Paraves).
- Highly efficient breathing (Ornithodira [see Chapter 13]).

Plainly, avialans inherited many features from their nonavian ancestors. Ironically, none of this is particularly new; the idea of birds-as-dinosaurs goes back as early as the discovery of *Archaeopteryx* in the mid 1800s (see Box 8.2).

Having a Purpose in Life

Pneumatic bones and feathers are commonly singled out as marvelous adaptations to maintain lightness and permit flight. They surely are marvelous, they maintain avian lightness, and there is no doubt that feathers work well for flight. But was the original purpose of pneumatic bones and feathers for lightness and flight, respectively?

In both cases, now that we have a sense of bird ancestry, the answer is probably not. Hollow bones are a saurischian character (even the names *Coelophysis* and *Coelurosauria* contain a reference to the hollow bones in these nonflying dinosaurs), and pneumaticity is likely related to efficient breathing (as implied above, we will see this again in immense, ponderous sauropods – Chapter 9 – which we can safely assume didn't fly). Pneumaticity likely developed long before there were things we would call birds. Even though these features were refined in modern birds, they inherited pneumaticity, unidirectional breathing, feathers, and a host of other "uniquely" avian features from nonavian theropods, who themselves inherited them from somewhere back in the "misty mist and the dusky dusk" of Ornithodira.

Most people would say that the purpose of feathers is flight. We agree; feathers are used for flight, but we doubt that they evolved for that purpose, particularly as many nonflying, nonavian dinosaurs had them. This, in turn, provides a real insight into what the origin of feathers was all about.

One striking feature of all feathered dinosaurs – flying or not – are skeletal designs suggesting very active, commonly predatory, lifestyles. With feathers appearing on nonflying theropods, many paleontologists infer that feathers likely provided the insulation that is a prerequisite for warm-bloodedness, allowing theropods to maintain high levels of activity for extended periods of time; an attribute that was eventually used for flight (see Chapter 14). It is perhaps no coincidence that the most highly evolved feathers are flight feathers; other feathers, such as monofilamentous feathers and down, likely represent earlier stages in feather development, and graced the coats of more primitive nonavian theropods.

But does that actually get us to the origin of feathers? Insulation for warm-bloodedness doesn't completely satisfy, either, because many of the early feathers that we've seen are monofilamentous, and may not have insulated very well or, as we saw in some large theropods, covered very well. But as we all know, modern birds are brightly colored, highly social, and visually acute (Chapter 6). We also know that theropods as a group – nonflying and flying – have distinctive patterns and colored markings. Could feathers – at first – have been at least as much for display as for insulation?

Recall that in Chapter 5 we discussed the sensory importance of feathers. Here, then, is a hypothetical, but plausible, scenario for the evolution of feathers: widely distributed (across the skin), monofilamentous feathers could have first arisen for tactile sensation. Color might have appeared as a byproduct – or a even a driving force – of evolving theropod sociality, associated with denser packing of the feathers. Still more densely packed, multifilamentous (downy) feathers might then (or concurrently) have evolved as a means of insulation, and been retained on the smaller, more active animals. And finally, as in Figure 5.16, pennaceous, flight feathers may have ultimately evolved in association with display, small size, and a very active lifestyle. That is simply a scenario – strengthened, we feel, by the fact that it exactly describes how feathers are used today. However, we can be sure that early feathers weren't originally about flight!

The Evolution of Controlled, Powered Flight

Today, two groups of animals have members that fly[6] – arthropods and vertebrates. The flying arthropods are insects; the flying vertebrates are bats and birds. Putting this in evolutionary language: mammals (synapsids) and birds (neornithine theropod dinosaurs) are the living flying vertebrates.

Primitively, neither insects nor vertebrates flew, and the common ancestor of the two didn't fly. This means that flight evolved independently in certain lineages of insects and vertebrates. Looking more closely at vertebrates, the common ancestor of mammals and theropod dinosaurs didn't fly. This means that flight evolved independently in each of these groups (mammals and theropods). So looking only among living animals, flight seems to have been invented three times: in insects, in bats, and in birds. Is that the total number of times that flight evolved independently?

Definitely not! Looking at archosaurs, recall the magnificent pterosaurs (Figure 4.13); not dinosaurs at all, but gorgeous flying nondinosaurian ornithodirans: the sister group to dinosaurs (Figure 4.11). So now we have flight being invented independently at least four times: twice in archosaurs (birds and pterosaurs), once in mammals (bats) and once in arthropods (insects). Is that all?

Not really; you haven't reckoned with the volant fertility of Paraves! The cladograms demonstrate that just in Paraves, that tiny corner of Archosauromorpha, powered flight was invented independently at least three times. Most famously there is the lineage that we have been tracking in this chapter: Aves. But there are also the so-called "four-winged" fliers, exemplified here by deinonychosaurs *Microraptor* and *Anchiornis* (see Figures 7.21 and 7.23, respectively). These animals had well-developed pennaceous feathers, but few of the other anatomical features for powered flight that have served ornithoracians so well. Perhaps arboreally based gliding flight, with a bit of power from the wings (not those on the legs, however), was primarily the norm for this group. Finally, a third independent attempt at flight occurred in the scansoriopterygids (Figure 7.25), the enigmatic, bat-like paravians that have been rewriting the book on what is possible in a theropod, or any dinosaur for that matter. Both these paravian lineages, basal paravians represented by Scansoriopterygidae, and Deinonychosauria, represented by the "four-winged flyers," were evolutionary dead-ends as far as flight is concerned: they never developed flying animals beyond these (or at least, this is what is understood at present!). Aves, however, starting in much the same place – a primitive nonvolant paravian theropod – went much further, reaching the undoubted flight capabilities seen in Neornithes today.

And these are minimum estimates: who knows how many faltering dead-end steps towards flight there were in insects? In just paravians alone the road to powered flight as we know it today is

---

[6] As distinguished from gliding.

littered with experiments and semi-successes. Likely, it would not have been otherwise for insects, bats, and pterosaurs.

A Closing Thought

Many readers may have heard the Mesozoic referred to as the "Age of Dinosaurs" and the Cenozoic called the "Age of Mammals." Dinosaurs are dominant in the Mesozoic; mammals in the Cenozoic, right?

Today, our best estimates of living Cenozoic mammal diversity are that there are about 4500 species; for living birds, there are about 10 000 species. That being the case, please allow us to welcome you to the Age of Dinosaurs.

SUMMARY

Birds *are* dinosaurs. We don't mean that they are *related* to dinosaurs – although, if they are dinosaurs, they must be related to them. We don't mean that they *come from* dinosaurs – although they obviously evolved from something that was itself a dinosaur. We mean that birds *are* dinosaurs, a statement that is no more radical than saying that humans are mammals.

Living birds have a suite of highly derived anatomical features that at first glance appear to be unique. In fact, a close look at Theropoda shows that most of the supposedly uniquely avian features of living birds are actually distributed throughout theropods: cladistic analysis demonstrates that many presumed "bird" characteristics are distributed in nonbird theropods, particularly within Tetanurae and Eumaniraptora.

The discovery in Bavaria of small, Late Jurassic theropod, *Archaeopteryx lithographica*, and the many discoveries of feathered, nonflying theropods from the Early Cretaceous of Liaoning Province of China blur the line between bird and nonbird. Clearly, although feathers are diagnostic for birds among living organisms, these fossils all demonstrate that presence of feathers does not guarantee that one is dealing with a bird. These discoveries reinforce the viewpoint that feathers had multiple uses, and were evidently invented as a form of insulation or perhaps display, only later to be coopted for flight purposes.

An important quality of Avialae appears to be flight. The origin of flight is shrouded in mystery, but the cladogram suggests that it was likely a multimillion-year process involving small size, arboreality, and weak flapping-powered gliding, followed by a variety of adaptations associated with a rigid trunk and a bony sternum, advanced arrangement of flight musculature, a pygostyle, alula, tail feather control, and some other flight-related features.

While *Archaeopteryx* was initially a key organism in identifying the relationship between Aves and dinosaurs, the wealth of new feathered fossils highlight and fill out the key events toward the evolution of modern birds (and flight). Despite their rarity as fossils, the fossil record of birds indicates the general order in which the evolutionary events leading to modern birds occurred.

The improvement of flight capability was a driving force in post-*Archaeopteryx* bird evolution. Pneumatic foramina became better developed, along with, sequentially, the pygostyle, a reduction in the number of trunk vertebrae, modifications of the shoulder, and the development of the carpometacarpus. Still, some primitive characters such as gastralia, were retained. All these features were present in Cretaceous ornithothoracan birds, including the small, comparatively common Enantiornithes, and the line leading to Aves, Ornithuromorpha.

Within Ornithuromorpha, several highly derived birds evolved, notably the diving Hesperornithiformes and the seagull-like Ichthyornithiformes. These birds, for all their advancement, were not exactly like living birds, lacking a number of features diagnostic for Neornithes (Aves), in particular the complete loss of teeth. The earliest fossil record of modern birds is very fragmentary and likely incomplete, but currently goes back into the Late Cretaceous. Nonetheless, the major evolutionary radiation of modern birds appears to have occurred after the end of the Mesozoic Era.

SELECTED READINGS

Benton, M. J. 2014. How birds came to be birds. *Science*, **345**, 508–509.

Brusatte, S. L., Lloyd, G. T., Wang, S. C., and Norell, M. A. 2014. Gradual assembly of avian body plan culminated in rapid rates of evolution across the dinosaur–bird transition. *Current Biology*, **24**, 2386–2392.

Carney, R. M., Vinther, J., Shawkey, M. D., D'Alba, L., and Ackermann, J. 2012. New evidence on the colour and nature of the isolated Archaeopteryx feather. *Nature Communications*. DOI: 10.1038/ncomms1642

Cau, A. 2018. Assembly of the avian body plan: a 160-million-year long process. *Bollettino della Società Paleontologica Italiana*, **57**, 1–25.

Chiappe, L. M. 1995. The first 85 million years of avian evolution. *Nature*, **378**, 349–355.

Chiappe, L. M. and Dyke, G. J. 2002. The Mesozoic radiation of birds. *Annual Review of Ecology and Systematics*, **33**, 91–124.

Chiappe, L. M. and Dyke, G. J. 2007. The beginnings of birds: recent discoveries, ongoing arguments, and new directions. In Anderson, J. S. and Sues, H.-D. (eds.) *Major Transitions in Vertebrate Evolution*. Indiana University Press, Bloomington, IN, pp. 303–336.

Chiappe, L.M. and Meng., Q. 2016. *Birds of Stone: Chinese Avian Fossils from the Age of Dinosaurs*. Johns Hopkins University Press, Baltimore, MD, 294 pages.

Chiappe, L. M. and Witmer, L. M. (eds.) 2002. *Mesozoic Birds*. University of California Press, Berkeley, 520 pages.

Clarke, J., Tambussi, C. P., Noriega, J. I., Erickson J. M., and Ketcham, R. A. 2005. Definitive fossil evidence for the extant radiation of Aves in the Cretaceous. *Nature*, **433**, 305–308.

Cracraft, J. and Clarke, J. 2001. The basal clades of modern birds. In Gauthier, J. and Gall, L. F. (eds.) *New Perspectives on the Origin and Early Evolution of Birds. Proceedings of the International Symposium in Honor of John H. Ostrom*, pp. 143–156.

Dial, K. 2003. Wing-assisted incline running and the evolution of flight. *Science*, **299**, 402–404.

Dingus, L. and Rowe, T. 1997. *The Mistaken Extinction*. W. H. Freeman and Company, New York, 332 pages.

Erickson, P. G. P., Anderson, C. L., Britton, T., *et al.* 2006. Diversification of Neoaves: integration of molecular sequence data and fossils. *Biology Letters*, **2**, 543–547.

Field, D. J., Bercovici, A., Berv, J. S., *et al.* 2018. Early evolution of modern birds structured by global forest collapse at the End-Cretaceous mass extinction. *Current Biology*, **28**, 1–7.

Gauthier, J. A. 1986. Saurischian monophyly and the origin of birds. In Padian, K. (ed.) *The Origin of Birds and the Evolution of Flight*. California Academy of Sciences Memoir no. 8, pp. 1–56.

Gauthier, J. A. and Gall, L. F. (eds.) (2001). *New Perspectives on the Origin and Early Evolution of Birds*. Peabody Museum of Natural History, Yale University Press, New Haven, CT, 613 pages.

Foth, C., Tischlinger, H., and Rauhut, O. W. M. 2014. New specimen of *Archaeopteryx* provides insights into the evolution of pennaceous feathers. *Nature*, **511**, 79–82.

Hecht, M. K., Ostrom, J. H., Viohl, G., and Wellenhofer, P. (eds.) 1984. The beginnings of birds. *Proceedings of the International Archaeopteryx Conference Eichstatt*, Freunde des Jura–Museums Eichstätt, Willibaldsburg, 382 pages.

Ji, Q., Currie, P. J., Ji, S., and Norell, M. A. 1998. Two feathered dinosaurs from northeastern China. *Nature*, **399**, 350–354.

Ostrom, J. H. 1974. Archaeopteryx and the origin of flight. *Quarterly Review of Biology*, **49**, 27–47.

Ostrom, J. H. 1976. Archaeopteryx and the origin of birds. *Biological Journal of the Linnaean Society*, **8**, 91–182.

Prum, R. O. and Brush, A. H. 2003. Which came first, the feather or the bird? *Scientific American*, **288**, 84–93.

Schweitzer, M. H., Suo, Z., Avci, R., *et al.* 2007. Analyses of soft tissue from *Tyrannosaurus rex* suggest the presence of protein. *Science*, **316**, 277–280.

Shipman, P. 1998. *Taking Wing*. Simon and Schuster, New York, 336 pages.

Stewart, T. A., Liang, C., Cotney, J. L., *et al.* 2019. Evidence against tetrapod-wide digit identities and for a limited frame shift in bird wings. *Nature Communications*, **10**, 3244. https://doi.org/10.1038/s41467-019-11215-8 www.nature.com/naturecommunications

Wagner, G. P. and Gauthier, J. A. 1999. 1, 2, 3 = 2, 3, 4: a solution to the problem of the homology of the digits in the avian hand. *Proceedings of the National Academy of Sciences*, **96**, 5111–5116.

Zhang, F., Kearns, S. L., Orr, P. J., *et al.* 2010. Fossilized melanosomes and the colour of Cretaceous dinosaurs and birds. *Nature*, **436**, 1075–1078.

Zelenitsky, D. K., Therrien, F., Erickson, G. M., *et al.* 2012. Feathered non-avian dinosaurs from North America provide insight into wing origins. *Science*, **338**, 510–514.

TOPIC QUESTIONS

1. What are the diagnostic characters of modern birds that might be preserved in the fossil record?
2. Compare these characters to those present in nonavian theropods.
3. What characters point to maniraptoran theropods as good candidates for the ancestry of modern birds?
4. Why did the ancestor of birds have to have been cursorial if one goes back far enough?
5. Why is it that we no longer think feathers evolved for flight? Does this mean that they are unrelated to flight?
6. How did bird flight evolve?
7. What were the evolutionary steps from *Archaeopteryx* to Aves?
8. What were the evolutionary steps from basal tetanurans to *Archaeopteryx*?
9. If we were unable to resolve the discrepancy between the paleontological and the embryological identifications of the fingers in a bird's carpometacarpus, would this in any way affect the hypothesis that birds are dinosaurs? Why?
10. During bird evolution, how did the wing evolve? Was a whole new wing structure required, or were all the pieces – as well as the correct proportions – there already?
11. If *Archaeopteryx* had never been found, would we be able to tell that birds are dinosaurs? Support your answer.
12. What characters suggest that considerable evolution occurred before the appearance of hesperornithiform birds?
13. What are the features of ichthyornithiform birds that indicate powerful, efficient flight?
14. Formulate a thoughtful answer to the question "What is a bird?".
15. What was the minimum number of times that powered flight must have evolved in vertebrates? Use a cladogram to explain your answer.

APPENDIX 8.1 WHAT MAKES A MODERN BIRD A MODERN BIRD?

Among living vertebrates, birds possess a remarkable and largely unique suite of diagnostic features (Figure A8.1.1). There are many; most of those discussed here are likely to be preserved in a fossil.

Feathers

All living birds have feathers – complex, distinctive structures that consist of a hollow, central shaft that decreases in diameter toward the tip. Radiating from the shaft are barbs, feather material that, when linked together along the length of the shaft by small hooks called barbules, form the sheet of feather material called the vane (Figure A8.1.1a). Feathers with well-developed, asymmetrical vanes are usually used for flight and are therefore called flight feathers. Feathers in which the barbules are not well developed tend to be puffy, with poorly developed vanes, and are called down; as we know from sleeping bags, comforters, and ski parkas, they are superb insulation.

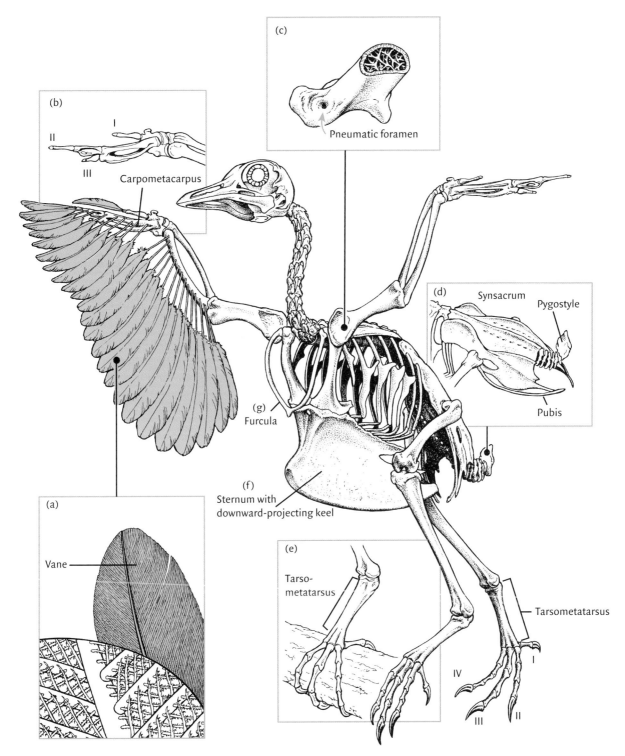

Figure A8.1.1. The skeleton of a pigeon, showing major features of its skeletal anatomy. (a) Detail of feather structure; (b) carpometacarpus with digits labeled; (c) hollow bone with pneumatic foramina; (d) synsacrum (fused pelvic bones) with pygostyle; (e) fused foot bones (tarsometatarsus); (f) sternum with large downward-projecting keel; and (g) furcula (fused clavicles).

Loss of Teeth

No living bird has teeth. The jaws of birds are covered with a rhamphotheca (beak).

Large Brains

Living birds have well-developed brains protected by a large braincase.

Carpometacarpus

The wrist and hand bones in the hand of modern birds are fused into a unique structure called the **carpometacarpus**[7] (Figure A8.1.1b). The carpometacarpus is composed of three fused fingers, generally identified as digits I, II, and III.

Legs and Feet

Birds are fully bipedal, and have an erect stance (see Chapter 4). The twin shin bones (tibia and fibula; together, the "drumstick" on the dinner table) are unequal: the tibia is large, but the fibula thins to a sliver close to the ankle.

The feet of all living birds are clawed and have three toes in front (digits II, III, and IV), and a smaller toe (digit I) at the back. The three central metatarsals (foot bones, to which the toes attach; in this case II, III, and IV) are fused together and with some of the ankle bones, to form a unique structure called a **tarsometatarsus** (Figure A8.1.1e).

Pygostyle

No living bird has a long tail skeleton. Instead, in most cases, most of the tail bones are lost, and those remaining are fused into a compact, vestigial bony structure called a **pygostyle** (*pygo* – rump; *stylus* – stake; Figure A8.1.1d). In living birds, the pygostyle is sandwiched by muscle and fatty tissue, the rectricial bulb, which allows movement of the tail feathers (so-called rectricial feathers); this tissue rejoices under the name of "Pope's [or Parson's, depending upon the Christian sect involved] Nose" at Thanksgiving time in the United States.

Pneumatic Bones

Living birds breathe unidirectionally with a complex system of air sacs (see Box 7.1). Their bones are thus pneumatic and have pneumatic foramina (Figure A8.1.1c).

Rigid Skeleton

Bird skeletons have undergone a series of bone reductions and fusions to produce a light, rigid platform to which the wings and the muscles that power them attach. Fused vertebrae in the back are connected with a well-developed breastbone, or sternum, by ribs with upper and lower segments. The sternum is large and, in flapping flyers, has a broad, deep **keel**, or downward-protruding bony sheet, for the attachment of flight muscles (Figure A8.1.1f). The pelvic region of the vertebral column is fused together into a **synsacrum**, a single structure consisting of many sacral vertebrae fused together (Figure A8.1.1d). The pubis is very slender and points posteriorly.

---

[7] Spiced and served with beer, we call them "buffalo wings"!

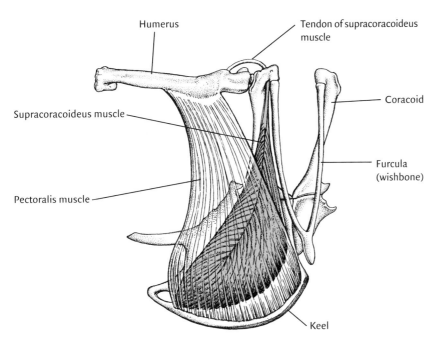

Humerus

Tendon of supracoracoideus muscle

Supracoracoideus muscle

Coracoid

Pectoralis muscle

Furcula (wishbone)

Keel

Figure A8.1.2. The two major muscles for flight: the pectoralis and the supracoracoideus. The pectoralis is the muscle used in the downward (power) stroke, while the supracoracoideus is used in the recovery stroke.

In the shoulder, pillar-like coracoid bones buttress against the front of the sternum, the shoulder blade (scapula), and against the furcula (Figure A8.1.1g). No *living* organism except birds has a furcula, although of course they are present in all neotheropods (e.g., most!).

Flight Musculature

In modern flying birds, the downward stroke of the wing is obtained by the pectoralis muscle, which attaches to the front of the coracoid and sternum, and to the furcula and humerus. The recovery stroke is carried out by the supracoracoideus muscle. The supracoracoideus attaches at the keel of the sternum, runs up along the side of the coracoid bone, and attaches via a tendon at the top of the upper arm bone through a hole (the triosseal foramen) formed by the coracoid, furcula, and scapula (Figure A8.1.2). This is an adaptation unique to living birds.

Figure 9.1. *Diplodocus*, one of the best-known sauropodomorphs, from the Late Jurassic of the Western Interior of the USA.

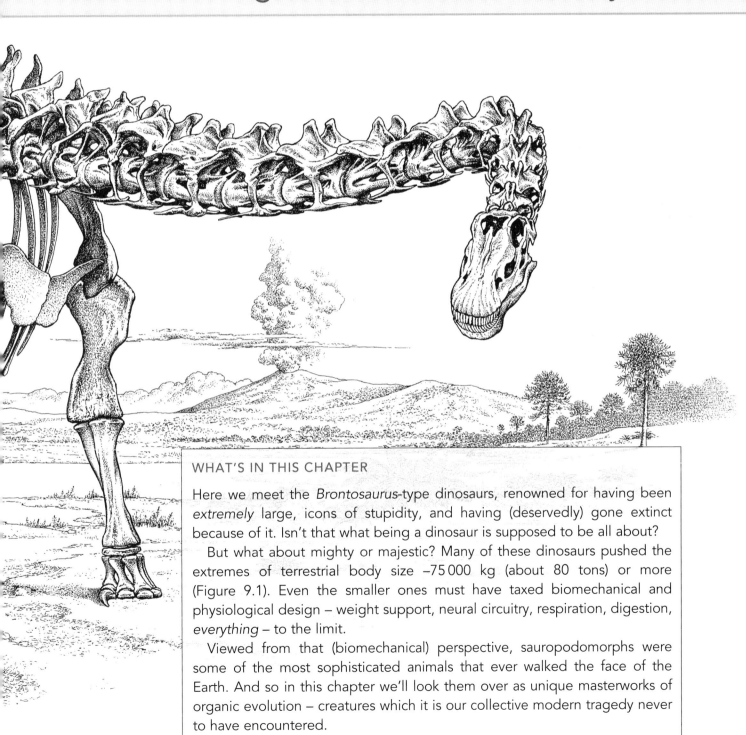

Chapter 9

Sauropodomorpha
The Big, the Bizarre, and the Majestic

WHAT'S IN THIS CHAPTER

Here we meet the *Brontosaurus*-type dinosaurs, renowned for having been *extremely* large, icons of stupidity, and having (deservedly) gone extinct because of it. Isn't that what being a dinosaur is supposed to be all about?

But what about mighty or majestic? Many of these dinosaurs pushed the extremes of terrestrial body size −75 000 kg (about 80 tons) or more (Figure 9.1). Even the smaller ones must have taxed biomechanical and physiological design – weight support, neural circuitry, respiration, digestion, *everything* – to the limit.

Viewed from that (biomechanical) perspective, sauropodomorphs were some of the most sophisticated animals that ever walked the face of the Earth. And so in this chapter we'll look them over as unique masterworks of organic evolution – creatures which it is our collective modern tragedy never to have encountered.

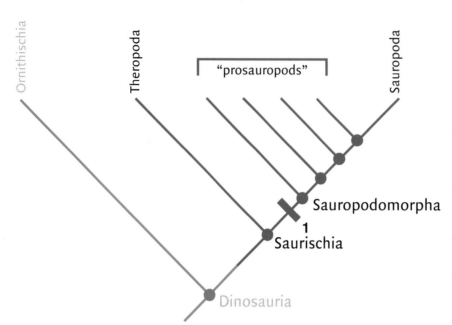

Figure 9.2. The current consensus of the relationships of "prosauropods" and Sauropoda. Derived characters include: at **1**, relatively small skull (about 5 percent body length), deflected front end of the lower jaw, elongate lanceolate teeth with coarsely serrated crowns, at least ten neck vertebrae that form a very long neck, dorsal and caudal vertebrae added to the front and hind ends of the sacrum, enormous thumb equipped with an enlarged claw, a very large obturator foramen in the pubis, and an elongate femur.

Sauropodomorpha

Sauropodomorphs (*saurus* – lizard; *pod* – foot; *morpho* – form) lived for more than 160 million years, from the very beginning of nonavian dinosaur history until its dramatic close. Over this long interval, sauropodomorphs managed to get to every continent (Figure 9.2), and spawned well over a hundred different species (with new ones being discovered all the time). They must have been doing a lot right.

Who are Sauropodomorphs?

Sauropodomorpha is a well-diagnosed group of saurischian dinosaurs (Figure 9.2). The dinosaurs that look like *Brontosaurus* – Sauropoda – are but one part of Sauropodomorpha; the other consists of a relatively short-lived group of primitive dinosaurs: basal sauropodomorphs called **prosauropods** (*pro* – before; see Figures 4.5 and 9.2).

While specialists have historically had some trouble figuring out what to do with "prosauropods," current thinking has them as a series of successively more derived, early sauropodomorphs (Figure 9.3a), moving closer and closer to Sauropoda. Here we use quotation marks around the term to denote the fact that although we refer to basal sauropodomorphs as "prosauropods," the group is not monophyletic.

Thinking about sauropodomorphs in terms of a formal definition, we might call Sauropodomorpha the group specified by the most recent common ancestor of *Plateosaurus* (a very primitive sauropodomorph) and *Saltosaurus* (a very derived sauropodomorph), and all of its descendants.[1]

---

[1] Another way to think about this definition might be: those sauropodomorphs closer in relationship to *Plateosaurus* than to *Saltosaurus*.

Figure 9.3. The derived prosauropod *Lessemsaurus,* from the Upper Triassic Los Colorados Formation, Argentina.

Sauropodomorphs: Among the Very First Dinosaurs

Sauropodomorphs are well represented among the very earliest dinosaurs (see also Chapter 5); indeed *Eoraptor* (Figure 5.7a), the earliest dinosaur known, is now thought to have been a sauropodomorph. They appeared early, and quickly diversified. The changes that they underwent were almost progressive: starting with a bipedal ancestor, heads became smaller, necks became longer, and, as we have seen, hind legs became closer in length to that of the front legs: the trend in these animals seems to

(a)

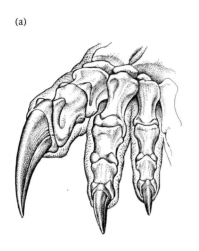

(b)

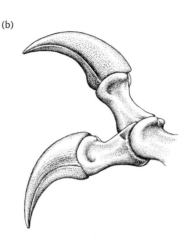

Figure 9.4. Left hand of the "prosauropod" dinosaur *Plateosaurus*, showing its well-developed thumb claw: (a) reconstructed hand; (b) thumb showing amount of movement permitted by skeleton.

have been becoming quadrupedal (see Figure 9.3). Paralleling this were increasing steps towards an unquestionably herbivorous lifestyle.

The link between herbivory and "prosauropod" evolution is underlined by the fact that the history of basal sauropodomorphs parallels the rise of gymnosperms – seed-bearing plants (see Figure 15.9). That is, as gymnosperms – particularly tall ones – became an important component of the land plant biota, "prosauropods" themselves became larger, and, as a group, constituted an increasingly important component of the terrestrial vertebrate fauna. "Prosauropods" were the first land creatures ever to take advantage of tall-growing plants.

"Prosauropods"

"Prosauropods" were a collection of relatively primitive dinosaurs with small heads, long necks, barrel-shaped bodies, and long tails, known from the Late Triassic through Early Jurassic, from all continents except Australia. In general, the front limbs were somewhat shorter than the hindlimbs, and all had five digits. "Prosauropod" hands were equipped with a large, half-moon-shaped thumb claw (Figure 9.4). Whether for food procurement, defense, and/or some unspecified social activity, the function of this claw remains unknown.

"Prosauropod" Lives and Lifestyles

Feeding

In the mood for food, sure, but which? The skulls show almost none of the advanced design features associated with chewing (see Part III Ornithischia); however, the jaw joint is slightly lower than the tooth row (Figure 9.5), a characteristic sometimes associated with chewing. The teeth are generally separated, leaf-shaped (Figure 9.6), and reveal few grinding marks, suggesting puncturing as the dominant tooth function.

Although paleontologists have traditionally considered sauropodomorphs to be herbivores, carnivory has been suggested in some of the basal forms because primitive sauropodomorph skulls, teeth, and general body proportions lack herbivore specializations. Yet, supporting herbivory in "prosauropods," the skull is proportionately smaller than that generally seen in carnivores. Recent treatments of the group split the difference, calling them predominantly herbivores that might have enjoyed an occasional meaty snack.

Once the food was past the mouth, grinding likely took place via gastroliths – which have been found in association with some "prosauropod" skeletons – and perhaps by stomach fermentation, to judge from their barrel-shaped torsos (see Chapter 10).

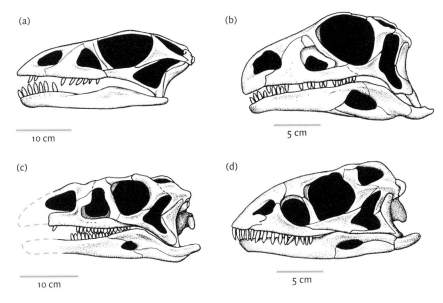

Figure 9.5. Left lateral view of the "prosauropod" skulls of (a) *Anchisaurus*, (b) *Coloradisaurus*, (c) *Lufengosaurus*, and (d) *Yunnanosaurus*.

(a)

10 cm

(b)

5 cm

(c)

10 cm

(d)

5 cm

Need for Speed?

In the most primitive of "prosauropods," the forelimbs are shorter than the hindlimbs and the trunk region is relatively short, suggesting that these animals walked principally on their hindlimbs rather than on all fours. Indeed, *Eoraptor* was surely a biped. However, the largest and most derived "prosauropods" (among them *Lessemsaurus* and *Melanorosaurus*; Figures 9.3 and 9.8) appear to have become fully quadrupedal. The early history of locomotion in Sauropodomorpha is consistent with the primitive condition for all dinosaurs: bipedality (see Chapters 4 and 5). This interpretation is supported by evidence from trackways as well; the trackway genus *Otozoum*, generally attributed to "prosauropods," is all but exclusively bipedal.

"Prosauropods" appear to have been quite slow. Calculations from trackways (see Chapter 13) suggest speeds of no more than 5 kmh, about the average walking speed of humans. Of course, those are just the tracks that they left one Sunday (?) back in the Late Triassic; in fact, their top speeds were likely much faster.

Socializing

Very little is known of "prosauropod" social habits. The famous *Plateosaurus* bonebeds in Germany and Switzerland, as well as others elsewhere, however, hint at herds; indeed, herds of prosauropods migrating across the European continent were proposed as early as 1915.

Detailed analyses of *Plateosaurus*, *Thecodontosaurus*, and *Melanorosaurus* (prosauropods for which large numbers of individuals are known) reveal sexual dimorphism in skull dimensions and in thigh bone (femur) size. Sexual dimorphism tends to be more pronounced in highly social animals, and thus there may be a connection between the likelihood of herding and sexual dimorphism.

Eggs, Nests, and Babies

Eggs and nests are known for the "prosauropods" *Mussasaurus* (Argentina) and *Massospondylus* (South Africa). The eggs were very thin-shelled; clutches tended to be small by dinosaur standards (about ten eggs), and the hatchlings proportionately large-headed, toothless, and quite small. Adult "prosauropods" are roughly 500 to 1000 times larger than the hatchlings, so the formative years of basal sauropodomorphs must have been times of radical growth and reorganization.

In the case of *Mussasaurus*, because hatchlings, juveniles, and adults are known, the changes undergone by young mussasaurs as they faced adult dinosaurhood have been investigated by Argentinian paleontologist A. Otero and colleagues (2019). First, they recognized three different levels of development based upon morphology: hatchling; juvenile, and adult. Then, they determined

(a)   (b)   (c)

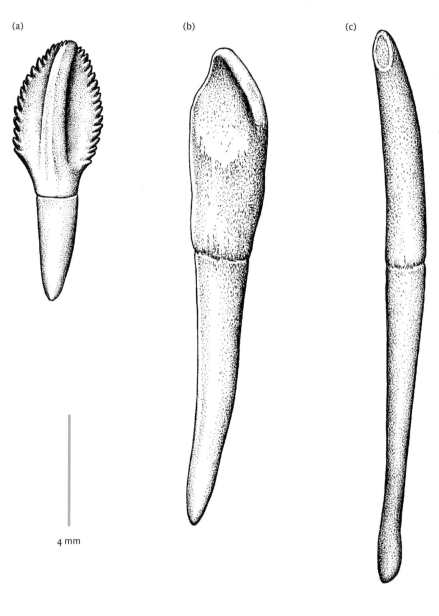

4 mm

Figure 9.6. Teeth in selected sauropodomorphs. (a) Leaf-shaped "prosauropod" tooth of *Plateosaurus*; (b) spatulate tooth of sauropod *Camarasaurus*; (c) pencil-like tooth of *Diplodocus*. The lower part of each tooth is the root.

the ages of the hatchling; juvenile, and adult to be 0, about 1, and 8 years or older, respectively. At birth, these dinosaurs weighed about 60 g; after 1 year, they weighed about 7 kg (15.5 lb), and as adults they weighed more than 1000 kg (approximately 2205 lb). Talk about a pre-teen growth spurt (about 143 kg [95 lb])/year)! Exactly how all this occurred metabolically is unclear (see Chapter 13), but crazy rapid growth rates are clearly indicated.

In this case, some other interesting observations were made. The hatchling likely began as a quadruped, forelimbs only slightly shorter than hind limbs. As the animal grew, the proportion of the tail to the total body weight (and size) increased, while that of the neck decreased. This caused the center of gravity to move from up in the chest, to just ahead of the pelvis, facilitating the ultimate adult bipedal stance (Figure 9.7).

"Prosauropods" were not particularly common beasts and did not hang around on Earth for a very long time, all dutifully going extinct in the Early Jurassic. Still, as the first tall-browsing herbivores, they represent the first appearance on Earth of the modern ecosystem that is with us today. Our hope

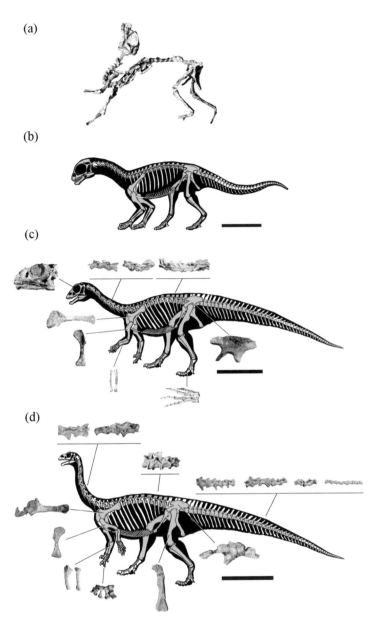

(a)

(b)

(c)

(d)

Figure 9.7. Growth in *Mussasaurus*. The hatchling and yearling show typical juvenile features (large head; large eyes, proportionately larger legs and feet). Aging involved changing the proportion of the neck and tail, and an associate shift of the center of gravity, all contributing to a bipedal stance. From Otero, A., Cuff, A.R., Allen, V., *et al.* 2019. Ontogenetic changes in the body plan of the sauropodomorph dinosaur *Mussasaurus patagonicus* reveal shifts of locomotor stance during growth: *Nature Scientific Reports*, **9**,7614. https://doi.org/10.1038/s41498–019-44037-1

is that, with more attention and finds, the future will bring more insights into this enigmatic and basal group of dinosaurs.

Sauropoda

Sauropods are the familiar *Brontosaurus*-type dinosaurs: quadrupedal, herbivorous, small-headed, long-necked and long-tailed. Although there has been considerable discussion about relationships within sauropods, most specialists would be comfortable defining Sauropoda itself as the monophyletic group comprising the most recent ancestor of *Vulcanodon*, *Saltosaurus*, and all of its descendants (Figure 9.8).

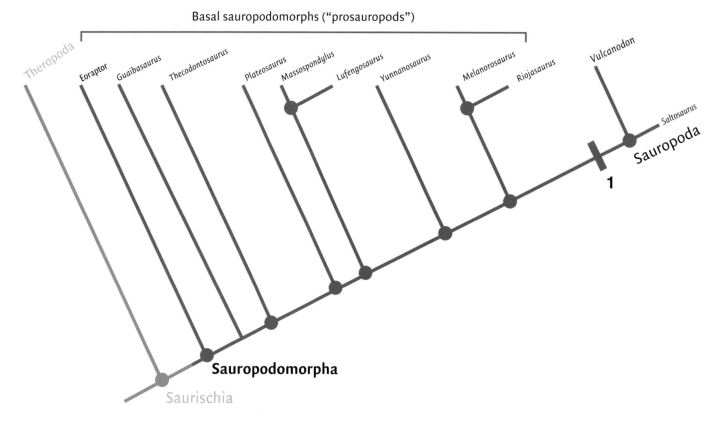

Figure 9.8. Cladogram of basal sauropodomorphs ("prosauropods") showing successive phylogenetic proximity to the clade Sauropoda. Derived characters of Sauropoda include, at **1**, special laminar system on forward cervical vertebrae, forelimb length greater than 60 percent hindlimb length, triradiate proximal end of ulna, subrectangular distal end of radius, length of metacarpal V greater than 90 percent that of metacarpal III, compressed distal end of ischial shaft, reduced anterior trochanter on femur, femoral shaft elliptical in horizontal cross-section, tibia length less than 70 percent femur length, metatarsal III length less than 40 percent tibia length, proximal end surfaces of metatarsals I and V are larger than those of metatarsals II, III, and IV, metatarsal III length is more than 85 percent metatarsal V length.

Design

Even if you begin with a basal sauropodomorph, getting *really* big takes some major evolution, and many sauropods were really huge dinosaurs. Yet, the sophisticated basic sauropod design, once it appeared, remained unique and only underwent fine tuning during their approximately 140 million years on Earth (Figure 9.9). It obviously worked!

The skulls of most sauropods were highly derived, in keeping with the rest of the animal. Although the tooth row was not deeply inset (indicating an absence of cheeks; see Part III Ornithischia), as one sees in mammalian and ornithischian herbivores, the teeth varied and, depending upon the sauropod, had simple crowns and were triangular, spatulate, or slender and pencil-like (see Figure 9.6). In some forms, there was a tendency to limit the teeth to the front of the jaws (Figure 9.10). From this we infer that what sauropods did with plants (besides digest them!!) must have varied; a mouth full 'o teeth like *Camarasaurus* (Figures 9.6b and 9.10c) must have handled some mighty tough (conifer?) vegetation, shearing it unceremoniously off the plant; while the almost delicate, pencil-like teeth restricted to the front of the jaw (*à la Diplodocus* – Figures 9.6c and 9.10d) would have been used for

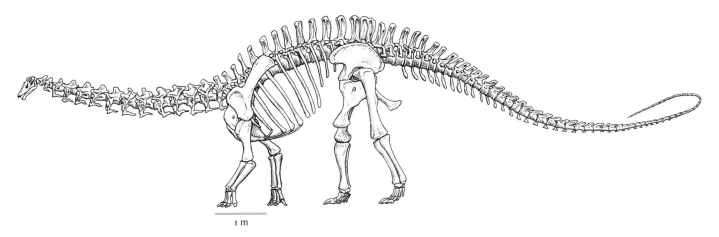

1 m

Figure 9.9. Left lateral view of the skull and skeleton of *Apatosaurus*.

stripping leafy vegetation – and then sending it back, essentially whole, to be processed in the stomach. Like their ornithischian brothers in herbivory, some advanced sauropods likely inclined towards some chewing (as we will see in Part III Ornithischia, chewing is a highly specialized behavior which most vertebrates don't do – even if mammals happen to!). A few sauropods may have had rudimentary cheeks; there is evidence for tooth occlusion in many, especially primitive forms, and in at least one case (*Nigersaurus*) there is even evidence for the development of a distinctive block of teeth: a kind of dental battery (see introduction to Part III Ornithischia and Chapter 12). As we shall see, these are all features that are characteristic of animals that chew. But, it is fair to say that in the main, sauropods were not chewing animals, and developed other means to digest their food once it was gotten off the plant.

Sauropod skulls tended to be delicately built, with large openings in the skull roof (Figure 9.10). They range from the relatively primitive (*Shunosaurus*, Figure 9.10a) to the more derived (such as *Brachyosaurus*, Figure 9.10b) and *Diplodocus* (Figure 9.10d). The external nares, instead of residing at the tip of the snout, had an as-yet unconvincingly explained phylogenetic tendency to migrate backward, toward the top of the head (Figure 9.11); this has led to speculation about whether they are located in that position to facilitate drinking at the end of a long neck; or whether perhaps although the nares are retracted towards the back of the skull, the soft-tissue nasal openings were farther forward, and were connected by soft-tissue canals to the narial openings (conveniently leaving no remnants for muddled sauropod paleontologists). But there must have been something special about the whole arrangement, because not only do the narial openings tend to be removed from the tip of the snout in most sauropods, but highly derived sauropods like *Nemegtosaurus* (Figure 9.10e) appear to have convergently evolved some of the same adaptations as *Diplodocus*. The skulls appear absurdly tiny (even if they are more than a meter long in a medium-to-large sauropod!) – until you realize that only an idiot would design a large, heavy skull at the end of an extremely long neck.

That "extremely long neck" was made up of a complex and very sophisticated system of girders and air pockets that maximized lightness and strength (Figure 9.11a). Unlike the vertebrae in mammals, most sauropod vertebrae have an internal system of pleurocoels, as well as pneumatic foramina (see Chapter 7 [Tetanurae] and Chapter 13). These same features are found in birds (both Mesozoic and modern) and many other theopods, all of which augment the volume of air in their lungs by a complicated series of interconnected auxiliary air sacs (see Figure 13.2). By analogy with

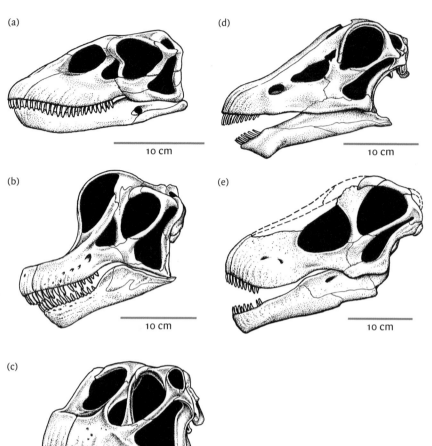

Figure 9.10. Left lateral view of the skull of (a) *Shunosaurus*, (b) *Brachiosaurus*, (c) *Camarasaurus*, (d) *Diplodocus*, and (e) *Nemegtosaurus*. Arrows indicate position of nares.

modern birds, auxiliary air sacs in sauropods reached (and expanded into) the internal cavities via the pneumatic foramina (Figure 9.11b).

Also distinctive in sauropods were the Y-shaped neural arches on the vertebrae. These held the **nuchal ligaments**, elastic ropes of connective tissue that ran down the back of the animal and supported the head and neck, so that it was not held up exclusively by muscles (Figures 9.11a and 9.12). Paleoartist Mark Hallett and sauropod specialist Mathew Wedel, in their wonderful book, *The Sauropod Dinosaurs* (2016), suggest that the ligament was actually a pair of ligaments, particularly in those sauropods that reared up (see below), providing strength, control, and stability. In their rendering, sauropod necks would have maintained a gentle "S"-shaped curvature, angling at a low angle away from the trunk, gaining a bit of elevation, and with the head, and perhaps several of the neck vertebrate closest to the head, turning back downwards as in Figure 9.12. The nuchal ligaments would have allowed the animals to maintain this curvature.

Sauropods were quadrupeds, having, as we have seen, secondarily evolved their stance from their bipedal, basal saurischian ancestors. The limbs were pillar-like, and would have done a Greek temple proud (Figure 9.13). The bones are composed of denser material than that found in the upper parts of the skeleton, an adaptation locating the weight and strength in the skeleton where it was most needed.

(a)

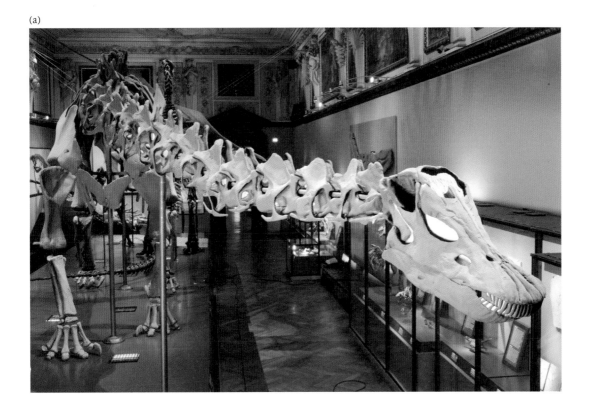

(b)

Figure 9.11. The neck of the *Apatosaurus* looking at its left side. (a) The complicated girder-like structure of the zygapophyses was used as a framework for muscle and ligament attachment, used to support the neck. Bifurcated neural spines indicate the positions of the paired nuchal ligaments along the neck. (b) In-pocketings in two vertebrae, marking the location of the pneumatic foramina and pleurocoels. White blobs are plaster filling pleurocoel entrances; not all pleurocoels, however, are plaster-filled. The long thin, bony projections at the base of each vertebrate are cervical ribs.

The hindlimbs articulated with an immense, robust pelvis. The connection between the thigh and the pelvis, the hip socket, is quite large and, of course, perforate (on display in Figure 5.9). The ilium is thick and robust; and not uncommonly a ridge on the ilium sits above the top of the thigh bone (femur). The large opening would have contained a chunk of cartilage that fit the opening in the pelvis, and formed a strong structure against which the femur articulated. In short, the whole arrangement was built for strength at the pelvis; the mighty central core of the animal.

Figure 9.12. Nuchal ligament region (shown in solid blue) running from the head, down the neck, and beyond.

The forefeet (the "hands," as it were, on the forelimb) were **digitigrade**, which means that the animal stood on its fingertips. The fingers were arranged in a nearly symmetrical horseshoe-shaped semicircle, and the first digit (the thumb) carried a large claw (Figure 9.14a). By contrast, the hindfeet were semi-**plantigrade**, which means that the animal's weight was supported along the lengths of its toe bones.[2] The foot was asymmetrical, and generally had three large claws (on digits I, II, and III; Figure 9.14b). Sometimes the trackways reveal the impression of a fleshy heel pad that nestled behind the claws of both the fore- and hindfeet, and supported the body as well (Figure 9.15). Sauropod footprints are immense, a single print not uncommonly spanning as much as 1 m (about 3')!

The whole thing has been likened to a kind of suspension bridge; four massive supporting pillars (the legs) from which the neck and head were suspended in front, and a long tail was suspended to the rear. Suspension of the neck would have been carried out by the nuchal ligaments described above; **caudal** (tail) ligament would have supported the tail from above, in both cases connecting to various projections from the vertebrae. From below, in the neck, long, thin extensions – actually **cervical** (neck) ribs – would have served as ligamentous attachments, supporting and connecting each vertebra. The cervical ribs are visible in Figure 9.11b.

Despite the fact that the trunk of sauropods was relatively broad, most sauropod trackways tend to be quite narrow, with the feet aligned toward the midline of the body. Most significantly, trackways never include a tail-drag mark, providing strong evidence that many sauropods carried their immense, whip-like tails – as long as 15 m (49') in the longest cases – entirely off the ground (Figure 9.15).

For many years, reconstructions commonly showed sauropods as swamp-dwellers, their great bulk buoyed up by water. In this way, so the story went, they could have remained deeply submerged, breathing with only their high nostrils poking out of the water, *à la* crocodile; chewing soft, water-logged swamp vegetation with their "weak" teeth. But sauropods don't exactly look like crocodiles. With four powerful pillar-like limbs, what is it about sauropods that says "amphibious"? In fact, close study of sauropod anatomy gives no evidence that sauropods whiled away their palmiest days in Mesozoic swamps, their great bulk buoyed up by the water. Our best evidence, supported by biomechanical studies of sauropod limbs, is that the dense-boned, massive, and pillar-like limbs were designed for fully terrestrial locomotion.

Thoughts of a Sauropod

Sauropods did not have obvious pretensions to deep thought, so this section will of necessity be brief. The head is proportionally small on the body, and the brain is proportionally small within the head;

---

[2] For reference, humans are supported along the lengths of their toe and foot bones, and are thus fully plantigrade.

the fact is that sauropods had the smallest brains for their body size of any dinosaur. Brain size is itself not a meaningful measure on intelligence; yet, using other, more sophisticated measures of estimating intelligence from brain morphology (see Box 11.1), sauropods do not evince great intellect. Yet their long, successful record of survival speaks volumes; their brains clearly did the basic functions that a brain is expected to do: coordinate; control; react to a wide range of stimuli. As we shall see, their behavioral repertoire may have been more sophisticated than a human might expect for an animal with proportionally so small a brain.

Lives of the Huge and Ancient

A Place to Roam

When we find the remains of these magnificent animals, they come from a myriad different environments, from river floodplains to sandy deserts. At Tendaguru (see Box 16.6) in southeastern Tanzania, as well as in the USA in northern Texas at the famous Glen Rose trackway sites, and in Maryland where the remains of the brachiosaurid *Astrodon* have been uncovered, there is strong evidence that these environments were once close to the sea and quite humid. Perhaps these were some of the conditions that sauropods found most congenial. Or perhaps they are just places where trackways preserve well, because gorgeous sauropod fossils have also been recovered from Argentina's Lower Cretaceous Neuquén, Mongolia's Upper Cretaceous Nemegt, and the United States' upper Morrison formations; riverine settings all. To judge from their global distribution, sauropods could live and thrive in a wide range of environments.

Reaching for the Stars

The unique sauropod design fairly screams for browsing in tall Mesozoic conifer (see Chapter 15) forests, and the superficial similarity of sauropod necks to those of giraffes, only reinforces this inference. Thinking superficially for just a moment longer, we can divide sauropods into three informal "types" – and these are *not* monophyletic groups – to gain insight into how they might have behaved. "Type I" sauropods are those with extremely long necks and tails; in general, elongate in body and limbs. *Apatosaurus* (Figure 9.9) epitomizes this body type.

Figure 9.13. The right forelimb of *Apatosaurus* on display in the imperial surroundings of the Naturhistorisches Museum Wien.

"Type II" sauropods are represented by brachiosaurs, in which not only was the neck very long, but the front limbs were longer than those in the rear (Figure 9.16). With this extra boost, its head could potentially have been raised to as high as 13 m. We'll lump other sauropods together as "Type III";

(a)

(b)

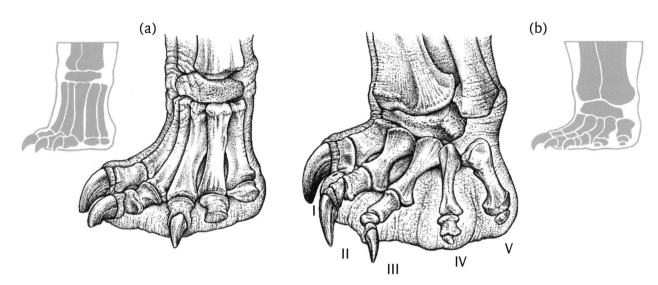

Figure 9.14. Sauropod left (a) forelimb and (b) hindlimb. The "hand" is far more digitigrade than the foot, which is nearly plantigrade (as shown in the limb cross-sections).

Figure 9.15. Five parallel trackways of Late Jurassic age, Morrison Formation, Colorado, USA. Tracks are thought to have been made by diplodocids walking alongside each other. Notice how close to each animal's midline each of the tracks is and the absence of any mark made by the tail.

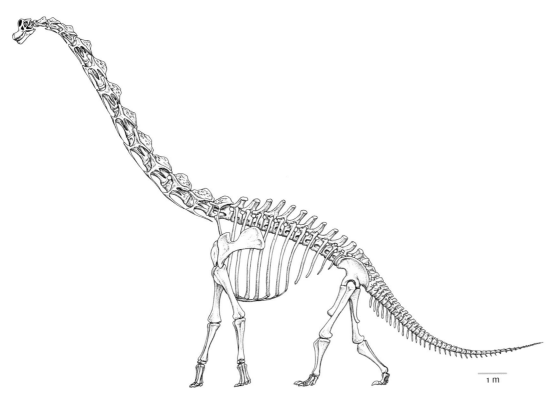

Figure 9.16. Left lateral view of the skull and skeleton of *Brachiosaurus*.

most would have been more stockily built, and not capable of tree-top feeding. Because the question here is about feeding by high-browsing, we'll not concern ourselves with the stockier, shorter types.

So it is pretty clear what "Type II" sauropods did: they used their long front legs to gain a substantial boost in elevation for the neck and head, allowing the animal to browse, with its long neck, high in the tree tops. But what of "Type I" sauropods?

Recent reconstructions of "Type I" sauropods, indicate that the head was generally held at or near the height of the shoulder, and that a giraffe-like, vertically oriented neck was, at best, a strain, and, at worst, not physically possible. It is very likely that most sauropods held their heads at, or very near, the level of the body and browsed downward. If this is true, why develop such extraordinarily long necks and tails in the first place?

But suppose these sauropods could adopt a tripodal stance, in which the animal reared back, pivoting at its pelvis, and the long, powerful tail acting as a kind of third leg (Figure 9.17)? By angling the body upwards from the pelvis, this would not require the neck to angle abruptly upwards from the body, and it would allow a 37 m (approximately 120') animal to reach very high into the trees indeed, secure in the length and strength of its tail acting as a third leg. A bit of evidence for the

Figure 9.17. A sauropod in reconstructed in a tripodal posture, using the tail as a "third leg."

possibility of the tripodal stance comes from the pelvis of these sauropods; many had longish neural arches on a particularly robust pelvis; the neural arches would have supported heavy musculature required to stabilize a head raised into tall treetops; the more robust pelvis would have functioned as a stable base for the tripod. Regardless, the possibility of a tripodal stance in sauropods remains a contentious subject, with some paleontologists questioning whether the hindquarters of the animal – long tail, robust pelvis, or no – could support the kind of weight and stress that rearing up would entail. Advocates respond with spectacular museum mounts like that shown in Figure 1.13. Our view is that all sauropods fed quadrupedally some of the time, but in case of those sauropods possessing a "Type I" morphology, the likelihood that they adopted a tripodal stance now and again – perhaps even as often as they did not – is great.

In high-browsing sauropods, if the head was truly perched so high, its brain would have towered about 8 m above its heart. So, for example, to push blood through the arteries up an 8.5 m long neck, the heart of a brachiosaur must have pumped with a pressure exceeding that known

Figure 9.18. Systolic blood pressures compared: (a) a sauropod (approximately 630 mm), (b) a giraffe (320 mm), and (c) a human (150 mm).

in any living animal – indeed, double that of a giraffe. It would take a very powerful heart – some estimate one weighing as much as 400 kg – to do the pumping (Figure 9.18). How fine capillaries in the brain might have withstood such pressures is again a matter for speculation. Perhaps strong valves maintained blood pressures up those long necks; these functioning in tandem might decrease the loads on the hearts. Here, unfortunately, the fossil record falls silent.

None of these considerations addresses yet another challenge posed by long-necked life: the extraordinary amount of unused, wasted, air contained in the neck if sauropods simply breathed in and out, bidirectionally, as do mammals and most tetrapods. If, however, sauropods used an avian style of respiration, with auxiliary air sacs (for which, as we have seen, there is significant evidence), almost all of the oxygen could be extracted from the inhaled air, and the problem of the amount of unused air contained within the neck would be diminished (Chapter 13; see also Figure 9.11).

With sauropods as well as theropods possessing pneumatic bone and pleurocoels (see above), this apparently highly derived character in modern birds (see Figure 13.2) might actually be a basal saurischian characteristic – reinforcing our point from Chapter 7 that birds, so apparently derived in the modern fauna, contain a lot of characters unexpectedly shared by their long extinct, "primitive" relatives.

Considerations of blood pressure, heart size, lung capacity, and breathing style remind us of how little we know about the way sauropods really functioned, but that in these respects at least, sauropods obviously were a very highly evolved group of animals.

Feeding

Tooth form and especially tooth wear indicate that sauropods stripped foliage, unceremoniously delivering a succulent vegetative bolus to the gullet, variably chewed, depending upon the sauropod. Some sauropods, such as *Camarasaurus* and brachiosaurs, had short snouts and proportionately deeper skulls with sturdy teeth, suggesting a more powerful bite. This idea is reinforced by considerable tooth wear that is preserved in these forms. By contrast, the longer-snouted diplodocoids, with their thin, pencil-like teeth restricted to the front of the jaws, may have stripped leafy plant material from the tree tops, and left the heavy food processing and bacteria in the stomach (see below). Diplodocoids evidently had some ability to move the lower jaw fore and aft relative to the skull (humans can do this too!) which may have contributed to the stripping process. Here, too, the well-developed thumb claw may have played a role, stripping vegetation off plants into bite-sized veggie filets. Swallowing with powerful neck musculature sped the bolus of food down its long journey through the esophagous (tube leading from the throat to the stomach) whereupon it entered the stomach.

The sauropod cross-section is wide and surely accommodated a capacious gut, even considering the forward-projecting pubis (in contrast to all ornithischians, which rotated the pubis rearward to accommodate an enlarged gut; see introduction to Part III Ornithischia). Sauropods likely had an exceptionally large fermentation chamber (or chambers) that would have housed endosymbionts; that is, bacteria that lived within the gut of the dinosaur. The endosymbionts would have chemically broken down the cell walls of the plant food, thereby liberating whatever nutrition was to be gotten. Considering the size of the abdominal cavity in sauropodomorphs, these animals probably fed on foliage with high fiber content (see Chapter 15); perhaps they also had slow rates of passage of food through the gut in order to extract a maximum of nutrients from such low-quality food. We can only conclude that these huge animals, with their comparatively small mouths, must have fed constantly to acquire enough nutrition to maintain themselves. The digestive tract of a sauropod had to have been a nonstop, if low-speed, conveyor belt.

Locomotion

The top speeds of *Brachiosaurus*, *Diplodocus*, and *Apatosaurus* have been calculated to be between 20 kmh and 30 kmh, a reasonable clip for animals the size of a house and weighing in excess of three to ten elephants. They undoubtedly walked a good deal more slowly most of the time, perhaps at rates of 20 to 40 km per day, as calculated from sauropod trackways (see Figure 9.15 and Chapter 13).

Hanging with the Big Dogs

One but needs to see the parallel trackways shown in Figure 9.15 to know that sauropods maintained some social relationships. These likely varied from large herds to small pods, to, perhaps, a solitary existence.

The many mass accumulations in the Morrison Formation in the USA (see Figure 9.15), the Tendaguru bonebeds of Tanzania (see Box 16.6), the Lower Jurassic sauropod sites of India, and most recently the Middle Jurassic of Sichuan, China, together with vast sauropod footprint assemblages, all speak to the existence of gregariousness in some sauropods, including *Shunosaurus*, *Diplodocus*, and *Camarasaurus*. But how gregarious is gregarious? Large bonebeds imply large herds. But just as we've seen that sauropod morphologies are not identical, sauropods likely did not socialize identically either. So in the cases of *Brachiosaurus* and *Haplocanthosaurus* from the Morrison Formation of the western USA, and *Opisthocoelicaudia* from Mongolia, isolated finds suggest that they were not so numerous and may thus have lived more solitary existences,

The image of large herds raises a separate issue: as we have seen, sauropods were very large animals that must have been eating constantly. If that is true, then large herds likely wreaked severe damage on local vegetation, both by stripping away all the foliage they could reach and by trampling into the ground all of the shrubs, brush, and trees that got in the way. The inference is that, in time, large herds would have depleted their food sources and needed to move on. Potentially, these sauropods would have migrated, therefore, to keep the fires stoked. And by comparison, the more solitary sauropods, neither functioning in herds, nor needing to keep traveling far for sustenance, may not have been migratory animals – or at least their migration was not driven by these factors. Other than these rough generalities, we know little about how sauropods communicated within herds, who ran the show, or what might have caught a sauropod's wayward eye.

Defense

Although breathless speculation over sauropod defense can be found on the internet, ultimately – and obviously – size must have been the best deterrent against an attack; since, depending upon the sauropod, they were between 50 percent and 300 percent larger than known coexisting predatory dinosaurs. Living in herds, healthy animals must have been nearly invulnerable. The tail was flexible and whip-like, and could have been employed to dust off would-be predators. In the case of an attack from a pack, both size and the tail may have been brought into play (but see Chapter 6 for how predators might have handled sauropod prey). Sauropod claws were not sharp like those of carnivorous theropods, but they could doubtless inflict significant damage, particularly when backed by the full weight and power of an adult sauropod. Like many anatomical features, sauropod claws likely served multiple functions. Were older and juvenile sauropods the most vulnerable, to be culled from the herd and set upon by heartless, drooling large theropods? Probably; but the young and the old are always the most vulnerable groups in most tetrapods. In short, beyond what are common-sense observations, we have very little insights on how sauropods protected themselves.

Growth and Development

For along time, we knew next to nothing about sauropod nesting and, indeed, even in the early 1990s it was seriously proposed that sauropods gave birth to live young (which, if it were true, would have made them unique among dinosaurs).

Sauropod eggs began to pop up as paleontologists looked for them and, in 1997, a stunning sauropod nesting ground was discovered in Patagonia. This site, known as Auca Mahuevo ("Auca more eggs"), consists of a massive nesting ground covering more than a square kilometer and littered with tens of thousands of large, unhatched eggs. Upon further investigation, four (or six, depending upon how you count them) layers of eggs were uncovered and, in each layer, the eggs were organized into clusters of 15 to 34 linearly paired eggs, thought to represent individual nests or clutches. Most spectacularly, a high proportion of these eggs contained embryonic skeletons, some with impressions of embryonic skin (Figure 9.19)!

The geographical extent of the nesting horizons confirms at least some rudimentary herding and/or social behavior in sauropods. Clearly several enormous colonies were preserved, to which mothers likely regularly returned. There, they dug crescentic trenches and gently deposited about 30 eggs into them. Possibly, the eggs were then covered by mounds of vegetation to keep them at optimal temperature and humidity.

For several reasons, the females are thought to have left the site after laying their eggs, leaving the eggs to their fates. First, consideration of the size disparity between an adult Auca Mahuevo sauropod and the juvenile. In large, similar dinosaurs, the hatchling would have weighed about 0.5 kg (4 lb) and the adult might have weighed some 45 000 kg (about 50 US tons); in short the adult weighed about 100 000 times more than the embryo (but see Chapter 13 on dinosaur weight estimates). Second, no adult bone has ever been found with these eggs. Thirdly, nothing other than embryos has ever been found at the site; no hatchlings or very young dinosaurs, whose presence would suggest that they stuck around for a bit of parental love. Finally, the embryonic teeth have

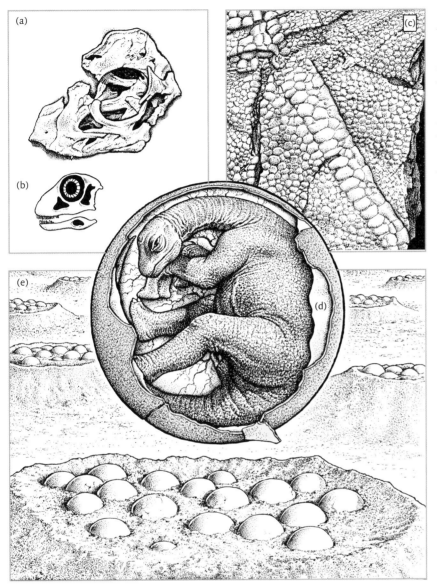

Figure 9.19. Titanosaurian remains from the Auca Mahuevo locality of Patagonia, Argentina. (a) Titanosaur skull (fossil); (b) reconstructed skull; (c) titanosaur skin (fossil) impressions; (d) reconstructed egg/embryo; and (e) schematic field of nests. The nests were not likely mounds as shown; more likely they were scooped-out troughs into which eggs were laid, and then covered lightly, perhaps with local vegetation.

small wear facets in them, not needed in the egg, but surely needed if the newly born (neonates) were on their own from the very beginning.

If our supposition is true, these sauropod babies pecked out of their shells (yes, there is evidence for an egg "tooth" in sauropods) and scampered sea-turtle hatchling style for a place to hide and food to eat. The Birth Day was probably the same as Feast Day for the local nasty theropods, but if the births were synchronous, a few of the hundreds (thousands?) of neonates might get past the theropod hazing to live (and eat and grow) another day.

Alternatively, the neonates potentially they may have been communally guarded by a few adult sauropods, who staked out the whole nesting area from its periphery; however, there is no evidence that bears on this idea, either for or against.

Since the find at Auca Mahuevo, eggs have been associated with particular sauropods in the Upper Cretaceous of southern France, Mongolia, and India, where, in 2007, a second, massive sauropod nesting ground was also uncovered. These vast clutches of eggs give us insights into dinosaurian life history strategies; that is, the ways in which particular organisms grow, reproduce, and die. One strategy, called the *r*-strategy, is to produce enormous numbers of eggs that result in thousands of offspring, the vast majority of which do not survive to reproduce during their relatively short lifespans. Think mayflies. Such offspring, born as near-adults, are called precocial. Not so much parental care here – too many children for serious parental investment.

This contrasts with the *K*-strategy, involving fewer offspring, lots of parental care, and longer lifespans. Think whales. Such offspring, born with a longer trek toward adulthood and requiring parental investment to get there, are called altricial.

Sauropods clearly laid a lot of eggs; thus, the resources expended by the parents on the offspring, as well as infant survival, may not have been so great. Sauropods evidently were *r*-strategists.

Beyond these few details, what do we know about the general aspects of sauropod reproduction, growth, and life histories? Sex in these animals undoubtedly involved a tripodal male and a quadrupedal female; however, beyond this most elemental of positions[3] all else remains speculative. For example, could the trenchant thumb claw have been used in this aspect of sauropod behavior as well? The question of the existence of a penis in male sauropods also looms large.

Once hatched, sauropodomorphs apparently grew at very high rates. New studies of the microscopic structure of sauropod bone indicate rapid and continuous rates in both prosauropods and sauropods. Studies of bone tissue (see Chapter 13) suggest that, during the time of its peak growth, *Apatosaurus* grew at the astounding rate of around 5.5×10^3 kg per year (see Chapter 13). The food consumption required to support such growth rates beggars description. Moreover, rather than imagining animals taking a leisurely 60 years to reach sexual maturity and having a longevity of perhaps 100 to 150 years, as had once been thought, the best current estimates are that it took about 20 years or less for a sauropod (and perhaps for a prosauropod as well) to become sexually mature. Similarly, lifespans for these animals were probably on the order of not much more than 30 years. Life for a sauropod may well have been in the fast lane.

The Evolution of Sauropodomorpha

Sauropodomorpha is easily diagnosed by more than a dozen derived features (Figure 9.8; see also Figure 9.2). With "prosauropods" as a grade of primitive sauropodomorphs, we move directly to Sauropoda. Sauropoda is supported by more than a dozen unique features, many of which relate to the attainment of great size and weight on land (Figure 9.8).

Sauropoda

There appears to be consensus on at least the larger contours of sauropod evolution. As currently understood, sauropods consist of several primitive taxa (among them *Blikanasaurus*, *Vulcanodon*, and *Kotasaurus*) on the one hand, and the more derived clade Eusauropoda on the other. Eusauropods are diagnosed by many features (Figure 9.20). The most primitive known member of the group, *Shunosaurus*, was a 9 m long sauropod from the Middle Jurassic of China (Figures 9.10a and 9.21). Its skull is relatively long and low, and vaguely reminiscent of the primitive sauropodomorph condition, with nostrils near the front of the snout and a mouth filled with many small and spatulate teeth.

---

[3] A detailed description of this, complete with illustrations – along with so much else sauropodish – can be found in Hallett & Wedel's (2016) book.

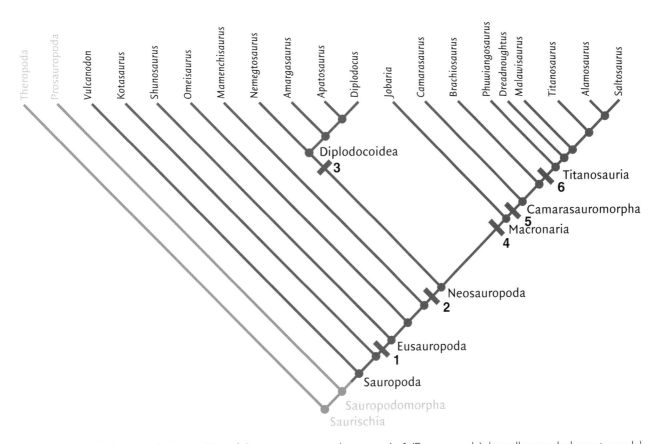

Figure 9.20. Cladogram of relationships of the major sauropod groups. At **1** (Eusauropoda), broadly rounded snout, caudal margin of external nares that extends behind the posterior margin of the antorbital fenestra, lateral plate on premaxillae, maxillae, and dentaries, loss of the anterior process of the prefrontal, frontals wider than length, wrinkled tooth crown enamel, most posterior tooth positioned beneath antorbital fenestra, at least 12 cervical vertebrae, neural spines of the cervical vertebrae that slope strongly forward, dorsal surface of sacral plate at the level of dorsal margin of ilium, block-like carpals, metacarpals arranged in U-shaped colonnade, manual phalanges wider transversely than proximodistally, two or fewer phalanges for manual digits II to IV, strongly convex dorsal margin of ilium, loss of the anterior trochanter of femur, lateral muscle scar at mid length of fibula, distally divergent metatarsals II to IV, three phalanges on pedal digit IV, ungual length greater than 100 percent metatarsal length for pedal digit I; at **2** (Neosauropoda), subnarial foramen on premaxilla–maxilla suture, preantorbital fenestra in base of ascending process of maxilla, quadratojugal in contact with maxilla, pedal digit IV with two or fewer phalanges; at **3** (Diplodocoidea), subrectangular snout, fully retracted external nares, elongate subnarial foramen, reduction of angle between midline and premaxilla–maxilla suture to 20° or less, most posterior tooth rostral to antorbital fenestra; at **4** (Macronaria), greatest diameter of external nares greater than that of orbit, subnarial foramen found within the external narial fossa; enlarged jaw muscles; at **5** (Camarasauromorpha), nearly vertical dorsal premaxillary process, splenial extending to mandibular symphysis, acute posterior ends of pleurocoels in anterior dorsal vertebrae, metacarpal I longer than metacarpal IV; and at **6** (Titanosauria), prominent expansion of rear end of sternal plate, very robust radius and ulna.

As sauropod evolution proceeded, various aspects of jaw mechanics and body form appear to have been linked. Sauropods evidently inherited a fully quadrupedal stance from their basal sauropodomorph ("prosauropod") ancestors. At the same time, there was a significant increase in body size; in fact, body sizes evidently increased throughout sauropod evolution, until the highly derived titanosaurids (see below), which in some cases appear to have been somewhat smaller.

With the advent of the great clade Neosauropoda, the snout broadened, the lower jaw strengthened, and wear facets from grinding appear on the teeth, indicating front and rearward

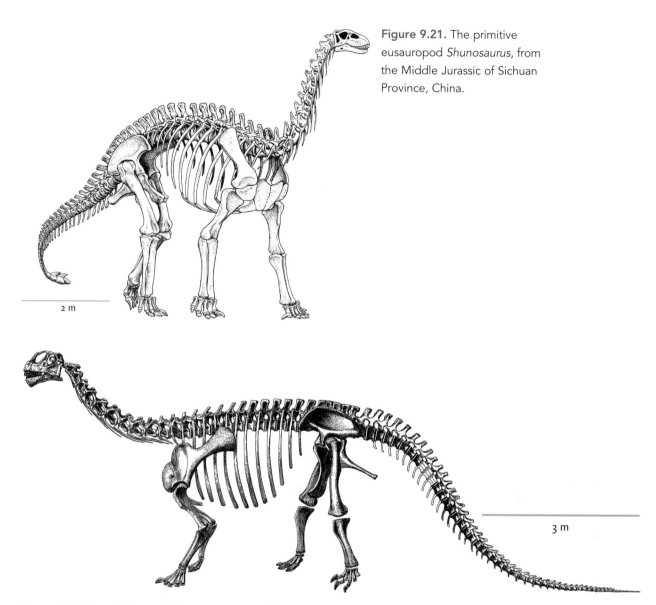

Figure 9.21. The primitive eusauropod *Shunosaurus*, from the Middle Jurassic of Sichuan Province, China.

2 m

3 m

Figure 9.22. The skeleton of *Camarasaurus*, a well-known macronarian from the Late Jurassic of the western United States. This somewhat dated drawing (note the dragging tail) dates back to its original description in the 1880s by legendary paleontologist O. C. Marsh (see Box 16.3).

movement of the jaws. The largest division within Neosauropoda is between the stocky, pumped Macronaria, and the elongate, ectomorphic Diplodocoidea (see Figure 9.20). These differences in body type match differences in the skulls: macronarians, including camarasaurs (Figure 9.22) and brachiosaurs (Figure 9.16), generally show a shortening and elevation of the skull, indicating a more powerful bite; by contrast, diplodocoids had lower, more elongate skulls, with the nostrils fully migrated away from the snout, and the teeth generally pencil-like and restricted to the front of the jaws (see Figure 9.10 for differences in the skulls). As we have seen, these very real differences in skull and tooth morphology likely translated into behavioral differences as well.

Our quick survey of Macronaria would not be complete without a stop at the top: titanosaurs (see Figure 9.20), an extremely diverse group of sauropods, known from all continents, and among the

Figure 9.23. A 37 m (122′) long cast of *Patagotitan*, a very large Argentinian Late Cretaceous titanosaur, standing guard at the January 15, 2019, opening of a permanent exhibit at the American Museum of Natural History (New York) called, fittingly, "The Titanosaur."

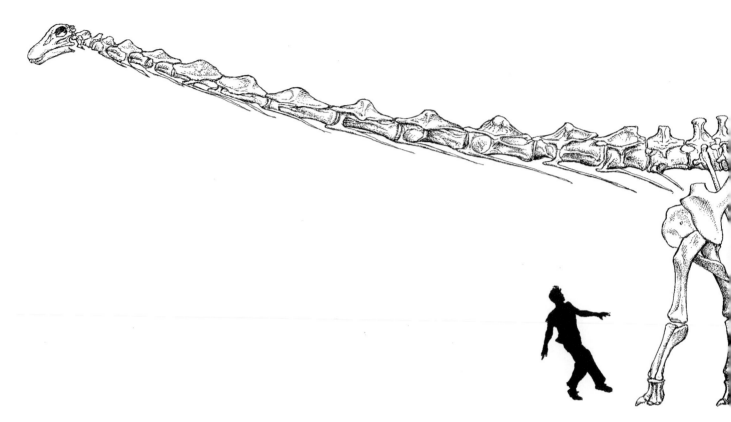

Figure 9.24. *Dreadnoughtus schrani*, an immense titanosaur from the Late Cretaceous of Argentina. *Dreadnoughtus* is among the largest known sauropods, as well as among the best preserved. Despite its immense size, it was likely still growing when it died. Human for scale.

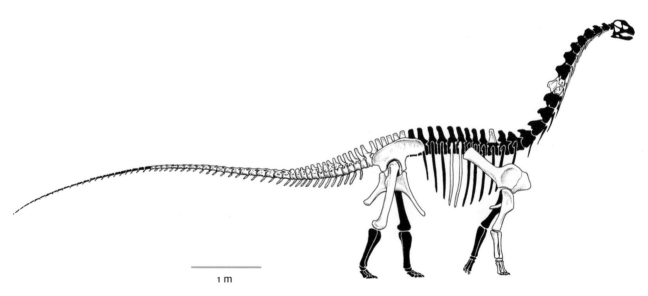

1 m

Figure 9.25. *Alamosaurus*, a poorly known Late Cretaceous sauropod from the southwest of the United States. The white sections are what is known from this dinosaur.

Figure 9.26. The diplodocoid sauropod *Amargasaurus*, from the Early Cretaceous of Argentina.

Box 9.1 On Again; Off Again: the Checkered Career of *Brontosaurus*

With the discoveries of dinosaurs in the Western Interior of the USA during the late nineteenth century, box-car-loads of brand new, but often incomplete, sauropod skeletons were shipped back east. Yale's O. C. Marsh (Box 16.3) described one of these new sauropods as *Apatosaurus* in 1877. With further shipments of specimens and more studies, Marsh found another "new" sauropod in 1879, which he named *Brontosaurus*.

Years went by and – thanks to the burgeoning popularity of many kinds of dinosaurs – *Brontosaurus* became as famous as any dinosaur ever discovered, while *Apatosaurus* languished in semi-obscurity. Nevertheless, there was the suspicion by many sauropod researchers that *Apatosaurus* and *Brontosaurus* were the same. In fact, this case was made as early as 1903 by E. S. Riggs, of the Field Museum of Natural History, Chicago, IL. Since then, most sauropod workers have regarded *Brontosaurus* as synonymous with *Apatosaurus*. If *Apatosaurus* and *Brontosaurus* are the same sauropod, the older name has priority, and thus *Apatosaurus* should be applied to this Late Jurassic giant.

Meanwhile, O. C. Marsh, lamenting as early as 1883 that no head was known for his *Brontosaurus* (now *Apatosaurus*), made his best guess as to the kind of skull this animal would have had; he concluded that it ought to have one like yet another sauropod, *Camarasaurus*. And so it was that *Brontosaurus/Apatosaurus* got the short-snouted type of skull of *Camarasaurus*.

Early in the twentieth century, H. F. Osborn, curator of vertebrate paleontology and powerbroker of the American Museum of Natural History in New York, and W. J. Holland, curator of fossil vertebrates at the Carnegie Museum of Natural History in Pittsburgh, each equally confident of his interpretations, skirmished over the issue of whose head should reside on the neck of *Apatosaurus* (for it was by then acknowledged to be such). Osborn followed Marsh and had his skeleton of *Apatosaurus* – in New York – topped with a camarasaur head (Figure B9.1.1a), while Holland, in Pittsburgh, was equally certain that *Apatosaurus* had a more *Diplodocus*-like head (based on a somewhat removed yet associated skull found near an otherwise quite complete skeleton at what is now Dinosaur National Monument in Colorado). But Holland gained no adherents and his mount of *Apatosaurus* in the Carnegie Museum remained headless in defiance of Osborn's dogma. After Holland's death, however, the skeleton was fitted with a camarasaur skull, almost as if commanded from Beyond, by Osborn himself.

Which head belongs to which dinosaur was finally resolved in 1978 by Carnegie Museum Curator of Paleontology D. S. Berman and the late sauropod authority J. S. McIntosh. Through some fascinating detective work on the collection of sauropod specimens at the Carnegie Museum, these two researchers were able to establish that *Apatosaurus* had a rather *Diplodocus*-like skull – long and sleek, not blunt and stout as had previously been suggested. As a consequence, a number of museums that

last sauropods to grace the planet Earth. Titanosaurs achieved incredible sizes: *Patagotitan* from Argentina, cast and now mounted at the American Museum of Natural history (New York) is 37 m (122') stem to stern (Figure 9.23).

Titanosaurs were also the heaviest animals ever known; some, such as *Argentinosaurus* (from Argentina) and *Paralititan* (from Egypt) may have exceeded 60 metric tons,[4] however, their fragmentary remains, along with the problems of estimating weight, make such estimates very crude approximations (Chapter 13). The mighty Argentine *Dreadnoughtus* (Figure 9.24), first described in 2014, is relatively complete (some fragmentary jaw material plus around 70 percent of the postcranial skeleton is represented) and, using a new computer algorithm for weight estimation based upon the circumference of the humerus and femur, the thing is estimated to have weighed an

---

[4] 1 metric ton = 1000 kg; 1 kg = 2.20462 lb, or 0.001 ton.

display *Apatosaurus* skeletons celebrated the work of Berman and McIntosh (and Holland) by conducting painless head transplants (Figure B9.1.1b).

Still, people were attached to the name "*Brontosaurus*," and nobody gets points for raining on parades. So an unofficial compromise was reached: the word "brontosaur" could be a common name like "dog" – you call a sauropod a brontosaur, and everybody knows what you're talking about. In any event, the scientific upshot of all this was "*Brontosaurus*" is actually *Apatosaurus*; *Apatosaurus* had an elongate, *Diplodocus*-like skull. Case closed, right?

Uh, no. In April 2015, three European specialists did an immense, exhaustive analysis of dinosaurs very closely related to *Diplodocus* (for example *Apatosaurus* and even *Brontosaurus*). Three hundred pages later, they concluded that *Brontosaurus* is very real (if not alive and well). In fact, there are actually three species of the thing: *B. parvus*, *B. excelsior*, and *B. yahnapin*. And everybody had *Diplodocus*-like elongate skulls.

This discussion has gone on for 135 years; do we really think we've heard the last word on these iconic sauropods? We hope so.

Figure B9.1.1. (a) "*Brontosaurus*" on display at the American Museum of Natural History in the early 1970s. This dinosaur was a chimera with a camarasaur-type head on an *Apatosaurus* body. (b) *Apatosaurus* being readied for display at the American Museum of Natural History in the mid 1990s, with its "new" *Diplodocus*-style skull.

unbelievable 59.3 metric tons.[5] Argentina seems to have been titanosaur heaven during the Cretaceous, and many new sauropods have been described from there in even the past five or so years. Ultimately, our understanding of who lived in the Mesozoic may be radically revised when the full extent of global titanosaur diversity is understood.

Yet sauropod evolutionary history is not entirely one of getting bigger. In Transylvania is found the titanosaur *Magyarosaurus*, a creature 5 to 6 m in length, much smaller than contemporary sauropods elsewhere in the world. These smaller forms were living on islands, a common phenomenon in the Mesozoic as now ("island dwarfing"). This dwarf sauropod was a "mere" 6 m long.

Several titanosaur forms, including *Saltasaurus*, from the Late Cretaceous of Argentina, and *Malawisaurus* (obviously from Malawi), are known to have been covered with special bones,

---

[5] But is this number right? More recent (and reasonable?) estimates of the weight of *Dreadnoughtus*, using different computer algorithms, put its heaviest weight at 30 to 40 metric tons (see Chapter 13). Nobody doubts that it was *very* big and *very* heavy, and, in the end, that is probably enough.

Figure 9.27. Dermal spines reconstructed along the back of the Late Jurassic North American diplodocoid, *Diplodocus*.

embedded in the skin, called osteoderms[6] (*osteo* – bone; *derm* – skin); in this case they are scattered, immense (up to 22 cm in length), and hollow. Why scattered? Paleontologist Kristina (Kristi) Curry Rogers proposed that in juveniles, the osteoderms were close together and served, as we will see in ankylosaurs (Chapter 10), as a kind of armor pavement. When the animals grew, the osteoderms, she proposed, grew apart, and then their hollow interiors potentially served as reservoirs for minerals and other nutrients during times of environmental stress.

Among the most famous of titanosaurs is, ironically, among the most poorly known: *Alamosaurus*, from the Late Cretaceous of the southwest of the United States (Figure 9.25). While Late Cretaceous sauropods are very well known from Europe, Asia, and South America, North American ones are quite rare; and it appears that they never got farther north than the American Southwest. What *that* is about is a story for another day – or at least for Chapter 15.

By contrast to macronarians, diplodocoids were elongate and slender animals with extraordinarily long necks and tails; some of the longest, though not the heaviest, animals that ever lived on land. The Late Jurassic *Diplodocus* extended about 30 m, yet its weight is supposed to have been about 15 metric tons (again, see Chapter 13 for the perils of dinosaur weight estimates).

As we have seen, the long-necked, long-tailed, gracile diplodocoids (can an animal whose weight is in double-digit tons *ever* be called "gracile"?) developed a distinctive skull: they restricted their dentition to just a series of highly evolved, pencil-like teeth to the front of the jaws. Tooth wear is apparent at the apices of the teeth (instead of along the side of the tooth), and front–back jaw movement is thought to have characterized the way these animals managed food in the mouth.

Some remarkable dinosaurs are among the diplodocoids, including *Amargosaurus*, with its extraordinary neural arches, from the Early Cretaceous of Argentina (Figure 9.26). The 21 m long *Apatosaurus* (Late Jurassic, Western Interior, USA) is best known by its incorrect name "*Brontosaurus*" (Box 9.1). *Diplodocus*, renowned for its long neck and tail, evidently carried dragon-like dermal spines along the length of its back (Figure 9.27).

Sauropodomorphs were the largest terrestrial vertebrate life forms of their time and of all time. *Brachiosaurus*, from the Late Jurassic of the western USA (see Figure 9.16), as well as the very similar *Giraffatitan* from Tanzania, captured several decades' worth of people's imaginations as the largest

---

[6] Although very rare in mammals, osteoderms are relatively common in tetrapods; we are familiar with them, for example, as the bony plates in crocodile skin. We'll see them in many different dinosaurs.

land-living animals of all time (measuring 23 m long and weighing in excess of 50 000 to 60 000 kg). Now supplanted by the likes of *Argentinasaurus* and *Dreadnoughtus, Brachiosaurus,* looming over the United Airlines Customer Service counter in Concourse B of Chicago's O'Hare International Airport, is still big enough.

Sauropods were the fast-growing, yet slow-paced high-browsing giants of the Mesozoic. Today we view them as evolutionary marvels, as they continue to baffle, surprise, and inspire with biomechanical and evolutionary consequences of "living large."

SUMMARY

Sauropodomorpha consists of the great herbivorous saurischian quadrupeds Sauropoda, and an early nonmonophyletic (e.g., paraphyletic) offshoot, called "prosauropods." "Prosauropods" were primitive large dinosaurs, appearing in the Late Triassic at the dawn of Dinosauria, and surviving through the Early Jurassic. Initially thought to be the monophyletic ancestors of Sauropoda, they are now considered to represent an early saurischian radiation, representing the world's first high-browsing herbivores.

Sauropoda were the largest land animals ever to walk the Earth, reaching ~40 m from the tip of the snout to the tip of the tail. These obligate herbivorous quadrupeds were highly evolved, with many biomechanical adaptations for large size and weight, including four pillar-like limbs and a massive pelvis, a tendency to lighten bones not immediately involved in bodily support functions, and a complex girder-like neck design, tipped by a small skull, to maximize leverage and lightness. Among the many striking features in the design of sauropod bones are the presence of pleurocoels, hollow spaces suggestive of avian-style air sacs. The likely presence of air sacs, as well as the inefficiency, in an animal of such great size, of mammalian-style bellows breathing, suggest that sauropods likely enjoyed avian-style breathing.

The skulls of sauropods were relatively small, and the group showed a general tendency toward migrating the nostrils or nares to the top of the skull, which is emphasized in diplodocoids. Dentition varied, from simple leaf-shaped teeth to pencil-like teeth restricted to the front of the mouth. Sauropods in general lack the chewing adaptations present in ornithischians (see Part III), and we may be confident that the stomach microflora (the bacteria) did much of the heavy lifting in the breakdown of food. Nonetheless for all sauropods, there appear to have been a variety of strategies for handling food at the mouth, reflected in jaw musculature, skull design, and tooth wear.

Some sauropods appear to have been social animals (as some do not), particularly as reflected in sauropod bonebeds and in trackways. The trackways clearly indicate that sauropods did not drag their tails. Sauropod gregariousness – for at least egg-laying – is also reflected in the recent discoveries of extremely large sauropod nesting grounds. Sauropod lifespans, once thought to be in the hundreds of years, are now thought to have been around 30 years, with extremely rapid growth of juveniles. The large number of eggs and babies associated with sauropod nesting grounds implies that these dinosaurs may have been *r*-strategists.

Sauropod defense was likely accomplished mainly by huge body size, with perhaps an assist from the whip-like tail and the broad, trenchant claw on the forefoot.

The immense size of sauropods makes analogizing them with living terrestrial vertebrates extremely difficult. Although not likely possessed of a "warm-blooded" metabolism – at least as adults (see Chapter 14) – they nonetheless must have consumed copious quantities of food, and must have required virtually incessant food consumption through the comparatively small mouth. If the number of individuals was large enough, a healthy herd might have have defoliated landscapes. This in turn leads to at least the possibility that sauropods were often on the go, searching out new foliage to consume.

SELECTED READINGS

Curry Rogers, K. and Wilson, J. A. (eds.) 2005. *The Sauropods: Evolution and Paleobiology*. University of California Press, London, 349 pages.

Curry Rogers, K., D'Emic, M., Rogers, R., Vickaryous, M., and Cagan, A. 2011. Sauropod dinosaur osteoderms from the Late Cretaceous of Madagascar. *Nature Communications*, **2** (564), 1–4. doi: 10.1038/ncomms1578

Galton, P. M. and Upchurch, P. 2004. Prosauropoda. In Weishampel, D. B., Dodson, P., and Osmólska, H. (eds.) *The Dinosauria*, 2nd edn. University of California Press, Berkeley, pp. 232–258.

Hallett, M. and Wedel, M. 2016. *The Sauropod Dinosaurs*. Johns Hopkins University Press, Baltimore, MD, 320 pages.

Henderson, D. M. 2003. Typsy punters: sauropod dinosaur pneumaticity, buoyancy, and aquatic habits. *Proceedings of the Royal Society of London, Series B (suppl.)*, 271, S180–S183.

Klein, N., Remes, K., Gee, C. T., and Sander, P. M. (eds.) 2011. *Biology of the Sauropod Dinosaurs – Understanding the Life of Giants*. Indiana University Press, Bloomington, IN, 331 pages.

Seymour, R. S. and Lillywhite, H. B. 2000. Hearts, neck posture and metabolic intensity of sauropod dinosaurs. *Proceedings of the Royal Society of London, Series B*, **267**, 1883–1887.

Tidwell, V. and Carpenter, K. (eds.) 2005. *Thunder-Lizards – The Sauropodomorph Dinosaurs*. Indiana University Press, Bloomington, IN, 495 pages.

Upchurch, P., Barrett, P. M., and Dodson, P. 2004. Sauropoda. In Weishampel, D. B., Dodson, P., and Osmólska, H. (eds.) *The Dinosauria*, 2nd edn. University of California Press, Berkeley, pp. 259–321.

Upchurch, P., Barrett, P. M., Zhao, X., and Xu, X. 2007. A re-evaluation of *Chinshakiangosaurus chunghoensis* Ye vide Dong 1992 (Dinosauria, Sauropodomorpha): implications for cranial evolution in basal sauropod dinosaurs. *Geological Magazine*, **44**, 247–262.

Wedel, M. J. 2003a. The evolution of vertebral pneumaticity in sauropod dinosaurs. *Journal of Vertebrate Paleontology*, 23, 344–357. doi:10.1671/0272–4634(2003)023[0344:TEOVPI]2.0.CO;2

Wedel, M. J. 2003b. Vertebral pneumaticity, air sacs, and the physiology of sauropod dinosaurs. *Paleobiology*, 29, 243–255. doi:10.1666/0094–8373(2003)029,0243: VPASAT.2.0.CO;2

Yates, A. M. 2012. Basal Sauropodomorpha: the "Prosauropods." In Brett-Surman, M. K., Holtz, T. M., and Farlow, J. O. (eds.) *The Complete Dinosaur*, 2nd edn. Indiana University Press, Bloomington, pp. 424–443.

Yates, A. M., Bonnan, M. F., Nwevelling, J., Chinsamy, A., and Blackbeard, M. G. 2009. A new transitional sauropodomorph dinosaur from the Early Jurassic of South Africa and the evolution of sauropod feeding and quadrupedalism. *Proceedings of the Royal Society of London, Series B*, **277**, 787–794. doi:10.1098/rspb.2009.1440

TOPIC QUESTIONS

1. What is Sauropodomorpha? How does it relate to Saurischia?
2. What are the diagnostic characters of Sauropodomorpha? Why are the quotation marks always around the word "prosauropods"?
3. Outline what is known of sauropod reproductive strategies. Would you classify them as *r*-selected, or *K*-selected? Why?
4. Describe some features that are associated with large size in sauropodomorphs.
5. What is the evidence that sauropods didn't really chew their food? How can sauropods have been herbivores, but not have developed much chewing ability?
6. Why are sexual dimorphism and gregariousness often linked?
7. What kinds of design constraints are associated with having an extremely long neck?
8. What is the evidence that indicates that sauropods did not dwell in swamps? Where, then, and in what kinds of environment did they live?
9. Why do we think that the quadrupedal stance of sauropods evolved secondarily?

PART III

Ornithischia: Armored, Horned, and Duck-Billed Dinosaurs

Figure III.1. Right lateral view of the ornithischian pelvis as exemplified by *Stegosaurus*. Note that the pubis has a rearward-facing splint of bone (arrow), lying underneath the ischium in what is known as the **opisthopubic** condition (see also Figures 5.9 and 5.10).

A long long time ago, and many saurischians away, we met Ornithischia, one of the two great branches of dinosaurs. Although Ornithischia was identified as early as 1887, nobody at that time had an inkling how diverse the group really was. Now we know a bit better. We'll check out individual ornithischians in Chapters 10 to 12, but first, let's introduce the group.

What Makes an Ornithischian an Ornithischian?

Ornithischians, as we said in Part II, show obvious family resemblance, and diagnostic characters abound. Among these, two stand out:

- In all ornithischians, at least a part of the pubis has rotated backward to lie close to and parallel with the ischium;[1] this orientation is called **opisthopubic** (Figure III.1).
- All ornithischians had a unique bone, the **predentary**, an unpaired, scoop-shaped element that capped the front of the lower jaws (Figure III.2).

Both of these adaptations were associated with food consumption and processing. The evolution of the rearward-directed pubis (recall that, primitively, the pubis points forward; see Figure 5.9) is believed to be associated with the development of a large stomach (or stomachs?) and intestinal region (gut), the better for extracting nutrients from plants. Accommodating the large gut was a barrel-shaped torso, recognizable from the shape of the ribs. The predentary supported the lower portion of a beak, a characteristic feature of all ornithischians (see below). Other ornithischian diagnostic characters are listed in the cladogram shown in Figure III.3. You can see that the monophyly of Ornithischia is *extremely* well supported, and recognised in all schemes of dinosaur relationships (see Box 5.1).

---

[1] In birds, the pubis is rotated backwards as well, and so, confusingly, ornithischian dinosaurs are called "bird-hipped," while birds, as we have seen, are undoubted saurischians. This means that birds belong to the "lizard-hipped" branch of Dinosauria while the "bird-hipped" branch – Ornithischia – seems to have given rise to nothing still alive; certainly not birds.

Figure III.2. Left lateral view of the skull of the ceratopsid *Triceratops*. The predentary caps the front of the lower jaw, and is outlined in white.

The discovery of one entirely unexpected attribute of ornithischians came in 2009, with the unearthing of a primitive (basal) ornithischian dinosaur called *Tianyulong* (Figure III.4). *Tianyulong* bears a row of long, monofilamentous structures – Stage 1 proto-feathers? – down its neck, back, and tail.

And lest that dinosaur was thought to be a fluke, in 2014 P. Godefroit and colleagues described *Kulindadromeus*, a relatively derived ornithischian (a neornithischian; see Figure III.3 and below) from Siberia (no less!), preserving both scales and three different types of feathers. Ornithischians were evidently, at least primitively, feathered creatures (see also Chapter 11).

This leads to the obvious question: Do feathers – or their precursors – go all the way back to basal Dinosauria? The short answer is "quite probably," pointing the way to very primitive monofilamentous feathers as a basal dinosaur - or even older – character (see also Chapter 5).[2]

[2] The question really hinges on whether or not feathers evolved more than once. In the past, received wisdom suggested that feathers only evolved one time; that being the case, feathers would be primitive for all ornithodirans, since they are present in at least pterosaurs, theropods, and ornithischians. If, however, they evolved more than once, then perhaps they occurred independently in these groups, and it cannot be said the Ornithodira is characterized by the possession of feathers. The jury is still out on this question.

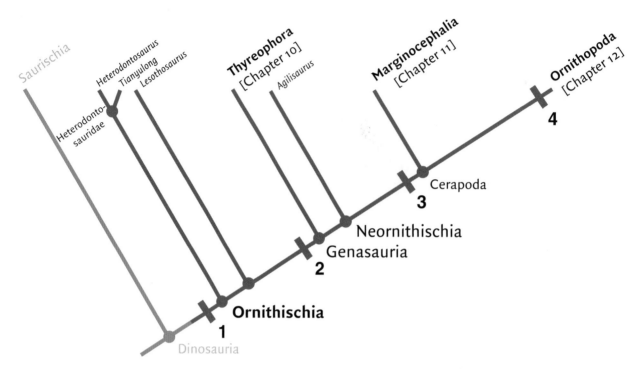

Figure III.3. Cladogram of Ornithischia. Derived characters include: at **1** (Ornithischia), opisthopubic pelvis, predentary bone, toothless and roughened tip of snout, reduced antorbital opening, palpebral bone, jaw joint set below level of the upper tooth row, cheek teeth with low subtriangular crowns, at least five sacral vertebrae, ossified tendons above the sacral region, small prepubic process along the pubis, long and thin preacetabular process on the ilium; at **2** (Genasauria), emarginated dentition (indicating large cheek cavities, and reduction in the size of the opening on the outside of the lower jaw (the external mandibular foramen); at **3**, gap between the teeth of the premaxilla and maxilla, five or fewer premaxillary teeth, finger-like anterior trochanter; at **4**, high-crowned cheek teeth, denticles on the margins restricted to the terminal third of the tooth crown, canine-like tooth in both the premaxilla and dentary.

Chew Chew Chew-Boogie!

One essential quality of ornithischians is that, to a greater or lesser extent, all apparently chewed their food. Because humans and many other mammals chew, we naturally assume that everybody does, and it can be surprising to learn that (as we noted earlier) most vertebrates don't chew; that teeth are really most commonly used only for biting off chunks of whatever is being eaten. The fundamental act of *chewing* – of grinding food down into a paste that can be digested relatively efficiently – is done by other organs in most vertebrates; generally, as in sauropods, a gastric mill; commonly with gastroliths. But ornithischians got into the chewing game in a big way, so a look at what chewing's all about is useful for understanding these dinosaurs.

Chewing in Mammals

We start by looking at the basics of chewing in a familiar living group: mammals. Among living herbivorous mammals, no matter what the size, the skull is generally divided into three sections (Figure III.5): at the front is the **cropping** part, where blade-like teeth (generally incisors) bite off chunks of food. Behind the cropping section is the **diastema**, or gap, a region that is toothless (or nearly so), likely used for the manipulation of the food by the tongue. Finally, further back in the

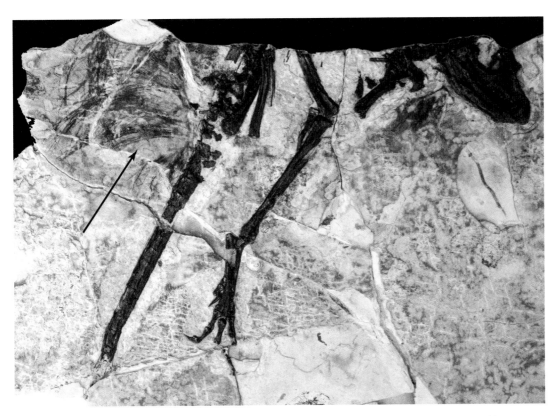

Figure III.4. The tail and left leg of *Tianyulong*, a basal ornithischian evidently bearing monofilamentous feathers (indicated by arrow). Head is to right of photograph. Because feathers are found in both primitive ornithischians as well as many saurischians (so far, theropods), the most parsimonious conclusion is that feathers must also have characterized the common ancestor of both groups: the earliest dinosaur.

mouth are the cheek teeth (molars in mammals) – in herbivores especially (which tend to chew their food more than carnivores), a block of teeth, relatively tightly fitted against one another, which are used to grind or shear plant material. In mammals the upper and lower cheek teeth occlude – or fit tightly against each other when the jaw is shut – which ensures that, as the chewing takes place, the grinding is efficient.

Two other features are associated with chewing in mammals. Firstly, toward the back of the lower jaw is a large expansion of bone, the coronoid process, which serves as an attachment site for strong jaw-closing muscles (Figure III.5). Secondly, the tooth row is deeply inset toward the midline of the skull. This makes room for cheeks, muscular tissues that play the obviously essential role of keeping food in the mouth while it is being chewed.

In mammals, then, chewing leaves a recognizable imprint on the design of the skull, the lower jaw, and the teeth. It's striking that, in almost all ornithischian dinosaurs, many of these same adaptations for chewing can be found, where they doubtless arose independently. We'll see this as we explore each of the ornithischian groups in Chapters 10 to 12.

Chewing in Dinosaurs

The primitive tetrapod condition, exemplified in theropods, is that the jaw joint is at the same level as the tooth row (Figure III.6b). The jaw thus functions like a scissors and the bite slices sequentially from the back of the jaw forward (we saw this in theropods, see also Figure 6.14).

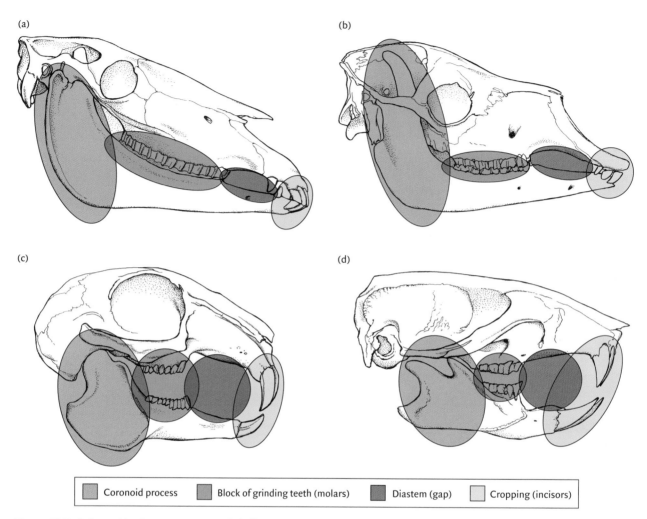

| Coronoid process | Block of grinding teeth (molars) | Diastem (gap) | Cropping (incisors) |

Figure III.5. Selected herbivorous mammal skulls (not drawn to scale). (a) Horse (*Equus*), (b) llama (*Lama*), (c) rabbit (*Lepus*), and (d) rat (*Rattus*). Divisions of skulls indicate: anterior cropping section (yellow), diastem (orange), block of grinding cheek teeth (light purple), and coronoid process (light blue). Despite the range of sizes and herbivorous behaviors, all skulls show the same basic organization.

By contrast, when the jaw joint is below the level of the tooth row, as it is in all ornithischians, the jaw functions a bit like water-pump pliers, in which the jaws close simultaneously along their entire length (Figure III.6a). The blocks of cheek teeth grind against each other simultaneously instead of shearing past each other sequentially. As it turns out, even the most primitive ornithischians had many of the required goods for serious chewing. And the many ornithischians that followed never kicked the habit.

In many dinosaurs, including all ornithischians, the cropping function of the mouth was carried out not by teeth, but by a beak, or **rhamphotheca**. Rhamphothecae were made of **keratin**, a protein-based substance that makes up horns, nails, hooves, and claws. It is also a key ingredient of hair, although you're not likely to find *that* in a dinosaur!

Ornithischia: the Big Picture

Most specialists agree that fundamental split within Ornithischia is between some very primitive ornithischians near the base, and then everything else (see Figure III.3).

(a)

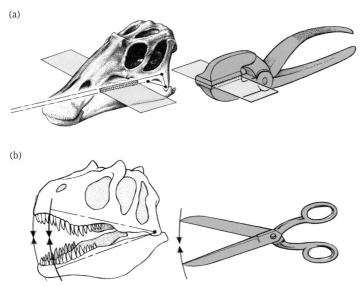

(b)

Figure III.6. Position of jaw joints. (a) In ornithischians, the jaw joint is below the level of the tooth row, and the blocks of cheek teeth *grind* against each other simultaneously, much like the water-pump pliers; (b) by contrast, the theropod condition (as we have seen) has the jaw joint at the same level as the tooth row; the dominant movement, therefore, is scissor-like *cutting*, back to front.

The Most Primitive Ornithischians: Heterodontosaurids, *Lesothosaurus*, and Their Friends

The most basal ornithischians are the Late Jurassic *Chilesaurus* from Chile, naturally enough, and the enigmatic, Late Triassic (?) Pisanosaurus, from Argentina. *Chilesaurus* was originally described in 2015 as a theropod (!), and reanalyzed two years later as a primitive ornithischian, so basal that the authors suggested that it might actually document the transition from theropods to ornithischians.

What transition from theropods to ornithischians? Here we get back to Seeley (1887) and the Ornithischia–Saurischia split: in that view of Dinosauria, there is no transition between theropods and ornithischians; once Saurischia and Ornithischia diverged, each lineage evolved separately. But, in the newly revised phylogeny of dinosaurs, in which Theropoda and Ornithischia are united within Ornithoscelida (Box 5.1), a primitive theropod could in fact be a precursor for an ornithischian – an ornithischian that just happened to be something like *Chilesaurus*.

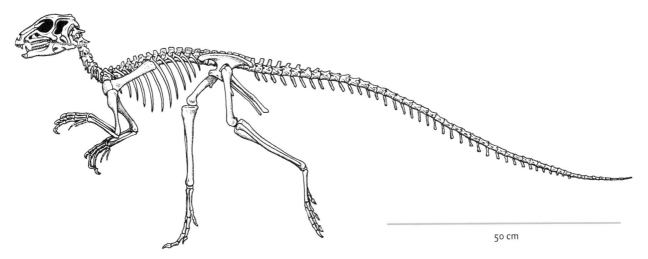

50 cm

Figure III.7. Left lateral view of the skeleton of *Heterodontosaurus*.

(a)

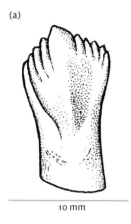

10 mm

(b)

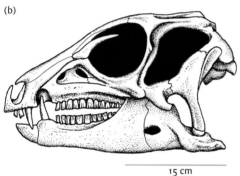

15 cm

Figure III.8. (a) Upper tooth of the heterodontosaurid *Lycorhinus*. (b) Left lateral view of the skull of *Heterodontosaurus*.

(a)

(b)

Figure III.9. Jaw mechanics in Heterodontosauridae, showing mobility of the lower jaws. (a) Skull of the *Heterodontosaurus*. (b) Cross-section through skull and mandible. As the lower jaw closed, each side rotated slightly outward along its long axis.

Pisanosaurus has a much less tantalizing story. Discovered in Argentina in the 1940s, it is so incomplete and poorly preserved, that it barely qualifies as an ornithischian. Yet, ornithischian it is, and the earliest one known at that. For all its antiquity, incompleteness, and poor preservation, however, it is apparently more derived than *Chilesaurus*. Here we reaffirm: older $\neq$ more primitive, a key distinction we made in Chapter 3 as well.

Heterodontosaurids were small, bipedal ornithischians (Figure III.7) of uncertain phylogenetic position; with a mix of primitive and advanced characters, some authors consider them to have been basal ornithischians, while others have them as the sister group Marginocephalia (Chapter 11), and still others as the sister group to Euornithopoda (Chapter 12). Because heterodontosaurids bear the inset tooth rows (and cheeks) diagnostic of genasaurs, some consider them to be basal genasaurs. We tentatively consider them to be basal ornithischians (Figure III.3).

In many respects heterodontosaurids are not exactly primitive; for example, they evolved teeth bearing a high, chisel-shaped crown ornamented with tiny bumps (Figure III.8a; correctly termed denticles) as well as a large canine-like tooth on both upper and lower jaws (the basis for the "heterodonto-" part of the name; Figure III.8b). Moreover, skull design and tooth-wear striations suggest that they chewed distinctively: they amplified the familiar vertical movement of the lower

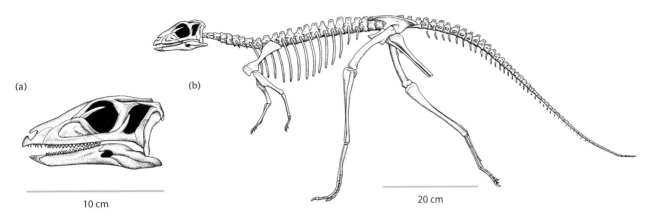

(a)

(b)

10 cm

20 cm

Figure III.10. Left lateral view of the skull (a) and skeleton (b) of the basal ornithischian *Lesothosaurus.*

jaws with slight outward rotations of each side of the mandible about its long axis (Figure III.9). This allowed them to get a bit more grind out of each bite. Powerful forelimbs and clawed hands likely were used to grab vegetation or to dig up roots and tubers.

Like many ornithischians, the evolution of canine-like teeth of heterodontosaurids likely was related to combat between animals of the same species (males?), ritualized display, social ranking, and possibly even courtship (see Chapters 10 and 11). A modern analog is found in the tusked deer-like tragulids, living mammals from southeastern Asia and Africa. In these mammals, tusk development is tied to sexual maturity, and is a dimorphic (differently expressed in the sexes) feature that is, as we propose for heterodontosaurids, used for intraspecific combat, ritualized display, and social ranking. Similarly, the development of a large knob of bone, a boss, in the cheek region (therefore called the jugal boss[3]) in heterodontosaurids might also be interpreted as a form of visual display.

Display in ornithischians may have come in another form as well. We've already met *Tianyulong,* the feathered heterodontosaur from western Liaoning Province, China (see Figure III.4). Depending upon the colors and patterns, these primitive feathers could have served a display function and may have been intimately associated with behaviors ranging from choosing mates (see Chapter 6) to marking territory, to camouflage, to keeping warm (Chapter 13).

Lesothosaurus[4] was a small, leggy Early Jurassic herbivore from South Africa (Figure III.10). It had a typical suite of diagnostic ornithischian characters including a jaw joint lower than the tooth row (see Figure III.6). That character hints at chewing; we'll have far more than hints for the rest of Ornithischia.

The Rest of Ornithischia

Genasauria

Besides the most primitive ornithischians (above), "everything else" is within the clade Genasauria (*gena* – cheek, a reference to their all having well-developed cheeks). Genasaurs include two great groups of ornithischians, Thyreophora (Chapter 10), and Neornithischia (including Cerapoda

---

[3] Named for the jugal, the skull bone on which it appears.

[4] Lesothosaurus has not gone quietly into the phylogenetic night. A major reanalysis of Ornithischia (Butler *et al.,* 2008) removed *Lesothosaurus* from its conventional basal position in Ornithischia and made it a basal thyreophoran (see Chapter 10), closely related to stegosaurs and ankylosaurs. This idea, however, does not appear to have been widely adopted; we therefore return it to its conventional spot at the base of Ornithischia pending a broader scientific consensus on this idea. The relationships of basal neornithischians are very much in a state of flux (see Box 12.1).

[Marginocephalia + Orthinopoda]; Chapters 11 and 12, respectively), as well as a few assorted forms with which we won't concern ourselves here. They all share the derived characters of muscular cheeks, indicated by the deep-set position of the tooth rows, away from the sides of the face, a spout-shaped front to the mandibles, and reduction in the size of the opening on the outside of the lower jaw (the external mandibular foramen), among others. Because it's hard to understand the point of cheeks without chewing, chewing should be thought of as a fundamental genasaur behavior. All of this may seem like an alphabet soup of dinosaur names, but the cladogram in Figure III.3 lays these relationships out clearly.

Thyreophora

Thyreophora (*thyreo* – shield; *phora* – bearer; a reference to the fact that these animals have dermal armor) consists of genasaurs in which there are parallel rows of keeled dermal armor scutes (or bony plates) on the back surface of the body. The most familiar thyreophorans are stegosaurs and ankylosaurs, but we'll encounter others along the way (Chapter 10).

Neornithischia

Moving up the cladogram is the slightly-more-derived Neornithischia (*neo* – new; see Figure III.3). Primitive neornithischians include a variety of small, bipedal forms – we'll call them *basal neornithischians* – whose relationships within Ornithischia are still very uncertain. Rather than getting mired here among the basal neornithischians, we revisit them again in Chapter 12.

Cerapoda

Cerapoda (*cera* – horn) are those genasaurs that share a pronounced diastem between the teeth of the premaxilla and maxilla among other derived characters (see Figure II.3). Marginocephalians (*margin* – margin; *cephal* – head), a group united by having, primitively, a distinctive, narrow shelf that extended over the back of the skull (see Figure II.3), consists of two well-known ornithischian taxa, the dome-headed pachycephalosaurs and ceratopsians (Chapter 11), the latter being the horned dinosaurs most famously known from the Late Cretaceous of North America.

Ornithopods, especially euornithopods, took chewing to new levels, possibly unmatched in the history of life. Within this group are found the familiar duck-billed dinosaurs, as well as one of the very first dinosaurs ever recorded, *Iguanodon*. We'll visit these animals in Chapter 12.

SELECTED READINGS

Baron, M. G. and Barrett, P. M. 2017. A dinosaur missing link? Chilesaurus and the early evolution of ornithischian dinosaurs. *Biology Letters*, 13, 20170220. http://dx.doi.org/10.1098/rsbl.2017.0220

Boyd, C. A. 2015. The systematic relationships and biogeographic history of ornithischian dinosaurs. *PeerJ*, 3, e1523. DOI: 10.7717/peerj.1523

Butler, R. J., Upchurch, P., and Norman, D. B. 2008. The phylogeny of the ornithischian dinosaurs. *Journal of Systematic Paleontology*, 6, 1–40.

Norman, D. B., Witmer, L. M., and Weishampel, D. B. 2004. Basal Ornithischia. In Weishampel, D. B. and Dodson, P. (eds.) *The Dinosauria*, 2nd edn. University of California Press, Berkeley, pp. 325–334.

Sereno, P. C. 1986. Phylogeny of the bird-hipped dinosaurs (Order Ornithischia). *National Geographic Research*, 2, 234–256.

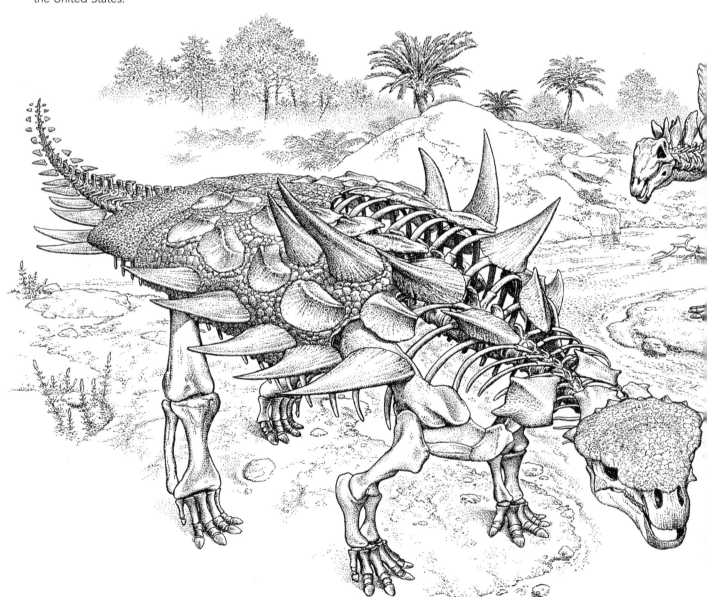

Figure 10.1. Representatives of the two great groups of thyreophorans: *Gastonia*, an ankylosaur from the Early Cretaceous of Utah, USA (left), and *Stegosaurus*, a stegosaur from the Late Jurassic of the United States.

Chapter 10

Thyreophora
The Armor-Bearers

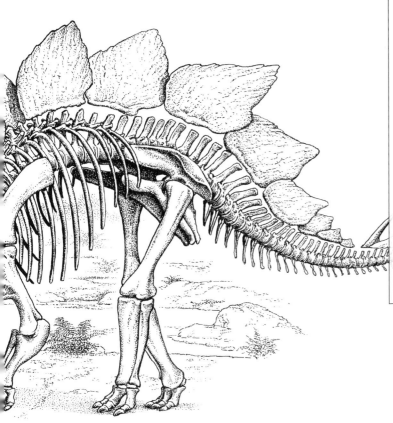

WHAT'S IN THIS CHAPTER

Here we meet our first ornithischians: the unpronounceable, but not unspeakable, Thyreophora. To understand them, we'll:

- Meet the two great clades of thyreophoran dinosaurs, Stegosauria and Ankylosauria.

- Learn about their strategies for fending off predators.

- Investigate life as a thyreophoran.

- Learn where they came from and who they are related to (we already know, unfortunately, what ultimately happened to them).

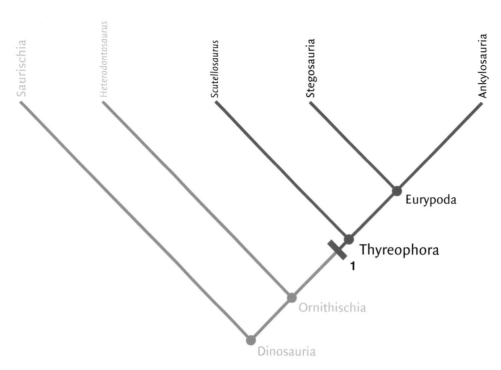

Figure 10.2. Cladogram of Thyreophora, emphasizing relationships within Ornithischia. Derived characters include: at **1**, transversely broad process of the jugal, parallel rows of keeled scutes on the back surface of the body.

Thyreophora

Thyreophorans were all about defense, evolutionarily opting for protection by armor. And it worked: these dinosaurs did very well during their approximately 100 million years on Earth, spawning upward of 50 species currently known.

Who are Thyreophorans?

All thyreophorans are characterized by parallel rows of osteoderms, which run down the necks, backs, and tails. The group is dominated by two great clades: Stegosauria (*stego* – roof) and Ankylosauria (*ankylo* – fused) (see Figure 10.1). Together, stegosaurs and ankylosaurs make up a monophyletic clade known as Eurypoda (*eury* – broad; *poda* – feet). Along with these two big groups, a few other miscellaneous, primitive Early Jurassic thyreophorans round out our story. Figure 10.2 lays out basic thyreophoran relationships.

Primitive Thyreophora

Primitive thyreophorans, outside of Ankylosauria and Eurypoda, are represented here by three forms: *Scutellosaurus*, *Emausaurus*, and *Scelidosaurus* (Figure 10.3).[1] Although primitive ornithischians in most respects, all have the diagnostic thyreophoran character of rows of bony elements embedded in

---

[1] And possibly a fourth: *Lesothosaurus*, if the proposal of Butler and colleagues (2008) is correct.

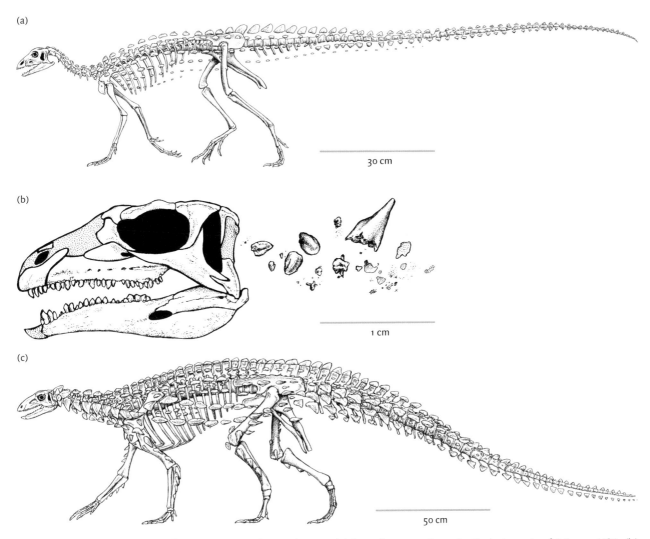

(a)

30 cm

(b)

1 cm

(c)

50 cm

Figure 10.3. Left lateral view of some primitive thyreophorans. (a) *Scutellosaurus*, from the Early Jurassic of Arizona, USA; (b) the poorly known *Emausaurus*, from the Early Jurassic of Germany; and (c) *Scelidosaurus*, from the Early Jurassic of England.

the skin – osteoderms – down the back and tail. Two were bipedal, reflecting the primitive condition in Dinosauria; *Scelidosaurus* was quadrupedal, foreshadowing the trend in the rest of Thyreophora.

Eurypoda

Eurypoda (*eury* – wide; *pod* – foot) is the clade of derived thyreophorans. It consists of two famous groups of ornithischian dinosaurs: Stegosauria and Ankylosauria (Figure 10.2). We'll look more closely at these one at a time.

Eurypoda: Stegosauria

Stegosaurs were medium-sized dinosaurs, 3 to 9 m in length and weighing 300 to 1500 kg (650 to 3500 lb), characterized by bearing osteoderms that developed into spines and plates, as well as by their quadrupedal stance (Figure 10.4). Their profiles sloped strongly forward and downward toward the ground as a result of the hindlimbs being substantially longer than the forelimbs (Figure 10.5). All

Figure 10.4. *Tuojiangosaurus*, a stegosaur from the Late Jurassic of Sichuan Province, China.

toes had broad hooves. They seem to have been relatively uncommon dinosaurs, yet it is becoming clear that they got around: to date they are known from China, Mongolia, India, the United States, Spain, France, and England. A generic thyreophoran known from South Africa has also been referred to stegosaurs. Stegosaurs occupied a narrow swath of time in the Mesozoic: the earliest is known from Middle Jurassic, and they seem to have petered out early in the mid Cretaceous.

Stegosaurian Lives and Lifestyles

Locomotion

The general body plan of stegosaurs does not suggest life in the fast lane (Figure 10.6). Indeed, with their long back legs and short front legs, stegosaurs must have had a locomotor conundrum: at the same cadence (the rate of feet hitting the ground), the hindlimbs would have covered much more distance than the forelimbs. At high speeds, therefore, the rear of the animal would have overtaken its head! This problem could be avoided in two ways: (1) by drawing the forelimbs up from the ground (that is, temporarily being bipedal while running) or (2) by limiting movement to a slow walking gait. Because of the mass distribution of stegosaurs, the first option is unlikely. Our best guess is that the pace of stegosaur life was leisurely, on the order of 6.5 to 7.0 kmh (4 to 5 mph) maximum speed (see Chapter 13). Chasing fleet prey may not have been too important to a hungry herbivore.

Neither, it would seem, was growing up quickly. Paleontologists can gain insights about the rate at which a dinosaur grew from juvenile to adult by the type of bone tissue it produces (Chapter 13).

(a)

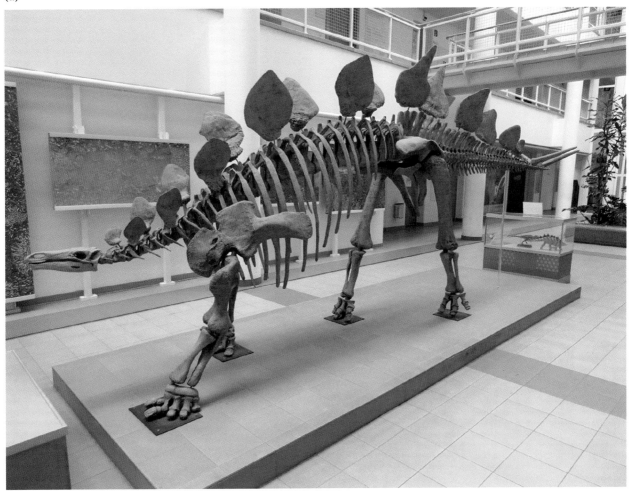

Figure 10.5. The best-known of all plated dinosaurs, the North American *Stegosaurus* from the Late Jurassic. (a) Front three-quarter view; (b) rear three-quarter view. This dinosaur is no longer posed with the old familiar hump-back, the anatomy of the ligaments and plates both support a straight tail.

Recent studies on stegosaur development, based upon the type of bone tissue they have, suggest that by the standards of most other dinosaur groups, stegosaurs generally took their time to grow, not getting too large too quickly. This has implications for how stegosaurs used their distinctive plates (see Spines and Plates below).

Dealin' with Mealin'

The business end of feeding in stegosaurs began at the rhamphotheca (beak), similar to those seen in modern turtles and birds, which covered the fronts of both the upper and lower jaws (Figure 10.7). The rhamphothecae were probably sharp-edged, and were used to crop and strip foliage.

Like all genasaurs, stegosaurs had an inset tooth row, implying cheeks, which in turn suggests chewing; however, exactly how that must have worked is baffling. The cheek teeth of stegosaurs were relatively small, simple, and triangular (Figure 10.8), and not tightly pressed together in a block for efficient grinding. Moreover, the teeth lack regularly placed, well-developed worn surfaces, features present in herbivores that chew by grinding. Furthermore, the coronoid process was low,

(b)

Figure 10.5. (*cont.*)

lending little mechanical advantage to the jaw musculature. Chewing? Perhaps, but not particularly efficient when compared with other chewing vertebrates.

So how else might stegosaurs have ground their food? Birds use gastroliths (see Chapter 6) to grind food. The problem is that gastroliths have never been found with stegosaur remains, as they have with other dinosaurs (prosauropods, sauropods, psittacosaurs, and ornithomimosaurs). Ultimately, however, the coexistence in stegosaurs of a small skull (in proportion to the size of the animal), simple, irregularly worn teeth, large gut capacity, cropping rhamphothecae, weak jaw musculature, and cheeks all conspire to make the business of dealin' with mealin' – beyond knowing that they were herbivorous – poorly understood in these dinosaurs.

What might stegosaurs have eaten? In most, the head was held near the 1 m level. Thus stegosaurs were likely low-browsers, consuming ground-level plants such as ferns, cycads, and other herbaceous gymnosperms (see Chapter 14).

Of course, the Mesozoic world of the low-browsers was not filled only with stegosaurs. It is very likely that stegosaurs competed with a variety of other dinosaurs, many of which appear to have been more efficient chewers. Could stegosaurs have used their narrow skulls to select only the most nutritious parts of the plant, while everybody else dined less discriminately?

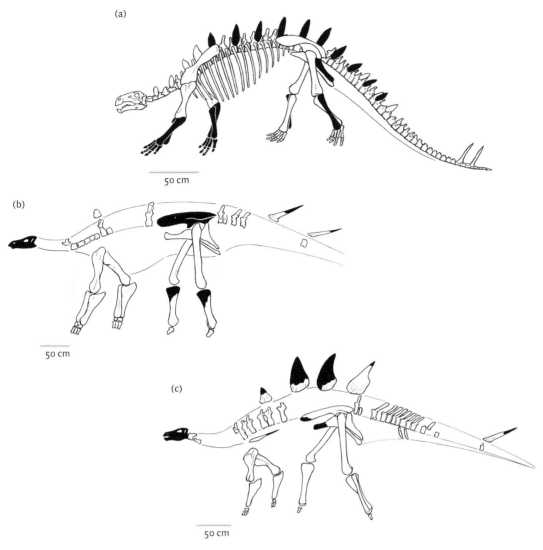

Figure 10.6. Left lateral views of the skeletons of (a) *Huayangosaurus*; (b) *Dacentrus*; (c) *Lexovisaurus*.

On the other hand, maybe stegosaurs weren't confined to low-browsing. Some paleontologists have argued that stegosaurs could have reared up on their hindlimbs in order to forage. Then the strong, flexible tail might have acted as a third "leg" (as we saw in sauropods) to form a tripod. If so, these animals could have reached as high as 6 m in the largest forms.

No Brains, One Brain, or Two Brains?

It is as clear as most anything can be at a distance of 150 million years that stegosaurs were just not all that bright. Their brains were an estimated 0.001 percent of the adult stegosaur body weight, putting them near the bottom of the dinosaur – for that matter, vertebrate – gray-matter scale (Figure 10.9). Brainy-ness must not have been part of the stegosaur largely defensive life strategy, as indeed they were so small-brained that early paleontologists felt compelled to assign them an extra brain: based upon an enlargement of the canal in the centra of the vertebrae (see inset to Figure 4.5) in the hip region, in which the spinal cord rests. Here began the legend of the dinosaur with two brains: a small one in the head and another in the pelvis, presumably to pick up the slack left by the

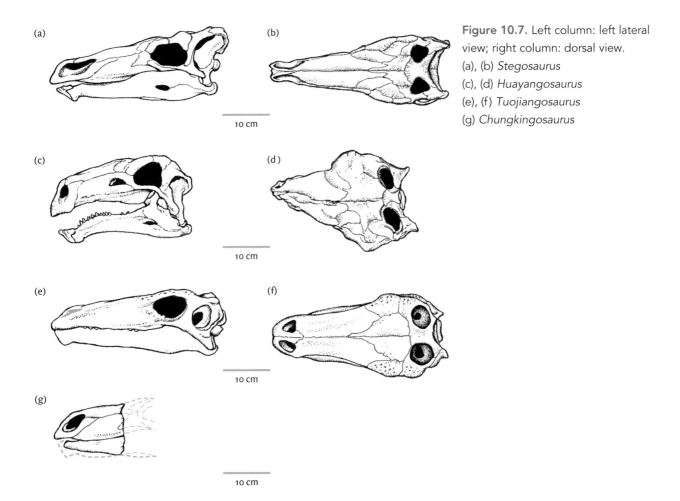

Figure 10.7. Left column: left lateral view; right column: dorsal view.
(a), (b) *Stegosaurus*
(c), (d) *Huayangosaurus*
(e), (f) *Tuojiangosaurus*
(g) *Chungkingosaurus*

first. All of this inspired literary effulgence, two samples of which we apologetically offer in Box 10.1.

The enlargement of the stegosaur spinal canal in the pelvic region, the putative "hind brain," is admittedly another stegosaur enigma. Many vertebrates have enlargements in the sacrum for nerves going to the hind legs, but the neural canal at the front of the stegosaur pelvis is upward of 20 times the volume of the brain. Some living birds have a similar enlargement that houses an organ whose function is thought to supply glycogen (a complex sugar-based molecule which the body stores, but can break down to obtain energy) to the nervous system. In the early 1990s, evolutionary biologist and anatomist Emily Bucholtz proposed that the enlargement in the stegosaur sacrum could have housed a glycogen body, an idea that still is perhaps the best explanation for the high degree of sacral enlargement.

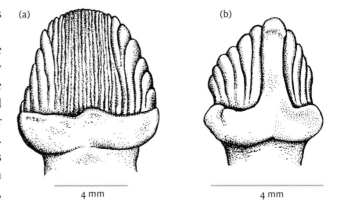

Figure 10.8. Inner views of an upper tooth of (a) *Stegosaurus* and (b) *Paranthodon*.

Speculation aside, the two stegosaurs in which the brain cavities are known – *Kentrosaurus* and *Stegosaurus* – suggest that stegosaur brains were relatively long, slightly flexed, and *small*

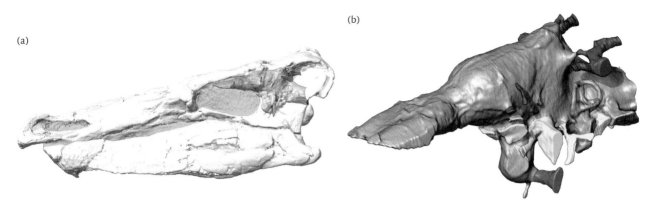

Figure 10.9. CT-scanned reconstruction of skull and brain of *Stegosaurus*. (a) cross-section through skull showing location of brain; (b) brain. Images courtesy of Larry Witmer and Witmerlab, Ohio University, USA.

(Figure 10.9). Only the olfactory bulbs, the portions of the brain that provide the animal with its sense of smell, are somewhat enlarged. Clearly stegosaurs, animals that had an unhurried lifestyle and possibly a relatively uncomplicated range of behaviors, would have had time and the hardware to stop and smell the roses... had there *been* any roses![2]

Social Lives of the Enigmatic

We don't have much of an idea about the social behavior of stegosaurs or know much about their life histories. No nests, isolated eggs, eggshell fragments, or hatchling material is yet known for any stegosaur. In fact, only a few juvenile and adolescent stegosaur specimens can tell us anything about the lives of subadult stegosaurs.

Among fully adult individuals, some workers have proposed that there is some sexual dimorphism; that is, differences between the sexes reflected in the morphology. It has been claimed that this shows up in, almost biblically, the number of ribs that contribute to the formation of the pelvis, as well as in the size of the thigh. While sexual dimorphism was never definitively demonstrated using these features, in 2016, differences in the size and shape of the plates were clearly shown for *Stegosaurus*. All of the appropriate criteria for the differences being related to gender were observed: there were only two varieties and no intermediate forms, one form was 45 percent larger and wider than the other; there were approximately equal numbers of each kind, and these differences occur at every plate position on the animal.

Little is known about the degree of sociality among stegosaurs. A mass accumulation of disarticulated, yet associated, *Kentrosaurus* material from Tendaguru in Tanzania (see Chapter 16) provides us with a hint that *Kentrosaurus* was gregarious (exhibited herding and other social behaviors), as does some *Stegosaurus* material from Montana. The combination of possible gregarious behavior (although historically stegosaurs were found as isolated specimens) and sexual dimorphism suggests that at least some stegosaurs may have interacted through display, something like we see in modern birds. Of course, feathers – or anything approaching them – are not known in any stegosaur, although the conditions of stegosaur preservation to date do not seem to suggest the likelihood of finding feather remains. Sexual signals – such as they were – likely came from the shapes and sizes of the spines and plates that graced the back of stegosaurs.

---

[2] Roses didn't appear until long after the last stegosaur went extinct – see Chapter 15.

Box 10.1 Dino Doggerel

OK. So we're not talking about Keats, Blake, Dickinson, or even Nash here; dinosaur poems never aspired to that. Most have contrasted dinosaurian enormity with their putative lack of brain power (and, no doubt, social grace), with few recent efforts to balance such dismal views.

The most famous dinosaurian poem celebrates the mental achievements of *Stegosaurus*, in particular the cerebral gymnastics powered by its double brains. The piece, by Bert L. Taylor, a columnist in the 1930s and 1940s for the *Chicago Tribune*, goes like this:

> Behold the mighty dinosaur,
> Famous in prehistoric lore,
> Not only for his power and strength
> But for his intellectual length.
> You will observe by these remains
> The creature had two sets of brains –
> One in his head (the usual place),
> The other at his spinal base.
> Thus he could reason a priori
> As well as a posteriori
> No problem bothered him a bit
> He made both head and tail of it.
> So wise was he, so wise and solemn,
> Each thought filled just a spinal column.

> If one brain found the pressure strong
> It passed a few ideas along.
> If something slipped his forward mind
> 'Twas rescued by the one behind.
> And if in error he was caught
> He had a saving afterthought.
> As he thought twice before he spoke
> He had no judgement to revoke.
> Thus he could think without congestion
> Upon both sides of every question.
> Oh, gaze upon this model beast,
> Defunct ten million years at least.

As a counterpoint to the power of Mesozoic intellects, we also provide some thoughts, entitled *The Danger of Being too Clever*, by John Maynard Smith, the famous English evolutionary biologist.

> The Dinosaurs, or so we're told
> Were far too imbecile to hold
> Their own against mammalian brains;
> Today not one of them remains.
> There is another school of thought,
> Which says they suffered from a sort
> Of constipation from the loss

Spines and Plates

In keeping with their thyeophoran heritage, the majority of stegosaurs had at least two rows of osteoderms down the back; one edge embedded deeply in the thick tough hide (Figure 10.10). These osteoderms generally take the form of spines, spikes (Figure 10.11), blunt cones, or plates. We've seen that they may have been dimorphic. Originally, the idea was that they were all about protection and defense. But defense, if it is any part of the story, isn't the whole story.

The shapes and patterns of plates and spines in stegosaurs are nearly always species-specific; that is, diagnostic for a particular species of stegosaurs). Moreover, they have their maximum visual effect when viewed from the side. These observations, in conjunction with those above, further suggest a display function – both for predators *and* for other stegosaurs. Predators surely would have seen a larger, more intimidating animal with the plates sticking up. If intraspecific display (display among members of the same species) was involved, it is likely that individual stegosaurs would have used these structures not only to tell each other apart, but also to gain dominance in territorial disputes and/or as libido-enhancers during the breeding season.[3]

---

[3] Which gender had the big plates and which had the narrower, smaller plates is an open question. If living birds and reptiles are any reasonable model it is likely that it was the males with the larger plates; but many exceptions surely exist.

Of adequate supplies of moss.
But Science now can put before us
The reason true why *Brontosaurus*
Became extinct. In the Cretaceous
A beast incredibly sagacious
Lived & loved & ate his fill;
Long were his legs, & sharp his bill,
Cunning its hands, to steal the eggs
Of beasts as clumsy in the legs
As *Proto-* & *Triceratops*
And run, like gangster from the cops,
To some safe vantage-point from which
It could enjoy its plunder rich.
Cleverer far than any fox
Or Stanley in the witness box
It was a VERY GREAT SUCCESS.
No egg was safe from it unless
Retained within its mother's womb
And so the reptiles met their doom.
The Dinosaurs were most put out
And bitterly complained about
The way their eggs, of giant size,
Were eaten up before their eyes,
Before they had a chance to hatch,

By a beast they couldn't catch.
This awful carnage could not last;
The age of archosaurs was past.
They went as broody as a hen
When all their eggs were pinched by men.
Older they grew, and sadder yet,
But still no offspring could they get.
Until at last the fearful time, as
Yet unguessed by *Struthiomimus*
Arrived, when no more eggs were laid,
And then at last he was afraid.
He could not learn to climb with ease
To reach the birds' nests in the trees,
And though he followed round and round
Some funny furry things he found,
They never laid an egg – not once.
It made him feel an awful dunce.
So, thin beyond all recognition,
He died at last of inanition.
MORAL
This story has a simple moral
With which the wise will hardly quarrel;
Remember that it scarcely ever
Pays to be too bloody clever.

The evidence that exists suggests that the idea is not completely out to lunch. The few juvenile stegosaurs known appear not to have had large spines or plates on their backs. The absence of these features in small, sexually immature individuals suggests that only when maturity was reached did looking big and sexy acquire importance. Here, then, is a strong argument for the importance of plates in display: if the plates develop only when you reach maturity, then they likely had little role to play in immature activities, and must have been associated with adult behaviors.

Thyreophorans as a group tend to be defensively armored, and stegosaurs were no exception. Along with the plates and spines, a flexible pavement of small osteoderms covered the bottom of the neck region.

Hot Plates?

The surfaces of the plates of *Stegosaurus* are covered with an extensive pattern of grooves, while the insides are filled with a honeycomb of channels (Figure 10.12). These external grooves and internal channels most likely formed the bony walls for an elaborate network of blood vessels. With such a rich supply of blood from adjacent regions of the body, could the plates have been used to cool the body by dissipating heat as air passed over them, or to warm the body by absorbing solar energy? In short, could the plates have been used for thermoregulation (temperature control)? As a test of this idea, paleontologist J. O. Farlow and colleagues tested the ability of the plates to radiate (or absorb)

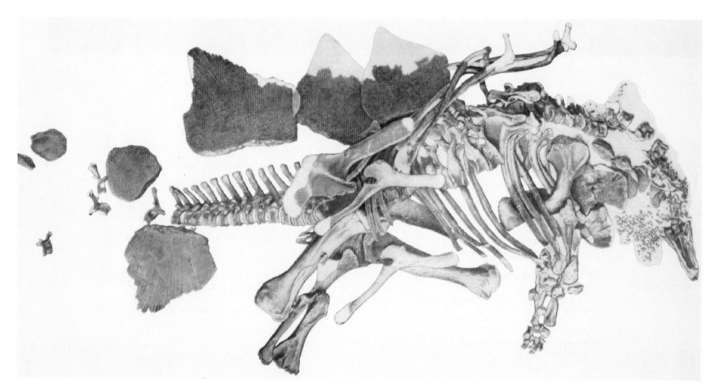

Figure 10.10. Perhaps the best specimen (NMNH 4349) of *Stegosaurus* as it was found in the field. Note that the plates do not articulate directly with the vertebrae; they were embedded in the skin; so exactly how they and the tail spikes were oriented on the living animal has been somewhat hard to determine. Paleontological consensus, however, now has them as two parallel, offset, vertically oriented rows down the back of the animal. The paleontologist K. Carpenter, of the Prehistoric Museum in Price, UT, USA, reconstructs the tail spines, however, oriented pointing outward; he noted, reasonably, that their evident functionality would have been impaired if they pointed directly upward like the plates. Image is the original drawing from 1887 publication by O. C. Marsh (see Chapter 16). Marsh, O. C. 1887. *American Journal of Science*, **34**, 413–417.

Figure 10.11. The skeleton of *Kentrosaurus*, a spiny stegosaur from the Late Jurassic of Tanzania.

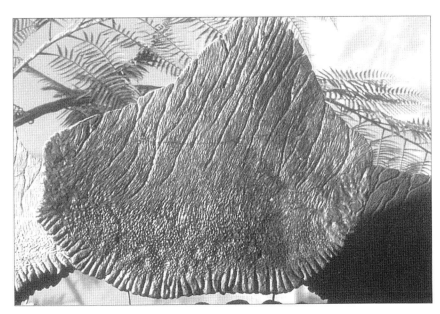

Figure 10.12. Lateral view of one of the dermal plates of *Stegosaurus*. Note the great number of parallel grooves, used to convey nutrients through blood vessels across the outer surface of the plate to the keratin covering the bony plate.

heat. Arranged in symmetrical offset pairs, as they are thought to have been in life, the plates provided significant heat dissipation, suggesting that thermoregulation may also have been a role of the plates.

Creative and attractive as this idea is, it was strongly critiqued by Harvard biologist R. P. Main and colleagues.[4] Their 2005 analysis of plate microstructure argues that no system of conduits exists that would have circulated blood in and out of the plates radiator-style; rather, the grooves preserved on the surface of the plates were likely used to bring nutrients to the keratinized sheath that covered the bony plate in life, much as one sees in the bony underpinnings of nails, claws, and beaks. And the tubes on the inside? Like part of the remodeling process associated with bone growth; in short, business as usual (see Chapter 14).

So we come back to where we began: as an appearance-enlarging device, the plates may have served an important role in defense. Yet, there has never been any doubt about the role of the long, pointed tail spikes! These were likely slashed from side to side on the powerful tail. Older reconstructions show these spikes pointing primarily upward; recent work by dinosaur specialist Ken Carpenter (see caption, Figure 10.10) suggests that the spikes actually splayed out to the sides, producing a much more effective defensive weapon.

So stegosaurs appear as a mass of contradictions: chewers that may not have chewed all that well; animals in which supposed sexual display functions were prominent but in which there is only a small amount of evidence for social behavior. Until very recently, we didn't even know the positions of the plates in *Stegosaurus*, the best-known genus. So much of what is apparently contradictory about stegosaurs is likely due to how little we know about them – a situation that we'd like to see changed!

---

[4] We will see several examples of creative and attractive, but potentially discredited ideas as this book unfolds. If science, as we asserted in Part I, is fundamentally a creative process, then these ideas are important not only for what they teach us about dinosaurs, but for what they teach us about the creative process of hypothesis-development and testing.

Eurypoda: Ankylosauria – Mass and Gas

Ankylosaurs were masters of the art of defense-by-hunkering. As their name implies, ankylosaurs were encased in a pavement of bony plates and spines – each embedded in skin and interlocked with adjacent plates – that formed a continuous shield across the neck, throat, back, and tail (Figures 10.13, 10.14, and 10.15). In many cases, it covered the top of the head, cheeks, and even eyelids.

Under all that armor, ankylosaurs were round and very broad (see Figure 10.13); clearly their girth accommodated a large gut. The head was low and broad (Figure 10.16), and equipped with simple, leaf-shaped teeth for pulverizing whichever plants an ankylosaur chose to eat (Figure 10.17).

Among the most striking features of the skulls of ankylosaurs are the complex internal nasal passageways, which have been revealed by CT scans (see Box 11.1) of the skulls. These passageways were elaborate, and in some cases, convolute (see below, Figure 10.18). Their precise function is uncertain, although they imply that olfaction, the sense of smell, was an important sense in this group of dinosaurs.

Figure 10.13. *Euoplocephalus*, the armored, club-tailed ankylosaur.

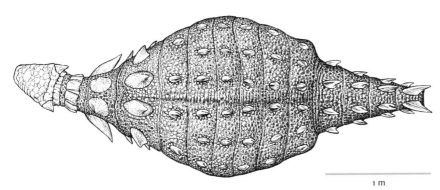

Figure 10.14. Dorsal view of the body armor of *Euoplocephalus*; reconstruction.

1 m

Figure 10.15. Illustration of the dorsal body armor of *Borealopelta*. The discovery of this extraordinary specimen was one of the paleontological highlights of 2017 (see text). From Brown, C. M. 2017. An exceptionally preserved armored dinosaur reveals the morphology and allometry of osteoderms and their horny epidermal coverings. *PeerJ*, DOI 10.7717/peerj.4066, 39 pages.

Ankylosaurs were mid-sized dinosaurs, rarely exceeding 5 m in length, although some (such as *Ankylosaurus*) ranged upward of 9 m. As in stegosaurs, the limbs were of different lengths, with the hindlimb exceeding the length of the forelimb by as much as 50 percent.

Like stegosaurs, ankylosaurs also had a global distribution, coming predominantly from North America and Asia, but also from Europe, Australia, South America, and Antarctica. As a group, they reached their peak diversity during the Cretaceous.

The best-preserved fossils come from Mongolia and China, where stunning specimens have been found inland, far from the ocean (either then or now), nearly complete and articulated, and in most cases preserved upright or on their sides. In contrast, in North America, partial skeletons are generally found, and these are often upside-down, sometimes in rocks deposited along the seashore or even in rocks representing open marine environments. The North American forms may have lived sufficiently close to the sea that their bloated carcasses might have been carried out with the tide, flipping upside-down because of their heavily armored backs. But as we shall see shortly, stunning exceptions to these beat-up wave-tossed, inverted North American forms surely exist!

Ankylosaur finds most commonly consist of individual skeletons or isolated partial remains; there is only one known ankylosaur bonebed – the *Pinacosaurus* (Mongolia) bonebed. Perhaps this

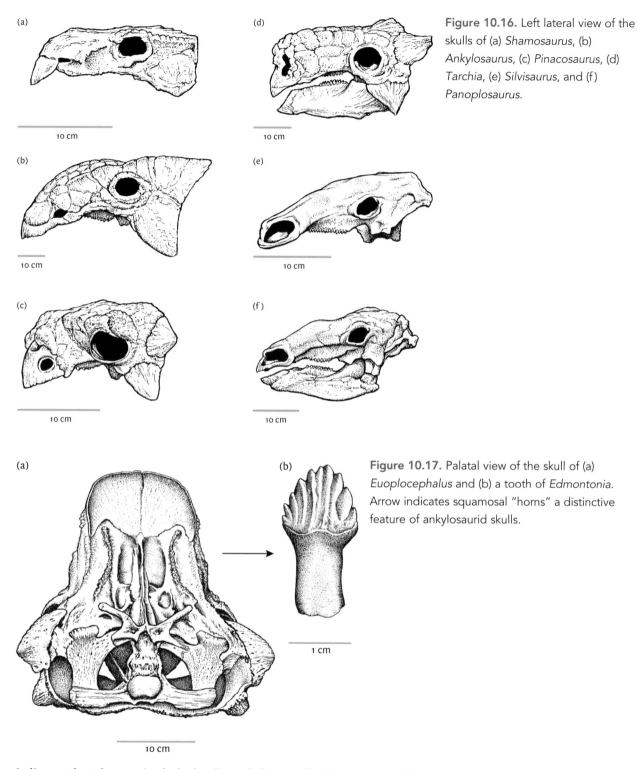

(a)

(b)

(c)

(d)

(e)

(f)

10 cm

Figure 10.16. Left lateral view of the skulls of (a) *Shamosaurus*, (b) *Ankylosaurus*, (c) *Pinacosaurus*, (d) *Tarchia*, (e) *Silvisaurus*, and (f) *Panoplosaurus*.

(a)

(b)

10 cm

1 cm

Figure 10.17. Palatal view of the skull of (a) *Euoplocephalus* and (b) a tooth of *Edmontonia*. Arrow indicates squamosal "horns" a distinctive feature of ankylosaurid skulls.

indicates that these animals had solitary habits or lived in very small groups. Even from our incomplete window on the past, it appears reasonably certain, and perhaps reassuring, that ankylosaurs did not enjoy the company of huge herds.

Ankylosauria consists of two great clades: Nodosauridae (*nodo* – knot; referring to the rounded osteoderms) and Ankylosauridae (Figures 10.18 and 10.19).

(a)

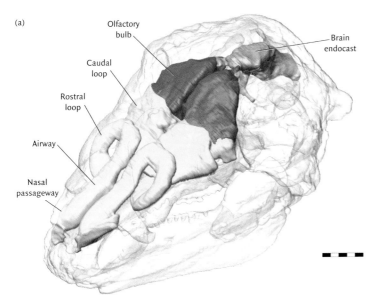

Figure 10.18. The skulls of nodosaurids (above) and ankylosaurids (below) compared. The skull of the ankylosaurid is broader, and was held at a more downward angle than that of the nodosaurid. Three-quarter view CT-scan reconstructions of (a) the nodosaur *Panoplosaurus* and (b) the ankylosaurid *Euoplocephalus*. In both cases, convolute nasal sinuses are revealed, running from the external nasal opening towards the braincase. These sinuses doubtless were involved in a heightened sense of smell; we don't know if ankylosaurs smelled good, but the evidence suggests that they smelled well. The small, darkened region to the rear of the skull is the brain. Scale bars = 5 cm.

(b)

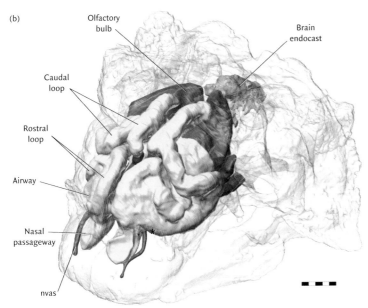

Nodosauridae

Nodosaurids had relatively long snouts and well-muscled shoulders, reflected by the presence of a large knob of bone on the shoulder blade, the **acromial process** (Figure 10.20), an attachment site for the heavy shoulder musculature that characterizes nodosaurs. Nodosaurs also had flaring hips, and pillar-like limbs. Many had tall spines at the shoulder (**parascapular spines**). Nodosaurids are known principally from the Northern Hemisphere (North America and Europe), although new discoveries in Australia and Antarctica have extended the geographical range of these animals deep into the Southern Hemisphere.

Ankylosauridae

Members of Ankylosauridae give the impression of impregnable, mobile fortresses. All are well armored (see Figures 10.15 and 10.21), but there are fewer tall spines along the body than in

(a)

(b)

Figure 10.19. (a) An ankylosaurid (*Euoplocephalus*) and (b) a nodosaurid (*Sauropelta*) compared. The ankylosaurid is stockier and has a tail-club. The nodosaurid is more lightly built and bears distinctive dermal armor.

nodosaurids. The tail ends in a massive bony club, in some instances with several paired knobs or triangular spikes along its length. The head is shorter and broader in ankylosaurids than in nodosaurids and there are large triangular plates attached to the rear corners of the skull (termed "squamosal horns;" see Figure 10.17a).

Ankylosaur Lives and Lifestyles

Mouths to Feed

Ankylosaurs likely had a very low browsing range, foraging for plants no more than a meter or so above the ground. The different beak shapes imply different feeding preferences. The narrow beak of nodosaurids may suggest selective feeding, plucking, or biting at particular kinds of foliage and fruits with the sharp edge of the rhamphotheca. In contrast, the very broad beak of ankylosaurids may imply less-selective feeding, in which plant parts were indiscriminately bitten off from the bush or pulled from the ground.

How the food was then prepared for swallowing is a bit of a mystery. Like stegosaurs (and pachycephalosaurs; see Chapter 11), the triangular teeth of both nodosaurids and ankylosaurids are small, not particularly elaborate, and less tightly packed than animals with well-developed chewing behaviors. However, the wear marks on the teeth indicate that grinding took place. In addition, it is likely that ankylosaurs had a long, flexible tongue (in their throats, they have large **hyoid bones** that support the base of the tongue) and an extensive **secondary palate**, which allowed them to chew and breathe at the same time. Moreover, deeply inset tooth rows suggest well-developed cheeks to keep whatever food was being chewed from falling out of the mouth (see Figure 10.17b). The jaw bones themselves were relatively large and strong (although lacking enlarged

Figure 10.20. *Edmontonia* on display in the Fukui Prefectural Dinosaur Museum (Japan). Note the projection on the scapula (outlined in white), known as the acromial process (indicated by red arrow), peeking out from under the carapace. The pronounced acromial process is one of the diagnostic characters of nodosaurs.

areas for muscle attachment). Indeed, most ankylosaur jaw features – except for tooth design and placement – suggest that ankylosaurs were reasonably adept chewers.

Perhaps the paradox of simple teeth in strong, cheek-bound jaws can be understood by looking not at how much chewing was done prior to swallowing (there obviously must have been some), but at the very substantial rear end of the animal, where the bones show a very deep rib cage circumscribing an enormously expanded abdominal region (see Figure 10.13). Commodious abdomens mean huge guts, and huge guts suggest that digestion took place in a very large, perhaps highly differentiated, fermentation compartment(s) in these armored dinosaurs. Bacteria likely lived symbiotically within the stomach (or stomachs, because ankylosaurs may well have had a series of them). The stomach(s) must have served as a great fermentation vat(s), bacterially decomposing even the toughest woody plant material. Among modern mammals, this method of breaking down tough plant material is well known in ruminants such as cows.

The combination of browsing at low levels and having anatomy indicative of chewing and fermenting places limits on what kinds of plants ankylosaurs may have fed on. At the levels where ankylosaurs could forage, the undergrowth consisted of low-stature ferns, gymnosperms such as cycads, and shrubby angiosperms (flowering plants), a rich array of plants to choose from within the first few meters above the ground (see Chapter 15).

Box 10.2 **The Year of the Ankylosaur (at Least in Canada)**

The year 2017 was *de facto* the Year of the Ankylosaur, as not one, but two spectacular specimens – one, a nodosaur, and the other an ankylosaurid – of these dinosaurs were recovered.

Borealopelta

The nodosaur was first found in 2011 in tar (oil) sands of northern Alberta, Canada; like all well-behaved North American ankylosaurs, on its back, in the fine sandy/silty sediment representing the floor of the seaway that bisected North America (the Western Interior Sea) during the Late Cretaceous. There, Shawn Funk, an astute workman who was clearly paying attention, encountered the thing while operating heavy equipment. Operations stopped, thankfully, and the dinosaur was extracted and shipped to the preparation laboratory of the Royal Tyrrell Museum, where for the next six years preparator Mark Mitchell labored over its preparation, racking up some 7000 hours on it (recall that in Chapter 1 we called preparators the "unsung heroes" of paleontology!). In 2017, postdoctoral researcher Caleb Brown published the formal description, now called *Borealopelta markmitchelli*.

The specimen consists of a skull, the front half of the bony dorsal shell, or **carapace**, of the nodosaur, with some posterior material (including stomach contents) appended (Figures 10.15 and B10.2.1). What makes this specimen

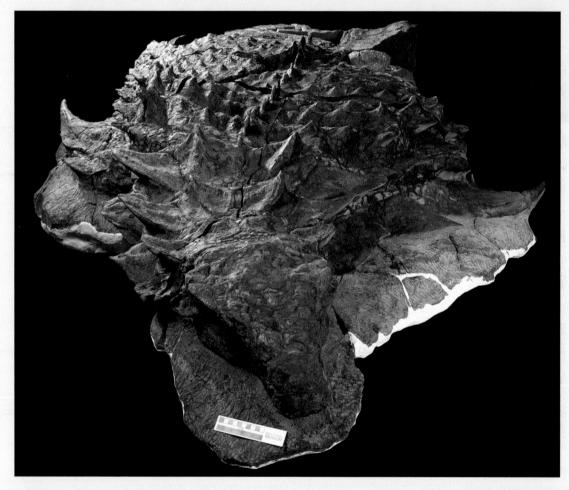

Figure B10.2.1. Three-quarter view of the fully prepared skull and carapace – now turned right side up – of *Borealopelta*.

Box 10.2 *cont'd*

spectacular is not just the completeness of the skull and carapace, but the fact that everything is preserved, including the keratin (proteins that form claws and horns) sheaths that covered all the bony material in life and, most remarkably, a dark-colored material, thought to represent original skin, still retaining the molecules that gave it its original coloration. "Remarkably," because it would seem that this dinosaur was a dark reddish-brown color dorsally, and a lighter greyish color ventrally, interpreted by Dr. Brown as countershading, a kind of camouflage in response to strong (large theropod?) predation. We revisit countershading – thought to occur in other dinosaurs – in Chapter 11.

A last question remains about *Borealopelta*: why is it so much better preserved than other ankylosaurs? Here the answer is in the preserving environment: the specimen was found in tar sands. The specimen was likely buried rapidly, before bacteria could decompose it. The oil that ultimately infiltrated the sands would have protected the original organic matter of the fossil by inhibiting anything from destroying its organic content, much as a telephone pole soaked in creosote (tar) is virtually impervious to weather.

Zuul

The other spectacular ankylosaur from 2017 was 35 million years younger than *Borealopelta*, but in many respects paralleled its discovery. Paleontologists prospecting well-studied outcrops in northern Montana in 2014 recovered *Zuul crurvivastator*, an ankylosaurid ankylosaur. Evidently, back in the latter part of the Late Cretaceous, an ankylosaur had died, turned turtle (another upside-down North American ankylosaur!), and floated into an estuary, where

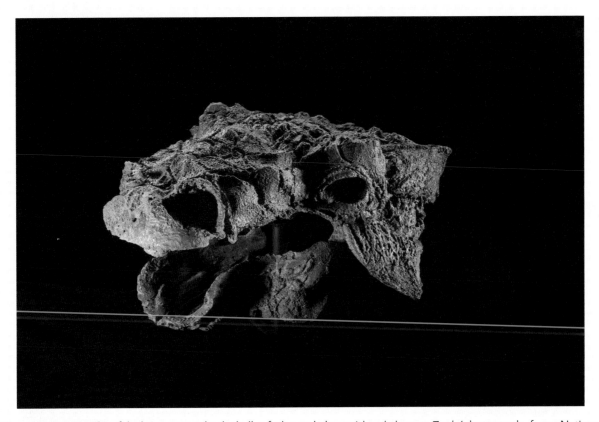

Figure B10.2.2. The fabulous, uncrushed skull of the ankylosaurid ankylosaur *Zuul* (photograph from National Geographic by Mark Thiessen.

Box 10.2 *cont'd*

it was buried before it could be disarticulated, dismembered, and crushed. Or rather, *completely* dismembered; the legs are missing from the specimen!

Zuul[a] was originally found in 2014 by a commercial collecting company, "Theropoda," while working on a specimen of *Gorgosaurus* to sell to the (wealthy) public. Its tail-club marked it for an ankylosaurid ankylosaur, and recognizing that it was something unusual that belongs in a museum for study, rather than in a rich person's living room for private display, "Theropoda" contacted paleontologist David Evans of the Royal Ontario Museum, which acquired the fossil in 2016 as a single 20-ton jacket. The specimen was sent to Research Casting International (RCI), a company that specializes in preparing museum exhibits.

By 2017 the skull and tail had been prepared, and the specimen was named and described. But beyond its gorgeous three-dimensional skull (Figure B10.2.2) and impressive tail-club, there is far more to *Zuul*, including a complete carapace, stem to stern, and beautiful skin impressions throughout. As this is being written, unsurprisingly, given the size of the dinosaur (approximately 6 m or 20'), preparator Amelia Madill and her team at RCI continues to remove sandy matrix uncovering the secrets of this extraordinary dinosaur.

Already the animal is thought to bear a cracked and healed carapace (a war wound from another angry ankylosaurid?) and has yet to be scoured for molecules that might indicate its color or other aspects of its paleobiology. In the end, this animal could be even more spectacular than the amazing *Borealopelta*!

[a] Here, we believe the authors' own words best explain the etymology of the generic name of this ankylosaur: "The generic name refers to Zuul the Gatekeeper of Gozer, a fictional monster from the 1984 film Ghostbusters." Dan Ackroyd, the star of Ghostbusters and a dinosaur enthusiast, loved it. You just can't make this stuff up! Arbour, V. and Evans, D. 2017. A new ankylosaurine from the Judith River Formation of Montana USA, based on an exceptional skeleton with soft tissue preservation. *Royal Society Open Science*, **4**, 161086. http://dx.doi.org/10.1098/rsos.161086

Brains and Senses

Ankylosaurs may not have been particularly adept at making foraging choices, however, because their brain power was close to the bottom of the dinosaur range (only sauropods had smaller brains for their size; see Chapter 13). Still, had there been roses (see footnote 2, this chapter), ankylosaurs, too, might have stopped to smell them; the skulls were equipped with enlarged, convolute nasal passages (Figure 10.18), suggesting that olfaction may have been an important sense for these dinosaurs. Hand in hand with slow thinking, ankylosaurs were among the slowest moving of all dinosaurs for their body weight. Estimates suggest that they were able to run no faster than 10 kmh (about 6 mph) and walked at a considerably more leisurely pace (about 3 kmh [less than 2 mph]). Ankylosaurs were built for digestion, not for speed.

Defense

Although ankylosaurs were slow on their feet, other aspects of the limb skeleton suggest that they could actively defend themselves against predators. In all ankylosaurs the entire upper surface of the body – the head, neck, torso, and tail – was covered by a pavement of bony plates (see Figures 10.13 to 10.15; see also Figure B10.2.1).

In the case of nodosaurids, the shoulder region, as we have seen, was exceedingly powerful, with tall spines. This pumped-up, well-defended front end, in conjunction with relatively long hindlimbs and wide stance, may have had the effect of dropping the center of gravity of the animal forward, providing nodosaurids with an aggressive frontal, spine-based defense.

(a)

(b)

Figure 10.21. (a) The bony tail-club of *Euoplocephalus*. (b) Tail spikes in the tail of *Pinacosaurus*. Visible along the length of the tail are the ossified tendons.

Ankylosaurids went even further, augmenting their armor by a powerful tail-club (Figure 10.21), constructed of paired masses of bone set at the end of a tail (itself about half the length of the body). In some cases the tail was also equipped with spikes along its length (Figure 10.21b).

Thanks to CT scans and careful muscle reconstructions, we now know a lot about the design of the ankylosaurid tail-club. The paired masses of bone varied in density, with the densest bone located outermost (Figure 10.22). While the tail was flexible at its base, the rear half was stiffened by modified, interlocking vertebrae, as well as by a series of tendons that had become ossified – turned to bone, and thus inflexible – running down its length. This part of the tail was not meant to bend, and it provided firm attachment for the powerful muscles. Using the flexibility of the tail vertebrae at the base and the stiffened tail farther out, the club could have been forcefully swung side to side like a baseball or cricket bat.

Just how forcefully? Paleontologist Victoria Arbour theoretically calculated the velocity and thus the power of the swing; her estimates of the velocity, depending upon the type of muscle reconstruction used, are as high as 18.9 m/s, producing bone-crunching impulses of 376 kgm/s. The idea, clearly, would be to avoid any such contact with an ankylosaur tail-club in the first place!

For nodosaurids, a proactive defense must have been a head-first (or shoulder-first) proposition,

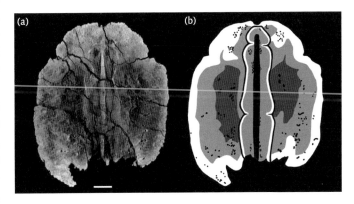

Figure 10.22. (a) CT-scan image through an ankylosaur tail-club. (b) Interpretation of (a), showing densest bone (white), bone of medium density (light gray), and bone of lowest density (dark gray). Scale bar equals 5 cm.

keeping the parascapular spines pointed toward the predator. Ankylosaurids, on the other hand, likely first planted their hindlimbs and then rotated their forequarters with their strong forelimb muscles, ever keeping watch on the threatening opponent. Then they would have wielded that massive club at their opponent's legs and feet.

In both ankylosaurids and nodosaurids, however, the last resort must have been to hunker down defensively. With its legs folded under its body, a 3500 kg ankylosaur would have been formidably immobile. Safe under protective armor, both nodosaurids and ankylosaurids were the best-defended fortresses of the Mesozoic.

The Evolution of Thyreophora

Thyreophora

We can be reasonably confident that the evolution of thyreophorans embodied increasing size and a return to quadrupedality, because the cladograms show that the quadrupedal eurypodan clades were all derived relative to primitive thyreophorans. Primitive thyreophorans suggest an evolutionary sequence from gracile, small, bipedal creatures like *Scutellosaurus* to larger, quadrupedal dinosaurs like *Scelidosaurus* (see Figures 10.3 and III.3).

Eurypoda

It is not difficult to imagine the evolution of an ankylosaur from an armored, primitive quadruped like *Scelidosaurus*. The cladogram (Figure 10.23) suggests that the basal eurypodan must have looked something like *Scelidosaurus*, and the step to a larger, more powerful, more heavily armored primitive ankylosaur or stegosaur is easy to conceive.

Stegosauria

Stegosauria is a monophyletic clade of ornithischian dinosaurs, diagnosed on the basis of a number of diagnostic features shown in Figure 10.24. The ancestral stegosaur must have been an animal with spine-shaped osteoderms and fore- and hindlimbs of not too dissimilar lengths. Within Stegosauria, the basal split is between *Huayangosaurus* and all the remaining species. This divergence took place sometime before the latter half of the Middle Jurassic. *Huayangosaurus* itself has a number of uniquely derived features shared by a more inclusive group of stegosaurs, Stegosauridae (Figure 10.24).

Within Stegosauridae, *Dacentrurus* represents the most basal form. The remainder of Stegosauridae includes *Stegosaurus*, *Wuerhosaurus*, *Kentrosaurus*, and *Tuojiangosaurus*. The evolution of this group was evidently characterized by an increase in the difference in length between the fore- and hindlimbs (Figure 10.24).

Finally, there is *Stegosaurus* itself, the best-known, most common stegosaur (see Figure 10.5). *Stegosaurus* must have evolved its distinctive plates from the spiny, conical osteoderms present in its ancestry. Plates, however, are only known in *Stegosaurus*, and their evolution occurred sometime during the Middle or early Late Jurassic.

Ankylosauria

Reflecting the importance of heavy armor to ankylosaurs and the ease of its preservation, it is not surprising that armor and/or its support comprise the majority of derived features uniting the clade

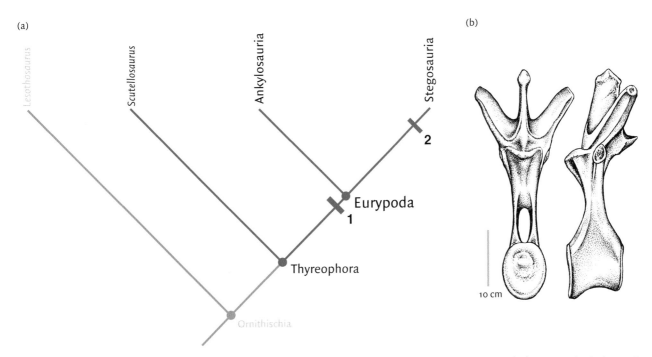

Figure 10.23. (a) Cladogram emphasizing the monophyly of Eurypoda and Stegosauria. Derived characters include: at **1** (Eurypoda), bones that fuse to the margins of the eye sockets, loss of a notch between the quadrate (see Figure 4.6) and the back of the skull, and enlargement of the anterior part of the ilium; at **2** (Stegosauria), back vertebrae with very tall neural arches and highly angled transverse processes, loss of ossified tendons down the back and tail, a broad and plate-like acromial process, large and block-like wrist bones, elongation of the pre pubic process, loss of the first pedal digit, and loss of one of the phalanges of the second pedal digit, and a number of features relating to the development of osteoderms, and formation of long spines on plates from the shoulder toward the tip of the tail. (b) The front and left lateral view of one of the back vertebrae of *Stegosaurus*. Note the great height of the neural arch; a diagnostic stegosaurian character (at **2**).

Ankylosauria (Figure 10.25). Ankylosaur evolution followed two principal pathways since the origin of the group sometime in the Jurassic: Ankylosauridae and Nodosauridae.

Primitively in all ankylosaurs, the beak was scoop-shaped but relatively narrow (although slightly broader than in stegosaurs) and remained so in the nodosaurids and in the ankylosaurid *Shamosaurus*. In all other ankylosaurids, by contrast, the beak became very broad, which matched the general broadening of the animal and the development of tail-clubs.

Ankylosauridae

Ankylosaurids share a suite of derived features. Figure 10.26 highlights the relationships among ankylosaur genera, as well as some of the key characters supporting these relationships.

Nodosauridae

Turning to the other great clade of ankylosaurs, nodosaurids share a number of derived features, as shown in Figure 10.27, particularly the well-developed acromial process. This musculature, and the parascapular spines that accompanied it, may have played a role in their defensive behavior. Nodosaurids changed little during their tenure on Earth; however, various diagnostic characters allow us to learn something of their relationships (Figure 10.27).

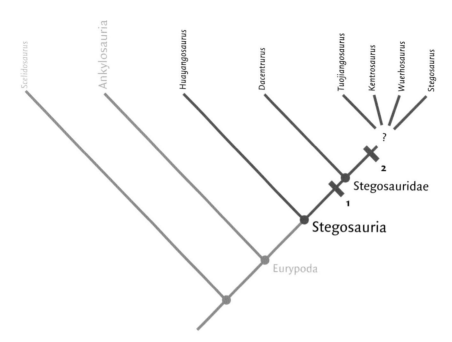

Figure 10.24. Cladogram of Stegosauria, with Ankylosauria and *Scelidosaurus* as successively more distant relatives. Derived characters include: at **1**, large antitrochanter, long prepubic process, long femur, absence of lateral rows of osteoderms on the trunk; at **2**, widening of the lower end of the humerus, an increase in femoral length, and an increase in the height of the neural arch of the back and tail vertebrae. Relationships of genera of the ultimate node on the cladogram remain uncertain.

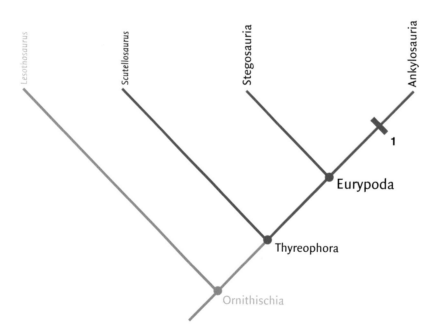

Figure 10.25. Cladogram of Eurypoda, emphasizing the monophyly of Ankylosauria. Derived characters include: at **1**, closure of antorbital and upper temporal fenestrae, ossification and fusion of keeled plate onto side of lower jaw, fusion of first tail vertebrae to sacral vertebrae and ilium, rotation of ilium to form flaring blades, closure of hip joint, development of dorsal shield of symmetrically placed bony plates and spines.

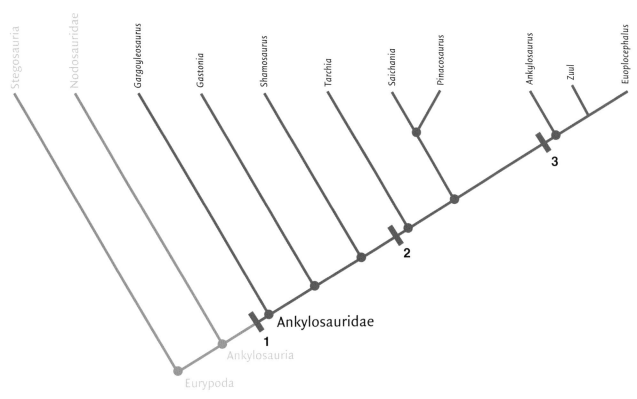

Figure 10.26. Cladogram of Ankylosauridae, with its two closest relatives, Nodosauridae and Stegosauria. Derived characters include: at **1**, pyramidal squamosal boss, shortening of premaxillary palate, presence of premaxillary notch, rostrolaterally directed mandibular ramus of pterygoid, tail-club; at **2**, rostrodorsal- and caudoventral-arching of palate, vertical nasal septum, rugose and crested basal tubera; at **3**, caudal end of postaxial cervical vertebrae dorsal to cranial end, fusion of sternals.

SUMMARY

Thyreophorans were small- to medium-sized quadrupedal ornithischians with rows of scutes down the back. Aside from some primitive forms, the two great groups of thyreophorans were Stegosauria and Ankylosauria, linked together within a monophyletic Eurypoda. Eurypodans, as a group, were not renowned for their intellects; some of the lowest brain:weight ratios known for dinosaurs come from Eurypoda.

Stegosaurs are relatively poorly known eurypodans with hindlimbs significantly longer than the forelimbs, and paired rows of plates or spines down the back, terminating in a tail with elongate spines: likely a defensive weapon. Their fossil record extends from the Middle Jurassic to the Early Cretaceous.

Stegosaurs are rare finds, and they likely lived in isolation rather than gregariously. The plates in stegosaurs were most likely associated with at least some aspect of display. Stegosaurs – like all eurypodans – had cheeks, which suggest chewing. The teeth, however, occluded relatively poorly, suggesting inefficient grinding. Owing, in part, to their rarity, the reproductive strategies and behavior of stegosaurs are still largely unknown.

Ankylosaurs were squat, armored quadrupeds, coated with a pavement of osteoderms, and bony protection everywhere. They lived from the mid Jurassic to latest Cretaceous. Two groups are known:

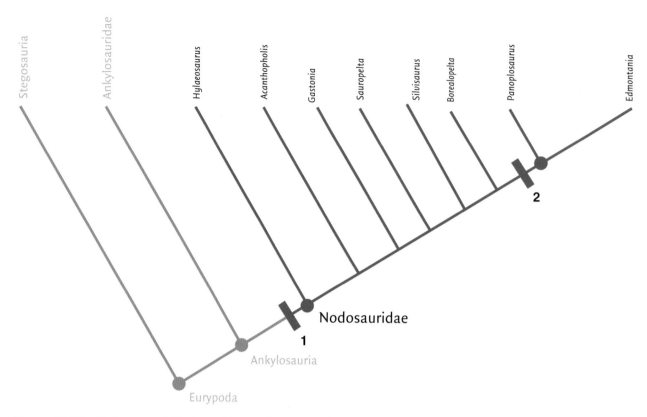

Figure 10.27. Cladogram of Nodosauridae, with its two closest relatives, Ankylosauridae and Stegosauria. Derived characters include: at **1**, knob-like acromion on scapula, occipital condyle derived from but one bone (the basioccipital), large knob (supraorbital boss) above the eye; at **2**, premaxillary teeth lacking, distinct rostrodorsal feature of secondary palate. After Brown *et al.* (2017).

Nodosauridae and Ankylosauridae. Nodosaurids were slightly more lightly built, with tall parascapular spines, while ankylosaurids were more squat, and equipped with a large tail-club.

Ankylosaurs were low-browsing animals. Their teeth and cheeks were rather like those of stegosaurs, but ankylosaurs had secondary palates, which may have aided in the efficiency of mastication. Unquestionably, though, much of their energy was obtained through gut fermentation, as suggested by the striking breadth of their girths. Ankylosaur bonebeds suggest that, unlike stegosaurs, these dinosaurs may have been gregarious animals. Little is known of their reproductive strategies and, by morphology, they were evidently animals that relied heavily upon defense, either by simply hunkering down or, in the case of ankylosaurids, by wielding tail-clubs.

SELECTED READINGS

Brown, C. M., Henderson, D. M., Vinther, J., *et al.* 2017. An exceptionally preserved three-dimensional, armored dinosaur reveals insights into coloration and predator–prey dynamics. *Current Biology*, **7**, 2514–2521. http://dx.doi.org/10.1016/j.cub.2017.06.071

Buffrenil, V. de, Farlow, J. O., and de Ricqlès, A. 1986. Growth and function of Stegosaurus plates: evidence from bone histology. *Paleobiology*, **12**, 459–473.

Carpenter, K. (ed.) 2001. *The Armored Dinosaurs.* Indiana University Press, Bloomington, IN, 512 pages.

Carpenter, K. 2012. Ankylosaurs. In Brett-Surman, M. K., Holtz, T. R., and Farlow, J. O. (eds.) *The Complete Dinosaur*, 2nd edn. Indiana University Press, Bloomington, IN, pp. 505–525.

Coombs, W. P., Jr. and Maryańska, T. 1990. Ankylosauria. In Weishampel, D. B., Dodson, P., and Osmólska, H. (eds.) *The Dinosauria*. University of California Press, Berkeley, pp. 456–483.

Coombs, W. P., Weishampel, D. B., and Witmer, L. M. 1990. Basal Thyreophora. In Weishampel, D. B., Dodson, P., and Osmólska, H. (eds.) *The Dinosauria*. University of California Press, Berkeley, pp. 427–434.

Galton, P. M. and Upchurch, P. A. 2004. Stegosauria. In Weishampel, D. B., Dodson, P., and Osmólska, H. (eds.) *The Dinosauria*, 2nd edn. University of California Press, Berkeley, pp. 343–362.

Giffin, E. B. 1991. Endosacral enlargements in dinosaurs. *Modern Geology*, 16, 101–112.

Hopson, J. A. 1980. Relative brain size in dinosaurs – implications for dinosaurian endothermy. In Thomas, R. D. K. and Olson, E. C. (eds.) *A Cold Look at the Warm-Blooded Dinosaurs*. AAAS Selected Symposium no. 28, pp. 278–310.

Jerison, H. J. 1973. *Evolution of the Brain and Intelligence*. Academic Press, New York, 482 pages.

Maryańska, T. 1977. Ankylosauridae (Dinosauria) from Mongolia. *Palaeontologia Polonica*, 37, 85–151.

Norman, D. B., Witmer, L. M., and Weishampel, D. B. 2004. Basal Thyreophora. In Weishampel, D. B., Dodson, P., and Osmólska, H. (eds.) *The Dinosauria*, 2nd edn. University of California Press, Berkeley, pp. 335–342.

Sereno, P. C. and Dong, Z.-M. 1992. The skull of the basal stegosaur Huayangosaurus taibaii and a cladistic analysis of Stegosauria. *Journal of Vertebrate Paleontology*, 12, 318–343.

Saitta, E. T. 2016. Evidence for sexual dimorphism in the plated dinosaur Stegosaurus mjosi (Ornithischia, Stegosauria) from the Morrison Formation (Upper Jurassic) of western USA. *PLoS ONE*, 10(4), e0123503. doi:10.1371/journal.pone.0123503

Vicaryous, M., Maryańska, T., and Weishampel, D. B. 2004. Ankylosauria. In Weishampel, D. B., Dodson, P., and Osmólska, H. (eds.) *The Dinosauria*, 2nd edn. University of California Press, Berkeley, pp. 363–392.

TOPIC QUESTIONS

1. What are the diagnostic characters of Thyreophora? How are thyreophorans related to other ornithischians?
2. What are the diagnostic characters of Stegosauria?
3. What are the diagnostic characters of Ankylosauria?
4. How do ankylosaurids differ from nodosaurids?
5. Describe the evidence for there being two brains in *Stegosaurus*. What is the prevailing view on this issue now?
6. Did ankylosaurs defend themselves passively, or could they actively defend themselves?
7. What are the apparent contradictions in the mouths of stegosaurs? That is, what features appear to be well designed for chewing and what features appear not to be so well designed for chewing?
8. Do ankylosaurs have the same apparent contradictions in their chewing mechanisms as stegosaurs? Elaborate.
9. What might the plates of *Stegosaurus* have been used for? Would this use apply to all stegosaurs? Why?
10. Do the small brains in stegosaurs appear to have hindered their evolutionary success? Why?
11. What is the evidence for intraspecific competition in stegosaurs?
12. What are some of the differences between ankylosaurid defense and nodosaurid defense?
13. Why do we think ankylosaurs potentially ruminated?
14. Summarize what is known about herding behaviors in thyreophorans.
15. How does the interpretation of thermoregulation in stegosaurs correlate with their inferred activity levels?
16. What is known about sexual dimorphism in thyreophorans?

Figure 11.1. Representatives of the two great groups of Marginocephalia. Left: *Homalocephale*, a pachycephalosaur; right: *Protoceratops*, a ceratopsian.

Chapter 11

Marginocephalia
Bumps, Bosses, and Beaks

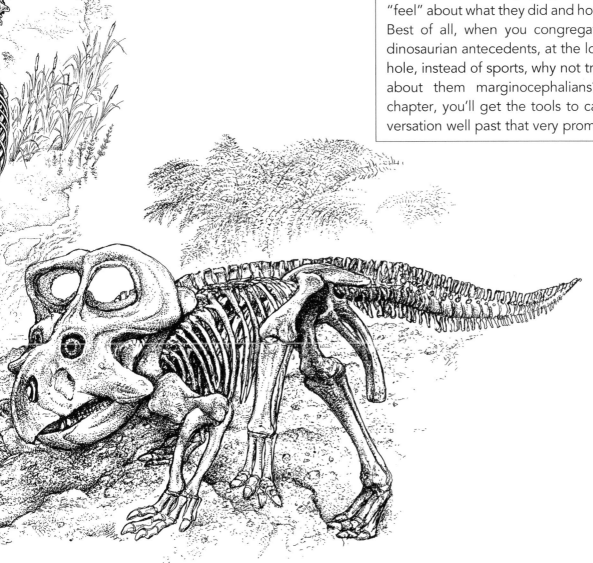

WHAT'S IN THIS CHAPTER

Here we meet Marginocephalia, a group of primarily Asian and North American Cretaceous ornithishcians that produced some of the most iconic dinosaurs that ever lived. We'll check out their phylogenetic fit and get a "feel" about what they did and how they did it. Best of all, when you congregate, like your dinosaurian antecedents, at the local watering hole, instead of sports, why not try, "Say, how about them marginocephalians?!"? In this chapter, you'll get the tools to carry the conversation well past that very promising start.

Marginocephalia

Who Were Marginocephalians?

Marginocephalia (margin – edge; *cephal* – head): not a name you'll hear from the local five-year-old dino-it-all. Yet, Marginocephalia reflects an important phylogenetic link between two major, somewhat different-looking, groups of dinosaurs: **Pachycephalosauria** (*pachy* – thick;) and **Ceratopsia** (*kera* – horn; *tops* – face). Together with **Ornithopoda** (Chapter 12), marginocephalians make up the taxon known as Cerapoda (Figure 11.2).

Marginocephalians all bear a ridge, or shelf, of bone running across the back of the skull. The size of this feature varies greatly, but in all cases, when viewed from above, it blocks from sight the bones at the back of the skull.

Although marginocephalians come in many shapes and sizes, they were restricted to the Northern Hemisphere during the Cretaceous Period and, if the fossil record is any indication, largely to Asia and North America.

Marginocephalia: Pachycephalosauria – in Domes We Trust

Pachycephalosaurs were bipedal ornithischians with thickened skull roofs (Figures 11.3 and 11.4). In the North American forms, this took the form of high domes, but in Asia, along with the high domes, certain varieties had flattened, thickened skulls (Figure 11.4). Originally recognized as separate genera (e.g., dome-headed and flat-headed forms), the evidence is accumulating that in many

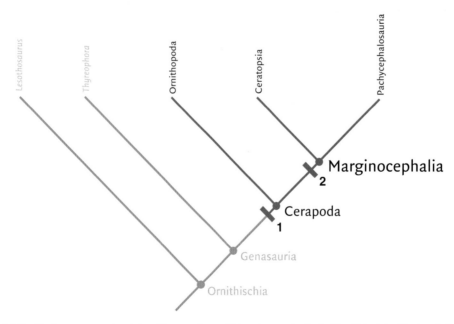

Figure 11.2. Cladogram emphasizing Cerapoda and Marginocephalia in Ornithischia. Derived characters include: at **1**, significant diastema between premaxillary and maxillary teeth, five or fewer maxillary teeth, finger-like anterior trochanter; at **2**, narrow parietal shelf obscuring occipital elements in dorsal view, lateral portions of shelf formed by squamosal.

Figure 11.3. The largest known pachycephalosaur, the North American Late Cretaceous *Pachycephalosaurus*. Note the welter of knobs and spikes, surrounding the large pachycephalosaur dome.

(all?) cases, the flat-headed forms are simply younger versions of the dome-headed forms, whose domes develop in the mature animal (see below; see also Chapter 13).

Pachycephalosaur Lives and Lifestyles

Where Did a Pachycephalosaur Call Home?

In Asia, pachycephalosaurs apparently lived in a Sahara-like desert, punctuated by ephemeral streams in small drainage basins. Their remains are commonly found as nearly complete skulls and beautifully articulated skeletons. These fossils show little evidence of transport, and were apparently fossilized close to where the living animal died.

In North America, by contrast, the environments were very different. The rocks where marginocephalian remains are found represent a broad, Cretaceous coastal plain in a then-temperate climate – built from sediment eroded as the Rockies mountain range rose to the west (Figure 11.5). The most common finds are isolated, thickened skull caps, whose water-worn appearance suggests that they were transported long distances in rivers before burial and fossilization (Figure 11.6). Indeed, many pachycephalosaurs are known from skull material only! The high proportion of skull

(a)

(b)

(c)

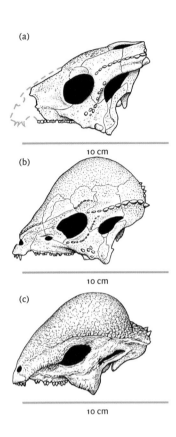

10 cm

10 cm

10 cm

(d)

(e)

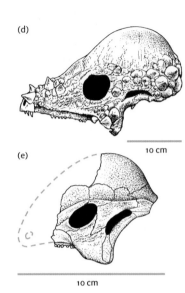

10 cm

10 cm

Figure 11.4. Left lateral view of (a) *Homalocephale*, (b) *Prenocephale*, (c) *Stegoceras*, (d) *Pachycephalosaurus*, and (e) *Tylocephale*.

cap finds suggest that only the most robust of bones – in this case, the skull caps – survived long journeys. The relative absence of articulated specimens in coastal plain sediments implies that, in life, North American pachycephalosaurs likely lived toward, or even in, mountains, where sediments were more likely to be eroded rather than deposited.

Feeding

The herbivorous habits of pachycephalosaurs are evident not only from their teeth, but also from the impressive volume of their abdominal regions. At the front of the mouth, the jaws contained simple, peg-like gripping teeth, the last of which were sometimes enlarged in a canine-like fashion (see Figure 11.4). A small rhamphotheca is thought to have been present. Further back, the cheek teeth of pachycephalosaurs were uniformly shaped with small, triangular crowns; the typical primitive ornithischian "leaf-shaped" tooth (Figure 11.7). The front and back margins of the crowns of such teeth bear coarse serrations, for cutting or puncturing plant leaves or fruits.

Turning to the opposite end of the gastrointestinal system, the rib cage of pachycephalosaurs was very broad, extending backward to the base of the tail. Its size suggests that it accommodated a large stomach (or stomachs) that broke down tough vegetation via bacterial fermentation (Figures 11.3 and 11.8).

Pachycephalosaur Brains

Pachycephalosaurs had typical ornithischian brains, and doubtless aspired to typical ornithischian thoughts. Atypical, though, were the enlarged olfactory lobes of the brain, suggesting a better-than-average sense of smell. What they smelled, however, is a secret that died with them.

For all its unremarkable-ness, the pachycephalosaur brain was uniquely oriented in its skull. The back half of the brain is angled downward, which is reflected in the rotation of the back of the skull

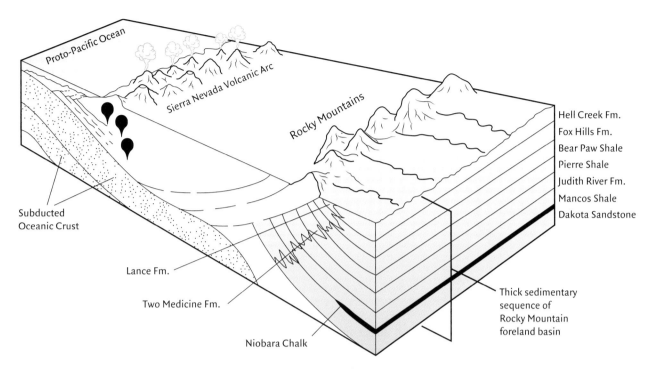

Hell Creek Fm.
Fox Hills Fm.
Bear Paw Shale
Pierre Shale
Judith River Fm.
Mancos Shale
Dakota Sandstone

Proto-Pacific Ocean

Sierra Nevada Volcanic Arc

Rocky Mountains

Subducted
Oceanic Crust

Lance Fm.

Two Medicine Fm.

Niobara Chalk

Thick sedimentary
sequence of
Rocky Mountain
foreland basin

Figure 11.5. Late Cretaceous paleogeographic block diagram of North America. As the ancestral Rocky Mountains were uplifted and then drained and eroded by rivers, a thick sequence of sedimentary deposits filled geological basins directly to the east of the rising Rockies. Fossil material was carried from the highlands by those rivers, and deposited onto the flat coastal plain to the east, producing some of the world's richest fossil deposits from this time.

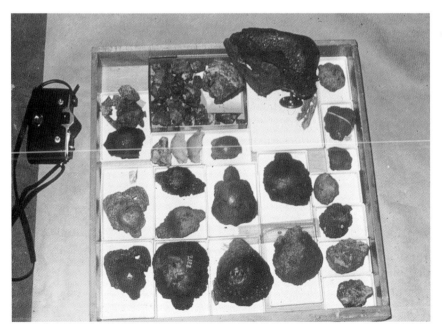

Figure 11.6. A museum tray filled with the isolated skull caps of pachycephalosaurs. Camera is 10 cm long.

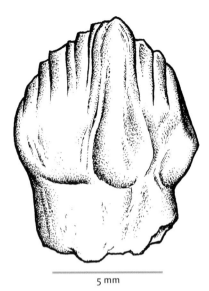

Figure 11.7. An upper cheek tooth of *Pachycephalosaurus*. This is an example of a typical primitive ornithischian "leaf-shaped" tooth; similar teeth are found in other ornithischians, including heterodontosaurids (Figure III.8a) and thyreophorans (Figure 10.17b).

5 mm

(the **occiput**) to face not only backward (as it does conventionally), but also slightly downward. The higher the dome, the more downward the orientation of the occiput. And that might be related to pachycephalosaurs using their heads for something other than profound thought.

Using Your Head… For a Battering Ram?

The morphology of the domes has long suggested to many paleontologists that, incredibly, pachycephalosaurs used their thickened skull roofs as battering rams (Figure 11.9). Internally, the structure of the dome is very dense, with the bone fibers oriented in columns approximately perpendicular to the external surface of the dome. Such an arrangement may be ideal for resisting forces that come from strong and regular thumps to the top of the head and for transmitting such forces around the brain, much as the helmet worn by sparring boxers is supposed to channel forces around the head.

Using special clear plastic cut to resemble a cross-section of the high-domed pachycephalosaur *Stegoceras* (Figure 11.10), paleontologist H.-D. Sues stressed the model in a way that simulated head-butting. The stress lines, seen under ultraviolet light, mimicked the orientation of the columnar bone, reinforcing the suggestion that the fibrous columns evolved to resist stresses induced by head-butting.

Building a Better Butt-er

If head-butting was the preferred means of pachycephalosaur expression, the body had to be set up to allow for it. Recall that the back of the pachycephalosaur skull is angled down and back when the head is held up. With the dome of the head pointing forward – the only position that makes sense for head-butting – rotation of the back of the skull minimizes the chance of violent twisting or even dislocation of the head on the neck.

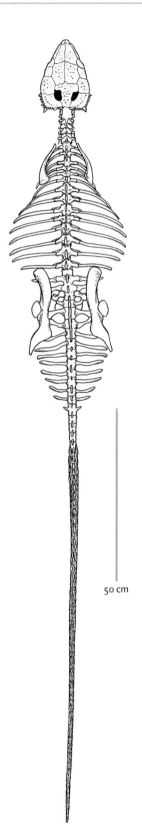

50 cm

Figure 11.8. Dorsal view of the skeleton of *Homalocephale*. Note the broadly expanded abdominal rib cage.

We might hope to see some protective measures in the neck as well; unfortunately, the neck of pachycephalosaurs is not well understood. Still, it is clear from the large muscle-attachment sites on the occiput (Figure 11.11) that the neck musculature was unusually well developed and very strong. We surmise that it was used to position and hold the head correctly for head-butting.

Further down the spinal column, and utterly unique to pachycephalosaurs, the vertebrae have been described as having distinctive tongue-and-groove articulations, which could have provided rigidity to the back. These articulations would have prevented the kinds of violent lateral rotations of the body that would otherwise have been suffered at the time of impact.

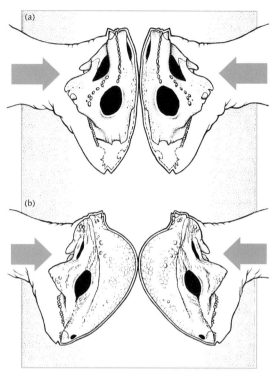

Figure 11.9. Head-on pushing and butting in (a) *Homalocephale* and (b) *Stegoceras*.

Conscientious Objectors?

But not so fast – or at least not so hard. The thickened skull cap of at least one North American pachycephalosaur – "*Stygimoloch*" – has been shown to contain abundant openings for blood vessels; the vessels and other soft tissues would have filled the spaces within the bone. With so much vascularization, the skulls of pachycephalosaurs may not have done well with either front or side impacts; this leaves the domes in this genus, and probably others, principally as display structures rather than WMDs.

Socializing Pachycephalosaur Style

Either way (or both!) – as display and/or as battering ram – social behavior is strongly implied, and some degree of sexual dimorphism might be anticipated. And so this idea was tested using a large sample of a single pachycephalosaur species, *Stegoceras validum*. It turns out that the domes of *Stegoceras* can be segregated into two groups on the basis of relative size and dome shape (Figure 11.12). One group had larger, thicker domes than the other. Strikingly, the ratio of larger-domed to less-large-domed individuals was one-to-one; exactly what you might expect if one population was male and the other, female. Which was which, however, is anybody's guess.

Sexual Selection

Pachycephalosaurs all had – along with, and perhaps including, the dome – a suite of features related to visual display. Firstly, there are the canine-like teeth. These could have been used in threat display or biting combat between rival individuals, much like pigs and some deer do today. Equally suggestive are the knobby and spiny osteoderms that covered the snout, the side of the face, and most extensively on the back of the marginocephalian shelf (see Figure 11.2, node 2; see also Figure 11.4). These distinctive features were likely all about showing off and establishing dominance.

The establishment of dominance gives one gender – males, if living reptiles are any guide – preferred reproductive access to the other gender (females, in that case), who then select with whom they'll mate. Recall from Chapter 6 that this practice is called sexual selection, selection by one gender (today generally, females), of the opposite gender. Sexual selection is quite distinct from

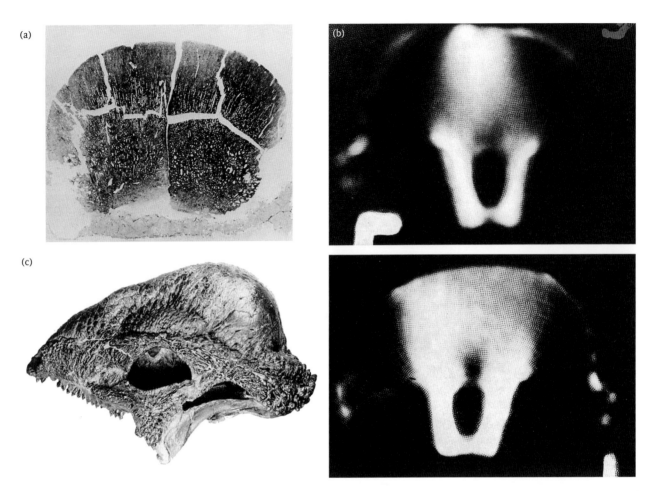

Figure 11.10. Stress patterns in pachycephalosaur domes during head butting. (a) Vertical section through the dome of *Stegoceras*. Note the radiating organization of internal bone. (b) Plastic model of the dome of *Stegoceras* in which forces were applied to several points along its outer edge and seen through polarized light. Note the close correspondence of the stress patterns produced in this model and the organization of bone indicated in (c) the left side of the skull of *Stegoceras*.

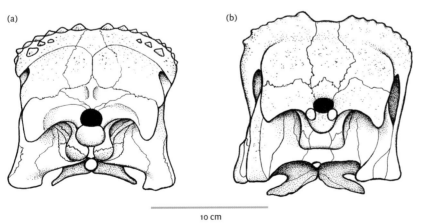

10 cm

Figure 11.11. Rear (occipital) view of the skull of (a) *Homalocephale* and (b) *Stegoceras*.

(a)

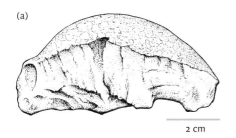

(b)

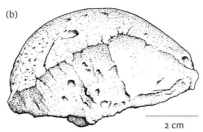

2 cm

2 cm

Figure 11.12. Two forms of the dome of *Stegoceras*. The shallower dome (a) is thought to pertain to a female, while the other dome (b) may pertain to a male.

natural selection, in which particular morphologies, unrelated to gender, are favored for reproductive success.

In pachycephalosaurs, domes, knobs, and spikes have been interpreted in the context of display: ritual, and/or potentially violent, clashes. The winner, likely the male with the best-fashioned cranial hardware, got the opportunity to be selected by a female. But he always had to be vigilant for other males that wanted to literally knock his block off – or at least knock him off the block.

The Evolution of Pachycephalosauria

Pachycephalosaurs share a host of derived features (Figure 11.13), most of them cranial and associated with the skull roof (which, as we have seen, is what preserves very well). For all that, however, our understanding of their evolution is very much in a state of flux (Figure 11.14). By some accounts, the most primitive pachycephalosaurs on the cladogram are Asian, suggesting an Asian origin for the group; yet, at least one recent analysis has the most "primitive" member of the group, *Wannanasaurus*, as the most derived!

Among the problems in pachycephalosaur phylogeny is the phylogenetic significance of the thickened, high skull roofs and the importance of the skull hornlets and decoration. These had been used as indicators of relationship and were used as a means by which general evolutionary trends were identified; if, as we report below, they are actually indices of maturity, many published phylogenies may be problematic. As we see in Figure 11.14, there's trouble a-brewin' in Pachycephalosaur Phylogenylandia.

The broad girth of pachycephalosaurs is clearly derived relative to the narrower more primitive girths seen in most other ornithischians. It suggests a backward migration and enlargement of the digestive tract to occupy a position between the legs and under the tail. As was the case for thyreophorans (see introduction to Part III Ornithischia; and Chapter 10), simple styles of chewing must have combined with fermentation-based digestion to increase the nutrition available to pachycephalosaurs from the plants they ate.

But is it Right?

It's all a good story; it falls together cohesively, and there is no reason why it isn't right. But *is* it right? Pachycephalosaurs are not common dinosaurs; as we have seen, they are not generally found as articulated specimens, but rather as skull-cap fragments; thick domes and wedges of bone, washed along and deposited in former river channels. These animals were obviously not living where they were found (within a river), so the conditions of preservation do not generally shed an ideal amount of light on their lives. And in the past ten or so years, what we thought we knew about them has had some serious challenges.

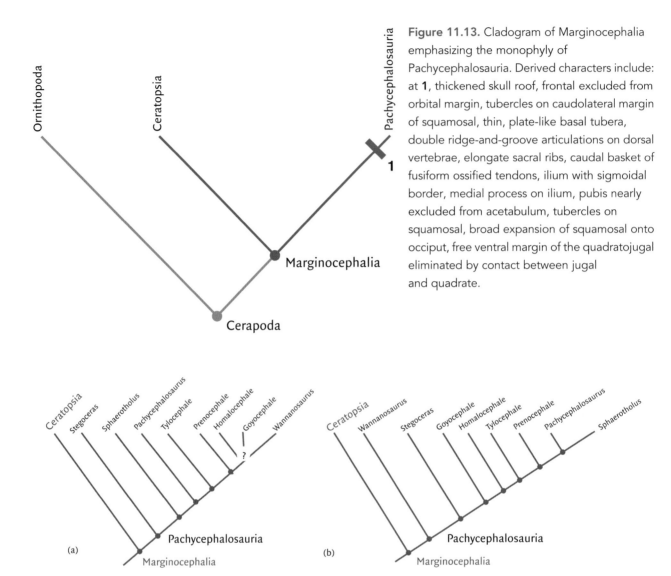

Figure 11.13. Cladogram of Marginocephalia emphasizing the monophyly of Pachycephalosauria. Derived characters include: at **1**, thickened skull roof, frontal excluded from orbital margin, tubercles on caudolateral margin of squamosal, thin, plate-like basal tubera, double ridge-and-groove articulations on dorsal vertebrae, elongate sacral ribs, caudal basket of fusiform ossified tendons, ilium with sigmoidal border, medial process on ilium, pubis nearly excluded from acetabulum, tubercles on squamosal, broad expansion of squamosal onto occiput, free ventral margin of the quadratojugal eliminated by contact between jugal and quadrate.

Figure 11.14. Two recent cladograms of the proposed relationships of pachycephalosaurs. (a) An abbreviated 2010 cladogram constructed by Longrich and colleagues; and (b) an abbreviated 2013 cladogram, constructed by Evans and colleagues. These cladograms are strikingly divergent, suggesting that considerable work yet remains before the definitive phylogenetic history of Pachycephalosauria is understood.

For example, work by J. R. Horner and M. B. Goodwin, as well as other colleagues, suggests that the Late Cretaceous North American genera *Dracocorex*, *Stygimoloch*, and *Pachycephalosaurus* may all be the same animal, with each genus representing a different stage of growth in its life. In this case, the juvenile form (*Dracocorex*) was a flat-headed creature, and with maturity came the dome. Along with the growth of the dome, these authors proposed that the spiky skulls of the juveniles became less spiky as the animals matured (Figure 11.15). The evidence that they have marshalled is impressive, and if all of this is correct, it brings up some uncomfortable second thoughts about our story:

(1) Horner and colleagues' model call for the development of the dome to go from flat to domed; doming occurring around the time of sexual maturation; a growth trajectory with which most paleontologists would agree. Does this mean that all flat-headed pachycephalosaurs were juveniles? Some? It certainly suggests that flat domes vs. domed domes don't automatically mean that they are different genera.

(2) Pachycephalosaurs were never a very diverse group, but perhaps some of even the known diversity is an overestimate, since the different types of pachycephalosaurs were identified by paleontologists by the size of their domes and the shape, position, and size of the hornlets, the very thing that appear to have changed with growth during an individual's lifetime.

(3) Almost 40 years ago, the late R. Chapman and colleagues believed that they had identified sexes from a statistically significant sample of two dome sizes in *Stegosceras*, work that we described above (Figure 11.12). Those dome sizes were interpreted as male and female, but could they have simply been two statistically distinct age populations?

(4) The argument for sexual selection in pachycephalosaurs was made by analogy to modern large herbivorous mammals like moose, deer, elk, and cows, which engage in ritualized intraspecific (and intragender) fighting as part of the sexual selection process (in mammals, generally, males perform; females choose); these presumed behaviors are enshrined in books like this and in evocative artwork such as R. T. Bakker's memorable image of pachycephalosaur head-butting (Figure 11.16).

The premise for head-butting – or any kind of butting – is based on the argument for sexual selection, and that argument is based upon the hornlets, ridges, and the dome which adorn the skulls. But even the word "adorn" is a judgment, because this word suggests that we already know what these features are used for: some kind of extraneous (generally sex-related) signaling device – like jewelry, or brand name clothes in humans. But they might be for something

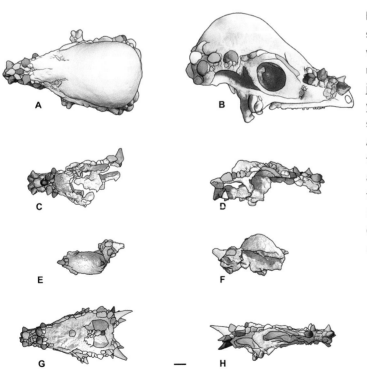

Figure 11.15. Growth in *Pachycephalosaurus* skulls. Left column: top view; right column: side view. E–H were originally named other generic names, even though they are simply likely to be juvenile forms of *Pachycephalosaurus*. The youngest specimen is "*Dracocorex*" (G, H). The subadult growth stage is "*Stygimoloch*" (E, F). A younger adult stage *Pachycephalosaurus* (C, D) is followed by an older adult stage (A, B). The horns are colored so as to be identifiable from specimen to specimen. From: Horner, J. R. and Goodwin, M. B. 2009. Extreme cranial ontogeny in the Upper Cretaceous dinosaur *Pachycephalosaurus*. *PLoS ONE*, **4**, e7626.

Figure 11.16. *Pachycephalosaurs* squaring off for some head-butting; at Dinopark Amersfoort Zoo, Netherlands.

else – not yet proposed – and if they are, then the entire sexual selection story for pachycephalosaurs is weakened.

(5) Unrelated to the trek towards maturity, head-butting as practiced in the head-to-head charges of two domed surfaces (Figure 11.16) seems to be a sure fire recipe for broken necks and significant bruising. Perhaps flank-butting, or pushing contests were the ways that pachycephalosaurs flaunted their desirability. Because unless the impact was precisely aimed – and how could it be? – the gender that didn't head butt would find itself in short supply of those that did.

The evidence for sexual selection in pachycephalosaurs is strong; of that there is little doubt. Even so, as is always with case with dinosaurs, what appears simple and obvious at first blush, might be true; or, there might be much complexity which lies beneath, that we just haven't yet thought carefully enough about, or fully understand.

Marginocephalia: Ceratopsia – Horns and All the Frills

Rhinos. Hippos. First discovered in the second half of the 1800s, paleontologists have been trying to stuff ceratopsian (*cera* – horn; *tops* – face) dinosaurs into some (any!) familiar category. But they're not rhinos and they're not hippos; they're ceratopsian dinosaurs, and there was never anything quite like them on Earth before or since. Superficially, a few late-model ceratopsians were admittedly rhino-like; powerful, large herbivores ranging upward of 6 or 7 metric tons (Figure 11.17) equipped with horns. But equally famous, and very un-rhino-like, is a host of smaller, lighter (25 kg and 200 kg) nonhorned Asian ceratopsians from slightly earlier in the Cretaceous (Figure 11.18).

We know a lot about ceratopsians: the fossil record from Asia and North America is one of the most outstanding of any dinosaur group. Primitively small bipeds, these animals evolved into powerful quadrupeds early in their history, developing thick hooves on all toes and reaching sizes to rival that of small tanks.

Figure 11.17. *Chasmosaurus*, a ceratopsian from the Late Cretaceous of the Western Interior of North America.

With or without horns, it is easy to recognize the stamp of the ceratopsian clan: ceratopsians all had skulls with a hooked beak in front and that flared deeply in the cheek region (Figure 11.19). And at the tip of the snout in the upper jaw was the uniquely evolved rostral bone (Figure 11.20).

In fact, while many ceratopsians had horns, some did not. Those ceratopsians that had them, though, grew some of the most impressive horns ever seen on any vertebrate (Figure 11.21). Like the horns of many mammals, the skulls only preserve the bony horn cores, covered by keratin sheaths that actually comprised the working end of the horn. The horn visible on the head of the living animal was therefore significantly larger than the horn core alone (Figure 11.22).

(a)

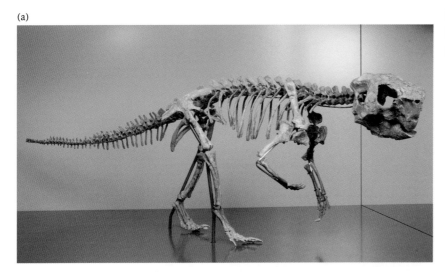

Figure 11.18. Two small, Asian, hornless ceratopsians. (a) *Psittacosaurus*; (b) *Protoceratops*.

(b)

Equally memorable is the ceratopsian frill, the marginocephalian shelf run amok. Extending from the back of the skull, frills vary considerably in almost every respect: size, ornamentation, and shape (see Figures 11.19 and 11.21). The largest reach 2 m in length.

Ceratopsian Lives and Lifestyles

Dressed and Ready to Chew

Ceratopsians chewed. And chewed. A hooked rhamphotheca, blocks of cheek teeth in both upper and lower jaws, a sturdy coronoid process, and inset tooth rows, evidence for the existence of fleshy cheeks, all scream CHEWING.

The business end of the ceratopsian mouth was the narrow, hooked, beak-tipped snout, suggesting the potential for careful selection of the plants for food. Individually the cheek teeth were relatively small, but they grew stacked and overlapping together into a large, single functional slicing block in each jaw, the dental battery (Figure 11.23). Unlike in mammals, in which there is only one episode of tooth replacement ("baby teeth" are replaced by "adult teeth"), in ceratopsians, worn teeth were

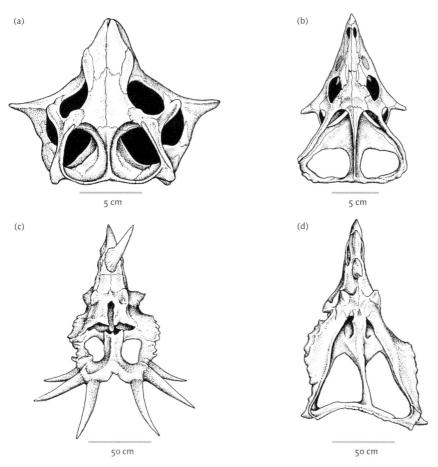

Figure 11.19. Dorsal view of the skull of (a) *Psittacosaurus*, (b) *Protoceratops*, (c) *Styracosaurus*, and (d) *Chasmosaurus*.

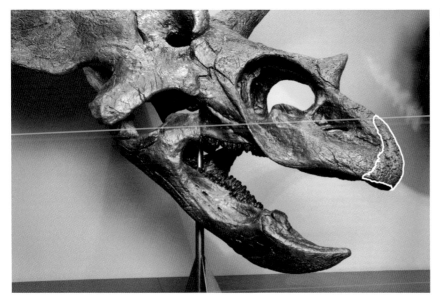

Figure 11.20. Ceratopsian snout; with diagnostic rostral bone highlighted. Photo from Museum of Natural History in Vienna.

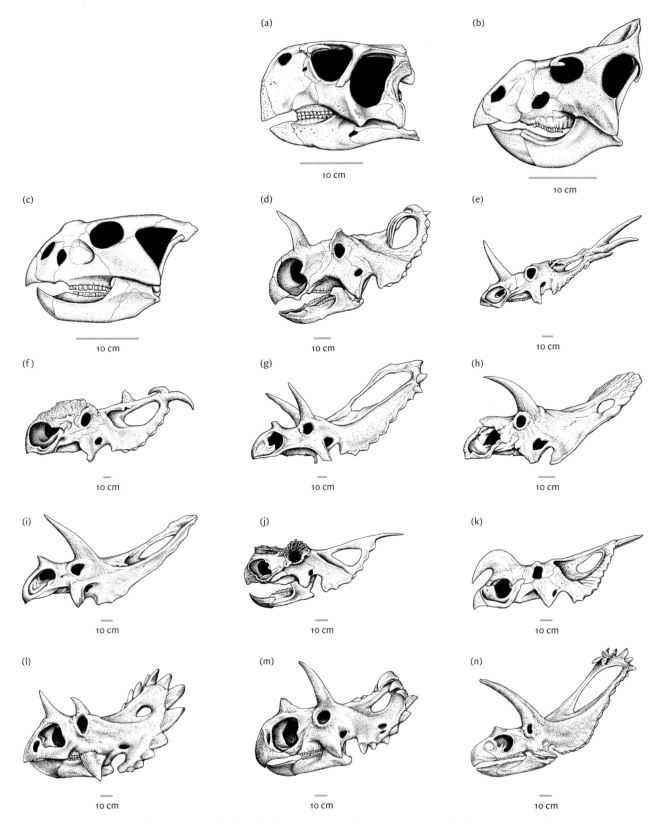

Figure 11.21. Left lateral view of the skull of (a) *Psittacosaurus*, (b) *Leptoceratops*, (c) *Bagaceratops*, (d) *Centrosaurus*, (e) *Styracosaurus*, (f) *Pachyrhinosaurus*, (g) *Pentaceratops*, (h) *Arrhinoceratops*, (i) *Torosaurus*, (j) *Achelousaurus*, (k) *Einiosaurus*, (l) *Regaliceratops*, (m) *Wendiceratops*, and (n) *Titanoceratops*.

Figure 11.22. The skull of *Triceratops* showing the horn core covered by the keratinized (finger-nail and claw material) horn as it would have been in life.

constantly replaced, so that the active chewing surface of each of the four dental batteries was continually refurbished. New ones coming in popped the old, worn ones out.

Inexplicably, and unique in the animal kingdom, the orientation of the grinding surfaces migrated during evolution, becoming more and more vertical, until, in the large, highly derived North American forms, they occurred nearly vertically along the *sides* of the teeth comprising the dental battery (Figure 11.23b).

The force behind this high-angle shearing mastication derived from a great mass of jaw-closing musculature, which crept through the upper temporal opening and onto the base of the frill. The other end of this muscle attached to a massive, hulking coronoid process on the mandible (Figure 11.24). All in all, the chewing apparatus in ceratopsians was among the most highly evolved in any vertebrate, ever.[1]

Beyond the Mouth

Based upon their narrow girth of the body (by comparison with thyreophorans and pachycephalosaurs), the digestive tract does not appear to have been disproportionately large in ceratopsians and may not have relied upon wholesale bacterial fermentation for extracting nutrients from plants. Nevertheless, it must have been big enough to accommodate the endless parade of foliage that these animals consumed.

Even the largest quadrupedal ceratopsians never browsed particularly high above the ground. The browse height of the largest was probably less than 2 m. Nevertheless, they may have been able to knock over trees of modest size in order to gain access to choice leaves and fruits.

---

[1] The primitive ceratopsians *Yinlong* and *Psittacosaurus* lacked the highly refined chewing specializations of the North American ceratopsians, but *Psittacosaurus*, at least, is known to have harbored a packet of gastroliths lodged in its gizzard, which would have doubly pulverized its meal. Gastroliths are known from no other ceratopsian.

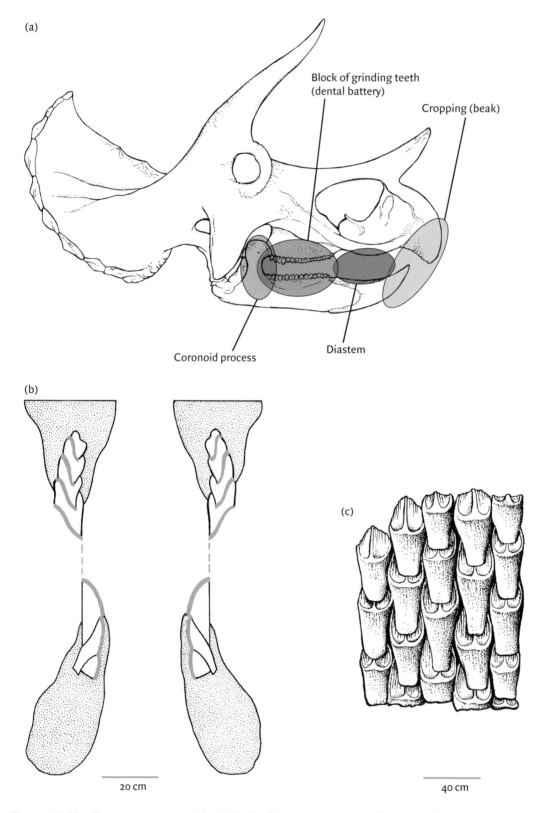

Figure 11.23. Chewing in ceratopsids. (a) Skull of *Triceratops*, exemplifying the divisions of the ceratopsid skull: anterior cropping section (beige); diastema (maroon); block of grinding cheek teeth (blue); coronoid process (brown). See also Figure III.5. (b) Cross-section through the upper and lower jaws of *Triceratops* showing high-angle grinding motion of the dental batteries; (c) internal view of the dental battery in the lower jaw of *Triceratops*.

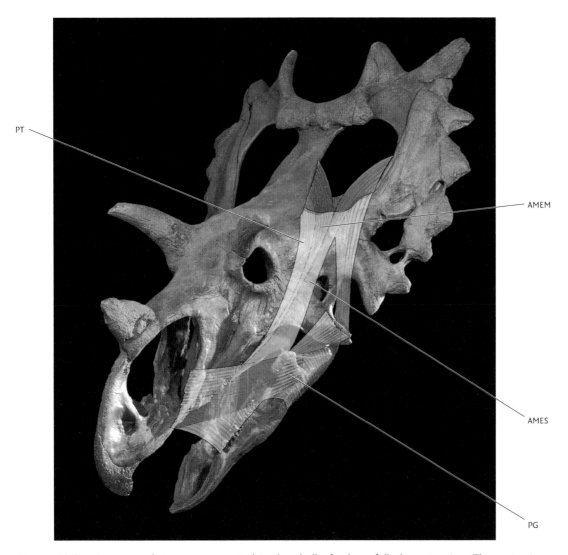

Figure 11.24. Jaw musculature reconstructed in the skull of a long-frilled ceratopsian. The major jaw closing (adductor) muscles are (1) the adductor mandibularis externus superficialis (AMES); (2) the adductor mandibularis externus medialis (AMEM); (3) the pseudotemporalis (PT); (4) the pterygoideus (PG).

Which plants were preferred by ceratopsians remains a mystery. The principal plants whose statures match browsing heights of ceratopsians were a variety of shrubby angiosperms, ferns, and small conifers (see Chapter 15).

Locomotion

The primitive ceratopsian *Psittacosaurus* appears to have been fully bipedal and must have walked in typical bipedal ornithischian fashion. But the rest were quadrupeds all, and while the back legs were fully erect, the orientations of the front limbs are somewhat controversial. Some have argued that the front limbs were directly under the body, as they are in mammalian quadrupeds. Others have argued, based on the shape of the bones of the front legs, that a more sprawling posture in the front legs is indicated (Figure 11.25).

These considerations naturally affect how we understand the way ceratopsians ran. The sprawling front legs would likely entail slower speeds, perhaps accompanied by a more unhurried lifestyle. The

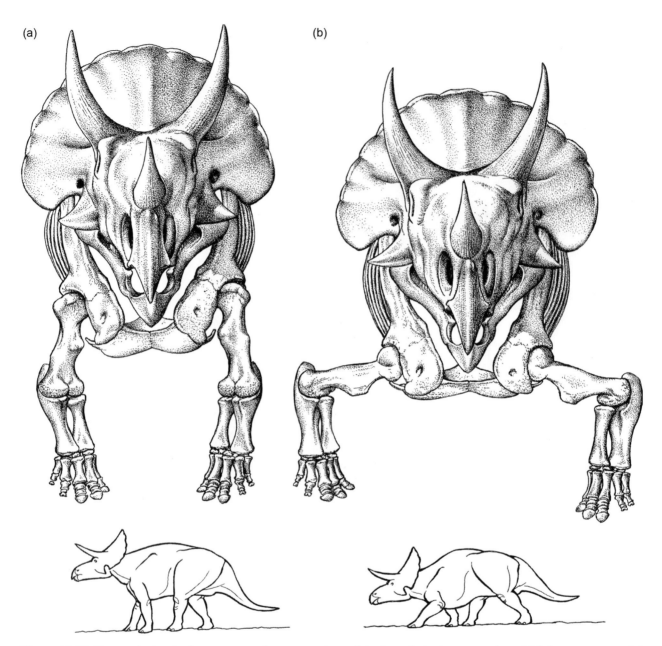

Figure 11.25. Two endpoints in the spectrum of reconstructions of the front limbs in ceratopsians: (a) fully erect stance; (b) semi-erect stance.

mammal-like reconstructions, with fully erect stances for both front and rear legs, suggest faster speeds, reminiscent of a large, Cretaceous rhinoceros. With such uncertainty about stance, tentative estimates for walking speeds range somewhere between 2 and 4 kmh (1.2 to 2.5 mph; see Chapter 13), while maximum running speeds might have ranged from 30 to 35 kmh (around 18 to 22 mph).

Bringing Up Baby

The first dinosaur egg nests ever found, in 1922, were thought to belong to the small Asian ceratopsian *Protoceratops*, and it was with this dinosaur that, for the next 70 years, the eggs were posed in museum displays all over the world. So it was, uh, *informative* to learn – after some

70 years – that the embryos inside those "*Protoceratops* eggs" actually belonged to the theropod *Oviraptor* (see Chapter 6; Figure 6.30).

Despite the confusion with *Oviraptor*, however, juvenile ceratopsians from Asia are now relatively well known. Hatchlings of *Psittacosaurus*, no more than 23 cm long (with tail!) have been found. And recently, complete skeletons of *Protoceratops* hatchlings – grown somewhat past the newly hatched stage – were found in nests (Figure 11.26). This means that some parental care at the nest after birth is strongly implied for *Protoceratops*: why, otherwise, would the juveniles congregate at the nest and grow there during the first year of life? Moreover, all of the growth stages from hatchling to adult have been documented in *Protoceratops*, making the ontogeny – or growth and development – of this dinosaur perhaps the best understood of all dinosaurs. Similarly, the ontogeny of a number of other ceratopsian dinosaurs is known, including that the primitive ceratopsian *Psittacosaurus* (Figure 11.27a) as well as that of the latest Cretaceous *Triceratops* (Figure 11.27b). And, as in pachycephalosaurs, we will see that the ontogeny of ceratopsians turns out to hold key clues about their behavior.

Horns, Frills, and Ceratopsian Behavior

Ceratopsian horns were once thought to be all about battling predators at close quarters. More recent interpretations have not completely ruled this out (see below) but, as in pachycephalosaurs, have instead focused on intraspecific behaviors such as display, ritualized combat, defense of territories, maturity and species identification, and establishment of social ordering.

The link between dominance, defense, and horns comes from studies of large horned (ungulate) mammals. Dominance in these mammals (and in other tetrapods) is accentuated by the development of structures that "advertise" the size of the animal; these obviously include horns and antlers, as well as the bony horn-like knobs (ossicones) of giraffes and the nasal horns of rhinoceroses. In short, the variety of horn and antler shapes in mammals are known to reflect (1) species-recognition mechanisms that aid in preventing interspecific (*inter* – between) attempted matings (that is, matings between different species), and (2) intraspecific (*intra* – within) differences so that displays and ritualized fighting behavior occur between appropriate individuals (e.g., equally eligible males, in the case of modern mammals).

Turning to ceratopsids and using modern horned mammals as analogs, current thought suggests that the large nasal and brow horns of ceratopsids functioned primarily during territorial defense and in establishing dominance. Similarly, the appearance and shape of elaborate scallops and spikes along the frill margin in many of the more highly derived ceratopsians separates one species from another, as well as juveniles from adults (see Figures 11.21 and 11.27). Thought of this way, the remarkable variations in the horns and frills in ceratopsians could be used for *interspecific* identification as well as the establishment of *intraspecific* dominance (Figure 11.28).

And Now the Data Become a Bit Murky...

With sexual selection indicated, we might expect sexual dimorphism in ceratopsians. And indeed, initially, statistical studies in the 1980s of *Protoceratops* showed two populations of adult frill and facial morphologies – strong evidence of sexual dimorphism. But in 2014, a well-supported, data-rich, catchily titled study ("Males resemble Females") concluded that there was no measurable difference between populations (presumably "males" and "females") of *Protoceratops*. Was the "two populations" idea just wishful thinking? The answer might be "yes." But it is worth remembering that the more recent authors may have failed to identify and measure the key sexually dimorphic features in this genus.

(a)

(b)

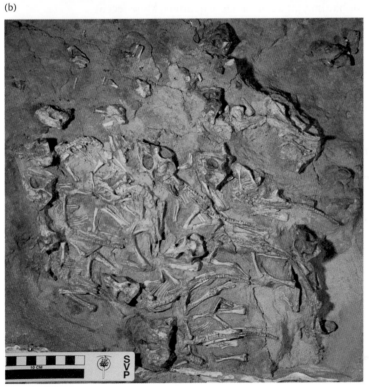

Figure 11.26. Two nests of juvenile ceratopsians from Asia. (a) A nest of 34 *Psittacosaurus* juveniles from the Lower Cretaceous Yixian Formation of Liaoning Province, China. An adult *Psittacosaurus* skull, presumably a parent, is visible on the left center of the photograph. (b) A nest with 15 juvenile *Protoceratops*, from the Upper Cretaceous Djadokhta Formation of the Gobi Desert, Mongolia. An adult *Protoceratops* is about 2 m long; each of these babies is about 10 cm, suggesting that they were perhaps within their first year of life. Their having at least a year of life together at the nest, before scattering into the wide world, implies the *Protoceratops* babies must have enjoyed parental care. But look at this image respectfully; it is actually a dramatic photograph revealing a very private moment: the last seconds before these baby dinosaurs' young lives were snuffed out. Ruler is marked in cm.

So, can we rule out sexual selection in ceratopsians? Hardly; frills don't appear too large in juvenile specimens – they only develop when the animals reach 75 percent of adult body sizes (Figure 11.27). This suggests that frill growth is coordinated with sexual maturity and therefore that there is a reproductive connection to frill size and shape. Doesn't *that*, at least, sound like sexual selection?!

Sexual selection is thought to also occur in other ceratopsians, among them *Centrosaurus* and *Chasmosaurus* (Figure 11.28). In many of these forms, the development of scallops and spikes on the frill margin could enhance the dimorphic nature of the frill; equally plausibly, however, such features

(a)

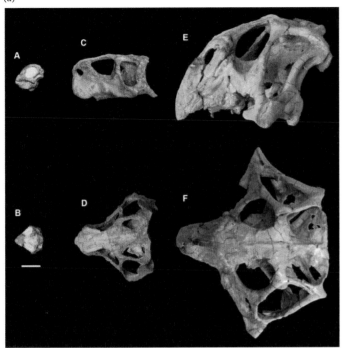

Figure 11.27. Ontogeny of ceratopsian dinosaurs. (a) *Psittacosaurus*, a primitive ceratopsian. Top row (A–C–E): side views; bottom row (B–D–F): top views (scale bar = 1 cm). Three growth stages are represented: baby, juvenile, and adult. From Bullar et al., 2019. (b) *Triceratops*, the most recent, most derived ceratopsian. A growth series for the Late Cretaceous ceratopsian *Triceratops*. Four growth stages are represented: baby, juvenile, sub-adult, and adult. The authors propose that the "adult" form might actually be *Torosaurus*, making this dinosaur the same animal as *Triceratops* (see text). From Goodwin and Horner, 2006.

(b)

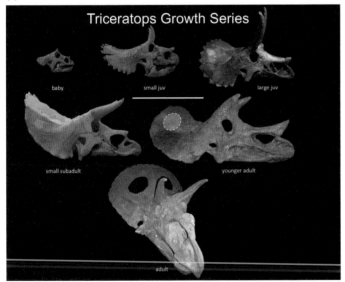

would also be useful for identification and indicators of maturity: who represents the competition for mating opportunities, and who does not. A close look at Figure 11.27b suggests an interesting phenomenon in *Triceratops*: as the animals aged, the frill, along with becoming more pronounced, become denuded of the bumps (properly termed epioccipital bones) along the frill margin. Paleontologists had always thought of these bones as indicative of maturity; the more bumps and hornlets, the more mature the animal. What if it was exactly the opposite?

Regardless, all of this implies social interactions, and it thus comes as no surprise to learn that ceratopsians lived in large herds. The evidence for this comes from an overwhelming catalog of

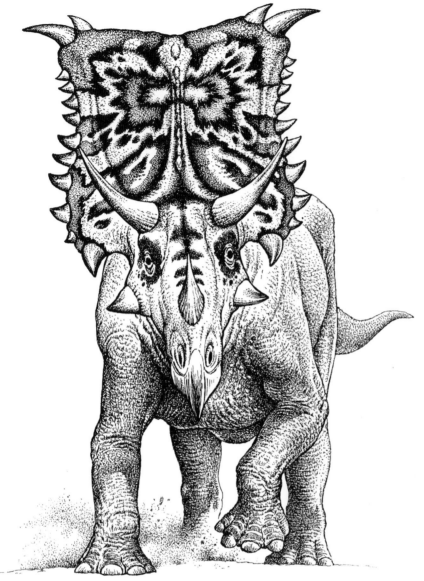

Figure 11.28. "Back off": frill display in *Chasmosaurus*. The very long frill could have provided a very prominent frontal threat display, not only by inclining the head forward but also by nodding or shaking the head from side to side.

ceratopsian **bonebeds**, mass accumulations of single species of organisms. But how does a herd get preserved? Easily, if, as in the case of the famous *Centrosaurus* Bonebed of Alberta, Canada, a herd of centrosaurs was crossing a flooding river, swollen with storm water. Members of the herd were evidently caught up, drowned, and then unceremoniously deposited downstream. Bonebeds are known for more than nine different species, including several bonebeds exceeding 100 individuals. Such gregariousness makes sense when putting frills and horns into their behavioral context: identification, territoriality, ritualized combat and display, and the establishment of dominance are to be expected in animals that come together in highly social circumstances such as herds.

These ideas suggest that we ought to find puncture wounds inflicted on faces, frills, and bodies of competing ceratopsians. In fact, such wounds are preserved in at least five different ceratopsians. In *Triceratops* and *Centrosaurus*, healed wounds on the cheek region and frill were correlated with the types of horns each animal bears. The horns of *Triceratops* (two large brow horns; a smaller nasal

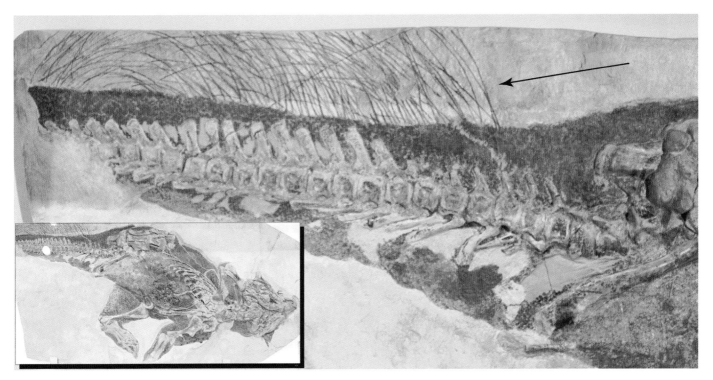

Figure 11.29. The Wonderful (and unexpected) World of *Psittacosaurus*. The tail region of the primitive ceratopsian *Psittacosaurus*, showing elongate (presumably) monofilamentous feathers (see arrow) along the back of the tail, interpreted as display features. Inset: complete specimen. Who knew!? Photographs taken from a cast of the specimen.

horn) are very different from those of *Centrosaurus* (a single large nasal horn), and if *intraspecific* competition was the main use of the horns, the wounds left on the frills and faces of these animals should look very different from one another. As it turned out, *Triceratops* had a significantly higher number of healed wounds on the frill than did *Centrosaurus*, suggesting that, given the mechanics of fighting, the damage was caused by other *Triceratops* with their massive, highly placed brow horns. These results provide strong evidence that blood-letting came from head-on engagements between competing members of the same species.

The argument for the importance of display in ceratopsians got an unexpected bump from the primitive ceratopsian *Psittacosaurus*. Feathers, or at least monofilamentous coats (as we learned in Part III), may have been primitive not just for theropods, but perhaps for all dinosaurs, because the basal ornithischian *Tianyulong* has them (see Figure III.4). *Psittacosaurus*, too, had some kind of monofilamentous bling, evidently restricted to the back of the tail (Figure 11.29). These structures certainly have the appearance of a display feature; reinforcing the importance of display in both the origin of feathers and sociality in Ceratopsia.

Thoughts of a Ceratopsian

We have a surprisingly good idea about the brains of dinosaurs, including ceratopsians, even though those brains are soft tissue, and were last seen on Earth millions of years ago (Box 11.1). Given the complex repertoire of inferred ceratopsian behaviors, it comes as a bit of a surprise that their brains were not particularly large (see also Chapter 13). Despite being near opposites in terms of body size and display-related anatomy, both *Protoceratops* and *Triceratops* had brains less than the size expected of a similarly sized crocodilian or lizard. Cerebrally, they were above sauropods,

Box 11.1 Dino Brains

Brains, of course, are soft anatomy, and so finding one preserved is next to impossible. Yet, we have good evidence for how brains looked in many dinosaurs and, considering areas of enlargement, we can suggest that a particular dinosaur had a well-developed sense of smell, or perhaps good vision. How can this be?

Historically, if the bones of the braincase were well preserved, paleontologists initially obtained **brain endocasts**, three-dimensional models of the brain. This was not so easy: all of the interior rock needed to be prepared out of the braincase, a process that generally involved breaking the braincase, and then the thing in effect reassembled, so that its inside could be cast. Reassembly took a *very* skilled preparator. Casting was accomplished by painting several layers of latex on the interior of the braincase, and then, once it had dried, pulling the flexible material through the foramen magnum (the opening through which the spinal cord contacts the brain) if the braincase was intact; peeling it off of the exposed interior if it was not. Latex cast in hand, a mold could then be prepared that faithfully recorded the contours of the inside of the braincase. From this mold were cast the brain **endocasts**; in plaster, or perhaps resin. Such heroic efforts gave a hint about dinosaur brains, but it turns out that, unlike mammals, dinosaur brains do not exactly match (fill up) the inside of the cavities (the braincases)

where they are located. So the major features of the brain can be deduced, but details are commonly hard to muster.

Very much more recently, paleontologists have begun to use computer-aided tomography (CT) scanning techniques (see Figure 6.20) to successfully image the brain; in so doing acquiring revolutionary insights about dinosaur brains. As in humans, the technique is noninvasive, and responds to density differences between the fossil bone and the rock that encases it. Preparation is therefore not necessary. The instrument obtains a sequence of two-dimensional "slices" – images, in fact – which can then be assembled into a computer-generated composite image, and be rotated and visualized from any angle. The advantages of this are multiple: aside from obviating the need for preparation – which, as we have seen, is costly, time-consuming, and potentially destructive – the CT scan provides as much detail as is preserved – the equal of, or better than, the very finest preparation jobs. The fit, as we have said, is not exact, but the amount of information that can be retrieved is still astonishing. For example, the CT-scanned image presented here (Figure B11.1.1) shows details of the cranial nerves (identified by Roman numerals) unavailable by conventional means. The CT dinosaur brain images we've shown in this book, as well as many others, are available on paleontologist and anatomist L. M. Witmer's laboratory ("Witmerlab") website: www.oucom.ohiou.edu/dbms-witmer/3D-Visualization.htm.

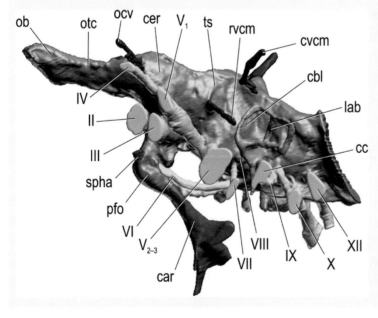

Figure B11.1.1. CT-scan reconstruction of the brain of the ceratopsid genus *Pachyrhinosaurus* (see Figure 11.21f). Scale bar = 4 cm. For abbreviations see Figure 6.20.

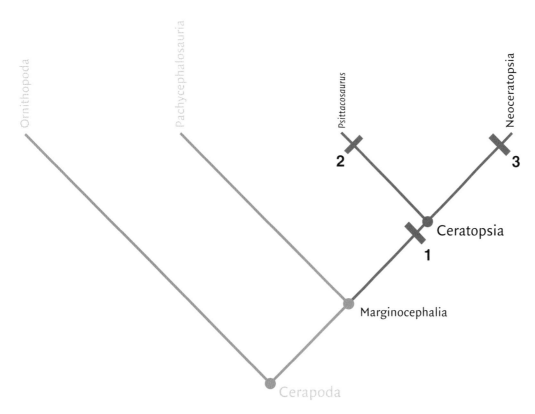

Figure 11.30.   Cladogram of Ceratopsia, emphasizing the monophyly of *Psittacosaurus* and Neoceratopsia. Derived characters include: at **1**, rostral bone, a high external naris separated from the ventral border of the premaxilla by a flat area, enlarged premaxilla, well-developed lateral flaring of the jugal; at **2**, short preorbital region of the skull, very elevated naris, loss of antorbital fossa and fenestra, unossified gap in the wall of the lacrimal canal, elongate jugal and squamosal processes of postorbital, dentary crown with bulbous primary ridge, manual digit IV with only one phalanx, manual digit V absent; at **3**, enlarged head, keeled front end of the rostral bone, much reduced quadratojugal, primary ridge on the maxillary teeth, development of humeral head, gently decurved ischium.

ankylosaurs, and stegosaurs, but commanded proportionally less cognition-related gray matter than either ornithopods or theropods. Regardless, the variety of exotic morphology in and around the head suggests that ceratopsians, large-brained or no, may have had a relatively complicated behavioral repertoire.

Color me… countershaded!

Until the discovery of melanosomes in dinosaur feathers (Figure 6.29), when paleontologists were asked what color dinosaurs were, we could answer, "Well, the bones are brown." Nobody ever asked the question, anyway, because everybody knew that dinosaurs were kind of a mossy dark green color! Of course the melanosomes started to provide clues about colors, and we've seen some striking patterns in *Anchiornis* (Figure 7.24), and *Archaeopteryx*, for example, is thought to have been black. Those are all feathered dinosaurs; how about one without feathers?

An answer to that came in 2010, when preserved colors were actually found a specimen of *Psittacosaurus*, the very one seen in Figure 11.29. Along with that mane of monofilamentous feathers on the tail, skin impressions are preserved. These show patterns of small, rounded pieces of dermal bone (ossicles), which in life would have given the animal a bumpy surface. These show a

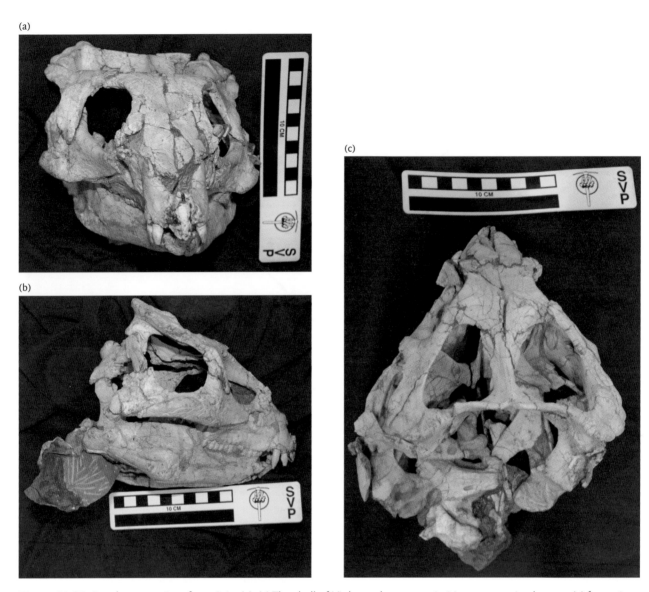

Figure 11.31. Basal ceratopsians from Asia. (a)–(c) The skull of *Yinlong*, the most primitive ceratopsian known: (a) front view; (b) right side; (c) top view. Photographs courtesy of J. M. Clark.

distinct network of dark and amber-brown coloration. Interestingly, the darkest network is reserved for the dorsal surface; the ventral surface of *Psittacosaurus* seems to have been much lighter. This is again an example of what is known as "countershading," in which the animal is colored as to make it more difficult to see from above. Recall that a similar example of countershading was reported in the ankylosaurs *Zuul* and *Borealopelta* (Chapter 10).

The Evolution of Ceratopsia

Ceratopsia is a monophyletic taxon, indicated by a rich array of derived features (Figure 11.30). Two ceratopsians that are thought to represent the primitive condition for the group are *Yinlong* (Figure 11.31) and *Psittacosaurus* (Figure 11.18), both small, Asian bipeds. All more-derived

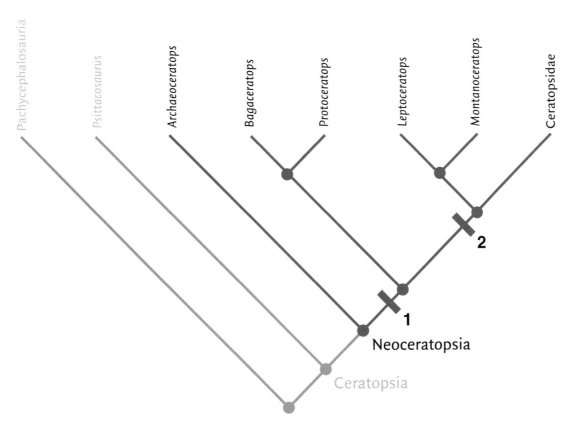

Figure 11.32. Cladogram of basal Neoceratopsia, with the more distantly related *Psittacosaurus* and Pachycephalosauria. Derived characters include: at **1**, elongated preorbital region of the skull, an oval antorbital fossa, triangular supratemporal fenestra, development of the syncervical (fusion of cervical vertebrae); at **2**, greatly enlarged external nares, reduced antorbital fenestra, nasal horn core, frontal eliminated from the orbital margin, supraoccipital excluded from foramen magnum, marginal undulations on frill augmented by epioccipitals, more than two replacement teeth, loss of subsidiary ridges on teeth, teeth with two roots, ten or more sacral vertebrae, laterally everted shelf on dorsal rim of ilium, femur longer than tibia, hoof-like pedal unguals.

ceratopsians – Neoceratopsia – are quadrupeds. This underlines an important evolutionary event that we can read from the cladogram (Figure 11.30): relatively early in their history, ceratopsians, for whatever their reasons, adopted a quadrupedal stance.

Those early days also brought with them evidence of a major ceratopsian migration. Neoceratopsia (Figure 11.32) consists of a series of small, relatively primitive forms such as the Asian *Protoceratops* and *Bagaceratops*; the somewhat younger, though still primitive North American *Montanaceratops* and *Leptoceratops*, as well as the more derived, exclusively North American clade Ceratopsidae, that group of large, familiar ceratopsians such as *Triceratops* and *Centrosaurus* (Figure 11.33), including newly discovered ones like *Wendiceratops* (Figure 11.34).

When we compare the geographical locations of various neoceratopsians, that is, their biogeography, with primitive and advanced ceratopsians on the cladograms shown in Figures 11.32 and 11.33 it becomes clear that, early in neoceratopsian history, a primitive neoceratopsian – looking perhaps a bit like *Protoceratops* – migrated to the New World. The route of choice would likely have been briefly exposed land across the Bering Straits. Recent work not only confirms this pattern, but also suggests that several such migrations took place during ceratopsian history (Figure 11.35).

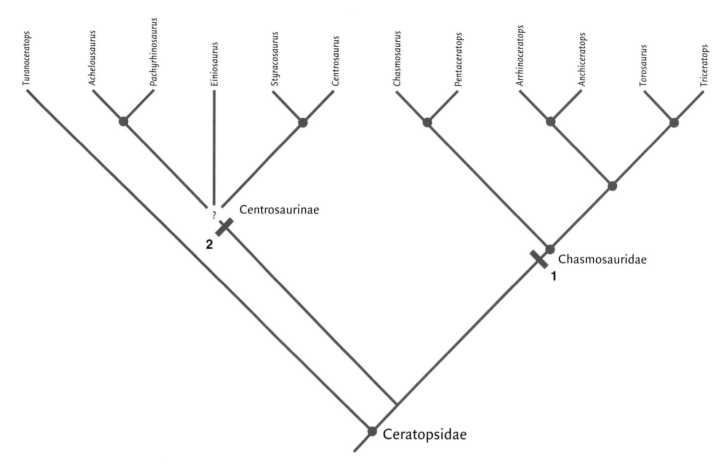

Figure 11.33. Cladogram of Ceratopsidae. Derived characters for Ceratopsidae. Derived characters (for chasmosaurines) include: at **1**, enlarged rostral, presence of an interpremaxillary fossa, triangular squamosal epioccipitals, rounded ventral sacrum, ischial shaft broadly and continuously decurved. Derived characters (for centrosaurines) include: at **2**, premaxillary oral margin that extends below alveolar margin, postorbital horns less than 15 percent of skull length, jugal infratemporal flange, squamosal much shorter than parietal, six to eight parietal epoccipitals, predentary biting surface inclined steeply laterally.

Once in North America, a few lineages retained the comparatively modest morphology of their ancestors. However, the clade radiated into two spectacular and diverse groups of much larger, flashier ceratopsids: chasmosaurines, after *Chasmosaurus*; and centrosaurines, after *Centrosaurus*. Chasmosaurines are generally called "long-frilled," after a tendency in the group to develop large, open frills, while centrosaurines are sometimes called "short-frilled," after a tendency in the group toward shorter frill lengths. Their postcrania (everything behind the skull and lower jaws) are very similar; but the skulls are remarkably distinct, with many different types of horns and facial features denoting identity and perhaps display behavior (Figures 11.19 and 11.21). Yet, the origin of all these big, flashy ceratopsids seems not to have been in North America; *Turanoceratops*, from Uzebekistan (Asia), is evidently a very primitive ceratopsid, indicating that the evolutionary roots of ceratopsids may extend earlier than the North American migrations (Figure 11.33).

In North America, the latest Cretaceous forms *Triceratops* and *Torosaurus* may yet hold a few surprises. These two latest, largest ceratopsian dinosaurs thundered across what is now the Great Plains of North America – Montana, North Dakota, South Dakota, Wyoming, and Alberta – just before the (nonavian) dinosaur extinction. Their postcranial bones are virtually identical; with these,

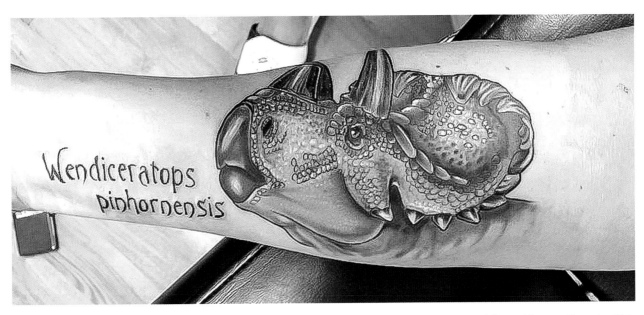

Figure 11.34. The recently (2015) discovered *Wendiceratops*, a centrosaurine ceratopsid from Alberta, Canada. This ceratopsian is named after its discoverer, Wendy Sloboda, who celebrated her find by tattooing its image on her arm (see also Figure 11.21m).

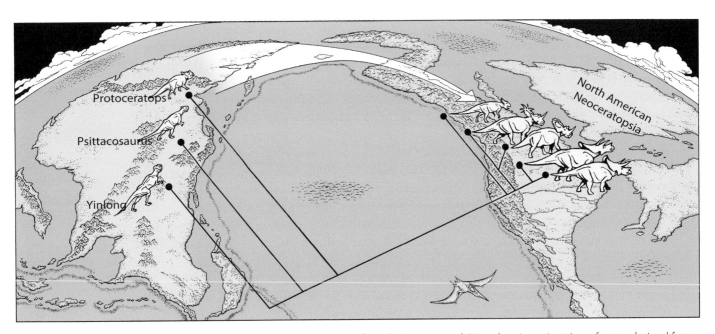

Figure 11.35. Cladogram of ceratopsians superimposed on a map of North America and Asia, showing migration of more derived forms to North America. Arrow shows the general route of migration.

as we have seen in many other ceratopsians, it is the skulls that are distinct. *Torosaurus* has two large openings in the frill, while *Triceratops* has no such openings. But work by paleontologists M. B. Goodwin, J. Scannella, and J. R. Horner suggests that these different-looking animals could be the same thing: as *Triceratops* got older, its frill became thinner and thinner in just the place where the opening in *Torosaurus* is found. Horner, Scannella, and Goodwin proposed that *Torosaurus* is simply

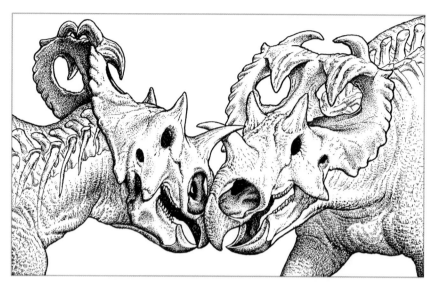

Figure 11.36. "Crossing of the horns": combat between (male?) *Centrosaurus*.

a more aged version of *Triceratops*! We show them as different animals in Figure 11.33; but in Figure 11.27 you can see the paleontologists model for how *Triceratops* ages until it is "*Torosaurus*." If Goodwin, Scannella, and Horner are correct, however, dinosaurs now called "*Torosaurus*" will end up being called *Triceratops* because that name came first.

The Evolution of Behavior

If there is some correspondence between morphology and behavior, then the morphological trends identified by all the ceratopsian cladograms should give us insights into the evolution of neoceratopsian behavior. In those ceratopsians with relatively modest frills and horns – forms such as the Asian *Protoceratops*, and the North American *Leptoceratops* and *Montanoceratops* – display could have involved swinging the head from side to side or up and down, a display seen, for example, in modern lizards.

The more derived ceratopsids share more elaborate frills and either nasal or brow horns. Among the long-frilled ceratopsians (for example, *Chasmosaurus*, *Pentaceratops*, and *Triceratops*), the display function of the frill may have been emphasized (see Figure 11.28). In contrast, most of the short-frilled ceratopsians (such as *Centrosaurus*, *Avaceratops*, and possibly *Pachyrhinosaurus*) were somewhat rhinoceros-like in their appearance, and perhaps tried to catch each other on their nasal horns, thus reducing to a degree the amount of damage inflicted on the eyes, ears, and snout (Figure 11.36).

The elaborate frill ornamentation; the variety of horns all seem to scream "sexual selection!!" and, as we've noted in the past, horns, frills, and chewing seem to drive the evolution of ceratopsian dinosaurs. In this diverse group, we think we're witnessing a world where display, recognition, and competition were all-important. However, the study of *Protoceratops* and the likely fate of *Torosaurus* remind us yet again that for all that we think we do know about nonavian dinosaurs, it is well to remember how vastly it is exceeded by what we don't.

SUMMARY

Marginocephalia consists of the bipedal Pachycephalosauria, the dome-headed ornithischians, and the quadrupedal Ceratopsia, the horned, parrot-beaked, frilled ornithischians. The group was largely restricted to the Cretaceous of Asia and North America, and is diagnosed by the presence of a variably sized, bony shelf that formed along the back of the skull.

Marginocephalians were gregarious animals, and sexual selection was likely a driving force in much of their behavior, a fact that is reflected in their morphology. The domes of pachycephalosaurs have been interpreted as structures designed for intraspecific competition: head- and flank-butting have been suggested. The striking morphological variety of horns and frill shapes, and cranial ornamentation, in ceratopsians suggests a high level of intraspecific competition. In both groups, sexual dimorphism has been recognized. Ceratopsian gregariousness is also reflected by the presence of large monospecific bonebeds, suggesting that herds of ceratopsians roamed what is now the Great Plains of Canada and the USA.

All marginocephalians are genasaurs, which means that they chewed their food to a greater or lesser extent. While pachycephalosaur teeth don't reveal evidence of remarkable chewing adaptations and pachycephalosaurs likely gut-fermented in their capacious stomachs, ceratopsians developed sophisticated chewing mechanisms including a robust coronoid process, dental batteries, a skull partitioned into cropping, diastema, and grinding sections, and powerful jaw adductor muscles that may have attached high on the frill.

Care of the young is known in ceratopsian dinosaurs. Nests of partially grown ceratopsians have been found, suggesting parental care at the nest.

Ceratopsian evolution was characterized by increasing size, as well as by one or more migrations across the Bering Strait land bridge, from Asia to North America. While intraspecific competition was likely an important behavioral aspect of even the earliest ceratopsians, later forms evolved elaborate frill or horn displays.

SELECTED READINGS

Bullar, C. M., Zhao, Q., Benton, M. J., and Ryan, M. J. 2019. Ontogenetic braincase development in *Psittacosaurus lujiatunensis* (Dinosauria: Ceratopsia) using micro-computed tomography. *PeerJ*, **7**, e7217. DOI: 10.7717/peerj.7217

Chapman, R. E., Galton, P. M., Sepkoski, J. J., and Wall, W.P. 1981. A morphometric study of the cranium of the pachycephalosaurid dinosaur *Stegoceras*. *Journal of Paleontology*, **55**, 608–618.

Dodson, P. 1996. *The Horned Dinosaurs*. Princeton University Press, Princeton, NJ, 346 pages.

Dodson, P. and Currie, P. J. 1990. Neoceratopsia. In Weishampel, D. B., Dodson, P., and Osmólska, H. (eds.) *The Dinosauria*. University of California Press, Berkeley, pp. 593–618.

Dodson, P., Forster, C. A., and Sampson, S. D. 2004. Ceratopsidae. In Weishampel, D. B., Dodson, P., and Osmólska, H. (eds.) *The Dinosauria*, 2nd edn. University of California Press, Berkeley, pp. 494–513.

Evans, D. C., Schott, R. K., Larson, D. W., Brown, C. M. and Ryan, M. J. 2013. The oldest North American pachycephalosaurid and the hidden diversity of small-bodied ornithischian dinosaurs. *Nature Communications*, **4**, 1828. DOI: 10.1038/ncomms2749 |www.nature.com/naturecommunications

Farke, A. A, Wolf, E. D. S., and Tanke, D. H. 2009. Evidence of combat in *Triceratops*. *PLoS ONE*, **4**, e4252. doi:10.1371/journal.pone.0004252

Farlow, J. O. and Dodson, P. 1975. The behavioral significance of frill and horn morphology in ceratopsian dinosaurs. *Evolution*, **29**, 353–361.

Goodwin, M. B. and Horner, J. R. 2006. Major cranial changes during *Triceratops* ontogeny. *Proccedings of the Royal Society of London, Series B*, **273**, 2757–2761. doi:10.1098/rspb.2006.3643

Horner, J. R and Goodwin, M. B. 2009. Extreme cranial ontogeny in the Upper Cretaceous dinosaur *Pachycephalosaurus*: *PLoS ONE*, **4**, e7626.

Lingham-Soliar, T. and Plodowski, G. 2010. The integument of *Psittacosaurus* from Liaoning Province, China: taphonomy, epidermal patterns and color of a ceratopsian dinosaur. *Naturwissenschaften*, **9**, 479–486. DOI 10.1007/s00114-010-0661-3

Longrich, N. R., Sankey, J., and Tanke, D. 2010. *Texacephale langstoni*, a new genus of pachycephalosaurid (Dinosauria: Ornithischia) from the upper Campanian Aguja Formation, southern Texas, USA. *Cretaceous Research*, 31, 274–284.

Maiorino, L., Farke, A. A., Kotsakis, T., and Piras, P. 2014. Males resemble females: Re-evaluating sexual dimorphism in *Protoceratops andrewsi* (Neoceratopsia, Protoceratopsidae). *PLoS ONE*, 10(5), e0126464. doi:10.1371/journal.pone.0126464

Makovicky, P. 2012. Marginocephalia. In Brett-Surman, M. K., Holtz, T. R., and Farlow, J. O. (eds.) *The Complete Dinosaur*, 2nd edn. Indiana University Press, Bloomington, pp. 527–549.

Sereno, P. C. 1990. Psittacosauridae. In Weishampel, D. B., Dodson, P., and Osmólska, H. (eds.) *The Dinosauria*. University of California Press, Berkeley, pp. 579–592.

Schott, R. K., Evans, D. C., Goodwin, M. B., Horner, J. R., Brown, C. M., and Longrich, N. R. 2011. Cranial ontogeny in *Stegoceras validum* (Dinosauria: Pachycephalosauria): A quantitative model of pachycephalosaur dome growth and variation. *PLoS ONE*, 6(6), e21092. doi:10.1371/journal.pone.0021092

Witmer, L. M. and Ridgely, R. C. 2008. Structure of the brain cavity and inner ear of the centrosaurine ceratopsid dinosaur Pachycephalosaurus based on CT scanning and 3D visualization. In Currie, P. J., Langston, W., and Tanke, D. H. (eds.) *A New Horned Dinosaur from an Upper Cretaceous Bone Bed in Alberta*. National Research Council of Canada Monographs no. 49729, pp. 117–144.

You, H.-L. and Dodson, P. 2004. Basal Ceratopsia. In Weishampel, D. B., Dodson, P., and Osmólska, H. (eds.). *The Dinosauria*, 2nd edn. University of California Press, Berkeley, pp. 778–793.

TOPIC QUESTIONS

1. Who are marginocephalians?
2. What are the diagnostic characters of Marginocephalia? How are marginocephalians related to other ornithischians?
3. What are the diagnostic characters of Pachycephalosauria?
4. What are the diagnostic characters of Ceratopsia?
5. Describe chewing as practised by ceratopsian dinosaurs.
6. What do we know about ceratopsian egg-laying and nesting?
7. Give a brief history of ceratopsian biogeography.
8. What is sexual selection? Intraspecific competition?
9. What is in the inferred function of the horns and frill in ceratopsians?
10. What is the inferred function of the dome in pachycephalosaurs?
11. How do marginocephalian features relate to intraspecific competition and sexual selection?
12. Why is it so difficult to determine pachycephalosaur phylogeny?

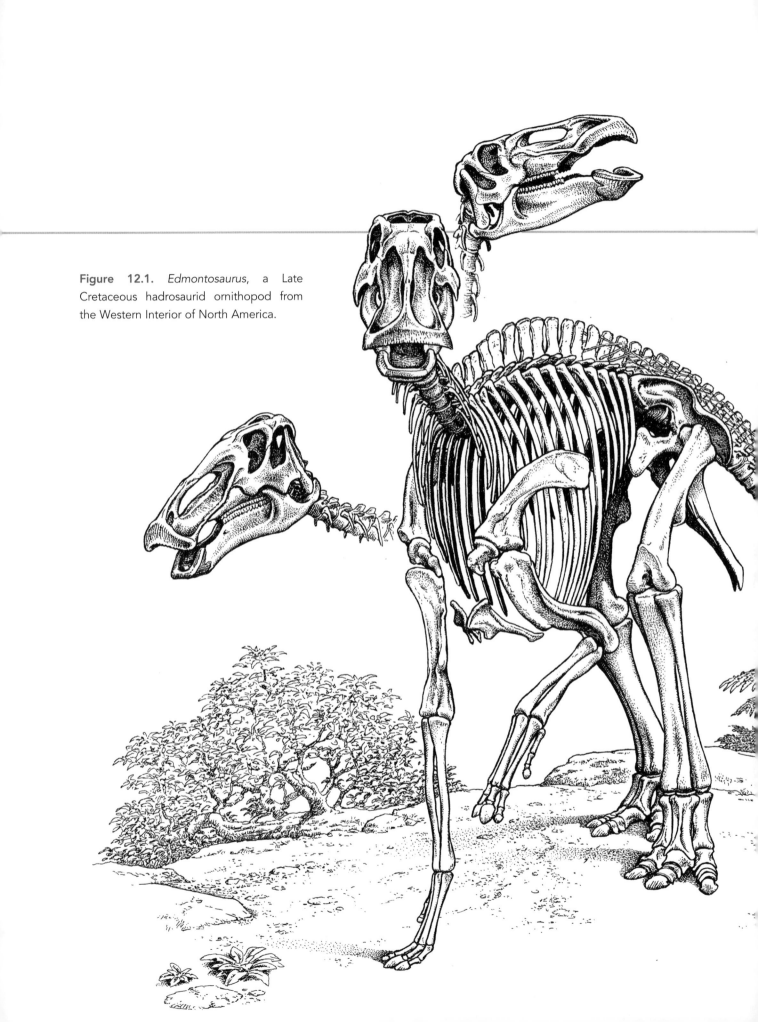

Figure 12.1. *Edmontosaurus*, a Late Cretaceous hadrosaurid ornithopod from the Western Interior of North America.

Chapter 12 Ornithopoda
Mighty Mesozoic Masticators

WHAT'S IN THIS CHAPTER

And now we meet Ornithopoda, the bipedal, herbivorous ornithischians. Here, deep into Righteous Ornithischia, we'll see powerful chewers, such as the world has never encountered before or since. Ornithopods were numerous and diverse, and have a great fossil record, which includes the preservation of "mummies," eggs, babies… what more can we possibly ask for?! We'll also see where ornithopods came from, and explore the adaptations that made them so successful. So bring your teeth and let's chew on Ornithopoda.

Ornithopoda

Ornithopods (*ornith* – bird; *pod* – foot) were the cows, deer, bison, wild horses, antelope, and sheep of the Mesozoic (Figure 12.1). Herbivores all, they were one of the most numerous, most diverse, and longest-lived groups in all Dinosauria. From the Middle Jurassic, when they first appeared, until the end of the Cretaceous, when they all went extinct, ornithopods evolved well over 100 species at present count.

Ornithopods lived on every continent. They ranged from near the then-equator to such high latitudes as the north slopes of Alaska, the Yukon, and Spitsbergen in the Northern Hemisphere, and Seymour Island, Antarctica, and the southeast coast of Australia in the Southern Hemisphere. Local conditions in these regions varied widely, so ornithopods lived in quite diverse habitats and in a wide range of climates.

They also evolved a range of sizes: early in their history, ornithopods were generally small (ranging from 1 to 2 m in length); however, later some members of the group got very big (upwards of 15 m; Figure 12.2).

We know as much about ornithopods as about almost any other group of dinosaurs: *Iguanodon* was a charter member of Sir Richard Owen's original 1842 "Dinosauria" (see Chapter 16). Hadrosaurids ("duckbills") are known from single bones to huge bonebeds. Their remains include skin impressions with patterns that have even been used to distinguish different species, and ossified tendons, as well as delicate skull bones such as sclerotic rings (that support the eyeball), stapes (the thin rod of bone that transmits sound from the eardrum to the brain), and hyoid bones (delicate bones that support the tongue). We have hadrosaurid eggs with all growth stages represented, from hatchling, to "teenager," to adult. Ornithopod footprints and trackways abound in many parts of the world.

Who Were the Ornithopods?

Ornithopods are genasaurian cerapodans as we can see in Figure 12.3 (see also Figure III.3). As ornithopod phylogeny is currently understood, there is a basic split between a few primitive ornithopods (generally small bipedal herbivores) such as *Orodromeus*, and the remaining ornithopods, Euornithopoda. Most of the dinosaurs we commonly think of as ornithopods, including *Iguanodon* and the duck-billed dinosaurs (hadrosaurids; see Figures 12.2 and 12.1,

Figure 12.2. *Iguanodon bernissartensis*, the magnificent Early Cretaceous ornithopod, as seen in the Museum of Natural History of Vienna in its original early 1880s mount. A more modern take on ornithopod posture has the backbone of the animal largely parallel with the ground.

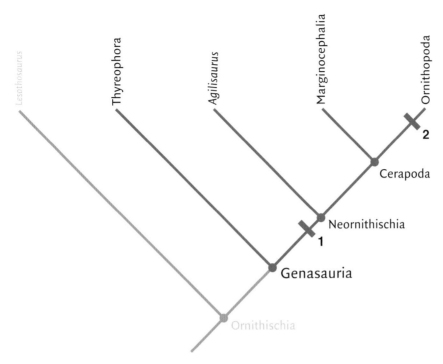

Figure 12.3. Cladogram of Genasauria (cheek-bearing ornithischians; see Part III), emphasizing the monophyly of Ornithopoda. Derived characters include: at **1**, pronounced ventral offset of the premaxillary tooth row relative to the maxillary tooth row, crescentic paroccipital processes, strong depression of the mandibular condyle beneath the level of the upper and lower tooth rows, elongation of the lateral process of the premaxilla to contact the lacrimal and/or prefrontal; at **2**, scarf-like suture between postorbital and jugal, inflated edge on the orbital margin of the postorbital, deep postacetabular blade on the ilium, well-developed brevis shelf, laterally swollen ischial peduncle, elongate and narrow prepubic process.

respectively), are euornithopods. The matter is somewhat complicated because a number of small bipedal herbivores, once considered ornithopods (such as *Agilisaurus* and *Hexinlusaurus*), are now thought to be outside of Ornithopoda, either more basal to it or immediately basal to Cerapoda as a clade within Neornithischia (Figures 12.3 and 12.4).

Ornithopod Lives and Lifestyles

Gettin' Around

Ornithopods, especially the larger ones, functioned as bipeds and quadrupeds, with both locomotor modes commonly occurring in the same beast. The smallest were predominantly bipedal, with gracile bones that suggest agility and speed. When eating or standing still, however, they also may have adopted a quadrupedal stance. Some of the larger ornithopods, however, such as *Iguanodon*, functioned more as full-time quadrupeds, only going bipedal when in a hurry. A number of larger ornithopods have a sturdy wrist and a hand with thickened hoof-like nails on the central digits, a hand that clearly was capable of considerable weight support. Indeed, some hadrosaurs are now known to have been obligate quadrupeds; that is, they were quadrupedal at all times (see Arms and Hands below). For all this, however, reconstructions of their hip musculature suggest more similarities with bipeds than with quadrupeds. Perhaps the strongly bipedal cast to their inferred hip

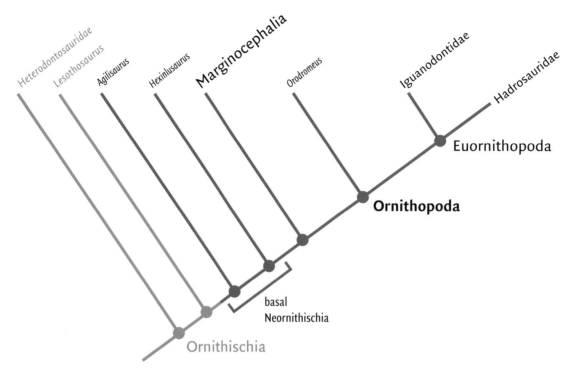

Figure 12.4. Cladogram showing the relationship of a few miscellaneous bipedal primitive ornithischian taxa, formally thought to be ornithopods. As we saw in Part III, primitive ornithischian relationships are in a state of flux; what is presented here is undoubtedly not the last word on them!

musculature is a holdover from their ancestry: like ceratopsians, they undoubtedly evolved from bipedal ancestors, but unlike ceratopsians, their hip morphology appears not to have evolved as distinctly quadrupedal. Juveniles may have been more bipedal than their fully grown, adult counterparts.

In all cases, the tail was long, muscular, strengthened by ossified tendons, that is, tendons that had mineralized and turned to bone, and held at or near horizontal, making an excellent counterbalance for the front of the animal. In the largest ornithopods, ossified tendons extended into the hip region, providing strength and rigidity at the pelvis, where considerable stress was placed, particularly during bipedal locomotion (Figure 12.5). In general, the powerful hindlimbs tend to be at least as long as, and in some cases more than twice the length of, the forelimbs.

How fast could these dinosaurs have traveled? Larger iguanodontians such as hadrosaurids may have been able to reach 15 to 20 kmh (9 to 13 mph) during a sustained bipedal run, but perhaps as high as 50 kmh (31 mph) on short sprints. Quadrupedal galloping appears unlikely given the rigidity of the vertebral column and the limited movement of the shoulder against the ribcage and sternum. For smaller ornithopods, running speeds could have been even higher (see Chapter 13).

Arms and Hands

Many ornithopods appear to have had gracile forelimbs, and likely used their hands to grasp at leaves and branches, bringing foliage closer to the mouth so that it could be nipped off by the toothed beak.

Most iguanodontians had very specialized fingers and hands, indicating multiple functions (Figure 12.6). In *Iguanodon*, for example, the first digit (thumb) was conical and sharply pointed. It has been suggested that it was used as a stiletto-like, close-range defensive weapon, or for breaking into seeds and fruits (or both, or... ?). The middle digits (II, III, and IV) were tipped with hooves, and

Figure 12.5. Ossified tendons (indicated by arrows) along the neural arches of *Iguanodon* at the Museum of Natural History in Vienna. This would have immobilized the backbone, providing strength in the pelvic region where the stresses were high during bipedal locomotion.

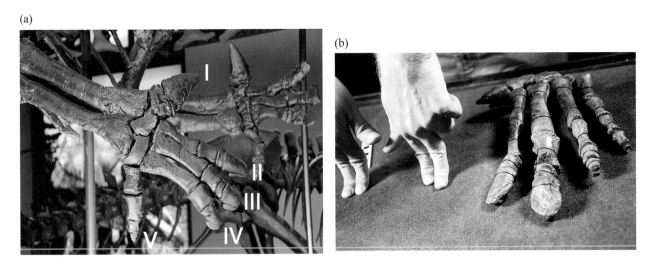

Figure 12.6. Hands in iguanodontians. (a) *Iguanodon*. Digits labeled on right hand. Note the spiked thumb (digit I). Photo from Museum of Natural History in Vienna. (b) *Mantelisaurus*. Here, the thumb spike (digit I) points left, while only digits II and III would have been hooved. Digits IV and V have many articulations in the phalanges, suggesting considerable flexibility. Human hands to the left of the image model how the stance might have looked.

were evidently weight-bearing; these would have been key players when the animal was in a quadrupedal stance. Finally, the outer finger (V) was highly flexible, and could bend across the palm, very much as the thumb does in humans: a grasping, opposable pinkie. The multifunctional hands suggest behaviors involving quadrupedal and bipedal stances, an option not available to most medium- to large-sized quadrupedal herbivorous mammals alive today.

Unlike other iguanodontians, in hadrosaurids digit I of the hand was lost, and digit V was relatively small. This left three main fingers, all tipped with hooves, with hardly any way to function other than to support the animal while standing. Hadrosaurids likely spent a lot of time on all fours;

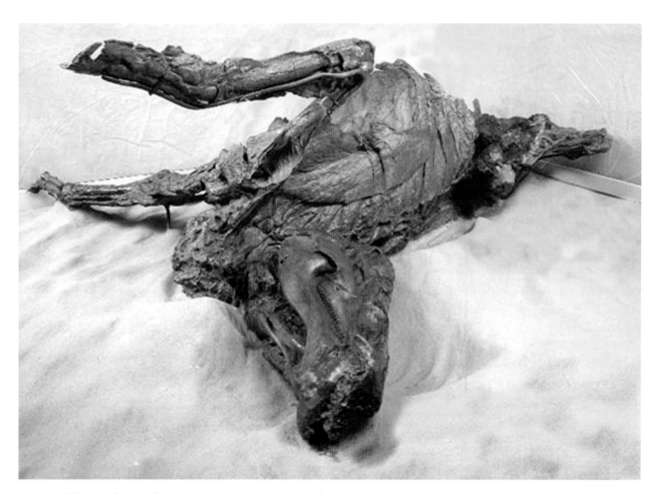

Figure 12.7. A hadrosaurid "mummy," complete with skin, ligaments, muscle tissue, stomach contents, and skeleton. The animal died on a floodplain, after which its carcass desiccated to dinosaur jerky. With burial, the dried dinosaur tissues were replaced with minerals, ultimately producing an exact replica of the original dried carcass.
Courtesy of the American Museum of Natural History, where this specimen is on display.

yet, as is plain from the fossils, the hands are far less stout than the feet, suggesting that they did not support the bulk of the animal, to the degree that the legs and feet must have.

Dietary Fiber

Fine dining, ornithopod style, is relatively well understood, for hadrosaurids at least. "Mummies" of hadrosaurids, complete with stomach contents, have been found. Figure 12.7 shows one of best examples of these fossilized "mummies." These spectacular specimens, though not true mummies in the sense that original soft tissue and bone are preserved, evidently died, and then dried into dinosaur jerky before decomposition and burial.[1] Eventually, the hardened, dried carcass will be buried. The toughened muscle and sinew didn't decompose, so the whole package – tissue *and* bones – was

---

[1] Even today, this process is not too uncommon; when an animal dies in a pasture (such as the horse in Figure 1.1), if conditions are relatively dry, bacterial action is not too great, and there are not too many scavengers around, the organs will decompose first, leaving the muscles, ligaments, tendons, and skin to dry, harden, and turn to jerky. Once dry – as long as it stays that way – it can last many years. Indeed, drying meat is a very good way to keep it from spoiling (which is what salting was originally used for).

replaced during burial (see Chapter 1). The startling result is preserved skin impressions, stretched tendons and muscles, and the last supper in the stomach and intestines. Hadrosaurids, we now know, ate twigs, berries, and coarse plant matter.

This selection of food correlates nicely with ornithopod height: they are thought to have been active foragers on ground cover and low-level foliage from conifers and in some cases from deciduous shrubs and trees of the newly evolved angiosperms; that is, the clade of all plants that bear flowers (see Chapter 15). Browsing on such vegetation appears to have been concentrated within the first meter or two above the ground, but the tallest animals may have reached vegetation as high as 4 m.

Eating coarse, fibrous, not particularly energy-rich food requires some no-nonsense equipment in the jaw to extract enough nutrition for survival, and ornithopods had the required goods. Like all genasaurs, ornithopods had a beak in the front for cropping vegetation, a diastem, a group of cheek teeth for shearing, and a large, robust coronoid process for serious mastication (Figure 12.8). A deeply inset tooth row indicates large fleshy cheeks, and powerful, closely pressed cheek teeth developed in hadrosaurids into true dental batteries (Figure 12.9), with smaller teeth, even more tightly packed, than those we saw in ceratopsids. But beyond these basics, different ornithopods had different modifications of the jaw, and different kinds of jaw motions are believed to have been used for the processing of food.

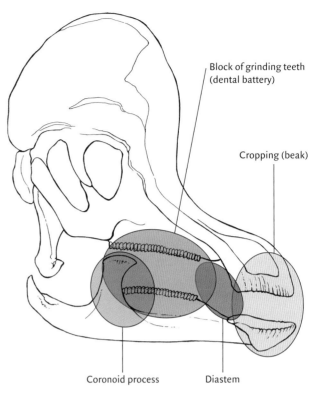

Figure 12.8. The skull of the Late Cretaceous hadrosaurid *Corythosaurus*, exemplifying chewing in derived ornithopods (hadrosaurids). The divisions of the skull are (as throughout this book): anterior cropping section (beige); diastema (maroon); block of grinding cheek teeth (blue); coronoid process (brown). See also Figures III.5 and 11.23.

Modern treatments of ornithopod jaw mechanics suggest some differences among ornithopod feeding behaviors. In basal ornithopods, the beak was relatively narrow, implying a somewhat selective cropping ability. Iguandontians, particularly the more highly evolved hadrosaurs, had broad snouts (Figure 12.10), and in some cases even developed a strongly serrated edge on the

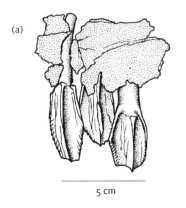

(a)

5 cm

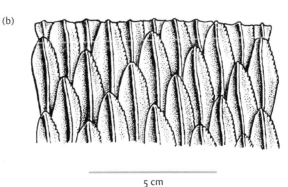

(b)

5 cm

Figure 12.9. Teeth in ornithopods. (a) three upper teeth of *Iguanodon* and (b) lower dental battery of *Lambeosaurus*. The teeth throughout ornithopod evolution become progressively more tightly packed, culminating in the lambeosaur dental battery.

rhamphotheca. They were likely not too selective; instead, they hacked and severed leaves and branches without much regard for what they were taking in, confident they had the equipment to handle whatever went in. Basal ornithopods were more likely careful nibblers, while iguanodontians were wood chippers.

Beyond the diastema, the chewing began. Here it was aided by something that is utterly foreign to humans (but not so, for example, to snakes). Our skulls and lower jaws are **akinetic**, meaning that, except for the vertical motion of the lower jaws, the bones in our skulls are solidly fused and locked together. Not so with ornithopods. Above and beyond the familiar up and down compression of chewing, slight movements of particular, individual bones within the skull and lower jaws allowed the cheek teeth to grind past each other from *side to side*. A skull in which individual bones move is called **kinetic**.

Ornithopods likely evolved a unique, kinetic skull, in which they mobilized their upper jaws. This kind of mechanism, called **pleurokinesis**, involved a slight outward rotation of portions of the upper jaw, especially the maxilla (the bone that contains the upper teeth), with each bite (Figure 12.11). When the upper and lower teeth were brought into contact on both right and left sides, the opposing surfaces of the dental batteries sheared past one another. Pleurokinesis was an important advance for ornithopods, especially iguanodontians, giving them the ability to chew the toughest, most fibrous plants.

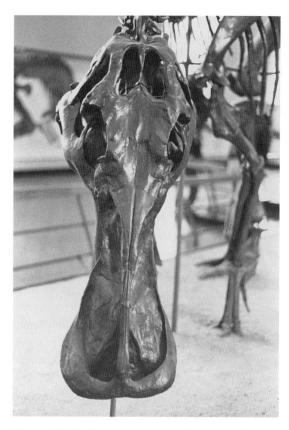

Figure 12.10. *Edmontosaurus*, a flat-headed hadrosaurid from the western USA, exemplifying the broad snouts of hadrosaurid ornithopods.

Chewing reached its most refined state in hadrosaurids, in which the cheek teeth were fitted tightly together into a dental battery, which effectively acted as a single shearing or grinding tool in each jaw (see Figure 12.11). We saw in ceratopsians that, unlike in mammals, for the full life of the animal new teeth constantly came in as older, worn teeth were popped out. Hadrosaurids had the same type of system, although perhaps even more efficient, since the grinding took place on the tooth crown and not along its side. Regardless, with constantly replacing teeth, tooth wear was never an issue (as it is in mammals, which only replace teeth once, so their adult teeth have to last their entire lives). The toughest, most fibrous plants undoubtedly succumbed to the hadrosaurid combination of powerful jaw muscles operating on a pleurokinetic skull equipped with constantly replaced grinding surfaces.

As in all of the other ornithischians that have been discussed, once ornithopods chewed their food, it was swallowed and digested in a capacious bacteria-filled gut. Unsurprisingly, given their sophisticated chewing abilities, the proportionately largest guts were found in the largest iguanodontians (including hadrosaurs); these ornithopods were all about extracting the most nutrition out of a low-quality, high-fiber, high-volume, conifer-based diet.

Thoughts of an Ornithopod

By dinosaur standards, ornithopods were smart – as smart or smarter than might be expected of living archosaurs if they were scaled up to dinosaur size (see Chapter 13). For example, *Leaellynasaura*, a basal ornithopod from Victoria, Australia, was apparently quite brainy and had acute vision, as suggested by prominent optic lobes in the brain.[2] In general, euornithopod smarts may be related to greater reliance on the senses – sight, smell, and hearing – for protection that, in the absence of other means of which we are aware, may have been their only defense. Moreover, brain size in these dinosaurs may have also been an integral part of a complex behavioral repertoire.

Socializing *à la* Ornithopoda

From the time of their discovery, ornithopods have attracted a good deal of attention, particularly for the extraordinary crests on the heads of hadrosaurids. These features, as well as their being found in large herds, hint at sophisticated social behavior.

Song of the Saurian

Hadrosaurids have attracted the most attention, in large part because some of them bore striking hollow crests (Figure 12.12). The hollow crest morphology was once thought to relate to putative aquatic habits of the group (as in the case of sauropods, nobody thinks they were aquatic any more!) or to increasing a sense of smell (still a possibility), but studies also suggest that the internal chambers of the crests would have made good resonating chambers, producing loud, low-frequency sounds[3] – a kind of Mesozoic *alphorn*. There is no point in making noise if there is nobody who can hear it; however, studies of the morphology of the hadrosaur hearing mechanism (inner ear) suggest that hadrosaurs could hear the low frequencies that may have been produced in the hollow crests.

With that insight, much of the discussion now centers on intraspecific competition (see Chapter 5) and sexual selection (see Chapters 6 and 11). To convey information about species, sex, and even rank, crests had to have been *visually* and, if part of their function was as a resonating chamber, *vocally* distinctive. Like Rod Stewart versus Placido Domingo. Only then can they have promoted successful matings between consenting adults. How strongly is the role of sexual selection implied by hadrosaurid crests?

Paleontologist J. A. Hopson made five predictions that test the hypothesis that hadrosaurid crest morphology was all about sexual selection.

(1) If communication and display were important, hadrosaurids must have had both good hearing and good vision.
(2) If the crest served the dual role of visual display and as a vocal resonator, then its external shape might not necessarily mimic the internal shape of the resonating cavities inside.

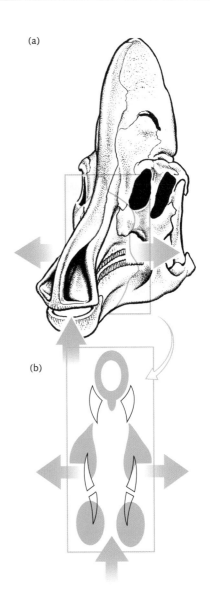

Figure 12.11. Jaw mechanics in Ornithopoda, showing lateral mobility of the upper jaws (pleurokinesis). (a) Skull of *Corythosaurus* showing movement directions. As the jaw closed, the bones carrying the dental batteries (the maxilla) in the upper jaw moved slightly outward, accentuating the grinding motion of the dental battery in upper jaw against that in the lower. (b) Cross-section cut across skull, showing pleurokinesis, in which the upper teeth move outwards relative to the lower.

---

[2] The animal had an estimated encephalization quotient (EQ; see Figure 13.3) of 1.8; J. A. Hopson estimated that the average EQ of other ornithopods is about 1.5.
[3] For you musicians: from a G two octaves below middle C to the F# above middle C.

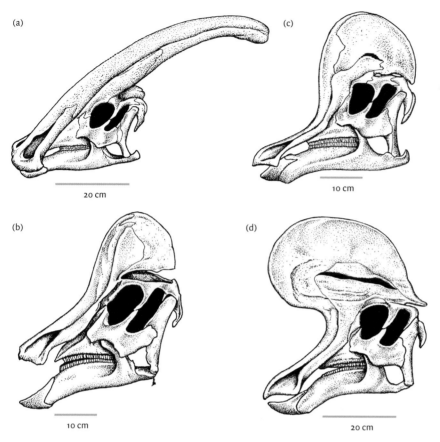

Figure 12.12. Left lateral view of the skull of some hollow-crested hadrosaurids. (a) *Parasaurolophus*, (b) *Hypacrosaurus*, (c) *Corythosaurus*, and (d) *Lambeosaurus*.

(3) If crests acted as visual signals, then they should be species-specific in size and shape, and they should also be sexually dimorphic.

(4) If the crests were a visual cue, they ought to be increasingly distinctive as the number of hadrosaurids living together increases.

(5) The crests should become more distinctive through time as a consequence of sexual selection.

How did these hypotheses fare? Hypothesis (1) is relatively well supported, in that hadrosaurids, to judge from their sclerotic rings (see Figure 4.6), had relatively large eyes, implying acute vision. Similarly, preserved middle and inner ear structures suggest that a wide range of frequencies was audible to these animals. Hypothesis (2) is upheld in virtually all cases, in that the profile of the crest is more elaborate or extensive than the walls of the internal plumbing (Figure 12.13). Hypothesis (3) is amply upheld in large part thanks to studies on the growth and development in lambeosaurine hadrosaurids, which show that crests become most prominent when an animal approaches sexual maturity. Moreover, adult lambeosaurines are known to be dimorphic, particularly in terms of crest size and shape (Figure 12.14). Could these "morphs" have been male and female?

Hypothesis (4) is based on the idea that distinctiveness would be an advantage during the breeding season. It was tested at Dinosaur Provincial Park in Alberta, Canada, where five distinctively equipped species of hollow-crested hadrosaurid and one species of solid-crested hadrosaurid all lived together in multi-specific bliss. In support of the hypothesis, elsewhere where hadrosaurid diversity is lower, the distinctiveness of the crests is decreased. Interestingly, however, hypothesis (5) is not well supported, for lambeosaurines' crests, at least, arguably become less distinctive over time.

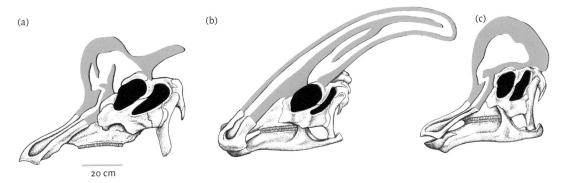

Figure 12.13. Highly modified nasal cavities housed within the hollow crests on the heads of lambeo-saurine hadrosaurids. (a) *Lambeosaurus*; (b) *Parasaurolophus*; (c) *Corythosaurus*.

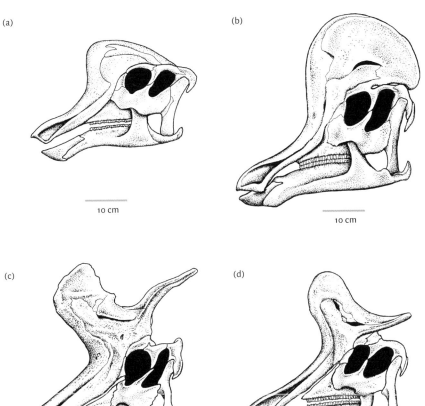

Figure 12.14. Growth and sexual dimorphism in lambeosaurine hadrosaurids. (a) Juvenile and (b) adult *Corythosaurus*. (c) Male (?) or older and (d) female (?) or younger *Lambeosaurus*. The more that we learn about the ontogeny of dinosaurs, the more difficult it becomes to be sure that we are not simply seeing the effects of maturity in the specimens.

Perhaps, then, as we saw in ceratopsians, some of this ornamentation was related to the identifica-tion of an individual's maturity, rather than his (her?) species.

If the crests were used for species recognition, ritualized display, courtship, parent–offspring communication, and social ranking, the accentuated nasal arch seen in genera such as *Gryposaurus*, *Maiasaura*, and *Brachylophosaurus* may have been used for broadside or head pushing

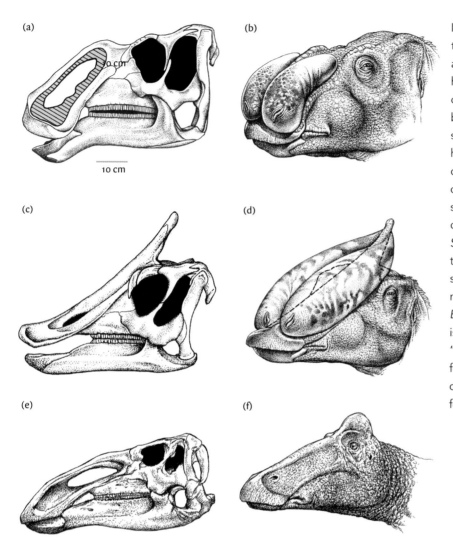

Figure 12.15. Some imaginary soft-tissue configurations that would accommodate skull morphologies in hadrosaurs. (a) The circumnarial depression in *Gryposaurus* (indicated by cross-hatched region) could have supported an inflatable flap of skin in hadrosaurines, as in (b). (c) The skull of *Saurolophus*, including the distinctive posterior-pointing solid spike; (d) speculative reconstruction of single or paired inflatable sac(s) in *Saurolophus*. Here the many soft-tissue features of modern bird heads serve as potential guides to what might have been present. (e) *Edmontosaurus*, which had no crest, is now known to have borne a fleshy "cockscomb" of skin a bit like that found on a rooster, perhaps brightly colored, and almost certainly used for display (f).

during male–male combat (Figure 12.15a). Inflatable flaps of skin possibly covered the nostrils and surrounding regions (Figure 12.15b); if present, these could have been inflated and used for visual display, as well as noise-making – more a Mesozoic bagpipe. In *Prosaurolophus* and *Saurolophus* (Figure 12.15c), in which a solid, spiky crest points posteriorly, a sac might have extended onto the solid crest that extended above the eyes (Figure 12.15d), while in *Edmontosaurus*, with neither hollow crest nor solid spike (see Figures 12.1 and 12.12), a broadly excavated nostril region may have housed an inflatable sac. Unfortunately, these soft-tissue-based hypotheses are all speculative. Not speculative, however, was the 2014 description of a "cockscomb"[4] preserved on a specimen of *Edmontosaurus*; here, then, was a hadrosaur without bony crests, that in life had a soft-tissue fleshy crest (Figure 12.15e, f).

In lambeosaurine (hollow-crested) hadrosaurids, the crests perched atop the head must have provided for instant recognition (Figure 12.12). This would have been by visual cues as well as by low honking tones produced in the large resonating chamber of the crest (see Figure 12.13).

---

[4] The fleshy red-colored skin tissues, used for male display, found on rooster heads.

Other Ornithopods

Other ornithopods show features potentially interpretable in terms of sexual selection and intraspecific competition. The low, broad bump on top of the head of *Ouranosaurus* and the arched snout of *Muttaburrasaurus* and *Altirhinus* may well have behavioral significance relating to intraspecific competition and sexual selection. *Ouranosaurus* was equipped with extremely tall neural spines, which formed a high, almost sail-like ridge down its back (Figure 12.16). What these did – or for that matter what the spines did in *Spinosaurus* (see Figure 6.24) – remains anybody's guess. It is possible that these long spines supported a membrane used by the animal to warm up and cool down, and/or the spines may have had a display function, providing the animal with a greater side profile than it would otherwise have had. We just don't know.

Display behavior in many ornithopods begins to make even more sense when considered in the context of the discoveries of bonebeds containing just one type of dinosaur. Monotypic bonebeds, that is, bonebeds containing only one type of animal, are known for *Dryosaurus*, *Iguanodon*, *Maiasaura*, and *Hypacrosaurus*, among ornithopods. In the case of hadrosaurids, at least, the evidence suggests that a single herd could have exceeded 10 000 individuals, rivaling the multiple mile-sized bison herds that roamed the Great Plains of North America before the unfortunate pairing of the transcontinental railroad with the .55 caliber Sharps rifle.

Bringing Up Baby II

The secrets of ornithopod reproductive behavior are just beginning to be told. For the small, basal ornithopod *Orodromeus*, hatchlings had well-developed limb bones, with fully formed joints, indicating that the young could walk, run, jump, and forage for themselves as well as any adult. We thus infer minimal parental care.

As first discovered by paleontologist J. R. Horner and wonderfully described in his and Gorman's book *Digging Dinosaurs*, hadrosaurids did not take a *laissez-faire* an attitude toward child-rearing. *Maiasaura*, *Hypacrosaurus*, and probably most others nested in colonies, digging a shallow hole in soft sediments and laying, in the case of *Maiasaura*, up to 17 eggs in each nest. These nests were separated by about a mother's body length, strongly suggesting that they were regularly tended by a parent. *Maiasaura* hatchlings (Figure 12.17) look like typical babies (think puppies): big limbs, feet, and hands; large heads and eyes; shortened snouts! The babies were found amid an abundance of eggshell fragments, implying an extended stay at the nest that wreaked havoc on the eggs that once housed them.

With under-developed joints, the offspring were helpless during the nest-bound time. They could hardly have foraged far from the nest and must have depended on their parents to provision and protect them. But to go from a 1 m hatchling to a 9 m adult, growth must have come hot and heavy: at approximately 12 cm per month, as fast as, or faster than, fast-growing mammals and birds (see Chapter 13). This means that hatchlings must have channeled into growth virtually all of the food that their parents brought them.

Recently the development of *Hypacrosaurus*, at least, has been taken to new levels with a detailed analysis of changes in the face as the animals grew up. A complete growth series of skull specimens representing embryos, nestlings, juveniles, subadults (teenagers!), and adults has been reconstructed (*à la Protoceratops*) and from this it is clear that these dinosaurs, like all juveniles, begin life relentlessly cute: short snouts, large eyes, and rudimentary crest development! Just as in all other vertebrates, these features change, becoming less cute, as adulthood is reached.

A new take on ornithopod child-rearing was provided by the hypsilophodont *Oryctodromeus*, a dinosaur that evidently raised altricial young in a burrow (see Figure 1.7). Only one specimen of the animal is known, but this reveals a burrow with an end chamber, in which were found the remains of

Figure 12.16. The Early Cretaceous iguanodontian *Ouranosaurus* from Niger. Mounted specimen on display at the Royal Ontario Museum, Ontario, Canada.

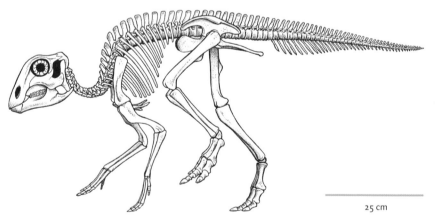

Figure 12.17. Left lateral view of the skull and skeleton of a hatchling hadrosaurid *Maiasaura*. Keeping in mind that the adult was about 4.5 m long, check out the scale bar on this figure (see also Figure 13.9).

25 cm

an adult and two juveniles. *Oryctodromeus* appears to have some digging specializations in its skull and thoracic region, suggesting a fossorial, or burrowing, lifestyle.

Ornithopod life, therefore, certainly involved much opportunity for interaction: within herds, as breeding pairs, and as families. All of this is part and parcel of the visual and vocal communication we postulated earlier, and affirms complex social behavior, particularly in hadrosaurids.

Box 12.1 Those Rascals, the Basal Ornithischians!

It used to be really easy: if it was bipedal, ornithischian, herbivorous, and three-toed, it was likely an ornithopod. But when paleontologists got to thinking about relationships a wee bit more carefully, and looked through the lens of cladistic analysis, who was a diagnosable ornithopod, and who was not, turned out to be a bit trickier.

The lid blew off the pot with heterodontosaurids: these three-toed, herbivorous bipedal ornthischians had appeared for years to be ornithopods. Careful cladistic analysis by R. J. Butler, P. M. Barrett, and colleagues (see Chapter 16), however, suggested that they properly belong way back at the base of Ornithischia.

Other supposed "ornithopods" also were problematic: *Othnielosaurus, Hexinlusaurus, Agilisaurus, Zephryrosaurus, Lesothosaurus, Pisanosaurus, Jeholosaurus, Yandusaurus, Stormbergia*, to name just a few! Worse yet, venerable

clades like Hypsilophodontidae were shown to be paraphyletic, so that perplexed researchers, as noted by North Dakota paleontologist C. Boyd (2015, p. 2) "chose conservatively to refer to all non-marginocephalian, non-iguanodontian…taxa as 'basal neornithischians.'"

Part of the problem stems from the strengths and weaknesses of phylogenetic systematics. As we've seen, the method depends upon synapomorphies for its power; thus, when dealing with primitive characters, it is at its weakest, because there are few synapomorphies to help distinguish the groups. The result is that it is hard to determine which of the many differing hypotheses of relationships – the cladograms – is the most parsimonious, and should thus be selected.

It turns out that most humans can live comfortable, satisfying lives without ever knowing the exact

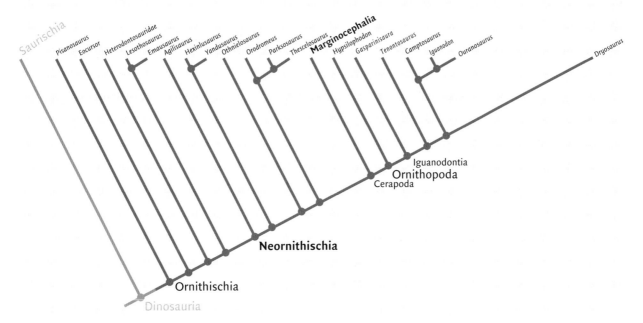

Figure B12.1.1. C. Boyd's 2015 proposal for the relationships among ornithischian dinosaurs. It is broadly comparable to the cladogram we show in Figure 12.18; however, it differs in some key aspects as well. For example, *Orodromeus*, considered by some to be a basal ornithischian, is here not even within the group. Likewise, *Hypsilophodon*, *Parksosaurus*, and *Thescelosaurus*, long regarded as fine, upstanding ornithopods, are here regarded as too primitive to be within Ornithopoda. *Gasparinisaurus*, considered by Butler and Barrett as rather close to *Hypsilophodon*, is here a member of Iguanodontia. It's bewildering, and the goal is not to give nonspecialists hives; it is just to convey a sense of the complexity and instability of our understanding of basal neornithischian relationships as of 2020! Figure modified from Boyd, C. 2015. The systematic relationships and biogeographic history of ornithischian dinosaurs. *PeerJ*, **3**, e1523; DOI 10.7717/peerj.1523.

Box 12.1 *cont'd*

phylogenetic position of, for example, *Gasparinisaurus*. That being the case, we offer two simplified examples of the many cladograms that have been proposed for these basal neornithischian dinosaurs. Figure 12.18, in the text, is a version of the 2012 cladogram developed by paleontologists R. J. Butler and P. M. Barrett. Figure B12.1.1, here, is a version of the cladogram suggested by C. Boyd in 2015. The take-home message in all this is that ornithopod – or perhaps we might better say neornithischian – evolution, was not a simple straight line, but rather a complicated bushy tree, with many dead ends and false starts on the way to the most highly evolved group of ornithopods: hadrosaurs.

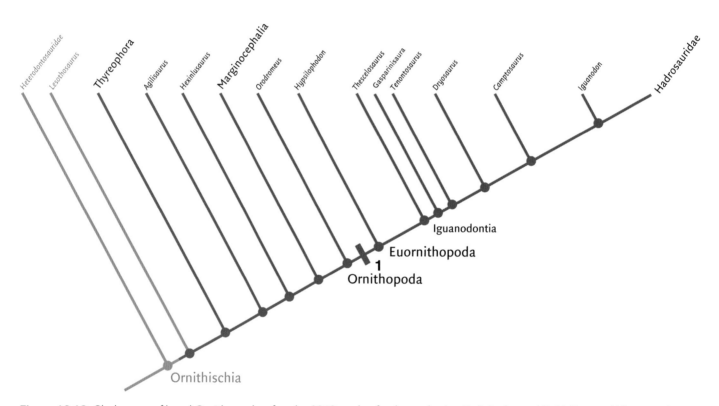

Figure 12.18. Cladogram of basal Ornithopoda, after the 2012 study of paleontologists R. J. Butler and P. M. Barrett. What is striking is that the majority of the most famous ornithopods are most precisely designated "euornithopods." We use the more general term "ornithopods" throughout this chapter, however, as a nod to convention. Derived characters include: at **1**, subcircular external antorbital fenestra, distal offset to apex of maxillary crowns, strongly constricted neck to the scapular blade, ossification of sternal ribs, hypaxial ossified tendons in the tail; at **2**, rectangular lower margin of the orbit, widening of the frontals, broadly rounded predentary, dentary with parallel dorsal and ventral margin, absence of premaxillary teeth, ten or more cervical vertebrae, six or more sacral vertebrae, presence of an anterior intercondylar groove, inflation of the medial condyle of the femur.

How do those ornithopods for which we have information conform to either of these two contrasting reproductive strategies outlined in Chapter 9? The small, primitive ornithopod *Orodromeus* seems to have been an *r*-strategist, an inference that is based on the precocial nature of the young. In contrast, *Maiasaura*, *Hypacrosaurus*, and perhaps other hadrosaurids had nest-bound, altricial hatchlings, and were likely *K*-strategists.

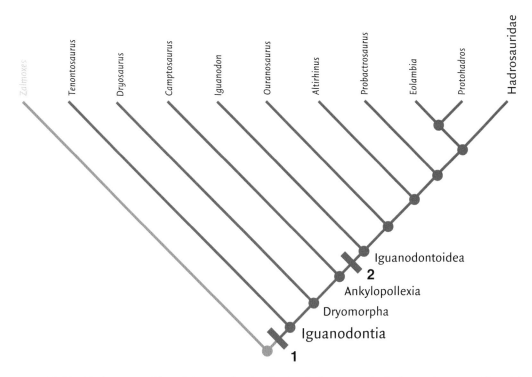

Figure 12.19. Cladogram of basal Iguanodontia. Derived characters include: at **1**, premaxilla with a transversely expanded and edentulous margin, reduction of the antorbital fenestra, denticulate margin of the predentary, deep dentary ramus, loss of sternal rib ossification, loss of a phalanx in digit III of the hand, compressed and blade-shaped prepubic process; at **2**, strong offset of premaxilla margin relative to the maxilla, peg-in-socket articulation between maxilla and jugal, development of a pronounced diastema between the beak and mesial dentition, mammillations on marginal denticles of teeth, maxillary crowns narrower and more lanceolate than dentary crowns, closely appressed metacarpals II–IV, deep triangular fourth trochanter, deep extensor groove on femur.

The Evolution of Ornithopoda

The basal split of Ornithopoda from the generalized cerapodan condition likely occurred in the latest Triassic or earliest Jurassic. The clade is diagnosed on the basis of a number of derived features (see Figure 12.3). As we've seen, an early split in Ornithopoda occurred between primitive ornithopods such as *Orodromeus* and *Agilisaurus*, and Iguanodontia, with iguanodontians containing much of the future diversity of the clade.

Ornithopoda, once the comfortable residence home of three-toed, bipedal ornithischians, has proven to be a rascally taxon under the rigorous scrutiny of cladistic methods (Box 12.1). It is best understood by its definition: all genasaurs more closely related to *Parasaurolophus* (a hadrosaur) than to *Triceratops*. On the cladogram, this means, in effect, all nonmarginocephalian ceropodans (Figure 12.18). It consists of a host of relatively small, agile ornithopods such as *Hypsilophodon* and *Gasparinisaura*, as well as a few somewhat larger, more robust forms (*Parksosaurus* and *Thescelosaurus*), and the diverse clade Iguanodontia (Figure 12.19), residence of such dinosaurian luminaries as *Camptosaurus* (Figure 12.20), *Iguanodon*, and all the hadrosaurids. In general, nonhadrosaur iguanodontians tended to reach their apogee in the Late Jurassic–Early Cretaceous interval.

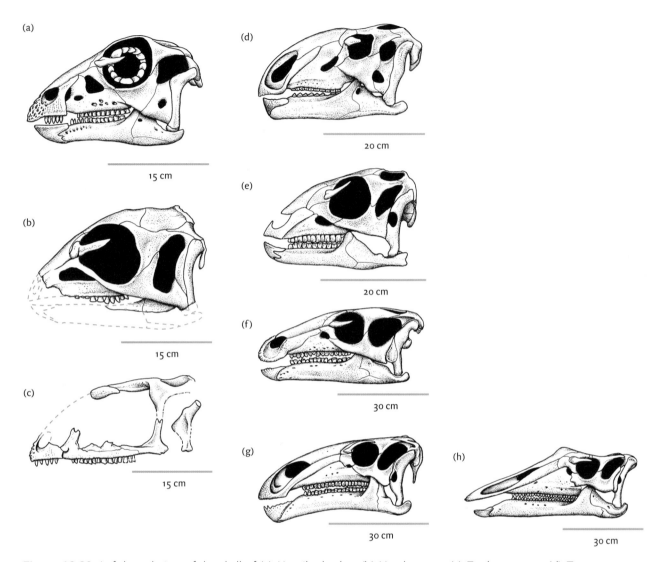

Figure 12.20. Left lateral view of the skull of (a) *Hypsilophodon*, (b) *Yandusaurus*, (c) *Zephyrosaurus*, (d) *Tenontosaurus*, (e) *Dryosaurus*, (f) *Camptosaurus*, (g) *Iguanodon*, and (h) *Ouranosaurus*.

Hadrosaurids are among the best known of all dinosaurs, highly evolved, Late Cretaceous ornithopods with a fabulous fossil record that allows, as we have seen, insights into their behavior. Two major clades within hadrosaurids – Lambeosaurinae and Hadrosauridinae – constitute most of Hadrosauridae, with a few forms left whose relationships within Hadrosauridae are uncertain (Figure 12.21). A variety of hadrosaurids is shown in Figure 12.22.

Several interesting evolutionary trends are present within Ornithopoda. It is perhaps no coincidence that ornithopod diversity seems to parallel gymnosperm (especially conifer) and angiosperm diversity (see Figure 15.8). It suggests that these dinosaurs and plants may have been involved in a kind of reciprocal *pas de deux*: as gymnosperms developed ways to discourage predation, ornithopods developed more and more efficient ways of extracting nutrients. The reciprocal evolution culminated in the highly efficient pleurokinetic hadrosaurid jaw, with its well-developed, integrated packages of teeth in continuously replacing dental batteries. Overall patterns within the Late

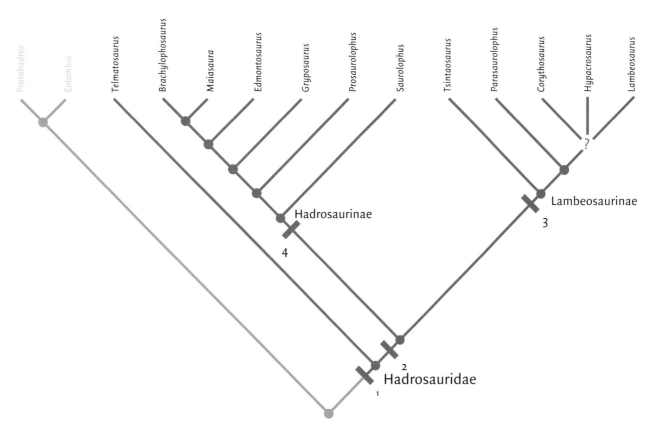

Figure 12.21. Cladogram of Hadrosauridae. Derived characters include: at **1**, three or more replacement teeth per tooth position, posterior extension of the dentary tooth row to behind the apex of the coronoid process, absence of the surangular foramen, absence or fusion of the supraorbital to the orbit rim, long coracoid process, dorsoventrally narrow proximal scapula, very deep, often tunnel-like intercondylar extensor groove; at **2**, absence of the coronoid bone, reduction in surangular contribution to coronoid process, double-layered premaxillary oral margin, triangular occiput, eight or more sacral vertebrae, reduced carpus, fully open pubic obturator foramen, absence of distal tarsals II and III; at **3**, maxilla lacking an anterior process but developing a sloping dorsal shelf, groove on the posterolateral process of the premaxilla, low maxillary apex, a parietal crest less than half the length of the supratemporal fenestrae; at **4**, presence of a caudal margin on the circumnarial fossa.

Cretaceous of North America and Asia, at least, suggest that hadrosaurids ecologically replaced large non-hadrosaurid iguanodontians, powerful chewers in their own right, but lacking the extraordinary specializations of hadrosaurs. Hadrosaurids arguably evolved the art of chewing to levels of sophistication unparalleled in the history of life; perhaps the price of a tough, fibrous gymnosperm diet.

Along with the specializations associated with chewing, ornithopod evolution may have also been characterized by a greater and greater investment by parents in their young. We have seen that *Orodromeus* produced relatively precocial offspring; its basal position within Ornithopoda (see Figures 12.4 and 12.18) suggests that precocity may be primitive for at least Ornithopoda. As the diversity of Ornithopoda increased, altricial behavior likely evolved in more derived ornithopods some time prior to the origin of Hadrosauridae, which all are thought to have given birth to altricial young.

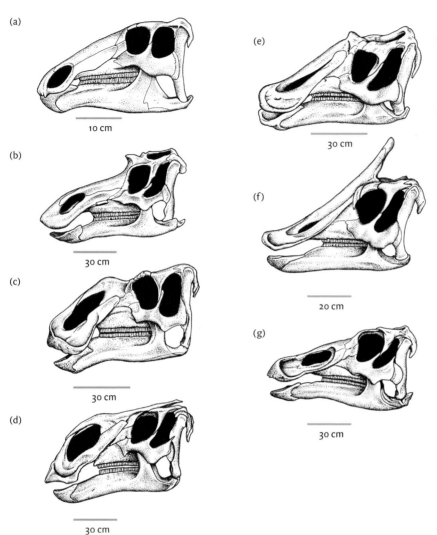

Figure 12.22. Left lateral view of the skull of (a) *Telmatosaurus*, (b) *Maiasaura*, (c) *Gryposaurus*, (d) *Brachylophosaurus*, (e) *Prosaurolophus*, (f) *Saurolophus*, and (g) *Edmontosaurus*.

SUMMARY

Ornithopods were the most numerous and diverse herbivores of Dinosauria, consisting of iguano-dontians (including hadrosaurids), and a cloud of more primitive forms. Although once thought to have included all bipedal ornithischians, it is now thought that many dinosaurs formerly thought to be ornithopods are more correctly regarded as outside of Ornithopoda (more basal ornithischians).

Ornithopods ranged in size from very small (less than 2 m) to rather large (more than 15 m), and evidently colonized virtually every inhabitable region of the globe. Like so many other dinosaurian groups, the most primitive forms were bipedal, although quadrupedal locomotion became increasingly dominant, particularly in larger forms. Basal iguanodontians appear to have mixed quadrupedal and bipedal locomotion, but the derived forms, the hadrosaurs, appear to have relied almost exclusively on quadrupedal locomotion. Even these, however, may have resorted to bipedal locomotion when running hard.

Ornithopods had very advanced chewing capabilities. The skull has an inset tooth row indicating cheeks, as well as the tripartite division of chewing herbivores (including a cropping beak, a diastem, and closely appressed cheek teeth for grinding), and in virtually all cases, a large coronoid process suggests strong jaw adductor muscles. Hadrosaurs took chewing to unprecedented heights with the evolution of a pleurokinetic skull combined with dental batteries. Coprolites and stomach contents suggest hadrosaurs needed all the teeth they had: their diet appears to have been coarse and fibrous.

Ornithopods were very social animals, none more so than hadrosaurs. The discovery of many bonebeds attests to this, as do a remarkable and complex variety of sexually dimorphic head and skull features found in many ornithopods, suggesting that life as an ornithopod involved intensive sexual selection. In hadrosaurs at least, communication may well have been enhanced by a variety of vocalizations.

The genus *Maiasaura* has given us a view of child-rearing, duckbill style. Complete growth series, from hatchlings at nests to adults, are known, and considerable evidence exists that hadrosaurs grew remarkably quickly. Nesting apparently was communal, and parental care was expended on altricial young: duck-bills were likely *K*-strategists. Interestingly, other ornithopods, such as the genus *Orodromeus*, may have raised precocial young and favored an *r*-strategy of child-rearing.

Ornithopod evolution was characterized by an increase in the sophistication of chewing specializations. Among the large ornithopods, it could be argued that globally, the highly advanced hadrosaurs ecologically replaced nonhadrosaurid iguanodontians.

SELECTED READINGS

Boyd, C. 2015. The systematic relationships and biogeographic history of ornithischian dinosaurs. *PeerJ*, **3**, e1523. DOI 10.7717/peerj.1523

Butler, R. J. and Barrett, P. M. 2012. Ornithopods. In Brett-Surman, M. K., Holtz, T. R., and Farlow, J. O. (eds.) *The Complete Dinosaur.* Indiana University Press, Bloomington, IN, pp. 551–556.

Eberth, D. A. and Evans, D. C. (eds.) 2014. *Hadrosaurs.* Indiana University Press, Bloomington, IN, 619 pages.

Hopson, J. A. 1975. The evolution of cranial display structures in hadrosaurian dinosaurs. *Paleobiology*, **1**, 21–43.

Horner, J. R. 1984. The nesting behavior of dinosaurs. *Scientific American*, **250**, 130–137.

Horner, J. R., Weishampel, D. B., and Forster, C. A. 2004. Hadrosauridae. In Weishampel, D. B., Dodson, P., and Osmólska, H. (eds.) *The Dinosauria*, 2nd edn. University of California Press, Berkeley, pp. 438–463.

Norman, D. B. 2004. Basal Iguanodontia. In Weishampel, D. B., Dodson, P., and Osmólska, H. (eds.) *The Dinosauria*, 2nd edn. University of California Press, Berkeley, pp. 413–437.

Norman, D. B., Sues, H.-D., Coria, R. A., and Witmer, L. M. 2004. Basal Ornithopoda. In Weishampel, D. B., Dodson, P., and Osmólska, H. (eds.) *The Dinosauria*, 2nd edn. University of California Press, Berkeley, pp. 393–412.

Ostrom, J. H. 1961. Cranial morphology of the hadrosaurian dinosaurs of North America. *Bulletin of the American Museum of Natural History*, **122**, 33–186.

Varricchio, D. J., Martin, J. A., and Katsura, Y. 2006. First trace and body fossil evidence of a burrowing, denning dinosaur. *Proceedings of the Royal Society Series B*, **274**, 1361–1368.

Weishampel, D. B. 1984. The evolution of jaw mechanisms in ornithopod dinosaurs. *Advances in Anatomy, Embryology and Cell Biology*, **87**, 1–110.

Weishampel, D. B. 1997. Dinosaurian cacophony. *BioScience*, **47**, 150–159.

TOPIC QUESTIONS

1. Who are the ornithopods? What are the diagnostic characters of Ornithopoda? How are ornithopods related to other ornithischians?
2. What are the major divisions of Ornithopoda, and what are their diagnostic characters?
3. How could a single ornithopod be both bipedal and quadrupedal?

4. Describe the hands of nonhadrosaurid iguanodontians. How did those differ from the hands of hadrosaurids?
5. Highlight what is known of nests and nesting in ornithopods.
6. What is pleurokinesis? How did it function in the jaws of hadrosaurids?
7. Name another vertebrate with a kinetic skull.
8. Why is it that most paleontologists now think that hadrosaurid head structures were related to intraspecific competition and/or sexual selection?
9. Give a nondinosaur example of a *K*-strategist, and an *r*-strategist.
10. How does ornithopod diversity parallel gymnosperm and angiosperm diversity?
11. Use a cladogram to make the argument that a *K*-strategy evolved *at least* two times in the history of vertebrates.

PART IV
Endothermy, Endemism, and Extinction

Nonprofessionals generally assume that paleontology is all about collecting fossils in exotic locales, and identifying and naming new and (to us) bizarre or unexpected life forms. And no active paleontologist would deny that these are all some of the best parts of the job.

But for most of us, good as these are, they really don't come close to the full intellectual richness of our work. It's not just about whether *T. rex* ate *Triceratops*, or whether an ornithomimid could run at 50 kmh. It's the *big* picture – the synthesis – where paleontology really makes its mark. Paleontology allows us to ask scientific, Earth-sized questions, such as how have entire ecosystems evolved over time, and whether evolution has any kind of intrinsic direction (like whether organisms tend to evolve towards greater intelligence or whether life through time is moving towards increasing diversity). The historical perspective allows us to investigate the great rhythms of life on time scales that can barely be grasped by a human mind.

Particularly relevant in a world in which dramatic climate change is causing rampant global extinction, paleontologists are uniquely positioned to ask questions like: Are we in the middle of a mass extinction right now? What happens then? Are some organisms more resilient to mass extinctions than others? Are some *ecosystems* more resilient to mass extinctions than others? How long does it take for things to return to normal (called the **recovery**)? We like to think of ourselves as unique, and living in unique times, but it turns out that the Earth has experimented with almost everything in the past (although, perhaps, not an industrial age), and the record of that experimentation is left for us to uncover in the Earth's rock record. That is what geologist M. Bjornerud meant when she subtitled her book *Reading the Rocks*, "*An autobiography of the Earth.*" It is by reading the rocks that we know that global climate change and rising sea levels have occurred many times in the past, and from the stories they tell, we have some ideas about how the life on Earth responds to these kinds of dramatic changes.

In fact, humans developed and have flourished in an unusually cold period in Earth history. In the past, things were generally warmer; we might ask, therefore, whether global warming is actually a return to normal. Paleontology, with its vast sweep of Earth history, is uniquely positioned to provide insights to these very timely questions.

In this section of the book, therefore, we take a small step towards thinking about larger synthetic questions that naturally arise from studies of the past.

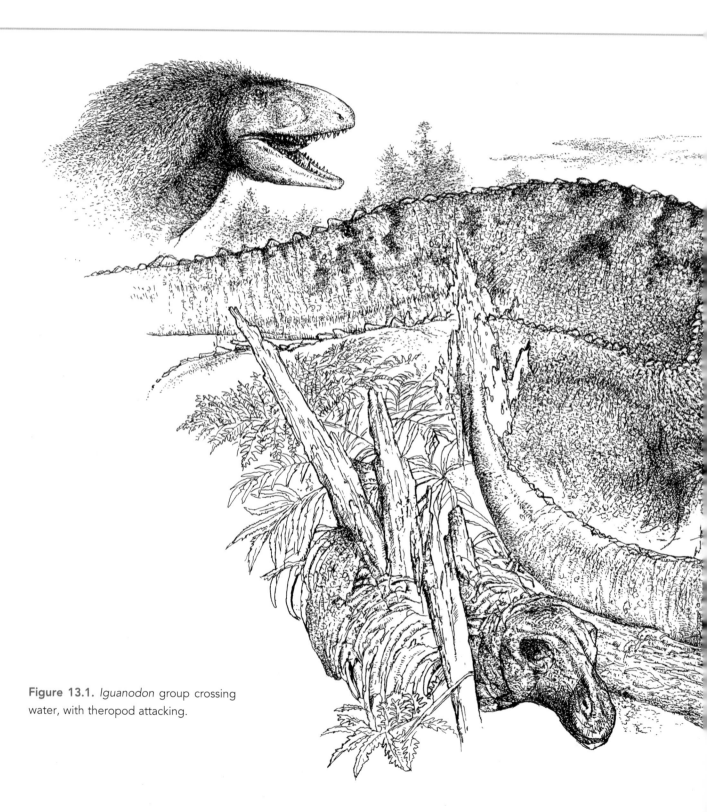

Figure 13.1. *Iguanodon* group crossing water, with theropod attacking.

Chapter 13

The Paleobiology of Dinosaurs I

WHAT'S IN THIS CHAPTER

Here we reach back, across tens of millions of years of time, to experience – to almost touch – dinosaurs as living beasts; breathing; fighting; loving; killing; wounding; wounded; feeding; pooping; *FEELING* animals, beyond the mere accumulations of fossil bones.

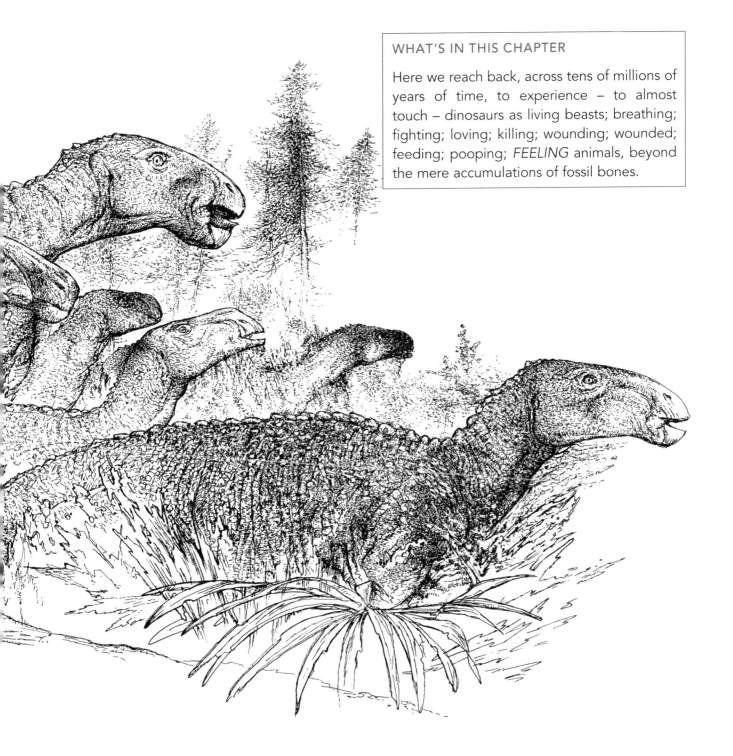

Paleobiology

In the late 1960s, vertebrate paleontologists began to recognize that even if the fossils they studied were (now) rocks found in rocks, it was *biology* of fossil organisms that they were really after: who these extinct organisms were in life, what they did, and how they did it. That recognition called for a new way of thinking, and that new way of thinking was celebrated by a new word, representing a new field: paleobiology, that is, the study of the biology of ancient organisms. What do we really know about nonavian dinosaurs as living beasts? More than we once dreamed possible, but still far less than we want to find out!

Dino Breath

When we discussed saurischian dinosaurs (sauropodomorphs and theropods, Chapters 6 to 9), recall that there was much loose talk about pneumatic bones and pleurocoels, both of which were adaptations for air sacs, features that we distinguished from the more familiar lungs. Now is the time to understand a bit better what all this was about.

Every Breath You Take

Mammals like ourselves breathe by passing air bidirectionally into and out of the lungs (that is, during inhalation and exhalation) like a bellows. Oxygen can be extracted from the air that reaches the lungs; however, air that stays in the trachea cannot be used as the trachea creates physiological dead space because some portion of the inhaled air never reaches the lungs. It is simply brought into the respiratory system and returned without being involved in oxygen–carbon dioxide exchange. In animals with long necks, like giraffes, this can be a lot of air. How much more, then, would it be in sauropods?

By contrast, birds, theropods, and sauropods have an extraordinary system called unidirectional air flow. When birds inhale air is drawn into posterior air sacs and lungs. Simultaneously, the oxygen-depleted air is pulled from the lung into the anterior air sacs. When the animal exhales, the now oxygen-depleted air is pumped out of the anterior air sacs, and the air in the posterior air sacs is brought into the lung. Because the air in the lung always moves in the same direction, the breathing is called "unidirectional." More significantly, the air moves unidirectionally while the blood circulation is in the opposite direction. This turns out to be the most efficient way that gaseous exchange (deoxygenated blood becomes oxygenated) takes place, and is called *countercurrent exchange.* Countercurrent exchange doesn't exist in nonavian respiration, and thus the avian system wrings more oxygen out of the air than the bidirectional, bellows-style lungs found in mammals (Figure 13.2). Pay attention, athletes!

Unidirectional breathing is to be expected in animals that have highly active lifestyles, in which the need to maximize the efficiency of oxygen consumption is very important. Along with birds, crocodiles are the only living unidirectional breathers, but crocodiles lack extensive air sacs, pleurocoels, and pneumatic foramina; by contrast, birds have a highly derived, complex system of air sacs, with approximately nine times the air storage capacity of the lungs, pleurocoels, and pneumatic bones. The air sacs don't get oxygen out of the air; but they surely make the unidirectional flow possible. All of these adaptations are associated with maximizing the efficiency with which birds can consume oxygen.

The presence of air sacs and pleurocoels leaves unambiguous marks on the bones, and so it is with confidence that we can be sure that most theropods and sauropods all breathed unidirectionally and had, to a greater or lesser extent, the complex suite of adaptations associated with it. This is yet another striking example of how a unique feature in living birds today – unidirectional breathing – is actually something that birds and nonavian theropod dinosaurs have in common, further supporting the evolutionary link between them.

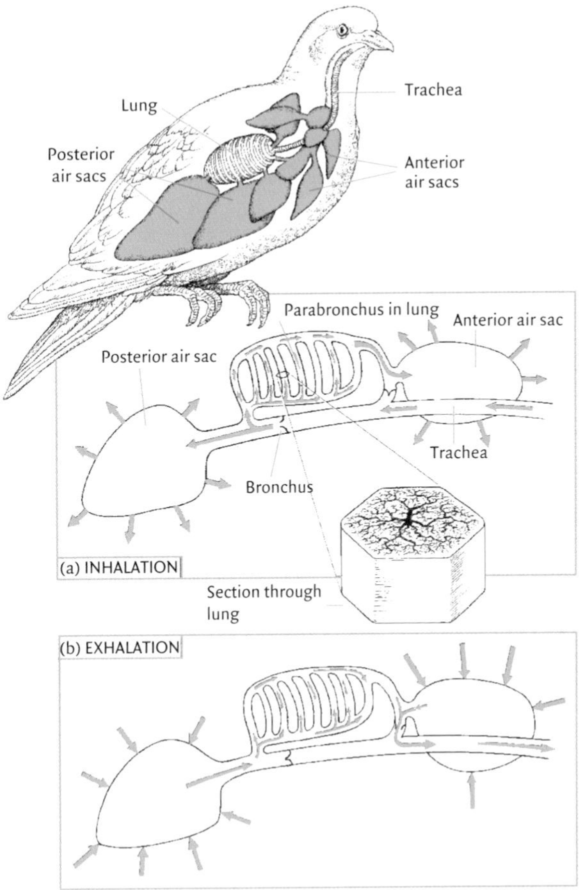

Dino Brains

How can we measure the intelligence of dinosaurs? In *living* animals, the very term "intelligence" is a term not easily defined. One might think of intelligence as some amalgam of problem-solving ability, behavioral flexibility, sensitivity, and perhaps the capability of abstract thought and/or self-consciousness. And so much of that, if any was ever there when they were alive, is *not* going to be accessible in extinction animals!

Yet, it is clear that, at a very crude level, there is a correlation between intelligence and brain:body-weight ratios. Brain:body-weight ratios are used because they allow the comparison of two differently sized animals (that is, brain:body-weight ratios theoretically, at least, allow for the comparison of chihuahua and St. Bernard dogs). The correlation suggests that, in a general way, the larger the ratio, more likely the organism will at least in a general way, contain some of the qualities enumerated above. Indeed, mammals have higher brain:body-weight ratios than fish and are generally considered to be more intelligent. But how smart could a very large dinosaur with a minuscule brain be (for example, see Figure 10.9 and Box 11.1)? Probably somewhat smarter than we might expect.

It is well known that organisms change proportions as they increase in size; this is termed allometry.[1] And it turns out that brain:body-weight ratios follow allometric principles as well: brains do not increase in size proportionally to the rest of the animal. For example, the brain of a 0.5 m rattlesnake is, in proportion to its body size, larger than the brain of a 3 m anaconda. Does this mean that the anaconda is significantly stupider than the rattler? Obviously not.

So, when considering how big or small a brain is in an animal, there has to be a way to compensate meaningfully for size. A quantitative method of doing this was first proposed by evolutionary psychologist H. J. Jerison, who, in the early 1970s, developed a measure called the

Figure 13.2. Unidirectional respiration, shown diagrammatically. As the animal inhales (a) air enters the trachea and posterior air sacs (here represented by a single sac), which expand. Air is then moved to the lungs, where it is deoxygenated, and then stored in the anterior air sacs (here represented by a single sac), which expand and fill with deoxygenated air. As the animal exhales (b), the deoxygenated air, in the lungs and anterior air sac, is expelled via contraction out of the trachea.

---

[1] We'll use ants as an easy way to understand allometry. Ants have particular proportions (legs; abdomen; thorax; eyes; all their features), which work very well at an ant's size. But, if you increased the size of an ant so that it could stare eye to (compound) eye with a sauropod, the ant would no longer be able to maintain the proportions of its formerly tiny self; it would be unable to function in even the most rudimentary way; its fundamental proportions would have to utterly change in order for it to do whatever giant sauropod-sized ants do.

encephalization quotient (EQ). Jerison constructed an "expected" brain:body-weight ratio for various groups of living vertebrates (reptiles, mammals, birds) by measuring many brain:body-weight ratios among these animals. Having constructed a range of expected brain:body-weight ratios, he could account for size in different organisms (and accommodate what might at first seem like an extraordinarily large or small brain). Noting that some organisms still didn't exactly fit in his groups (by virtue of having brains either larger or smaller than expected for that group), he measured the amount of deviation, and then termed this EQ.

Paleontologist J. A. Hopson, now knowing what he could expect for living vertebrates, measured how much the estimated brain:body-weight ratio EQ of extinct vertebrates deviated from the expected brain:body-weight ratios of their living counterparts. Figure 13.3 shows the EQs for several major groups of dinosaurs as reconstructed by Hopson. Because dinosaurs are "reptiles," he measured the deviation of various groups relative to a "reptilian" norm, in this case a living crocodile. Significantly, many ornithopods and theropods show a brain:body-weight ratio that is significantly larger than would be expected if a modern reptilian level of intelligence is being considered. The implication is that these dinosaurs were well above the crocodile "level" of intelligence.

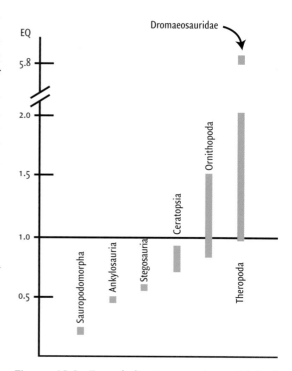

Figure 13.3. Encephalization quotients (EQs) of dinosaurs compared. The line at 1.0 represents the crocodilian "norm," and suggests that many groups of dinosaurs had larger brains than would be predicted from a conventional reptilian model (the crocodile). Note also the break between 2.0 and 5.8; if these measures mean anything, apparently coelurosaurs significantly outdistanced other dinosaurs in brain power. (Data from Hopson, 1980.)

Dino Bones

Once paleontologists began to look carefully, it became clear that the interiors of the bones, not just their external shapes, had a lot to tell us about dinosaur growth, about age, and even physiology. Fossil bone can preserve fine anatomical details. To see these, a thin slice of bone can be cut, mounted on a glass slide, and ground down to approximately 30 microns (μm), so thin that light can be transmitted through it. The slide with the slice of bone can then be studied under a microscope – or, without destroying the bone (as occurs with cutting and grinding), can be CT scanned. Either way, then the party begins.

Haversian Bone

A bone is a mineralized (e.g., hardened) piece of tissue; how then do bones grow? The answer is that they grow by **remodeling**, which involves the resorption (or dissolution) of bone first laid down – **primary bone** – and redeposition of a kind of bone called **secondary bone**. Secondary bone is deposited in the form of a series of vascular canals called **Haversian canals**, and resorption and redeposition of secondary bone can occur repeatedly during the process of secondary bone formation, termed "remodeling". The canals contain the blood vessels that supply nutrients to the remodeling bone; the living bone cells themselves are sequestered in dark spots throughout the remodeling bone. Remodeled bone has a distinctive look about it (Figure 13.4), visible in modern as well as fossil bone.

Now for some oversimplifications about living tetrapods, which nevertheless will help us understand the outlines of this process: birds and mammals tend to grow until they reach sexual maturity, after which they continue life (and sexual reproduction) as adults, with little growth. Such growth is

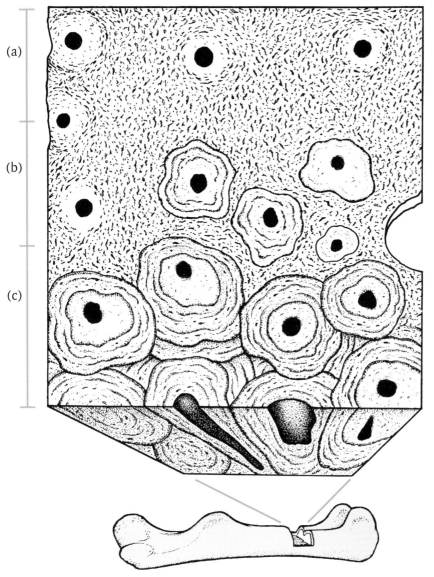

Figure 13.4. Primary bone in the process of being replaced by Haversian bone in the leg of a hadrosaurid. Longitudinal canals (at the top of the figure) in primary lamellar bone (a) are resorbed (b) and then reconstituted as Haversian bone (c).

called determinate, because it goes to a point (sexual maturity) and then it largely stops. The size of the adult of such species can therefore be predicted. Many reptiles (snakes, lizards, crocodiles, and turtles) and amphibians (salamanders and frogs) tend to continue growing through life, only stopping with death. Such growth is called indeterminate, since the final size of the species is not determined.

Because mammals and birds grow through a juvenile phase, generally to the onset of sexual reproduction, or thereabouts, and then stop, growth is compressed into the first part of these animals' lives. This means that growth rates tend to be much faster in birds and mammals. And faster growth rates require more intensive remodeling, which means that the Haversian systems are far more extensive in mammals and birds, than they are in reptiles and amphibians.

By and large – another generalization – the faster the growth rates, the denser the concentration of Haversian canals, reflecting the amount of energy and cellular activity required for the remodeling. Thus, the development of extensive Haversian bone occurs much more quickly in mammals and birds than in reptiles and amphibians, which would need to reach very old ages to show very extensive

Haversian bone remodeling. And between birds and mammals, birds show more dense Haversian remodeling than do mammals, reflecting the generally faster growth rates of birds than mammals. Among *extinct* vertebrates, Haversian bone has been observed in dinosaurs, pterosaurs, and **therapsids** (including Mesozoic and Cenozoic mammals).

Dinosaurs had determinate growth. Moreover, a look at their bones shows **dense Haversian bone** remodeling (Figure 13.5), suggesting that growth in dinosaurs occurred at rates somewhat comparable to those in modern birds. We'll talk about that those speedy growth rates mean for dinosaur "warm-bloodedness" in Chapter 14.

So if we know the rates at which dinosaurs grew, can we determine how long they lived? The answer to that comes from something called "LAGs."

LAGs

Concentric growth rings have been observed in the bones and teeth of dinosaurs. The growth rings represent times of slowed or stopped growth; for example, an extended (several year) drought or cold spell. Among modern tetrapods, such growth rings typically represent seasonal cycles. During times of slowed metabolism (such as dry seasons in the tropics, or cold seasons in more temperate latitudes), growth is stymied – hence the term **lines of arrested growth**, or **LAGs** (Figure 13.6).

LAGs as Time Markers

But are the LAGs truly seasonal? Our best evidence suggests that indeed they are. Of course, there is no numerical way to date them (see Chapter 2), so the evidence is circumstantial. Yet, it comes from several sources and those sources all produce a coherent story. For one thing, comparable lines known to form annually are found in a wide variety of groups closely and not-so-closely related to Dinosauria. Moreover, when growth rates are estimated based upon the density of Haversian canals in the remodeled bone, the amount of bone that forms between the lines is consistent with annual (seasonal) cycles. Finally, the spacing of the lines shows a steady decrease, consistent with a decrease in growth rate as the animals get older. Indeed, LAGs have never been found to come from anything other than annual growth.

And so if we accept that LAGs are annual signals, they suddenly give us insights into a heretofore elusive aspect of dinosaur paleobiology: how fast they grew, and how long they lived.

Age

How, then, does one determine the age of a dinosaur? Paleohistologists A. de Ricqlès, J. Horner, and K. Padian describe their technique:

(1) Count LAGs, which yield a number presumably corresponding to the age of the dinosaur specimen. Generally, long bones are used for this, because the features being studied, tend to be clearly exposed in them;[2]

(2) measure bone thickness between each LAG, and divide that thickness by the number of days/year, which should produce the amount of bone deposited per day (e.g., the rate of deposition); and

(3) measure total bone thickness and develop an estimate of how much time it would reasonably take to form.

From those steps, they infer the age of the particular specimen in question.

---

[2] Still, counting LAGs can be slightly tricky, because as the bones grow, the central area – the cortex – gets larger. As it enlarges, it remodels and destroys a few of the earliest (oldest) LAGs. Therefore accommodation must be made for this problem; usually by counting LAGs on several bones, which allows us to approximate the number of LAGs on a particular bone that might have been destroyed by cortical growth.

(a)

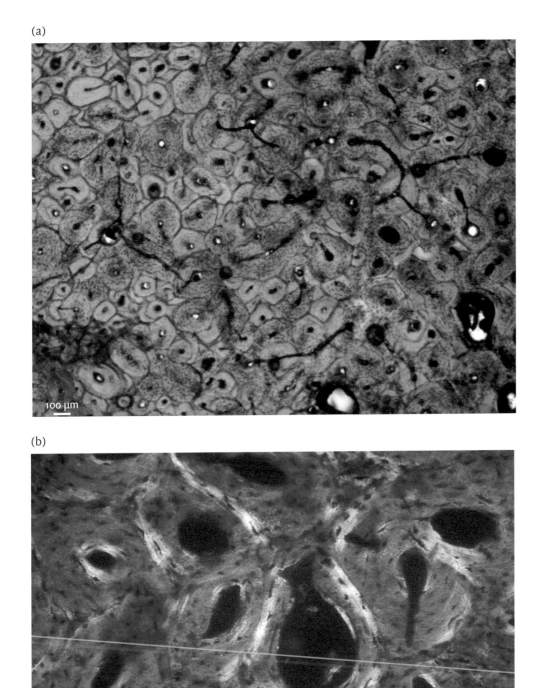

100 µm

(b)

126 µm

Figure 13.5. (a) Dense Haversian bone in *Tyrannosaurus*. (b) Magnified view of dense Haversian bone in *Archaeornithomimus*.

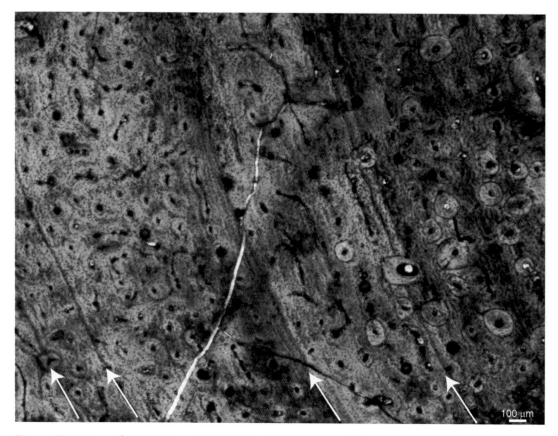

Figure 13.6. Lines of arrested growth (LAGs) in a *Tyrannosaurus* fibula. Arrows indicate LAGs.

Growth

With enough specimens of different ages, it is possible to put together entire growth trajectories of different species, that is, the pattern of growth through a particular organism's life. To take a familiar example, human growth trajectories aren't linear: we start with rapid growth, decreasing gradually until the onset of sexual maturity, after which growth, if it occurs at all, slows dramatically. By the time most humans have completed the first quarter of their lives, growth has stopped. By contrast, many animals, such as crocodiles, have indeterminate growth throughout their lives (although growth rates are fastest when they are juveniles; Figure 13.7).

How about dinosaurs? Two approaches can be used to learn about the growth of a dinosaur. The lengths of the long bones – femur or tibia, generally – can be measured as representative of growth. Alternatively, the animal's weight can be estimated (see below) and considered as representative of its growth. Neither of these methods would show the significant changes that, as we have seen, occur during growth (Figures 11.15 and 11.27), but at least the general growth trajectories could be known. Strikingly, the timing of those trajectories can vary significantly among different dinosaurs. Yet, all are constructed of the same basic "S"-shaped curve: slow-ish growth at birth, rapidly accelerating to fast growth, and then finally a slowing of growth with maturity and old age.

Figure 13.8 shows estimated growth curves for several dinosaurs, including the sauropod *Janenschia*, several tyrannosaurs; the avialan *Archaeopteryx*, the primitive ceratopsian *Psittacosaurus*, and the hadrosaur *Maiasaura*. Unsurprisingly, these dinosaurs do not really all share the same growth rates. *Tyrannosaurus*, so much larger than *Gorgosaurus*, *Albertosaurus*, and

Daspletosaurus, has a much more distinct, rapid, and sustained "growth spurt" than those contemporaries; indeed, its entire growth trajectory is shaped by its literally off-the-scale size (Figure 13.8b). *Archaeopteryx* (Figure 13.8c) if one would term it a "bird" at all, was very primitive and conveniently among the oldest known avialan as well. As it turns out, there is much about its physiology as well that would also be unfamiliar to a modern ornithologist, as the data suggest that it grew much more slowly than modern birds grow, although its growth rates appear to have been comparable to similarly sized maniraptoran theropods. So physiologically, compared to its near relatives, it was likely not unusual, although next to a modern bird it was quite primitive. *Maiasaura* (Figure 13.8e) is thought to have grown at an average rate of something like 500 kg (1100 lb)/year, an astounding rate, taking the animal from a very few kilograms as a hatchling, to around 2500 kg (5500 lb) as an adult in about five years. How this looks from the stand point of the fossils is shown in Figure 13.9. *Tyrannosaurus rex*, at the height of its growth spurt, would have been packing on up to 2.3 kg (5 lb)/day for nearly 10 years. With the exception of *Tyrannosaurus*, many of these dinosaurs seem to have reached their full size within 10 to 15 years. Presumably sexual maturity was reached somewhat earlier; at three (*Maiasaura*) to 10 (tyrannosaurs and the sauropod) years. An exception, tiny *Archaeopteryx*, reached its full size

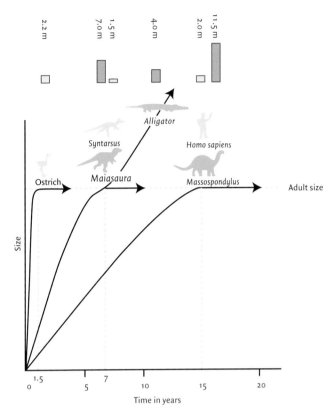

Figure 13.7. Estimated growth rates of an ostrich, *Alligator*, and a human. Unlike the other tetrapods presented, *Alligator* has indeterminate growth; hence, it has no fixed "adult size." Sexual maturity usually comes within six to eight years.

within about two years; sexual maturity would presumably have been somewhat earlier if, as we suspect, its development paralleled those of other theropods.

Our ability to infer the age of some dinosaurs has given us another insight as well. Most of the specimens that have been found turn out to be youngsters, somewhere in the range of 15 to 20 years old. For example, the mighty Sue, all $8.3 M of her, was only around 20 years old. The oldest specimens known reached – maybe – around – 30 years old. Despite the fact that *T. rex* was the "tyrant king," it would appear that life for *T. rex* was neither cossetted nor regal, but was instead hard and fast.

Tonnage

We said earlier that we preferred to use dinosaur lengths to capture the quantity of size because length can be measured; however, you may have observed that most researchers measure size by estimated weight. Estimating weight, however, is a tricky business; so why do we keep resorting to it? The answer is a practical one: weight summarizes size, and it reduces that complex quantity (size) to a single number. That in turn allows weight to be used in a variety of calculations. So here we think a little bit about how weights are actually estimated in animals whose flesh is long gone, and whose bones have turned to stone.

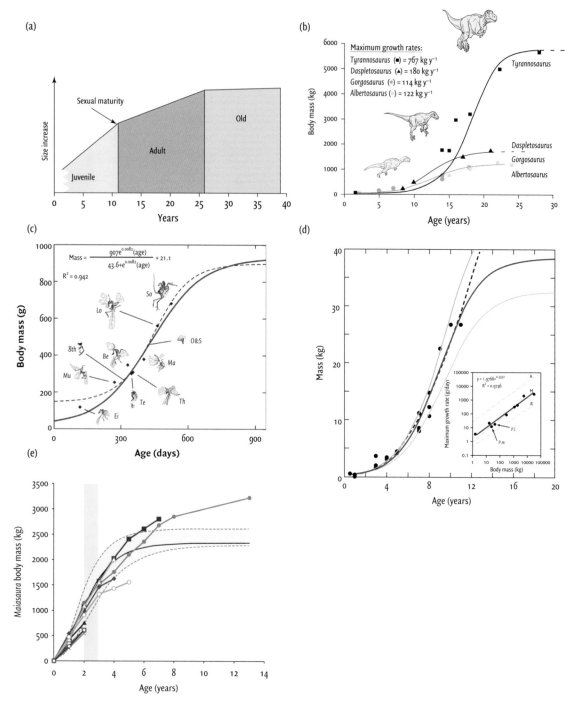

Figure 13.8. Inferred growth trajectories in some dinosaurs based upon LAGs and other features as described in the text. (a) *Janenschia* (sauropod); (b) Late Cretaceous tyrannosaurs; (c) *Archaeopteryx* (avialan theropod); (d) *Psittacosaurus* (primitive ceratopsian); (e) *Maiasaura* (hadrosaur). In each case, the curve is constructed from a series of specimens ranging from juveniles to adults. The specimens are indicated as dots; the resultant "S"-shaped curve is constructed from the positions of the dots. Parts (a) and (b) from Erickson, G. M. 2005. Assessing dinosaur growth patterns: a microscopic revolution. *Trends in Ecology and Evolution*, **20**, 667–684; (c) from Erickson, G. M., Rauhut, O. W. M., Zhou, Z., *et al.* 2009. Was dinosaurian physiology inherited by birds? Reconciling slow growth rates in *Archaeopteryx*. *PLoS ONE*, **4**(10), e7390, doi:10.371/journal.pone.0007390; (d) from Erickson, G.M., Makovicky, P.J., Inouye, B.D., Zhou, C.-F., and Gao, K.-Q. 2009. A life table

Back in the day, when people really didn't understand these things very well, it used to be easy to estimate the weight of a dinosaur. Just as American paleontologist E. H. Colbert did in the early 1960s, you made a scale model of the dinosaur, and then calculated its displacement in water (Figure 13.10).

That displacement was then (a) multiplied by the size of the scale model (for example, if the model were 1/32 of the original, the weight of the displaced water would have to be multiplied by 32) and (b) further modified by some amount using a fudge factor corresponding to the density of tetrapod bodies.

But what is the density of a tetrapod? Based upon studies with a baby crocodile, Colbert determined that baby crocodiles, at least, have a specific gravity of 0.89 (e.g., slightly less dense than the water they displaced). Studies with a large lizard (*Heloderma*) showed that the specific gravity of that lizard was 0.81. Among mammals, it would not be surprising to find the density of a whale differing from that of a cheetah, which might in turn differ from that of a gazelle. Full disclosure: there is no uniform density shared by all tetrapods.

What if you had a fossil represented by a big thigh bone, say, but not much else? Your plans are not quite DOA, because there turns out to be a relationship between limb cross-sectional area and weight. This relationship makes intuitive sense, because obviously, as a terrestrial beast becomes larger, the size (including cross-sectional area) of its limbs, to support its ever-increasing weight, must also

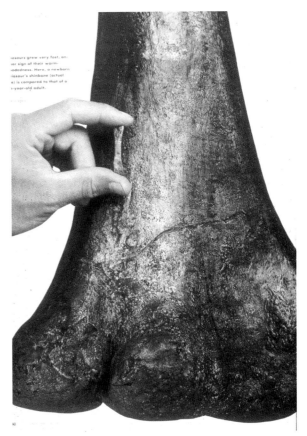

Figure 13.9. Femur of the hadrosaur *Maiasaura* hatchling compared to that of an adult. Note the size of the human hand.

increase. The question is, does it increase in the same manner for all tetrapods? If so, a single equation could apply to all. No surprise: it does *not*!

As noted by J. O. Farlow, weight is dependent upon muscle mass and muscle mass is really a consequence of behavior. Therefore weight estimates of dinosaurs are in part dependent upon presumed behavior. For example, reconstructing the weight of a bear would involve assumptions of muscle bulk and gut mass very different from those used in reconstructing the weight of a deer (Figure 13.11).

Indeed, the cross-sectional area of their limb bones may be identical, but they may weigh very different amounts; likewise, two animals might have the same weight, but be very differently sized. Moreover, our knowledge of dinosaurian muscles and muscle mass is incomplete. So the cross-sectional area method, although convenient and used by a number of workers, has the potential for serious mis-estimations of dinosaur weights.

Caption for Figure 13.8. (*cont.*) for *Psittacosaurus lujiatunensis*: initial insights into ornithischian population biology. *The Anatomical Record*, **292**, 1514–1521; and (e) from Woodward, H. N., Freedman-Fowler, E. A., Farlow, J. A., and Horner, J. R. 2015. *Maiasaura*, a model organism for extinct vertebrate population biology: a large sample statistical assessment of growth dynamics and survivorship. *Paleobiology*, **41**, 503–527, https://doi.org/10.1017/pab.2015.19.

In these days of CT scans, computers, and sophisticated algorithms, can we do better? We can, as demonstrated by the University of Liverpool's K. T. Bates and colleagues. These scientists begin with digitized, complete skeletons and have digitally reconstructed the fleshed-out bodies using laser-scanning techniques (Figure 13.12).

Among the most extraordinary features of these models were reconstructions of real internal anatomy, such air sacs, features that doubtless took up considerable space in nonavian dinosaurs but would cause the density of their owners to diverge significantly from that hopeful (but meaningless) quantity, the "specific gravity of tetrapods." The use of computers allowed the researchers to make innumerable other decisions as well; generally, how much "meat" to put on any given series of bones. Different parts of the body were accorded different densities, in accordance with their biology.

They began with work in the mid 2000s with five dinosaurs: three large theropods, an ornithomimid, and a duck-billed dinosaur (hadrosaur). Ultimately, the mass estimates for these dinosaurs varied significantly, "emphasizing the high levels of uncertainty inevitable in such reconstructions," as Bates and colleagues put it. Still, some "best estimates" were produced: 7655, 6071, and 6177 kg (= 8.44, 6.69, and 6.81 tons, respectively) for the three large-bodied theropods, 423 kg (= 932 lb) for the ornithomimid, and 813 kg (= 1792 lb) for the duckbill. Science is supposed to be testable; how could these scientists figure out whether or not their final numbers had any meaning?

One test was to use the method on the skeleton of a living ostrich. If the method had any validity, it should predict from the skeleton the correct weight of the modern bird, in which the actual mass is known. In fact, the ostrich's predicted weight closely matched that of the estimate calculated from the skeleton, suggesting that the method has some validity.

Ultimately, the most striking thing about the research was what it said about the weight distribution in these creatures. In all cases, the center of mass was consistently located low on the body, in front of the hip. This reinforces the idea, based upon the design of the bones, that these animals all tended to keep the backbone parallel, or nearly so, with the ground (as we saw in Chapter 4).

In 2015, Bates and some other colleagues turned their attention to *Dreadnoughtus* and two other big sauropods (*Apatosaurus* and *Giraffatitan*). *Dreadnoughtus*, they concluded, likely weighed between 30 and 40 metric tons, still in their view (and ours!) an extraordinary amount for a land-dwelling creature. But as for sauropod weight estimates of upwards of 60 tons, they concluded, "We find that 59 300 kg (59.3 metric tons) for *Dreadnoughtus* is highly implausible and demonstrate that masses above 40 000 kg require high body densities and expansions of soft tissue volume outside the skeleton several times greater than found in living quadrupedal mammals."[3]

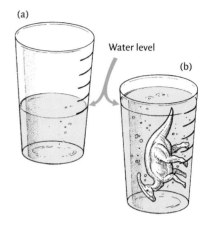

Figure 13.10. Estimating the weight of a dinosaur using displacement. For explanation of (a) and (b), see the text.

Figure 13.11. Estimating the weight of a dinosaur by comparing the cross-sectional areas of bones.

[3] Quote from Bates, K. T., Falkingham, P. L., Macaulay, S., Brassey, C., and Maidment, S. C. R. 2015. Downsizing a giant: re-evaluating *Dreadnoughtus* body mass. *Biology Letters*, **11**, 2015. dx.doi.org/10.1098/rsbl.2015.0215. See also, Bates, K. T., Manning, P. L., Hodgetts, D., and Sellers, W. I. 2009. Estimating mass properties of dinosaurs using laser imaging and 3D computer modeling. *PLoS ONE*, **4**(2), e4532.

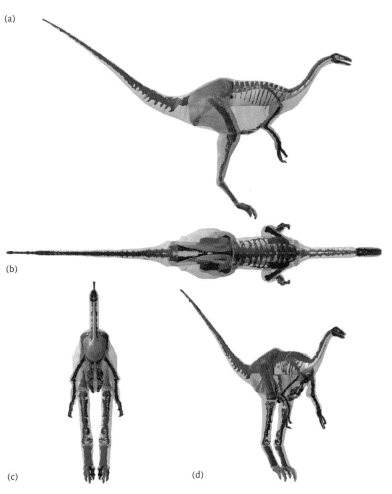

(a)

(b)

(c)   (d)

Figure 13.12. The "best estimate," fleshed-out reconstruction of the ornithomimid *Struthiomimus* in (a) right lateral, (b) dorsal, (c) anterior, and (d) oblique views (not to scale), using laser-scanning technology.

Fleet-Footed... or Flat-Footed

Trackways, the most tangible record of locomotor behavior, provide evidence for one aspect of an animal's walking and running capabilities, and the only independent test of anatomical reconstructions. When footprints are arranged into alternating left–right–left–right patterns, they demonstrate that all dinosaurs walked with a fully erect posture. But how can trackways also give us an indication of locomotor speed?

We begin with stride length; that is, the distance from the planting of a foot on the ground to its being planted again. When animals walk slowly, they take short strides, and when animals are walking quickly or running, they take considerably longer strides. This much is intuitive for anyone trying to catch a bus about to pull away from the curb. Now, consider the situation when you are being chased by something smaller than you. The creature chasing you must take long strides for its size, and more of them too, just to keep up. So there is clearly a size effect during walking and running, and these will likely be different for different kinds of animals under consideration.

How, then, to relate stride, body size, and locomotor speed? British biomechanist R. M. Alexander provided an elegant solution to this problem by considering dynamic similarity. Dynamic similarity is a kind of conversion factor: it "pretends" that all animals are the same size and that they are moving their limbs at the same rate. With these adjustments for size and footfall, it doesn't matter if you're a small or large human, a dog, or a dinosaur. All will be traveling with "dynamic similarity"; only speed will vary. That variable Alexander terms "dimensionless speed."[4] It is dimensionless speed that has a direct relationship with relative stride length. Stride length, of course, can be measured from trackways, which in turns allows us, for the first time, to calculate locomotor speed in dinosaurs.

To see how all this works, let's use Alexander's example of the trackway of a large theropod from the Late Cretaceous of Queensland, Australia. The tracks are 64 cm long, which Alexander, from other equally sized theropods, estimated must have come from a theropod with a leg length of about 2.56 m. The stride length of these tracks is 3.31 m, so the relative stride length (stride length:leg length) is 1.3. The dimensionless speed for a relative stride length of 1.3 is 0.4. And from all these measures, this Australian theropod must have been traveling reasonably quickly, at about 2.0 m/s, or 7.2 kmh.

As complicated as this approach appears, it represents the best method for estimating the actual speeds implied by trackways. But what about the fastest speeds a dinosaur might have been capable of? In 1982, R. A. Thulborn of the University of Queensland developed a method by which absolute locomotor abilities could be calculated. Thulborn's work relied heavily upon Alexander's slightly earlier studies on speed estimates from footprints and the relationship between body size, stride length, and locomotor speed among living animals. For both approaches, Thulborn determined that relative stride length has a direct relationship with locomotor speeds at different kinds of gaits (for example, walking, running, trotting, galloping).

Explicitly (for the quantitative among you):

$$\text{Locomotor velocity} = 0.25 \,(\text{gravitational aceleration})^{0.5}$$
$$\times (\text{estimated stride length})^{1.67}$$
$$\times (\text{hindlimb height})^{-1.17}.$$

Thulborn used this equation to estimate a variety of running speeds for more than 60 dinosaur species. The first group of estimates were for the walk/run transition, where stride length is approximately two to three times the length of the hindlimb. A potentially more important estimate – especially for dinosaurs fleeing certain death or pursuing that all-important meal – is maximum speed, which Thulborn calculated using maximum relative stride lengths (which range from 3.0 to 4.0) and the rate of striding, called limb cadence (estimated at $3.0 \times$ hindlimb length$^{-0.63}$).

Although we report some of these speeds elsewhere in this book, it is of some value to summarize the overall disposition of Thulborn's estimates. For small bipedal dinosaurs – which would include certain theropods and ornithopods – all appear capable of running at speeds of up to 40 kmh. Ornithomimids, the fastest of the fast, may have sprinted up to 60 kmh. The large ornithopods and theropods were most commonly walkers or slow trotters, probably averaging no more than 20 kmh. Thus the galloping, sprinting *Tyrannosaurus*, however attractive the image, did not impress Thulborn (or us) as likely: *T. rex* couldn't catch the Jeep.

Then there were the quadrupeds. Stegosaurs and ankylosaurs walked at no more than a pokey 6 to 8 kmh. Sauropods likely moved at 12 to 17 kmh. And ceratopsians – galloping along full throttle like enraged rhinos? Thulborn estimated that they were capable of trotting at up to 25 kmh.

Are these estimates accepted uncritically by all? P. Dodson has argued that these calculations would suggest that humans can run as quickly as 23 kmh, which in life they cannot. So are other means of determining dinosaur speeds possible?

---

[4] Dimensionless speed may appear oxymoronic, but in fact is equivalent to real speed divided by the square root of the product of leg length and gravitational acceleration.

Enter the computer. Analogous with the computerized weight estimates, vertebrate biologist W. I. Sellers and colleagues developed a series of complex computer models, using anatomy and what they called "mechanically and physiologically plausible" gaits to model locomotion in a duck-billed dinosaur (*Edmontosaurus*). An interesting aspect of their model was, given the parameters of the duck-bill and its gait, the generation of virtual trackways, which could then be compared with existing trackways. Their model suggested that hopping was the fastest gait (17 m/s), followed by quadrupedal galloping (16 m/s), and finally bipedal running (14 m/s), a result that is contrary to the widely held impression that when duck-bills wanted to go fast, they went bipedal. Clearly there's lots more to be done in this research direction.

Pathologies

One might suppose that fossil bones don't talk; that they're all, uh, stonyfaced. However, there have been a number of studies about dinosaur pathologies: the recognition of wounds – including bone breaks – that occurred in life, but were not fatal; and of pathogens, such as those shown in Figure 6.26. In these matters, the bones have much to say.

Bone breaks and lesions are perhaps the most common pathology recorded for dinosaurs. These are usually attributed to wounds obtained during fighting (defensively or offensively; for example, bite marks – see Figure 6.25). The wounds must not have been fatal if the bone shows growth around the break. And while fractures due to fighting are likely the most common damage seen to bones, now and then something different materializes: for example, a healed wound on an ornithopod (*Camptosaurus*) ilium originally attributed to overzealous sexual activity.

In general fractures obtained in life – as recognized by healing – are more common in larger dinosaurs than in smaller ones; this might due, partially, to the difficulty in inflicting fatal damage to a very large animal. Regardless, the list of dinosaur specimens with healed bone fractures is skewed toward large size, and includes *Allosaurus, Gorgosaurus, Albertosaurus, Iguanodon, Tyrannosaurus, hadrosaurs*, and *Pachyrhinosaurus*. Healed fractures are often associated with a "callus," a large bulb of bone that grows around the original break, protecting and strengthening it, which ultimately can remodeled with Haversian bone, processes suggestive of an endothermic ("warm-blooded") metabolism (see Chapter 14).

Other pathologies recognized in dinosaur bones include arthritis, bone infections (osteomyelitis), vertebral fusions, and cancers. Gout has even been (controversially, it is true) claimed for "Sue," the spectacular *Tyrannosaurus* now at the Field Museum in Chicago (see Box 1.1). Unsurprisingly, the precise agents of the infection are more difficult to determine. Joint-related traumas commonly (but not exclusively) associated with repetitive movements are known from a variety of dinosaurs, large and small. While a perhaps surprising amount is known of the physical traumas associated with being a dinosaur, a lot remains to be discovered; as a general rule, however, dinosaurs seem to have been hard-living creatures.

Zombies

Zombie-like, could dinosaurs ever come back (but perhaps better socialized)? Could scientists ever revive a (very) dead, (very) fossilized nonavian dinosaur?[5]

---

[5] In western African lore, a *bokor*, or wizard, can revive the dead (the word "zombie" is thought to have came from *nzumbe*, a word in the North Mbundu language of western Africa). Scientists aren't exactly wizards, but after Mary Shelley's *Frankenstein* (1818), the idea of scientific revival of the dead was firmly embedded in western thought as well.

The idea of creating extinct animals was vetted by Michael Crichton in *Jurassic Park* (1990). In that yarn, dinosaur blood, sucked by mosquitoes, was extracted from the insects when they themselves were trapped in amber. The dinosaur blood contained dinosaur blood cells; the blood cells were cloned, and Sam Neill faced off against *Velociraptor*. At the time, cloning dinosaurs from preserved cells received a bit of play in the serious scientific community, but in the end, DNA was found to degrade within a million or so years, and amber turned out to be semi-permeable, so that things could contaminate the samples. And the idea was buried, until the late 1990s when, zombie-like, it was re-exhumed in an entirely different form.

North Carolina State University biology professor Mary H. Schweitzer (see Figure 16.17), at the time beginning a PhD at Montana State University, observed red structures, each with a black dot in the center, in a blood vessel channel in a slice of bone from a particularly well-preserved *T. rex*. Her thought was – counter to everything that everybody "knew" about the process of soft-tissue (not bone) preservation – these looked like red blood cells; they were located in the right place; could they actually *be* red blood cells from the dinosaur? As time progressed, she ended up recovering a lot of soft tissue: blood vessels, bone cells, and keratin from the claws of *Rahonavis* (Chapter 8) and from the feathers of *Shuvuuia* (Chapter 7).

Could this stuff be *real*? One way to test it was by checking immune responses. Living bodies react to foreign objects by creating antibodies. So Schweitzer and her colleagues used mice to generate antibodies to the supposed dinosaur proteins she had extracted. Sure enough, when the antibodies that had been generated were exposed to hemoglobin from rats and turkeys, the antibodies bound to the hemoglobin, suggesting to Schweitzer and her colleagues that the proteins they had extracted from the dinosaur were something very much like hemoglobin. The keratin, too, was tested in the same way: keratin-specific antibodies were generated using the modern claw and feather proteins; these then bound to the ancient keratin. It appeared Schweitzer had actually isolated soft tissue from very extinct, long-gone dinosaurs. It got even crazier when another *T. rex*, astoundingly well preserved over the 66+ million years that had elapsed since the animal's death, was found. Schweitzer, who had by then moved to North Carolina State University, found medullary tissue in the fossil bones – tissue that is diagnostic of an animal that is producing eggs. Today medullary tissue is produced only by ovulating female birds; a clear indication that this *T. rex*, at least, must have been female – and ovulating (see Chapter 6). Schweitzer's reaction to her discovery, quoted in the pages of *Scientific American*, was priceless: "'Oh, my gosh, it's a girl – and it's pregnant!' I exclaimed to my assistant, Jennifer Wittmeyer. She looked at me like I had lost my mind."[6] From this *T. rex*, she extracted hollow, transparent, flexible tubes, some filled with some red material, which she interpreted as blood vessels with dinosaur blood (Figure 13.13).

Schweitzer's work was not accepted uncritically, and some authors have suggested that she is looking at modern invading microbes or bacterial biofilms growing on the fossil bone, rather than actual dinosaur tissue. That possibility became less likely in 2016, however, when collagen polypeptide sequences were obtained from a hadrosaur (*Brachylophosaurus*) and subjected to the same rigorous suite of tests undergone by the *T. rex* polypeptides. Indeed, recent work on fossil collagen has begun to elucidate aspects of the structure of this molecule that led to its unusual resistance to long-term degradation. Yet, even as recently as 2019, scientists have argued that bone is a particularly felicitous habitat for microbes, and that tissues heretofore ascribed to ancient, degraded

---

[6] Schweitzer, M. H. 2010. Blood from a stone. *Scientific American*, December, 62–69. As this book went to press, medullary bone was diagnosed unequivocally by the presence of a compound that is not unstable in the way that DNA is, the sulfated glycosaminoglycan keratan sulfate. The presence of keratan sulfate, therefore, may be a game-changing key to recognizing the sex of extinct dinosaurs, at least the pregnant ones! See Canoville, A., Zanno, L. E., Zheng, W., and Schweitzer, M. H., 2021. Keratan sulfate as a marker for medullary bone in fossil vertebrates, *Journal of Anatomy*, DOI: 10.1111/joa.13388.

collagen, could not be that. In this, as in so much, the devil is in the details, and it will likely be some years before Schweitzer's work can be accepted uncritically.

Meanwhile, molecular evidence has also been used in an entirely different context. A study published in April 2008 compared proteins from 21 different living creatures, including an alligator, an ostrich, a chicken, and two extinct creatures: *T. rex* and a mammoth. The *Tyrannosaurus* proteins were types of collagen extracted from a presumably unaltered femur. The results were unequivocal: the proteins showed that the mammoth and an elephant were phylogenetically close and, more important for our story, that the *Tyrannosaurus* and the birds were close – closer to each other than either was to an alligator.

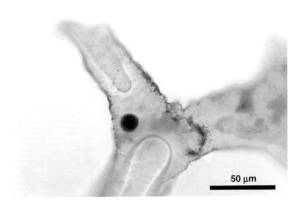

Figure 13.13. Red blood cells recovered by Mary Schweitzer from *Tyrannosaurus rex* long bones.

A deeper dive into the molecules of extinct organisms was provided by an international coalition of scientists, led by paleohistologist Alida Bailleul, who identified collagen, other proteins, and material derived from cell nuclei suggesting the presence of DNA, in a juvenile specimen of the hadrosaur *Hypacrosaurus*. This work, as well as other work published by Schweitzer and various colleagues, speaks to the reality of Schweitzer's and her colleagues' claim that they have isolated genuine molecules and soft tissues from nonavian, very extinct dinosaurs.

As for resurrecting an actual dinosaur, it won't end up happening quite as Michael Crichton described it, but the possibility of building backwards from the small fragments of DNA preserved in these fossils has become the kind of thing that some paleontologists dream about. Jack Horner, for one (see Figure 16.14), teaming up with molecular biologist colleagues, attempted to grow the tail of a nonavian dinosaur from long-dormant genes in modern birds (a chicken). His idea was that the ancestral dinosaur-tail genes, left over from when theropods regularly grew tails, are still there, waiting to be activated. They never quite got their tail, but their tale, and the dream that motivated it, are described in Horner's book *How to Build a Dinosaur* (2009). Even more tantalizing, as early as 1980, scientists were able to induce teeth in modern chicken embryos. Nobody, other than perhaps Warner Brothers or Disney, wants to see birds with teeth, but the point was clearly made that those long-unused tooth genes aren't lost in modern birds; they're just not activated during the development of a bird from embryo to adult.

And just when you thought it was safe to assume we'll never see a living dinosaur other than a bird, molecular biologist D. K. Griffin and colleagues published a paper in which the proposed "a glimpse into prehistoric genomics." The idea was a variant on phylogenetic bracketing: recognizing within the extraordinary diversity in modern birds some genetic commonalities, they attempted to reconstruct what they called the "overall karyotype [chromosomal content] of the Theropod (*sic*) lineage."[7] In a sense, this was a theoretical exercise, but if many of the long-unused banks of genes residing in nonavian dinosaurs, remain in existence in modern birds, and those genes can be identified, then perhaps the authors could, as they claimed, determine the likely karyotype of the avian ancestor.

With the ethically charged likelihood of seeing much more recently extinct animals – passenger pigeons; dodos; moas; mammoths, mastodons, marsupial tigers – brought back into existence, and

---

[7] Griffin, D. K., Larkin, D. M., and O'Connor, R. E. 2020. Time lapse: a glimpse into prehistoric genomics. *European Journal of Medical Genetics*, **63**; https://doi.org/10.1016/j.ejmg.2019.03.004. They believe that their method ought to take them close to the nonavian dinosaur ancestor of modern birds, described by them as "most likely a chicken-sized, two-legged, feathered, land dinosaur from the Jurassic period."

the recognition that ancient molecules, whose presence we had hardly guessed at, still exist, the possibility of coming closer to a living nonavian dinosaur becomes less and less fantastic.

SUMMARY

Here we have tried to focus on the biological aspects of dinosaurs as a group, rather than on the individual features of particular dinosaurs. We termed that type of study "paleobiology."

Dinosaurs – both avian and nonavian – had unidirectional breathing, a kind of breathing that differs markedly from bidirectional breathing, which is found in mammals. Unidirectional breathing uses a system of air sacs to store and direct air in one direction into the lungs. With the air moving in a single direction and opposite the direction of blood flow, countercurrent gaseous exchange takes place between the blood and the air; O_2 is removed from the inhaled air, and CO_2 is removed from the lungs. Unidirectional breathing as practiced in modern birds maximizes both the volume of inhaled air and the efficiency of gaseous exchange in the lungs.

Dinosaurs are famously thought to have been unintelligent, but when their EQs are considered relative to a crocodile (representing a modern reptile), many are seen to be somewhere towards the bird range of intelligence, if EQ – itself simply a variant on brain:body-size ratios – is a valid measure of intelligence.

Dinosaur bone, studied microscopically in thin section or by CT scan, shows evidence of extensive remodeling, such as is seen in modern birds. This suggests growth trajectories similar to those of birds; involving relatively fast growth as a juvenile; and perhaps somewhat slower growth after sexual maturity has been reached. The age of the dinosaur at each of the growth stages can be measured by using LAGs, which are believed to represent yearly growth. The ages, measured in combination with the particular development of dinosaurs at that age, has allowed paleontologists to develop "S"-shaped growth curves for many genera, reflecting rates of growth after birth, at adolescence, and adulthood.

Estimating the weights of dinosaurs can be difficult, but weight is an important measure of size. Various techniques for estimating weight are available; perhaps the simplest still-valid one is based upon cross-sectional area of the femur, which is approximately proportional to the overall size and weight of any tetrapod. We say "approximately," because different tetrapods of similar sizes have different amounts of muscle mass, which in turn affects their weight. More precise estimations of weight can be obtained using CT reconstructions of dinosaurs. These "ground-truthed" against a living dinosaur (ostrich), whose weight can be checked, have produced reasonably good estimates.

The speed at which dinosaurs could have moved is of much popular interest, but remains complicated to definitively ascertain. The most direct evidence of dinosaur velocities are the trackways that dinosaurs leave; however, while mathematical means of estimating the velocity of the trackmakers have been developed, most known trackways represent a leisurely stroll, rather than dinosaurs running to or from something at full tilt. Still, a combination of trackways and morphology allow some estimates; the fastest dinosaurs were likely cursorial bipeds like ornithomimosaurs; farther back in the pack would have been big carnivorous dinosaurs like *T. rex*. Quadrupeds like hadrosaurs and ceratopsians appear to have been capable of relatively high speeds, although some of the methods used to determine those speeds may lead to overestimations.

The question of whether or not any dinosaur soft tissue survived, in particular DNA, is addressed. With a shelf-life of a million years before degradation, the likelihood of finding nonavian dinosaur DNA in usable form is very slim. Nonetheless, some workers have identified soft tissues – red blood cells, collagen fibers, for example, and they have interpreted them as original (and millions of years old). These interpretations have met with considerable criticism, but there are enough data to support them, that they cannot be dismissed out of hand.

In the meantime, genes, such as those that were responsible for teeth in dinosaurs, have been identified in living birds, and those ancient genes reactivated to produce teeth. Some other efforts have been less successful, but the idea that the old sequences of genes are still present in living dinosaurs, is beginning to receive some scientific support.

SELECTED READINGS

Bailleul, A. M., O'Connor, J., and Schweitzer M. H. 2019. Dinosaur paleohistology: review, trends and new avenues of investigation. *PeerJ*, **7**, e7764. DOI 10.7717/peerj.7764

de Ricqlès, A., Horner, J. R., and Padian, K. 2006. The interpretation of dinosaur growth patterns. *Trends in Ecology and Evolution*, **21**, 596–597.

Erickson, G. M. 2005. Assessing dinosaur growth patterns: a microscopic revolution. *Trends in Ecology and Evolution*, **20**, 677–684.

Erickson, G. M., Makovicky, P. J., Inouye, B. D., Zhou, C.-F., and Gao, K.-Q. 2009. A life table for Psittacosaurus lujiatunensis: initial insights into ornithischian population biology. *The Anatomical Record*, **292**, 1514–1521.

Hopson, J. A. 1980. Relative brain size: implications for dinosaurian endothermy. In Thomas, R. D. K. and Olson, E. D. (eds.) *A Cold Look at the Warm-Blooded Dinosaurs*. AAAS Selected Symposium 28, Westview Press, Boulder, CO, pp. 287–310.

Horner, J. R., de Ricqlès, A., and Padian, K. 2000. Long bone histology of the hadrosaurid dinosaur *Maiasaura peeblesorum*: growth dynamics and physiology based upon an ontogenetic series of skeletal elements. *Journal of Vertebrate Paleontology*, **20**, 115–129.

Kaye T. G., Gaugler G., and Sawlowicz, Z. 2008. Dinosaurian soft tissues interpreted as bacterial biofilms. *PLoS ONE*, **3**, e2808. DOI: https://doi.org/10.1371/journal.pone.0002808, PMID: 18665236

Padian, K., and Lamm, E.-T. 2013. *Bone Histology of Fossil Tetrapods: Advancing Methods, Analysis, and Interpretation*. University of California Press, Berkeley, CA, 285 pages.

Rega, E. 2012. Disease in dinosaurs. In Brett-Surman, M. K., Holtz, T. R., Jr., and Farlow, J. O. (eds.) *The Complete Dinosaur* (2nd edn). Indiana University Press, Bloomington, IN, pp. 667–711.

Saitta, E. T., Liang, R., Brown, C. M., *et al.* 2019. Creatceous dinosaur bone contains recent organic material and provides and environment conducive to microbial communities, *eLife*. DOI: https://doi.org/10.7554/eLife .46205.001

Schweitzer, M. H. 2010. Blood from a stone. *Scientific American*, December, 62–69.

Sellers, W. I., Manning, P. L., Lyson, T., Stevens, K., and Margetts, L. 2009. Virtual palaeontology: gait reconstruction of extinct vertebrates using high performance computing. *Palaeontologia Electronica*, **12**(3), 11A, 26 pages. palaeoelectronica.org/2009_3/180/index.html

Spotila, J. R., O'Connor, M. P., Dodson, P. and Paladino, F. V. 1991. Hot and cold running dinosaurs: body size, metabolism, and migration. *Modern Geology*, **16**, 203–227.

Thomas, R. D. K. and Olson, E. C. (eds.) 1980. *A Cold Look at the Warm-Blooded Dinosaurs*. AAAS Selected Symposium no. 28, 514 pages.

Woodward, H. N., Freedman-Fowler, E. A., Farlow, J. A., and Horner, J. R. 2015. *Maiasaura*, a model organism for extinct vertebrate population biology: a large sample statistical assessment of growth dynamics and survivorship. *Paleobiology*, **41**, 503–527. https://doi.org/10.1017/pab.2015.19

TOPIC QUESTIONS

1. What is paleobiology?
2. What are remodeled bone, Haversian canals, LAGs, and respiratory turbinates?
3. What are some of the basic issues with determining the weight of dinosaurs? Why is it that cross-sectional area of a long bone has anything to do with weight?

4. Describe the relationship between Haversian canals and growth.
5. Please explain "dynamic similarity." What is meant when it is said that this quantity is "dimensionless" and why is it "dimensionless"?
6. How can it be that the genes responsible for teeth in dinosaurs have been identified?
7. Go to Figure 13.8 in this chapter and determine the maximum growth rates of the dinosaurs represented there. Which has the highest growth rate? Lowest?
8. How are rates of growth calibrated in fossil dinosaurs?
9. Suppose we wrote: "Isn't it possible that unidirectional breathing should be described as saurischian and not 'avian'?". Can you think of some reasons why this statement might be true?

Figure 14.1. David and Goliath upended. Two powerfully built, endothermic, intelligent, active ornithomimids set their sights upon a single, furry tiny mammal in the wrong place at the wrong time.

The Paleobiology of Dinosaurs II

Dinosaur Metabolism – Some Like it Hot

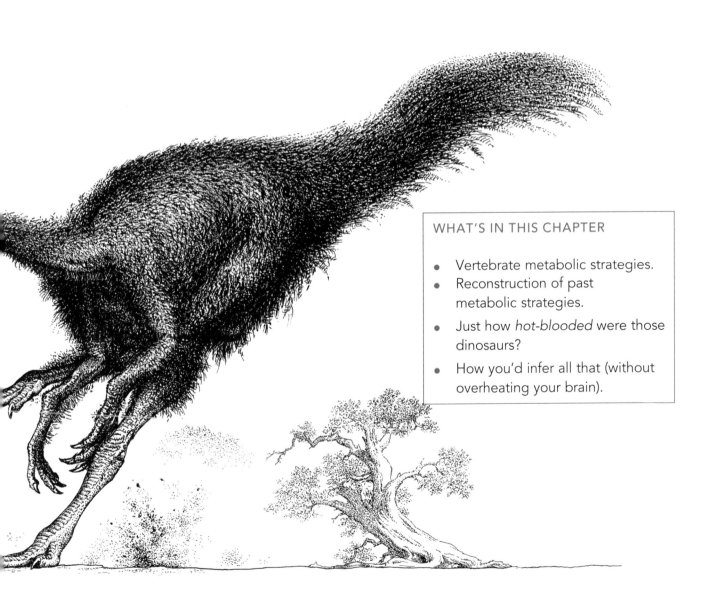

WHAT'S IN THIS CHAPTER

- Vertebrate metabolic strategies.
- Reconstruction of past metabolic strategies.
- Just how *hot-blooded* were those dinosaurs?
- How you'd infer all that (without overheating your brain).

The Way They Were

Since almost the beginning, paleontologists have been of two minds about dinosaurs. In the 1840s, brilliant (and nasty) English anatomist Sir Richard Owen originally described them as warm-blooded and mammal-like (see Chapter 16). Yet within 50 years, viewpoints had changed: since dinosaurs were reptiles, and reptiles, being more primitive than mammals, are cold-blooded, dinosaurs must have been cold-blooded. And so, for much of the twentieth century, dinosaurs were thought of as dolled-up crocodiles: green and scaly, and intellectually, not to say physically, a bit slow. That perspective was revisited with the revolutions of the 1960s and early 1970s and things have gotten a lot more interesting ever since. Perhaps Mesozoic dinosaurs are birds of a different feather?

Physiology: Temperature Talk

We often hear expressions like "warm-blooded" and "cold-blooded"; but it turns out that these terms don't mean much. For example, most "cold-blooded" vertebrates have warmed blood when they are active. A more meaningful distinction is between **endotherms** (*endo* – inside; *therm* – heat), organisms that regulate their temperature internally, and **ectotherms** (*ecto* – outside), organisms that use external sources of heat to regulate their temperatures.

Considered from another perspective, in some organisms, called **poikilotherms** (*poikilo* – changing), temperature fluctuates, but in others, called **homeotherms** (*homeo* – same), the temperature remains constant. Humans are *endothermic homeotherms*: when we are unable to maintain our body temperature, we get sick. Ectotherms, such as lizards, can tolerate decreases in core temperature, while endotherms must internally regulate their core temperatures.

Ectothermy and endothermy are two biochemically and biophysically different methods of obtaining heat. The terms poikilotherm and homeotherm, however, are endpoints in a spectrum that runs from maintaining a constant temperature to having a fluctuating temperature.

All of this is about **metabolism**, the chemical reactions that an organism uses to obtain energy and put it to work (Appendix 14.1). While many organisms cluster at the familiar metabolic endpoints (endothermic homeotherm; ectothermic poikilotherm), many do not (Box 14.1).

These metabolic endpoints translate into very distinct muscle capabilities. The muscles in both endotherms and ectotherms both achieve their maximum output almost immediately after being activated; some have even suggested that initial muscular energy output in ectotherms exceeds that of endotherms. Ultimately, however, endotherms produce more energy than ectotherms; it takes longer, during heavy use, for the muscles of endotherms to enter an **anaerobic** physiological state, characterized by **lactic acid** production (Appendix 14.1). This means that, generally, endothermic tetrapods are capable of higher levels of activity sustained over longer periods of time than are ectothermic tetrapods. And that's why most predatory ectotherms must kill by ambushing their prey, whereas endotherms have the metabolic option – not utilized by all endotherms, however – of running prey down, or running away from predators (Figure 14.2).

So if that is the case, why is anything still ectothermic? One might at first think that it's because ectotherms just haven't evolved the endothermic system, but it turns out that things are far more nuanced than that: there is a simple reality of energy budgets, based in modern animals: *endothermy is much more costly in terms of energy use than ectothermy.* It costs much more, energetically speaking – 10 to 30 times more – to maintain an endothermic metabolism as to maintain an ectothermic metabolism. So in fact, it's metabolically very expensive to be endothermic, and if you can control your behavior to accommodate this aspect of your metabolism, you can get through

Box 14.1 Warm-Bloodedness: to Have and to Have Hot

Although endothermy is *characteristic* of birds and mammals, it is by no means restricted to these groups. For some time, physiologists have known of plants (!) that can regulate heat in a variety of ways, the most common being to decouple metabolism (described in Appendix 14.1) from respiration, so that energy from the breakdown of ATP is simply released as heat. Several snakes are known to generate heat while brooding eggs, although this is accomplished by muscle exertion. Certain sharks and tunas can retain heat from their core muscles by counter-current circulation, and a variety of insects, including moths, beetles, dragonflies, and bees, are known to regulate their body temperatures.

Endothermy is not characteristic of these groups of organisms. Simply, it is known that some of them maintain temperatures warmer than those of the medium (air or water) in which they are living. Indeed, it has been estimated that maintaining a temperature against an external gradient has evolved independently at least 13 different times.

This differs from the idea that endothermy is *diagnostic* of a particular group. Indeed, endothermy is characteristic of but two groups: birds and mammals. And even in these, the familiar human (or cat/dog) endothermic homeothermy is not always what one finds. For example, many endotherms maintain basal resting rates, which are significantly lower than their active rates. Familiar examples would include hummingbirds, bats, and hibernating bears. These animals are all endotherms, but they are not properly considered homeotherms. Many other strategies and varieties abound, and it is the case that metabolic strategies are in reality far more complex and varied than the simplified endpoints that are most familiar to us.

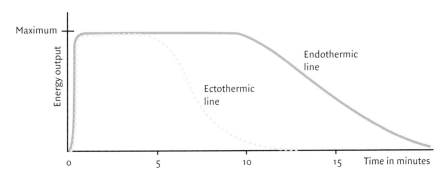

Figure 14.2. Energy output versus time in ectotherms and endotherms. The curves show that the muscles of both endotherms and ectotherms achieve their maximum energy output virtually instantaneously. In general, however, an endotherm *sustains* maximum energy output for more than twice as long as an ectotherm does.

life far more economically than any endotherm. And from an adaptation and survival standpoint, that's a very good place to be.

So What About Dinosaurs?

The *Jurassic Park*s and *Jurassic World*s notwithstanding, we can't stick a thermometer into the vent of a Mesozoic dinosaur and read the temperature, so how can we tell? As you will see, it will be by proxies; features that represent other features, just as "voting by proxy" means having someone represent you when a vote is taken, and you are not there to cast your own ballot.

First we cut to the chase: By now, evidence has become overwhelming that dinosaurs had some kind of endothermic metabolism(s). As we will see, the exact form that these took may have varied – perhaps from one type of dinosaur to another – but rather than think of dinosaurs as reptiles (and thus cold-blooded), we'll use the evolutionarily more informative term "amniotes" (see Chapter 4) which liberates us to think about the full range of metabolic strategies available to them. On to the evidence (by proxy).

The Cladogram

Birds are dinosaurs and modern birds are surely endothermic. So we ask the simple question: at what point during dinosaur evolution – e.g., how far down the cladogram – did "avian" endothermy evolve?

An important clue comes with insulation. All small- to medium-sized modern endotherms (avian and mammalian) are insulated with feathers or fur. There is a certain sense to this; if an ectotherm depends upon external sources for heat, why develop a layer of protection (insulation) from that external source? *Archaeopteryx*, with its plumage, is therefore usually considered to have been endothermic.[1] The discovery of nonavian, feathered (insulated) theropods from China (see Chapters 6 to 8) gives us a clue that endothermy must have occurred in theropods more primitive than Coelurosauria, and perhaps at an even more basal level than Theropoda (Figure 14.3).

Of course, using just this criterion, it's looking more and more like a lot of dinosaurs besides theropods could have been endotherms. We saw that the primitive ornithischian, *Tianyulong*, was likely feathered (see Part III Ornithischia and Feathers Without Flight" in Chapter 5). If the mono-filamentous feathers found in ornithischians are *homologous* with those found in theropods (sauris-chians), then the chances are that feathered insulation (and the metabolism type that it implies) could characterize all Dinosauria. Using this same logic, if the monofilamentous feathers of dinosaurs are homologous with those in pterosaurs, feather coats might have been invented as early as the first ornithodirans! But recall that the first feathers were likely sparse; not true coats, but display or sensory features. If so, those first feathers would hardly have served to insulate the creatures that bore them.

Air sacs and unidirectional breathing in living organisms are indicative of high metabolic output and point to endothermy. As we've seen, pneumaticity – indicative of expanded air sacs – and thus unidirectional breathing, is well developed in saurischian dinosaurs as a group (Chapters 8 to 10). These features are indicative of highly evolved respiratory systems, and suggest higher metabolic rates than commonly associated with ectotherms, as well as at least incipient endothermy.

Early Evidence for Dinosaur Endothermy that Fueled the Dinosaur Renaissance

Some of the earliest evidence that dinosaurs might be endothermic was marshaled by paleontologist R. T. Bakker (see Chapter 16), inspired by J. H. Ostrom's discovery of *Deinonychus* and redescription of *Archaeopteryx* (Bakker was Ostrom's student). Very much ahead of the wave, Bakker suggested (rather loudly, as it happened) that dinosaurs must have been warm-blooded. Along with paleon-tologist P. M. Galton, he even went as far as to propose a whole new *class* of vertebrates: Dinosauria (including birds)! Most paleontologists of the time saw the idea as fanciful, going against popular

---

[1] In 1992, J. A. Ruben suggested that *Archaeopteryx* could have been an ectotherm. His idea was based upon the amount of energy needed for flight, and upon the amount of energy available from an ectothermic metabolism. Since the bones of *Archaeopteryx* show limited adaptations for sustained, powered flight, Ruben argued that an ectothermic metabolism would have been more than sufficient for the kind of limited flight that apparently characterized *Archaeopteryx*. While powered flight may be possible in an ectothermic tetrapod, none (save perhaps *Archaeopteryx*) ever evolved it. Moreover, it seems to us that the presence of insulation (feathers) in *Archaeopteryx* is incompatible with an ectothermic metabolism.

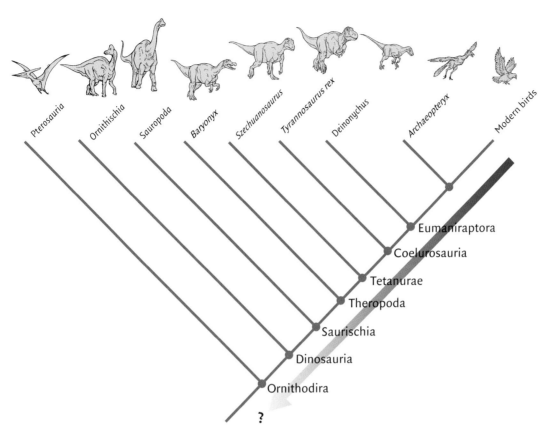

Figure 14.3. Cladogram showing the inferred "depth" within Theropoda of endothermy. While we can be certain the avialans were all endotherms, it is not clear how far back – or how deep within the cladogram – endothermy extends. With the discovery of the feathered *Yutyrannus* (Chapter 7), we now know that it goes back through all coelurosaurs. With monofilamentous-feathered ornithischians, perhaps it goes back as far as basal Dinosauria. Pushing such logic farther (and a cladogram like that in Figure 5.4 will help you visualize this), if they are present in pterosaurs, as they are now thought to be, then feathers could be primitive for all Ornithodira.

orthodoxy, but within 12 years, phylogenetic systematics demonstrated that Bakker and Galton's insight was correct. With all this activity, dinosaurs suddenly went from a curios to subjects of considerable public and professional interest – Bakker and Ostrom together fueled the "dinosaur renaissance" of the 1970s (see Chapter 16), and undoubtedly laid much of the groundwork for *Jurassic Park*, the book (1990) and the first installment of the *Jurassic Park* movie franchise (1993). And what were observations and discoveries that set the thing in motion?

Stance

A simple observation was that all nonavian dinosaurs had a fully erect stance. Among living vertebrates, a fully erect stance occurs only in birds and mammals, both of which are endothermic. The fully erect limb position in dinosaurs was therefore suggestive to him of endothermy.

Great coincidence; but is there a causal relationship between stance and endothermy? The original idea was that the fine neuromuscular control necessary to maintain a fully erect stance would only be possible within the temperature-controlled environment afforded by an endothermic metabolism. Since Bakker, further work (see also Box 4.3) suggests that, when an animal with sprawling stance moves, the trunk of the organism flexes from side to side as the animal walks. Such flexion reduces the amount of air that can fill the lungs on the scrunched side (Figure 14.4), so that when the animal needs the most air, it gets the least. The evolution of a fully erect stance may thus have been a means by which lung volume could be maximized during high-speed locomotion.

Two other simple correlations have since been noted between anatomy and endothermy. The first is that long-leggedness is characteristic of living endotherms while living ectotherms possess relatively stubby limbs. Certainly many dinosaurs possessed rather long limbs. The second correlation is the observation that, among living tetrapods, the only bipeds are endotherms. A recent study reconstructed the energy required for bipeds to walk and slow run. The amount of energy required for maintaining these gaits was theoretically calculated for both large and small dinosaur bipeds, and the somewhat surprising conclusion – surprising because it ran counter to prevailing wisdom – was that the amount of energy required by large bipeds, such as hadrosaurs and tyrannosaurs, to walk and slow run was significantly greater than that which would be available to ectotherms. The data were more equivocal for smaller bipeds, but even these pushed the maximum energy available with an ectothermic metabolism to its limit. Large bipedal dinosaurs, the authors concluded, must thus have been endotherms.

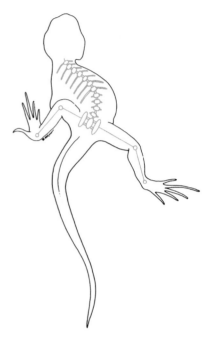

Figure 14.4. A sprawling vertebrate running quickly. The trunk alternately compresses the lung capacity on each side as the animal runs.

A study in 2009 used models that related locomotor anatomy to metabolic rates as measured by inferred oxygen consumption, and then applied the locomotor parameters of thirteen extinct bipedal species to determine their oxygen consumption. In all cases, including a nondinosaur dinosauromorph (*Marasuchus*; see Chapter 5), an oxygen consumption rate calling for an endothermic metabolism was indicated, particularly so in five large dinosaurs (*Tyrannosaurus*, *Allosaurus*, *Plateosaurus*, *Dilophosaurus*, and *Gorgosaurus*) and in small active bipeds such as *Heterodontosaurus*, *Compsognathus*, and *Velociraptor*. This, of course, was modeling, not data that were measured directly from living dinosaurs.

Legs

A variety of small- to medium-sized bipedal dinosaurs such as dromaeosaurids and ornithomimids are characterized by gracile bones, in which the thigh is short relative to the length of the calf. This, in turn, suggests high levels of sustained running – behavior generally not characteristic of modern ectotherms.

And what of the larger dinosaurs, especially those that were not bipedal? Here the issue becomes murkier. As we discussed in Chapter 13, the walking speeds of all tetrapods can be calculated from a combination of footprint spacing (stride length) and the length of the hindlimb. But, of course, the walking that produced most trackways was generally not pedal-to-the-metal running.

Could quadrupedal dinosaurs have run like the fastest mammals today? Ancestry gives a hint. In mammals, a fully erect posture evolved in a quadrupedal ancestor; however, in dinomorphs the fully erect posture evolved in a biped. Quadrupedal dinosaurs likely evolved their four-legged stance secondarily (see Chapters 10, 11, and 12) and thus the front limbs of dinosaur quadrupeds look, and likely functioned, differently from those of mammals (see Figure 11.25). Yet, as we have seen, quadrupedal gaits modeled in *Edmontosaurus*, at least, were evidently zippy.

Of Predators and their Prey

Perhaps Bakker's most brilliant insight was his inference of metabolism from the numbers of predators and prey found in the fossil record.

Given the fact, Bakker reasoned, that if predators are endothermic, they should require more energy than if they were ectotherms, and this should be in turn reflected in the weight proportions of predators to prey, or **predator/prey biomass ratios**. He estimated that predator/prey biomass ratios for *ectothermic* organisms are around 40 percent, while predator/prey biomass ratios for *endothermic* organisms are 1 percent to 3 percent. Here, then, was an order of magnitude difference in the **biomass** ratios, which ought to be recognizable in ancient populations. Now, by counting specimens of predators and presumed prey in major museums and by estimating the specimens' living weights, Bakker was armed with a tool from modern ecosystems that he believed could reveal the energetic requirements of ancient ecosystems.

His results seemed unequivocal: among the dinosaurs, the predator/prey biomass ratios were very low, ranging from 2 percent to 4 percent. He interpreted this low number to indicate that predators and prey in dinosaur-based food chains were endothermic (Figure 14.5).

This study, for all its originality, had some problems. For example, the assumption that prey are approximately the same size as the predators is clearly not correct (consider a bear eating a salmon), and when predator and prey are not similarly sized, this has drastic effects on the resultant ratio. Most significantly, the predator/prey biomass calculation assumes that all deaths are the result of predation; that there can be no mortality due to other causes. This is simply not the case in modern ecosystems.

There are other problems associated with using fossils. Most obvious are difficulties in estimating dinosaur weights (see Chapter 13). Moreover, the preservation of dinosaur material is subject to a variety of biases. How can one ever be sure that the proportions of the living community are represented? Because we can't, paleontologists commonly talk about fossil **assemblages**, which are simply the groups of fossils that we have collected in a particular locality, or representing a time interval or geographic area, may have nothing to do with the proportions of the same animals in the living *community* in which those animals lived.

Finally, Bakker obtained his data by counting specimens in museum collections, specimens that were likely collected because they were rare or particularly well preserved. Museum collections thus tend to represent assemblages of well-preserved organisms with a higher percentage of rare organisms than was present in the original fauna.

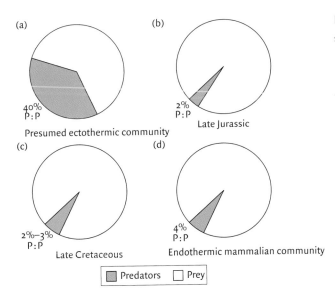

Figure 14.5. The proportions of predators to prey (P:P) in selected faunas in the history of life, as reconstructed by R. T. Bakker. Predators are shaded; prey are unshaded. (a) Early Permian of New Mexico; (b) Late Jurassic of North America; (c) Late Cretaceous of North America; (d) Middle Cenozoic of North America. The Cenozoic fauna (d) is mammalian and obviously endothermic, providing clear guidelines on what predator/prey ratios are in an endothermic fauna. As reconstructed by Bakker, the predator/prey ratios of the dinosaur communities clearly are closer to those seen in the known endothermic mammalian community (Middle Cenozoic) than in the inferred ecothermic community (Early Permian).

Several other studies have since attempted to duplicate Bakker's study, with mixed and/or equivocal results. Ultimately, too much uncertainty for the results to be definitive was introduced through the brilliant, but likely flawed, idea of predator/prey biomass ratios. But Bakker's evidence and advocacy kindled the flame of endothermic dinosaurs, and they would never be thought of as tricked-out crocodiles again!

Hearts and Minds

All living endotherms possess four-chambered hearts. The four-chambered heart system, in which the oxygenated blood is completely separated from the deoxygenated blood, may be a prerequisite for endothermy. Endothermy requires relatively high blood pressures in order to perfuse complex, delicate organs such as the brain with a constant supply of oxygenated blood. Such high blood pressures, however, would "blow out" the alveoli in the lungs. For this reason, mammals and birds separate their blood into two distinct circulatory systems: the blood for the lungs (pulmonary circuit) and the blood for the body (systemic circuit). The two separate circuits require a four-chambered heart – a pump that can completely separate the circuits.

Would such a heart be possible in dinosaurs? The nearest living relatives of dinosaurs, birds and crocodiles, possess four-chambered hearts; thus it is likely that a heart with a double-pumping system was present in basal Dinosauria.

This idea was strongly reinforced by the discovery in 2000 of what was controversially inferred to be a four-chambered heart with an aorta. The "heart" was preserved as an ironstone mass within the thoracic cavity of the basal ornithopod *Thescelosaurus*, and identified using computed tomography (a CT scan). Doubters doubted; advocates advocated; and in the end, it appears that doubters have taken the field.

In the late 1970s, attempts were made to assess the intelligence of dinosaurs using EQ (see Chapter 13). As part of his analysis of EQ, Jerison (who, it will be recalled, devised the concept) noted that, on the basis of EQ, living vertebrates cluster into two groups, endotherms and ectotherms. The idea was that living endotherms (birds and mammals) have significantly higher EQs than do living ectotherms (reptiles and amphibians), presumably because their more refined levels of neuro-muscular control require stable temperatures, such as are found with an endothermic metabolism. Based upon EQ, coelurosaurs were likely as active as many birds and mammals, while large theropods and ornithopods were somewhat less active than birds and mammals, but more active than typical living reptiles. Using EQ as a proxy for activity levels, other dinosaurs appear to have been in the range of living reptiles (see Figure 13.3).

The Nose Shows, and the Head Shows (Once You Know What to Look For)

Endothermy requires the lungs to replenish their air (ventilate) at a high rate. And high rates of ventilation lead to water loss, unless something is done to prevent it. What modern mammals and birds do is to grow convoluted sheets of delicate, tissue-covered bone, called respiratory turbinates, in the nasal cavities. The mucus-covered surfaces of the turbinates have dual functions: they pull moisture out of the air before it leaves the nose, thus conserving water, and they both heat the incoming air and cool the outgoing air, thereby stabilizing temperatures in the brain (Figure 14.6).

Biologist John Ruben wondered if the fact that modern endotherms require a means to conserve moisture during respiration and stabilize brain temperatures might provide a clue about dinosaur metabolism. Although a number of dinosaurs appear to have had convolute olfactory passageways, indicative of a well-developed sense of smell, apparently none – as far Ruben knew – had turbinates to allow them to recoup respired moisture. Considered exclusively on this basis, Ruben concluded that dinosaurs could not have been endothermic in the way that most mammals and birds are today.

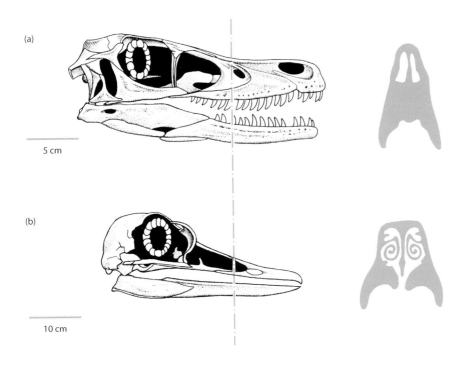

(a)

5 cm

(b)

10 cm

Figure 14.6. Cross-sections (solid shading) through the nasal regions of (a) an extinct dinosaur (*Velociraptor*) and (b) a living bird (*Rhea*); skulls and positions of the cross-sections are shown to the left. The nasal cavity of the bird shows convoluted respiratory turbinates, while that of *Velociraptor* does not.

In the meantime, however, the full details about nasal turbinates appear to be more nuanced than Ruben's original conclusions. First, the delicacy of the turbinates being what it is, a number of workers have pointed out that their absence doesn't mean that they weren't there; they simply might not have been preserved. And – more significantly – respiratory turbinates are known from two dinosaurs: the pachycephalosaurs *Stegoceras* and *Sphaerotholus*. But even more significantly, recent work suggests that you (or at least dinosaurs – you still do!) don't need respiratory turbinates to maintain brain temperature stability and retain moisture (see Air Conditioning below). There are other ways to stay cool – or at least homeothermic.

Where the Wild Things Are

The distribution of dinosaurs around the globe far exceeds the current distribution of modern ectothermic vertebrates, which are generally not found above and below, respectively, latitudes 45° north and 45° south. Large modern ectotherms rarely occur above latitude 20° north and below latitude 20° south (Figure 15.5).

Correcting for continental movements, Cretaceous dinosaur-bearing deposits have been found close to latitudes 80° north and 80° south of the equator. Both the northern and southern sites experienced extended periods of darkness, and at least occasionally, freezing temperatures.

The Arctic assemblage, from North America, includes hadrosaurids, ceratopsids, tyrannosaurids, and troodontids. The Antarctic dinosaur assemblage, from Australia, includes a large theropod and some basal euornithopods (including many juveniles). Along with these dinosaurs are fish, turtles, pterosaurs, plesiosaurs, birds (known solely from feathers), and, incredibly, a improbable late-surviving temnospondyl (an amphibian group that apparently went extinct in the Early Jurassic everywhere else in the world; see Figure 15.5).

This mix of animals is not particularly easy to interpret in terms of endothermy or ectothermy. Several members of the southern dinosaur assemblage had large brains and well-developed vision, for example, the basal neornithischian *Leaellynasaura*, potentially useful during periods of extended

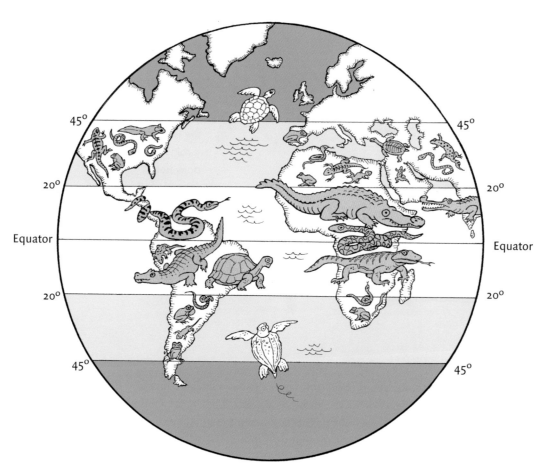

Figure 14.7. The latitudinal distribution of ectothermic tetrapods on Earth today. The larger terrestrial tetrapods do not get much beyond about latitude 20° north and south. These include large snakes and lizards, crocodilians, and tortoises.

darkness. Others, however, were not so well equipped. Burrowing may have been an option for a few, but surely not for all.

In the case of the Northern Hemisphere faunas, only *Troodon* is of a size that could make burrowing feasible. Migration was potentially a solution to inclement winter weather, although the dinosaurs would have had to migrate large distances before temperatures warmed sufficiently. There is evidence neither for nor against this possibility.

Bone Histology

Given the distribution of dense secondary Haversian bone in vertebrates, it is not too great a leap to suppose that dinosaurs, too, must have been endotherms.

Although this idea was initially promising, dense secondary Haversian bone forms due to a variety of factors, only one of which is endothermy. Secondary Haversian canals correlated with size and age, and possibly with the type of bone being replaced, the amount of mechanical stress undergone by the bone, and **nutrient turnover** (the metabolic interaction between soft tissue and developing bony tissue). The last two, however, are still related to metabolism.

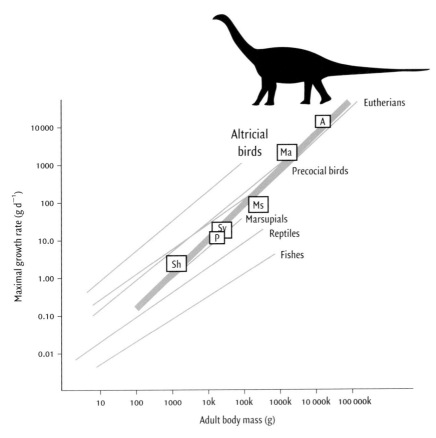

Figure 14.8. Maximum composite growth rates of various groups of animals, across a broad range of sizes. Composite rates constructed from growth rates of several animals in each group. In the case of the dinosaurs (**bold** line), it is constructed from *Shuvuuia* (Sh), *Psittacosaurus* (P), *Syntarsus* (Sy), *Massospondylus* (Ms), *Maiasaura* (Ma), and *Apatosaurus* (A). The difference in composite slope between dinosaurs and other vertebrate groups in the graph suggests that dinosaurs as a group grew at different rates from these other groups. Redrawn from Erickson, G. M. 2005. Assessing dinosaur growth patterns: a microscopic revolution. *Trends in Ecology and Evolution*, **20**, 677–684.

As we have seen, most dinosaurs – with the exceptions of some theropods, hypsilophodontids, iguanodontids, and sauropods – show well-developed LAGs, suggesting to researchers that growth in dinosaurs might have been more susceptible to external climatic influences than had been predicted by the simple homeothermic endothermic view of dinosaur metabolism. In two small flyers and one large flightless bird (*Patagopteryx*; see Figure 8.10), the presence of LAGs led to the conclusion that the early birds' metabolism(s) were subject to seasonal growth, even though the birds clearly bore feathers. The presence of LAGs in these early birds could mean that, ultimately, they had not quite attained the level of endothermy seen in living birds (see below).

Enantiornithine bone histology contrasts with that of ornithurine birds (for example, *Hesperornis* and *Ichthyornis*). In enantiornithines, the bone tissue looks very similar to that in modern birds. So too does the bone tissue of the primitive, Early Cretaceous *Confuciusornis* (see Figure 8.7).

So in the end, what does all this say about dinosaur endothermy? It has become well established that metabolic rates are well correlated with maximal growth rates. Fast growth rates – such as are found in modern birds – are highly suggestive of an endothermic metabolism. Figures 13.8 and 14.8 contrast the collective growth rate of dinosaurs as a group with those of some familiar groups of living animals. While some larger dinosaurs may have undergone very rapid growth to attain large size (Figure 13.8b), the growth rates of dinosaurs as a group were generally similar to modern endothermic growth rates. This in turn suggests that their metabolism may have been more closely allied to those seen in birds and mammals, than living reptiles likes snakes, lizards, and crocodiles (Figure 14.8).

Air Conditioning

As we have seen, the use of CT scanning has been extraordinarily useful in understanding the paleobiology of dinosaurs; used skillfully, CT scans have revealed an extraordinary amount of anatomy that otherwise would have gone unremarked. Among the most striking features that CT-scanning has allowed us to understand more fully are the elaborated nasal passages of many dinosaurs, such as *T. rex* (Figure 6.20) and ankylosaurs (Figure 10.18). In both of these cases, we noted that such nasal passages must reflect a heightened sense of smell, which is surely true, but surely also not the whole story. In 2018 Jason Bourke and colleagues performed a "fluid dynamic analysis" of the air passages in the ankylosaurs *Euoplocephalus* and *Panoplosaurus*. In essence, what they did was use digital models to reproduce air movement in ankylosaur sinuses, also modeling the heat exchange associated with that air movement. In an extraordinary bit of evolutionary engineering, the authors demonstrated that the ankylosaur nasal passages effectively modified the temperature of the air within them, serving to significantly warm air that was being drawn in, and cool air that was breathed out. Moreover, it appears that the elaborated nasal passages also helped exhaling ankylosaurs retain water in much the same way as the respiratory turbinates found in mammals. Bourke and colleagues concluded that the convolute nasal passages in ankylosaurs must relate to stabilizing brain temperatures in the living animal.

In sort, although it at first appeared that an absence of respiratory turbinates reflected an ectothermic metabolism, it is starting to become clear that convolute nasal passages, as well as respiratory turbinates, can serve the twin roles of maintaining both heat stability and reducing water loss.

Did other dinosaurs have built-in air conditioners? Recall (or maybe not!) that in Chapter 4 we talked about the vascularization of soft tissues in the upper temporal opening. In a 2019 study, anatomist Casey Holiday and colleagues proposed that the uppermost parts of the upper temporal opening, for many years thought to have been filled with jaw adductor muscles, actually contained fatty vascularized tissues that could be used for a kind of incipient temperature regulation: blood would be pumped into the region, either cooling or warming as best suited the animal. Those tissues could also support display structures – not preserved, of course – whose purpose could then be dual: display and thermoregulation. Did *T. rex* and other large theropods, for example, have some kind of soft tissue display feature, like a rooster's comb, sticking out of the back of its head? We don't know, although the evidence that the upper temporal opening was involved in temperature control is becoming overwhelming. Perhaps the soft-tissue display idea should not be dismissed too disdainfully, especially in light of fancy crests like those seen *Dilophosaurus* and in the high-latitude *Cryolophosaurus*, for example; both animals that lived in relatively extreme climatic settings (hot and cold, respectively, and likely relatively dry, both).

Stable Isotopes

Conveniently, fossil vertebrates carry around their own thermometers. These come in the form of stable istopes; isotopes that, unlike their unstable brethren, do *not* spontaneously decay. Of particular interest to us are the isotopes of the element oxygen. There are three: ^{16}O, by far the most common, ^{17}O, and ^{18}O.[2] The last, ^{18}O, is particularly interesting, because its proportion to ^{16}O varies as temperature varies. This process involves fractionation, which is the separation or division of the

---

[2] The isotope ^{16}O comprises 99.763 percent, ^{17}O 0.0375 percent, and ^{18}O 0.1905 percent of total atmospheric oxygen.

heavy and light isotopes during a phase transition.[3] Therefore, if a substance contains oxygen, one can learn something about the temperature at which that substance formed by the ratio $^{18}O/^{16}O$. In the case of bone and teeth, the oxygen is contained in phosphate (PO_4) that forms part of the mineral matter in the bone. Thus, knowing the oxygen isotopic composition of the bone, one can learn something of the temperature at which the bone formed. A number of different approaches bearing on the question of dinosaur endothermy have been tried, including core-to-extremities measurements, direct temperature estimates using isotope ratios, measurements of temperatures using latitudinal gradients, and "clumped isotope geothermometry."

Core-to-Extremities

If dinosaurs were poikilotherms, reasoned paleobiologist R. Barrick and geochemist W. Showers, there should be a large temperature difference between parts of the skeleton located deep within the animal (that is, ribs and trunk vertebrae) and those located toward the exterior of the animal (that is, limbs and tails; Figure 14.9).

If, however, dinosaurs were homeothermic, there should be little temperature difference between bones deep within the animal and those more external, because the body would be maintaining its fluids at a constant temperature. The difference in temperatures – or lack thereof – should be reflected in the proportions of ^{18}O to ^{16}O (that is, $^{18}O/^{16}O$).

Barrick and Showers showed that bones from the cores of some of the dinosaurs tested (*Tyrannosaurus*, *Hypacrosaurus*, *Montanoceratops*, *Orodromeus*, and a juvenile *Achelousaurus*) showed a $\pm 2\,°C$ temperature variation, suggesting that they were formed under near homeothermic conditions. A nodosaurid ankylosaur that they tested, on the other hand, had a large isotopic variation (an inferred temperature variation of around $11\,°C$) that is well beyond the limits of conventional homeothermy. Barrick and Showers concluded that all of the dinosaurs they tested, except nodosaurs, were homeotherms that experienced some "regional heterothermy" (Figure 14.9).

Direct Temperature Measurements

As we have seen, Barrick and Showers' work depended upon isotope ratios, but there was a catch. The calculation of temperature from isotopic ratios in bone requires knowing the isotopic composition of the environments in which the animal lived; usually, a function of the temperature of the water that it drank. But how could one ever know either the temperature of the isotopic composition of the water an extinct animal drank? Barrick and Showers, however, had designed a very clever experiment: they didn't need to worry about this because they were comparing *relative* temperatures – variations between cores and extremities.

A number of later studies by a variety of paleontologists and geochemists attempted, however, to obtain the actual temperatures at which various dinosaurs functioned. This was done by calibrating the estimated temperatures of known ectotherms to obtain those of dinosaurs. So, for example, crocodiles were an important feature of many dinosaur ecosystems. Knowing the temperatures at

---

[3] In this case, the phase transition in which the fractionation occurs is from liquid (the water that the organism ingests) to the solid (incorporation of the oxygen molecules into the bones and teeth). We will also see fractionation occurring during the phase transition from a gas (the atmosphere) to a liquid (rainwater, condensed from the atmosphere).

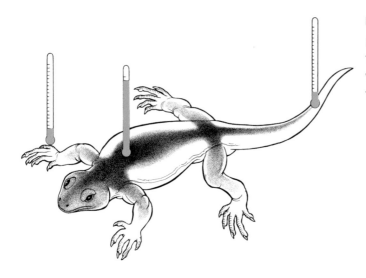

Figure 14.9. Core-to-extremity temperatures in a poikilothermic ectotherm. Because this tetrapod's temperature fluctuates with the ambient temperature, when it's cold outside, its extremities are much colder than its core.

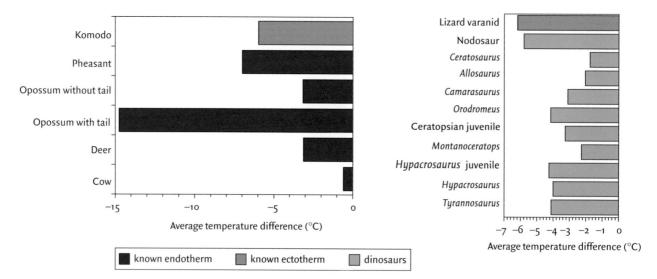

Figure 14.10. Estimated maximum temperature variations between bones located in the core of the body and those located at the extremities in living and extinct vertebrates, reconstructed with the use of oxygen isotopes. Living vertebrates are represented by the Komodo dragon (an ectothermic lizard), a bird, and a selection of mammals (endotherms). Note that the greatest variation between core and extremities occurs in the opossum (conventionally considered to be an endotherm), reinforcing the idea that even endotherms can undergo significant fluctuations between core and extremities (in this case the long tail). The researchers concluded that all dinosaurs tested, except the ankylosaur, matched the definition of homeotherms.

which modern crocodiles function, one could calibrate the estimated isotopic composition of the water, until the calculated temperature of fossil crocodiles was within a couple of degrees of the known temperatures of modern crocodiles. Once the isotopic composition of the water was properly estimated, the real temperatures of dinosaurs could then be determined.

The results of these types of studies have been consistent. For Cretaceous theropods, sauropods, ornithopods, and ceratopsians, the very considerable body of data collected suggests that these animals maintained body temperatures of between 30 °C and 37 °C, well within the range of modern endotherms, at least (Figure 14.10).

Latitudinal Gradients

There is yet another way to skin this dinosaur – isotopically speaking. Geochemist H. Fricke and paleontologist R. Rogers first – and other workers later – looked at the isotopic patterns of teeth across a latitudinal (e.g., north–south) gradient. They used teeth because teeth are tough and they preserve original isotopic signatures better than bone. And they (the scientists, not the teeth) looked across latitudinal gradients because temperature differences between the equator and the poles cause the ^{18}O:^{16}O ratio in rain to decrease from the equator to the poles. This occurs through the fractionation of the oxygen isotopes as they go from a gaseous state in the atmosphere, to a liquid state, as rain (via condensation). So, as one goes northward (or southward) from the equator, rain – and thus the water that fills the lakes and streams – becomes "lighter"; that is, relatively more enriched in ^{16}O than water in lakes and streams closer to the equator (which, being more enriched in ^{18}O, is said to be "heavier"). Lake and stream water, of course, gets drunk by terrestrial organisms – including dinosaurs – and contributes its share to the isotopic content of the organisms that drank it. Fricke and Rogers called such water "body water."

As we said, the body water is incorporated into the tissues of the living animal, including its teeth. Now, the step of incorporation into the bones and teeth involves a *second* fractionation step: from the liquid of the body water to the solid of the teeth. In the case of an endothermic homeotherm, the amount of the second fractionation will not change, because the temperature of the animal does not change regardless of the latitude in which it lives. In the case of the ectothermic poikilotherm, however, the temperature of the animal changes, and thus the amount of second fractionation changes: in lower latitudes, with higher temperatures, there is less fractionation; in higher latitudes, with lower temperatures, there is more fractionation.

All of that gives us a very powerful tool with which to recognize endothermic homeotherms and ectothermic poikilotherms (Figure 14.11). And, in fact, a very large study, involving about 100 dinosaurs, including sauropods, theropods, ornithopods, and ceratopsians, used the same technique to arrive at the same conclusion that had first been reached by Fricke and Rogers: dinosaurs gave a very different slope from that of ectothermic poikilotherms (crocodiles) when the isotopic compositions of the teeth are plotted against latitude. These results signal clearly that dinosaurs were fractionating oxygen isotopes in a manner closer to modern endothermic homeotherms, and quite differently from Cretaceous crocodiles, for example, which produced a signal that is comparable to their modern counterparts today.

Clumped Isotopes

The Achilles heel, as we have seen, of direct isotope-based temperature measurements is that isotopic composition of the water that the animals drank has to be estimated. Since 2010, however, a new technique – clumped isotopes – has been applied to the problem of the body temperatures at which ancient bone and teeth formed, as well as the temperatures at which eggs were incubated. Clumped isotopes refers to the attraction – the "clumping" – of ^{18}O and ^{13}C at different temperatures. The amount of clumping is temperature-dependent, and so with this technique, by measuring the degree of clumping, there is no need to estimate the isotopic composition of the water. Geochemist R. A. Eagle and colleagues applied this technique to a number of extinct and recent vertebrates, including the teeth of Jurassic sauropods and the eggs of oviraptorosaurids. Having calibrated this geothermometer with the modern vertebrates, they determined that the sauropods operated at temperatures between 31 °C and 38 °C, while the oviraptorosaurid eggs were formed at lower temperatures than those found in modern birds. These temperatures were suggestive of an

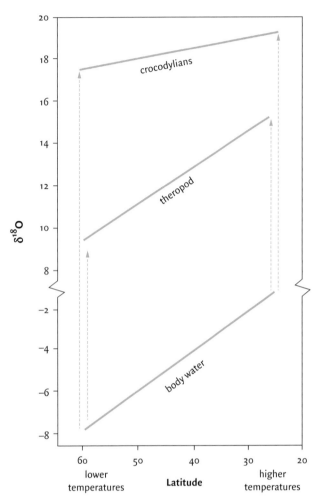

Figure 14.11. Using latitudinal gradients and oxygen isotopes to determine dinosaurian metabolic strategy. Latitude is plotted along the *x*-axis and oxygen stable isotopic content is plotted along the *y*-axis. The line labeled "body water" is the oxygen isotopic content of the water that organisms drink. Its oxygen isotopic content varies with latitude, as a result of fractionation due to latitudinal temperature differences. The lines labeled "theropod" and "crocodylians" are the oxygen isotopic content of the teeth for these animals along the same latitudinal gradient. The dotted arrows represent the isotopic fractionation that occurs during the formation of tooth enamel from body water. In the dinosaurs, regardless of latitude (outside temperature), the same amount of fractionation occurs, whereas in the crocodylians, the amount of fractionation that occurs as the oxygen is incorporated into the teeth is directly related to external temperature – and thus latitude. For this reason, the isotopic content relative to latitude of the crocodylians is a distinctly lower slope than that of the theropods – suggesting that the theropods' metabolism was that of an endothermic homeotherm, while that of the crocodylians is that of an ectothermic poikilotherm. Redrawn (and simplified) from Fricke, H. and Rogers, R. 2000. *Geology*, **28**, 799–802.

endothermic metabolism (but see below) in the sauropods, but not so obviously one in oviraptorosaurids.

Different Strokes for Different Folks?

Where does all that leave us? What is unimpeachable at this point is that dinosaurs were not ectothermic poikilotherms *à la* crocodile *mode*. They were something else; the question is, "what?" Several different, potentially coexisting, general classes of ideas have been proposed, including:

(I) *Different endothermic strategies could have been adopted by different dinosaurs, including, of course, birds.* The disparities in size and in dinosaur behavior make it difficult to support a metabolic "one-size-fits-all" model. Some scientists have argued that large sauropods could have relied upon a strategy called **gigantothermy** – small surface:volume ratios (resulting from large size) retained core heat, allowing sauropods to maintain a homeothermic metabolism without the metabolic cost of being truly endothermic. Yet the results of the clumped isotopic studies suggest that sauropod body temperatures may have been too *cool* for a passive strategy like gigantothermy. To those researchers, sauropods must have actively maintained their temperatures: they must have been true endotherms. By contrast, considerable evidence suggests that small- to medium-sized theropods, and perhaps similarly sized ornithopods, were likely homeothermic endotherms throughout their active lives; not perhaps to the extent that a modern bird is, but perhaps somewhat like the kind of endothermy exhibited by animals like an opossum. The recent discovery of feathers in large, as well as small, theropods, as well as the pervasive presence of extensive secondary (Haversian) bone, more closely resembling those of birds than any other living animal, supports this view.

(II) *Dinosaurs maintained an "intermediate" metabolic strategy, that is, a metabolism that was somehow in between modern ectothermic poikilotherms and endothermic homeotherms.* The late dinosaur paleophysiologist R. E. H. Reid argued for some years that dinosaurs had an "intermediate" metabolic strategy. In this view, dinosaurs would have been possessed of a four-chambered heart and a low metabolic rate, used gigantothermy and blood circulation to stabilize their temperature, and been tolerant of a range of temperatures. Reid saw the fast growth rates as *concordant with* endothermy, but not necessarily *requiring* endothermy, and pointed out in support of this idea that, unlike birds and mammals, nonavian dinosaurs never really produced truly tiny forms (which cannot rely at least in part upon their bulk for temperature stability). On the basis of this and other, generally histological, evidence, he argued that endothermy, as seen in modern birds and mammals, likely did not exist in nonavian dinosaurs. In this viewpoint, "intermediate" does not imply that the metabolism is on the way to evolving into modern-style endothermy; it is to be regarded as a fundamentally different and unique metabolic strategy that characterized nonavian dinosaurs. A similar take on this perspective was offered in 2014 by University of New Mexico physiologist J. M. Grady and colleagues, who concluded that reconstructed dinosaur growth rates matched neither modern endothermic homeotherms nor ectothermic poikilotherms, but rather were intermediate, a range of metabolic strategies that they characterized as "mesothermy."

(III) *Different metabolic strategies were in place during different ontogenetic stages of a dinosaur's development.* Some paleontologists suggest that some dinosaurs – particularly large ornithopods and theropods – maintained something close to endothermic homeothermy as fast-growing juveniles, but may have become closer to homeothermic ectotherms (via gigantothermy) as adults. While this may be true, the elevated temperatures consistently recorded by geochemists suggest that, as adults, these animals maintained a temperature different from that of the air that surrounded them: the hallmark of an endothermic metabolism.

(IV) *Endothermy was primitive for Dinosauria, and some, but not all, dinosaurs lost this trait, reverting back to ectothermic metabolisms.* Considering this from a slightly different perspective: because dinosaurs are a monophyletic group, a basal dinosaurian metabolic strategy must have existed. Moreover, we know that Saurischia and Ornithischia were feathered, suggesting that basal dinosaurs were feathered. If that is so, and if feathers are monophyletic (an assumption for which we have seen there is evidence), then it is possible that the earliest members of Dinosauria bore feathers. Based upon the feather coatings known from primitive ornithischians and the small sizes (relatively large surface:volume ratios) of the earliest dinosaurs, those feather coverings likely insulated the animals that had them. Because it makes no sense for an ectotherm to insulate itself, full feather coverings suggest endothermic metabolisms for the first dinosaur and its early descendants. However, it is quite possible that as dinosaurs diversified, full feather coats, and quite conceivably, endothermy, were lost in some lineages; for example sauropods, who perhaps sacrificed endothermy to gigantothermy: their size would allow them to be homeothermic, without the metabolic "price" of endothermy. Large dinosaurs such as hadrosaurs, whose "mummies" (Figure 12.7) preserve skin impressions that show no evidence of feathers, might also have used related strategies.

(V)   *All – or some – of the above.* This – the "I don't know/default option" – and thus our best call, is that dinosaurian physiology was likely a complex mix of various endothermic strategies, relating to size, behavior, and, potentially, environment. These strategies, however, could have evolved in different directions in different groups, according to behavior-based need.

Regardless, after this chapter, it should be clear that although the question of dinosaur metabolism has lost some of its immediacy and, uh, heat, the last word on this still simmering topic is very far from being said.

SUMMARY

Physiologists avoid terms like "warm-blooded" and "cold-blooded" and instead replace them by more meaningful terms like endothermy and ectothermy, which describe the heat source, and homeotherm and poikilotherm, which describe the degree to which temperatures fluctuate in organisms. All of these terms should be regarded as endpoints on spectra of metabolic strategies.

Anatomical indicators in dinosaurs suggestive of an endothermic metabolism include: the erect stance, which among living vertebrates is exclusively possessed by endotherms; the inferred presence of a four-chambered heart, necessary to sustain the higher blood pressures associated with an endothermic metabolism; and the relatively high EQs of some dinosaurs.

Each of these indicators, however, is inconclusive: the erect stance argument has been criticized as being purely coincidental (and not causal) and the high EQs of some dinosaurs are matched by strikingly low EQs in others. Moreover, the absence in some nonavian dinosaurs of respiratory turbinates has suggested that they perhaps didn't maintain the high rates of ventilation seen in living endotherms, although compensating features, such as elaborated nasal passages, may have taken their place.

The presence in dinosaurs of Haversian canals appears to suggest endothermy. These, timed by LAGs, suggest rapid growth rates that today are known only in endotherms. These suggest the possibility that, as juveniles at least, many dinosaurs may have possessed endothermic metabolisms.

Because endothermy is, in terms of energy, quite costly to maintain, it was suggested that the ratio predators/prey in endothermic ecosystems ought to be significantly smaller than the ratio of predators/prey in ectothermic ecosystems. Several attempts were made to calculate such ratios for dinosaurs. None ultimately proved definitive for a variety of reasons, including the unreliability of museum collections as accurate indicators of ancient communities, the fact that endothermic predators sometimes eat ectothermic prey (and vice versa), and the fact that the limiting factor on prey populations is not generally predation.

The existence of polar-dwelling dinosaurs has been interpreted as suggestive of an endothermic metabolism, since today large ectotherms don't get much above 20° north or below 20° south latitude. Yet, a high-latitude temnospondyl amphibian also preserved suggests that, perhaps as a result of a warmer Earth, the past distribution(s) of ectotherms was greater than today.

With insulation in the form of feathers potentially going as far back as Ornithodira, there is considerable evidence that endothermy may go back equally far within archosaurs. Regardless, as we saw in Chapter 7, *Archaeopteryx* and its avialan cohort enjoyed slower growth rates that we see in modern birds, suggesting that the insulation in basal ornithodires may be indicative of early flirtations with endothermy.

Ratios of $^{18}O/^{16}O$ are temperature-sensitive, and have been used in a variety of ways as paleothermometer in well-preserved fossil bone and teeth. The idea was that ectotherms would show a greater range of temperature fluctuations from core to extremities than endotherms. In fact, dinosaurs (and even some mammals) produced a somewhat mixed signal: while some dinosaurs, such as hadrosaurs, showed little temperature variation, ankylosaurs and two of the large theropods tested showed ectotherm-like variations. Reinforcing the point that endothermy and ectothermy are actually endpoints in a spectrum and that metabolisms among vertebrates are not easily predictable, an opossum, a living marsupial (mammal), also showed the kind of variation expected in an ectotherm. Latitudinal gradient studies suggest that dinosaurs maintained body temperatures that are significantly higher than ambient environmental temperatures. Clumped isotopic studies, although in the early phases, suggest similar conclusions in those dinosaurs in which they have been attempted.

While it is clear that the old "cold-blooded" lizard or crocodile model of dinosaur metabolism is defunct, the record suggests that dinosaurs potentially enjoyed a range of metabolic strategies, most or all of them rooted in some kind of at least primitive type of endothermy, perhaps intermediate between, and surely different from, the ectothermic poikilotherms and endothermic homeotherms of today.

SELECTED READINGS

Amiot, R., Lècuyer, C., Buffetaut, E., *et al.* 2006. Oxygen isotopes from biogenic apatites suggest widespread endothermy in Cretaceous dinosaurs. *Earth and Planetary Science Letters*, **246**, 41–54.

Amiot, R., Wang, X., Wang, S., *et al.* 2017. $\delta^{18}O$-derived incubation temperatures of oviraptorosaur eggs. *Palaeontology* **60**, 633–647.

Bakker, R. T. 1975. Dinosaur renaissance. *Scientific American*, **232**, 58–78.

Bakker, R. T. 1986. *The Dinosaur Heresies*. William Morrow and Company, New York, 481 pages.

Bakker, R. T. and Galton, P. M. 1974. Dinosaur monophyly and a new class of Vertebrates. *Nature*, **248**, 168–172.

Barrick, R. E., Stoskopf, M. K., and Showers, W. J. 1997. Oxygen isotopes in dinosaur bone. In Farlow, J. O. and Brett-Surman, M. K. (eds.) *The Complete Dinosaur*. Indiana University Press, Bloomington, IN, pp. 474–490.

Bourke, J. M., Porter, W. R., and Witmer, L. M. 2018. Convoluted nasal passages function as efficient heat exchangers in ankylosaurs (Dinosauria: Ornithischia: Thyreophora). *PLoS ONE*, **13**(12), e0207381. https://doi.org/10.1371/journal.pone.0207381

de Ricqlès, A., Horner, J. R., and Padian, K. 2006. The interpretation of dinosaur growth patterns. *Trends in Ecology and Evolution*, **21**, 596–597.

Desmond, A. 1975. *The Hot-Blooded Dinosaurs*. The Dial Press, New York, 238 pages.

Eagle, R. A., Tütken, T., Martin, T. S., *et al.* 2011. Dinosaur body temperatures determined from isotopic ($^{13}C-^{18}O$) ordering in fossil biominerals. *Science*, **333**, 443–445.

Eagle, R.A., Enriquez, M., Grellet-Tinner, G., *et al.* 2015. Isotopic ordering in eggshells reflects body temperatures and suggests differing thermophysiology in two Cretaceous dinosaurs. *Nature Communications*, **6**, 8296.

Farlow, J. O. 1990. Dinosaur energetics and thermal biology. In Weishampel, D. B., Dodson, P., and Osmólska, H. (eds.) *The Dinosauria*. University of California Press, Berkeley, pp. 43–55.

Fricke, H. and Rogers, R. 2000. Multiple taxon–multiple locality approach to providing oxygen isotope evidence for warm-blooded theropod dinosaurs. *Geology*, **28**, 799–802.

Grady, J. M., Enquist, B. J., Dettweiler-Robinson, E., Wright, N. A., and Smith, F. A. 2014. Evidence for mesothermy in dinosaurs. *Science*, **344**, 1268–1272.

Holliday, C. M., Porter, W. R., Vliet, K. A., and Witmer, L. M. 2019. The frontoparietal fossa and dorsotemporal fenestra of archosaurs and their significance for interpretations of vascular and muscular anatomy in dinosaurs. *The Anatomical Record*, **303**, 1060–1074. DOI: 10.1002/ar.24218

Hopson, J. A. 1980. Relative brain size: implications for dinosaurian endothermy. In Thomas, R. D. K. and Olson, E. D. (eds.) *A Cold Look at the Warm-Blooded Dinosaurs*. AAAS Selected Symposium 28, Westview Press, Boulder, CO, pp. 287–310.

Horner, J. R., de Ricqlès, A., and Padian, K. 2000. Long bone histology of the hadrosaurid dinosaur *Maiasaura peeblesorum*: growth dynamics and physiology based upon an ontogenetic series of skeletal elements. *Journal of Vertebrate Paleontology*, **20**, 115–129.

Padian, K. and Lamm, E.-T. 2013. *Bone Histology of Fossil Tetrapods: Advancing Methods, Analysis, and Interpretation*. University of California Press, Berkeley, CA, 285 pages.

Reid, R. E. H. 2012. "Intermediate" dinosaurs: the case updated. In Brett-Surman, M. K., Holtz, T. R., and Farlow, J. O. (eds.) *The Complete Dinosaur*, 2nd edn. Indiana University Press, Bloomington, IN, pp. 873–921.

Rich, P. V., Rich, T. H., and Wagstaff, B. E. 1988. Evidence for low temperatures and biologic diversity in Cretaceous high latitudes of Australia. *Science*, **242**, 1403–1406.

Sellers, W. I., Manning, P. L., Lyson, T., Stevens, K., and Margetts, L. 2009. Virtual palaeontology: gait reconstruction of extinct vertebrates using high performance computing. *Palaeontologia Electronica*, **12**(3), 11A, 26 pages; palaeoelectronica.org/2009_3/180/index.html

Spotila, J. R., O'Connor, M. P., Dodson, P., and Paladino, F. V. 1991. Hot and cold running dinosaurs: body size, metabolism, and migration. *Modern Geology*, **16**, 203–227.

Thomas, R. D. K. and Olson, E. C. (eds.) 1980. *A Cold Look at the Warm-Blooded Dinosaurs*. AAAS Selected Symposium no. 28, 514 pages.

TOPIC QUESTIONS

1. What are meant by the terms endothermy, ectothermy, poikilothermy, and homeothermy?
2. What are remodeled bone, Haversian canals, LAGs, and respiratory turbinates?
3. Was a four-chambered heart evidence for endothermic dinosaurs? What was it used to support?
4. Give some anatomical evidence that dinosaurs were endothermic. Critique it.
5. What is a predator/prey biomass ratio? How were these used in assessing dinosaur metabolism?
6. Critique the predator/prey biomass ratios.
7. How would the distribution of animals on Earth have anything to do with their metabolisms? What kinds of strategies were available to dinosaurs for protection against long, cold winters?
8. What is ^{18}O? What is meant by the statement "Fossil vertebrates carry around their own paleo-thermometers"?
9. Why would comparison of core temperatures to those of the extremities be useful in determining whether an animal had an endothermic or an ectothermic metabolism?
10. How are rates of growth calibrated in fossil dinosaurs?

APPENDIX 14.1 CHAIN OF FUELS

We all know that somehow we get energy from the family of carbon-based molecules called **carbohydrates** (think "Candy bar"!). Not quite so familiar, perhaps, is how this works. Simply put, the energy stored in the bonds of carbohydrates is transferred to energy stored in the bonds of a molecule called **ATP** (adenosine triphosphate). Then we – and all living organisms – access that energy by breaking the ATP molecule to produce a closely related molecule called **ADP** (adenosine diphosphate) and thereby releasing some of that energy.

Cellular Respiration

In cellular respiration, chemical bonds in carbohydrates (such as the sugar glucose) are broken via a type of reaction called oxidation. These reactions occur as complex, linked, series involving a number of intermediate steps. The breakdown of a single molecule of glucose (a simple carbohydrate) through this suite of reactions can produce 36 new molecules of ATP through a series of reactions called the "citric acid cycle," so named because citric acid is produced as an intermediate step in the carbohydrate breakdown (Figure A14.1.1).

This type of metabolism, called aerobic (involving oxygen), however, is not 100 percent efficient: ATP production captures about 40 percent to 60 percent of the energy of the bonds of the carbohydrates. The remainder is released as heat.

Organisms respire oxygen because energy storage as ATP involves, as we have seen, oxidation reactions. As the energy output of the organism is increased, the amount of ATP needed is increased, and hence more oxygen is consumed and more heat produced. This is why breathing, heart rate, and temperature increase when we exercise: we are using more energy, requiring more ATP to be generated, and thus we need more oxygen. And give off more heat!

There is a point, however, at which the volume of oxygen supplied by breathing is insufficient. Under such conditions, a different reaction path called *glycolysis* is followed. The process of generating energy through glycolysis is a type of metabolism called anaerobic (without oxygen). Glycolysis bypasses the citric acid cycle, and instead directly produces two three-carbon molecules called pyruvic acid. The pyruvic acid in turn generates lactic acid, which accumulates in the muscles and causes the familiar ache after extreme exercise (see Figure A14.1.1). After hard exercise, we breathe heavily to replenish our depleted oxygen supply, and eventually the lactic acid is removed from the muscles.

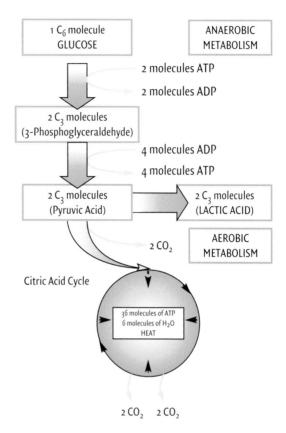

Figure A14.1.1. Cellular respiration consists of the breakdown of carbohydrates to produce energy that is stored in ATP. In this example, the six-carbon molecule glucose is broken down. Two pathways are shown: the aerobic path, in which ATP is produced via the citric acid cycle, and the anaerobic path, in which lactic acid is ultimately produced via glycolysis.

The Flowering of the Mesozoic

Chapter

15

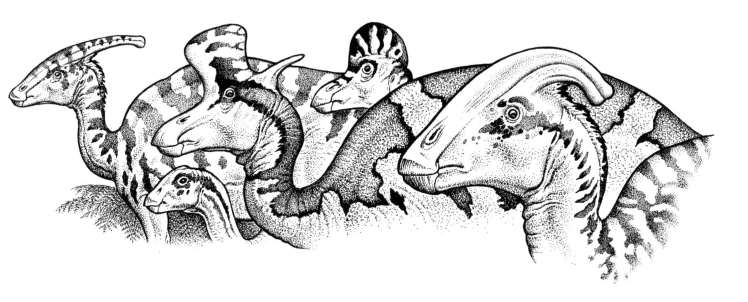

Dinosaurs in the Mesozoic Era

Throughout much of this book, we have considered dinosaurs as individuals; who they were, what they did, and how they did it. Now we'll step back and take a look at the large-scale ebb and flow of dinosaur evolution, and take a dip into dinosaur ecosystems. Before we can do that though, we need to think about the sedimentary record itself: what's there and what might be *missing*.

Limitations of the Sedimentary Record

Fossils, obviously, are generally found in sedimentary rocks. Sedimentary rocks are rocks that have been *deposited*; the type of sedimentary rocks in which dinosaur fossils tend to be found are called clastic rocks, rocks that are composed of sand, silt- and clay-sized fragments of rocks, and minerals that have been weathered, transported, and deposited by wind and water. It will also come as no surprise that water and wind ultimately deposit these rock fragments in topographically low areas, where they can accumulate (the fragments, not the water and wind). So in fact all sedimentary rocks – accumulations of sediment later lithified by burial to become sedimentary rocks – only represent topographically low areas: zones of sediment accumulation. The fossils that are found in those sedimentary rocks either lived in that very place or, like other bits of mineral and rock, were deposited there (and lived somewhere else). So, the sedimentary rock record is a highly selective thing: it only preserves ancient environments (represented by sedimentary rocks) and fossils (unless bits have been carried from somewhere else) from what are effectively *lowlands* – environments where sediment once accumulated.

Let's think about what this means: animals and plants that lived in mountainous areas won't get seen, unless somehow their hardparts – e.g. bones and teeth – get carried into lowland settings. As you see, the sedimentary and fossil records, like the Gentlemen's Clubs in Victorian London, are highly selective.

Preservation

Table 15.1 shows the distribution of dinosaurs among the continents through time. The paucity of dinosaur remains from Australia and Antarctica, however, is surely more a question of *local* preservation and inhospitable conditions today for finding and collecting fossils than defining where dinosaurs actually lived. Into this mix must be factored *geological* preservation; some time intervals simply contain more rocks than others. For example, the terrestrial Middle Jurassic of North America, particularly, is not well represented by rocks, with the result that it artificially appears to have been a time of very low tetrapod diversity (Box 15.1). The Late Cretaceous is rather the opposite, with the happy result that we have a rich record of Late Cretaceous dinosaurs. Several methods of estimating the completeness of fossil preservation have been developed to mitigate these problems (Box 15.2).

Dinosaurs Through Time

We can think of these geographical and temporal distributions as pages in a notebook, in which each succeeding page represents a new time interval with different continental arrangements and new and different assemblages of dinosaurs populating the continents. Considered in this way, the sequence of dinosaurs through time is like a grand pageant through Earth history, in which each interval of time has a characteristic fauna that gives that time a characteristic quality (Figure 15.1).

Table 15.1 **Distribution of dinosaurs on continents during the Mesozoic Era. Crosses indicate dinosaurs known.**

	Asia	Africa	South America	North America	Europe	Australia	Antarctica
Late Triassic	X	X	X	X	X	X	
Early Jurassic	X	X	X	X	X		X
Middle Jurassic	X	X	X	X	X	X	
Late Jurassic	X	X	X	X	X		
Early Cretaceous	X	X	X	X	X	X	
Late Cretaceous	X	X	X	X	X	X	X

Box 15.1 **The Shape of Tetrapod Diversity**

For at least 35 years, the University of Bristol's M. J. Benton has been compiling a comprehensive list of the fates of tetrapod families through time. We see several interesting features of the curve that result from this compilation. Note the drop in families during Middle Jurassic time (Figure B15.1.1). This, as we have seen, is an **artifact**; that is, a specious result. This particular one comes from the lack of finding localities more than from a true lack of families during the Middle Jurassic. Then, notice the huge rise in families during the Tertiary. Some of this may be real, and perhaps attributable in part to Tertiary birds and mammals (both of whom are very diverse groups), but some of it might be another artifact, due to what is called the "**pull of the Recent**." The pull of the Recent is the inescapable fact that, as we get closer and closer to the Recent, fossil biotas become better and better known. This is because more sediments are preserved as we get closer and closer to the Recent, and a greater amount of sedimentary rocks preserved means more fossils. The big spike at the end of the Jurassic is the Morrison Formation of the US Western Interior, a unit that preserved an extraordinary wealth of fossils.

So a curve like Benton's requires skill to understand and to factor out the artifacts. Nonetheless, we can see that,

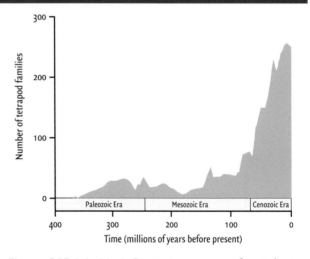

Figure B15.1.1. M. J. Benton's estimate of vertebrate diversity through time. On the x-axis is time; on the y-axis is diversity as measured in numbers of tetrapod families.

generally, dinosaurian diversity increased throughout their stay on Earth and, as they progressed through the Cretaceous, dinosaurs continued to diversify. The increase in diversity shown in Benton's diagram may reflect the increasing global endemism of the terrestrial biota, itself driven by the increasing isolation of the continental plates.

Figure 15.1. The great historical pageant of dinosaurs through time. The paleoenvironments of each time interval – and the dinosaurs that populated them – all have different qualities that characterized each successive ecosystem.

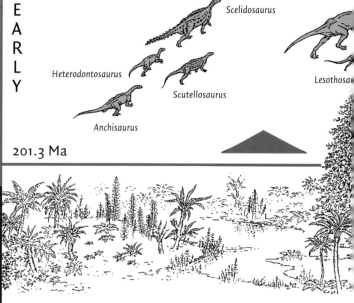
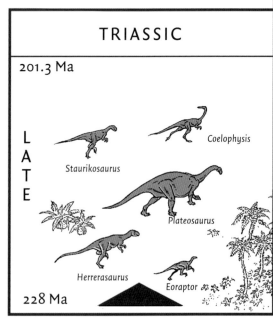

CRETACEOUS

66.1 Ma

L A T E

Maiasaura
Edmontonia
Oviraptor
Parasauralophus
Triceratops
T. rex
Gallimimus
Euoplocephalus
Troodon
Edmontosaurus
Protoceratops
Saltasaurus
Velociraptor
Centrosaurus
Pachycephalosaurus
Therizinosaurus
Alamosaurus
Carnotaurus

100.5 Ma

E A R L Y

Carchardodontosaurus
Psittacosaurus
Suchomimus
Deinonychus
Utahraptor
Ouranosaurus
Protarchaeopteryx
Polacanthus
Amargasaurus
Sinosauropteryx
Baryonyx
Wuerhosaurus
Iguanodon
Hypsilophodon
Caudipteryx
Brachiosaurus

145 Ma

Apatosaurus
Ornitholestes
Diplodocus
Archaeopteryx
Cetiosaurus

Box 15.2 Counting Dinosaurs

How many dinosaurs have ever lived? What percentage of those that have lived do we actually know about? In this box, we introduce several of the ways that diversity can be estimated. We also look briefly at the inverse: how many are missing? We start with the idea of **completeness**, a word one sees often in this context, and an important concept in the geosciences: completeness is the measure of how much of a particular thing is preserved. Sedimentary rocks, as we have seen, record time, and so the completeness (or incompleteness) of the sedimentary rock record is related to the completeness (or incompleteness) of the temporal record. Fossils, as we also saw, are preserved in sedimentary rocks, and so the completeness of the sedimentary rock record is also related to the completeness of the fossil record. How to figure out what's there and what's missing?

In the past 30 or so years, veteran sedimentary geologist P. M. Sadler and several colleagues, with the help of computers and piles of data, have taken on this problem. Because it all boils down to the completeness of the sedimentary record, Sadler and sedimentologist D. J. Strauss first constructed estimates of sedimentary completeness based upon a compilation of 25 000 rates of sediment accumulation (that is, rates of sediment deposition) in modern and ancient environments. Completeness, they determined, is ultimately a function of the age of the rocks, the long-term depositional rate, and the steadiness of the sedimentation over time. In a very general way, the larger the time scale under consideration, the more complete the sections. Later work by other authors refined some of these basic ideas. Sadler and paleontologist L. Dingus also attempted to apply the technique to fossils, estimating the completeness of particular intervals of time, such as blocks of time during which key evolutionary events, like mass extinctions, took place. They correctly reasoned that if not enough time was preserved, it would be rather difficult to learn about the events that took place (see Chapter 17).

Other Ways

Sophisticated statistical treatments have opened up other ways to count dinosaurs as well. Dodson (1990) used a published compilation, and simply counted. Fastovsky et al. (2004) used an updated 2004 version of the compilation used by Dodson (1990), and applied a statistical technique called **rarefaction** to the data. This technique allowed them to compare different sized samples to determine whether the diversity of dinosaurs actually changed through time, or whether just the samples varied because of preservation.

Another interesting statistical approach was applied by Wang and Dodson (2006), who developed a method for estimating the number of fossils for particular groups *that have yet to be discovered*! They introduced a metric called "coverage," which statistically assesses exactly how closely the *known diversity* from a locality (that is, what has been collected) conforms to its *actual diversity* (that is, a complete inventory of what theoretically ought to be preserved in the locality). Using the coverage metric, Wang and Dodson were able to take the total number of currently known dinosaur genera – 527 genera as of the year 2006 – and estimate the total number of dinosaur genera that ever existed: approximately 1850 genera.

Cladistic Estimates

We have heretofore emphasized the use of cladograms for reconstructing evolutionary relationships; however, they can also be used to get a sense of the completeness of the fossil record of a particular group. Recall that, along with character distributions, cladograms also portray the relative sequence of the appearance of organisms; who comes before whom. For with a fossil record (like dinosaurs!) this relative sequence from the cladogram can be compared with the real sequence of appearance that comes from the geological record of the same dinosaurs. The cladogram-based sequence ought to compare well with the sequence of stratigraphic occurrence of the fossils themselves. Suppose, however, that the fossil record is not complete; even if an ancestor is not preserved, the cladogram allows us to *infer* when it must have existed. To understand this, we turn to an example:

Suppose Dinosaur X and Dinosaur Y were each other's closest relative; thus they share a unique common ancestor. If Dinosaur X is known from rocks dated at 100 Ma and Dinosaur Y came from 125 Ma rocks, then this ancestor had to be at least 125 million years old (that is, the age of the older of the two dinosaur species). And if this is true, then there must be some not-yet-sampled history between this ancestor and Dinosaur X – to the tune of 25 million years – all because of phylogenetic continuity calibrated through the use of stratigraphy. Such a 25 million year gap can be referred to as a **minimal divergence time (MDT)**; it can be calculated for any two taxa so long as their phylogenetic relationships and stratigraphic occurrences are known and is an estimate of the completeness of the fossil record. Lineages must have existed that have so far not left us a physical record (through fossils) of their existence; these are called **ghost lineages**. MDTs are measures of their duration (Figure B15.2.1).

Completeness of the Ceratopsian Fossil Record

We exemplify how this method can give us an estimate of the quality of the fossil record of a particular group with Ceratopsia. It has sometimes been claimed that ceratopsians have one of the best fossil records among all dinosaur groups; that is, it is the most complete. How can we test this? There are about 32 ceratopsian species known, less than are found in theropods, sauropodomorphs, and ornithopods, yet more than in ankylosaurs, stegosaurs, and pachycephalosaurs. Averaged over their total time on Earth, ceratopsians apparently produced new species at the rate of one every 1.9 million years, as compared with a high of one new species per 1.4 million years for sauropodomorphs and a low of one per 5.6 million years for stegosaurs. By this reckoning, ceratopsians had relatively high rates of speciation.

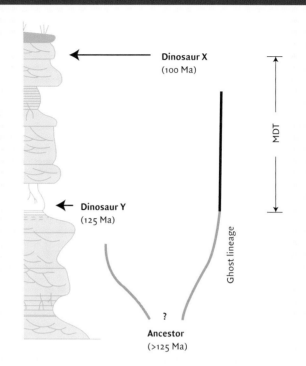

Figure B15.2.1. Ghost lineages and MDTs. Dinosaur X and Dinosaur Y are preserved 25 million years apart. If they are closely related, they both share a common ancestor that is at least as old as Dinosaur Y. Thus an estimate of the minimum divergence time (MDT) of the two lineages is as old as, or older than, Dinosaur Y (125 Ma). The record of that divergence is unpreserved and is therefore called a ghost lineage (curved gray lines on the figure).

To estimate the total diversity of a group, however, we turn to ghost lineages. For ceratopsians, MDT values range from 0 to nearly 30 million years, with an average of just over 5 million years. These are among the smallest MDTs for all Dinosauria, which suggests that the fossil record of this group is comparatively pretty well represented. Furthermore, actual ceratopsian species counts are nearly 70 percent of the total after ghost lineages have been added to the diversity total. On these measures, ceratopsians do indeed have one of the best records of all of the major dinosaur groups.

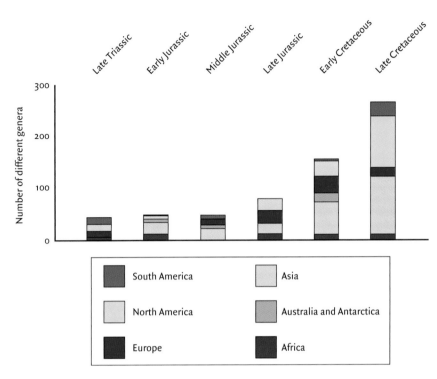

Figure 15.2. Changes in dinosaur diversity by continent measured through the Late Triassic–Late Cretaceous time interval. Each vertical bar shows the total number of different genera known from that particular time interval. Viewed from this perspective, dinosaurs appear to have steadily increased in diversity as the Mesozoic progressed.

Now let's look at this information in a more quantitative way: the **diversity**, that is, the number of different *types*, of nonavian dinosaur genera over time (Figure 15.2), through the approximately 164 million years that they were on Earth.

In the Beginning... the Late Triassic (237 Ma to 201.3 Ma)

Recall that dinosaurs radiated quickly in the Late Triassic (Figures 5.13 and 15.2). Unambiguous dinosaurs first appear on Earth at around 231 Ma,[1] but exactly how dinosaurs came to be the dominant terrestrial vertebrates in the Late Triassic remains tantalizingly shrouded in misty antiquity. As we implied in Chapter 5, our best data now suggest that dinosaurs likely moved quickly into a world *abandoned* by other vertebrates. Even the most ardent dinosaur advocates have failed to demonstrate that dinosaurs possessed a superior "secret weapon" that somehow allowed them to *outcompete* pre-existing tetrapods (mainly therapsids and primitive archosaurs, and nondinosaur dinosauromorphs; see Figures 15.3 to 15.6) and take charge. In the following section, we expand upon this interpretation.

The Rise of Dinosaurs – an Ecological Perspective

In Chapter 5, we discussed some of the basal dinosauromorphs closest to Dinosauria, and showed the transition, from nondinosaur archosaurs to dinosaurs, from a phylogenetic perspective. Now we take a look at this transition from an ecological perspective.

---

[1] There are some preservational and dating uncertainties.

Figure 15.3. Late Triassic therapsids ("mammal-like reptiles") and a very early mammal. (a) A large, 2.5 m herbivore (the dicynodont *Kannemeyeria*); (b, c) two carnivorous cynodontians (*Cynognathus*); and (d) an early mammal, the tiny (approximately 5 cm) *Eozostrodon*.

From its outset 251 million years ago, the Triassic was dominated on land by therapsids. Among these, the sleek, dog-like cynodonts were the chief predators, while the rotund, beaked, and tusked dicynodonts were the most abundant and diverse of herbivores (see Figure 15.3a). From the middle and toward the end of the Triassic, therapsids shared the Earth with squat, armored, plant-eating, archosauromorphs called aetosaurs and some carnivorous crocodile-like archosaurs (see Chapter 5). Dinosaurs appear not to have been important constituents of these Late Triassic ecosystems. Other organisms of the day included primitive turtles

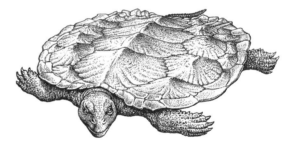

Figure 15.4. The primitive Late Triassic turtle *Proganochelys*.

(Figure 15.4), some crocodile-like amphibians called temnospondyls (Figure 15.5), and pterosaurs (Figure 15.6). Oh yes, and the very earliest mammals, tiny, shrew-sized, insectivorous creatures, were also present (see Figure 15.3d). As it turned out, their appearance on Earth was approximately coincident with – or even slightly preceded – that of dinosaurs. The other Late Triassic terrestrial vertebrate group, whom we met in Chapter 5, were the nondinosaur dinosauromorphs (see Figure 5.5). These were not numerically abundant, but they were very dinosaur-like. In short, Late Triassic terrestrial vertebrate faunas were not dinosaur-dominated; rather they were an eclectic mixture.

At the end of the Triassic, approximately 201 Ma, there was a great change in the fortunes of amniotes.[2] The majority of therapsids went extinct (one highly evolved group of therapsids, mammals, of course survived). And it was only then that dinosaurs somehow rose to become the **dominant** terrestrial vertebrates, by which it is meant that they became the most abundant, diverse, and probably visible group of tetrapods.

Puzzling, though, is *why* dinosaurs prevailed. Ideas, as we have seen, have boiled down to two contradictory theories:

(1) Dinosaurs out-competed their contemporaries, earning the right, as it were, to be the dominant terrestrial vertebrates.
(2) Dinosaurs somehow survived, because their nondinosaurian contemporaries went extinct, leaving a vacated planet open to dinosaur colonization.

Figure 15.5. A temnospondyl (an archaic amphibian) grabbing a snack.

Out-competition – Dinosaurs were Just Superior

In the late 1960s and early 1970s, A. Charig, Curator of Lower Vertebrates at what was then called the British Museum of Natural History, argued that those archosaurs that had the new, "improved"[3] **parasaggital**, that is, fully erect stance (see Chapter 5) were then able to out-compete contemporary predatory, semi-sprawling therapsids for their food sources (see Figure 5.2). The immediate descendants of these flashy new parasaggital archosaurs were the dinosaurs. The inevitable consequence of such progressive improvements in limb posture, Charig argued, was the pattern of faunal succession at the end of the Triassic. Dinosaurs prevailed, according to Charig, by virtue of having better-designed limbs and therefore more efficient terrestrial locomotion.

Figure 15.6. The primitive rhamphorynchid pterosaur *Dimorphodon.*

At nearly the same time, R. T. Bakker was making similar arguments about the competitive superiority of endothermy in dinosaurs (see Chapter 14). He believed that, instead of limbs, it was the achievement of internally produced heat that gave dinosaurs a competitive edge over contemporary and supposedly cold-blooded therapsids and rhynchosaurs. The same conclusions applied – dinosaurs won, therapsids lost, and the truth of the competitive superiority of endotherms over ectotherms could be read directly from the pattern of faunal succession at the end of the Triassic.

Filling the Vacated Planet?

M. J. Benton of the University of Bristol was not convinced that competition explained the Middle to Late Triassic fossil record of the earliest dinosaurs and their predecessors. According to Benton, it did *not*. Instead, he proposed that the fossil record of the last part of the Triassic is marked by not one but two mass extinctions. The earlier one, he argued, was the more extreme and ultimately most relevant

---

[2] This is the notorious Triassic–Jurassic mass extinction, an event that caused a considerable perturbation in all life forms, not just land-dwelling amniotes. How long this extinction took and precisely what caused it remain subject of considerable contention; however, the data are starting to point to a geologically rapid event (lasting, perhaps, a few tens of thousand years), driven by a mid-Atlantic submarine volcanic outpouring of basaltic lava and the wholesale emission into the atmosphere of a variety of not-so-healthy greenhouse-inducing volcanic gases.

[3] Rather transparently, Charig's own preferred term to describe the parasaggital stance of dinosaurs.

to the rise of dinosaurs. As Benton saw it, this earlier Late Triassic extinction completely decimated rhynchosaurs and nearly obliterated dicynodont and cynodont therapsids, as well as several major groups of large, predatory archosaurs.

Likewise, there was a major extinction in the plant realm. The important seed-fern floras (the "*Dicroidium* flora," which contained not only seed-ferns, but also horsetails, ferns, cycadophytes, ginkgoes, and conifers; see below) all but went extinct as well, to be replaced by other conifers and bennettitaleans (see Figures 15.7 and 15.8 later in this chapter). Dinosaurs began to dominate land vertebrate faunas only after the disappearance of therapsids, archosaurs, and rhynchosaurs. Thus the initial radiation of dinosaurs, according to Benton, occurred in an ecological near-vacuum, with the rapid loss of the dominant land-dwelling vertebrates setting the stage for the *opportunistic* evolution of dinosaurs. No competitive edge, because there was no competition.

The proposal of a double extinction remains controversial. The best biostratigraphic and geochronologic data available have narrowed the window for the end-Triassic extinction, so that a single event with a single cause is suggested: the massive Triassic–Jurassic episode of submarine basaltic volcanism known as CAMP (Central Atlantic Magmatic Province). Regardless, about the overall pattern of the extinction, Benton may still be right: and the dinosaurs and other organisms may have inherited a brave new Jurassic world, largely depopulated by noxious events associated with the end-Triassic extinction. If so, then perhaps dinosaurs may have just squeaked by, survivors not because they were somehow superior to the presumed competition, but because they happened to inherit a deserted Earth. Instead of survival having been something intrinsic to dinosaur superiority, it may have been that they simply had better luck.

As we saw in Chapter 5, since the days of Charig and Bakker, the more that we have learned about true dinosaurs by comparison to their very closely related nondinosaur dinosauromorph antecedents (e.g., *Marrasuchus*, *Silesaurus*, *Pseudolagosuchus*, and their cohort; see Figure 5.5), the less unique the features that true dinosaurs had to distinguish themselves. One option, never voiced earlier, was that maybe it was by pure dumb luck that the true dinosaur, barely differing from its nondinosaur dinosauromorph contemporaries, managed to survive and thrive. At that point in the Late Triassic, there may have been no evidence that the thing was going to spawn perhaps the greatest lineage of terrestrial vertebrates the world has ever seen; after all, it is only in hindsight that dinosaurs appear as an important group of animals; they surely didn't appear so in the Late Triassic. There at the beginning, here was this dinosauromorph with a perforate acetabulum; nothing inherently superior, but a character that got carried into Saurischia and Ornithischia, and all that this book is about. In short, perhaps that subsequent post-Late Triassic history is the thing that made the ancestor of all dinosaurs at all noteworthy.

Ironically (as we shall see in Chapter 16), some 164 million years later the tables again turned, and mammals inherited an Earth this time deserted by the very dinosaurs who had taken it from them all those millions of years earlier.

Continental Distributions and the Late Triassic Fauna

What kinds of evolutionary forces might have been driving the distinctive Late Triassic faunas? To start with, the very distributions of the continents likely played a role in the composition of global faunas.

Consider this example from the modern world: the large herbivore fauna of Africa (hippos; rhinos; wildebeests; zebras; African elephant) is rather different from that of North America (Bison, elk, moose). And both differ from that of India (Indian elephants; Bactrian camels; water buffalo; mountain goats). There are no land connections among these continents that would allow the fauna of one to spread to the other. Each of these faunas – in fact, the ecosystems of which they are a part – developed in relative isolation, and is therefore distinct. This type of distinctness is called endemism. A region that is populated by distinct faunas unique to it is said to show *high endemism*. High endemism is caused by evolution on widely separated continents, because there is no opportunity for faunal interchange.

Alternatively, if faunas of two continents appear very similar to each other, then it is likely that some land connection was present to allow the fauna of one continent to disperse to the other continent. Thus we can imagine a region characterized by *low endemism*; because the continents are closely allied with each other, and there are extensive opportunities for faunal interchange.

It is now clear that, during the Triassic and Early Jurassic, global terrestrial vertebrate faunas were characterized by unusually low endemism. The supercontinent of Pangaea still existed during this time, and land connections were more or less continuous among all of today's continents (see Figure 2.5). The interesting mixtures of faunas outlined here look similar on a global scale during these time intervals. While Pangaea remained united, land connections existed and endemism was low. Here, then, is an excellent example of large-scale, geological (abiotic) events driving and modifying large-scale patterns of biological evolution.

For all that, however, we should note that the first dinosaur-bearing faunas are rather patchily distributed, and somewhat enigmatic. The earliest dinosaurs, as we have seen, are known from South America (Argentina and Brazil). These include very primitive saurischians, whose affinities are likely primitive Theropoda and/or Sauropodamorpha (see Chapter 5). In some of those South American faunas, virtually no nondinosaur dinosauromorphs occur, sending the very misleading signal that once the first dinosaurs appeared, they took charge. Yet, in somewhat younger (around 7 myr) North American (US southwest) deposits, relatively advanced theropods are found (suggesting significant dinosaur evolution), happily(?) cohabiting with nondinosaur dinosauromorphs. Did these groups of animals coexist as mixed faunas or did they not? In short, these very earliest records of dinosaurs do not give a definitive signal about the nature of early dinosaur ecosystems; here, then, the incompleteness of the early dinosaur record is surely hampering its interpretation.

Jurassic (201.3 Ma to 145 Ma)

The Early Jurassic (201 Ma to 174 Ma) was the first time on Earth when dinosaurs truly began to dominate terrestrial vertebrate faunas. Indeed, although dinosaurs appeared in the Late Triassic, UC Berkeley paleontologist K. Padian marks the Early Jurassic as the real beginning of the "Age of Dinosaurs." Most of the players in the terrestrial game were now dinosaurs, although they continued to share the limelight with some relict nonamniotes (see Chapter 4; Figure 15.5), a few of the very highly derived, mammal-like therapsids (including some puny mammals; Figure 15.3), turtles (who, by now, appeared very nearly like their modern descendants), pterosaurs (Figure 15.6), and the newly evolved crocodilians. Interestingly, the Early Jurassic faunas retained much of the low endemism that characterized the Late Triassic world. The unzipping of Pangaea was in its very earliest stages, and it had not gone on so long that the fragmentation of the continents was yet reflected through increased global endemism.

The Middle Jurassic (174 Ma to 164 Ma) has historically been an enigmatic time in the history of terrestrial vertebrates. As noted earlier, Middle Jurassic terrestrial sediments are quite uncommon, particularly in North America. When we look at the total diversity of tetrapods through time (Box 15.1), the curve all but bottoms out during the Middle Jurassic. Did vertebrates undergo massive extinctions at the end of the Early Jurassic? Almost certainly not. More likely the curve is simply reflecting the serendipitous absence of terrestrial Middle Jurassic sediments on Earth. Without a good sedimentary record to preserve them, we can have little knowledge of the faunas that came and went during that time interval.

Regardless, the Middle Jurassic must have been a very important time in the history of dinosaurs. With the dismemberment of Pangaea well underway by this time, dinosaurs had diversified, and endemism was on the rise. Many of the nondinosaurian tetrapods that characterized earlier faunas – advanced therapsids, for example – were largely out of the picture. The insignificant exceptions to this, of course, were mammals, hanging on by the skin of their multi-cusped, tightly occluding teeth. The Middle Jurassic must have been a kind of pivot point in the history of dinosaurs, because it was then that most of the major dinosaur groups – sauropods, large theropods, thyreophorans, and ornithopods – assumed their familiar forms and consolidated their hold on terrestrial ecosystems. It's a shame that we cannot know more of this crucial time.

By the Late Jurassic (164 Ma to 145 Ma; see Figure 2.6), global climates had stabilized and were generally warmer and more equable (less seasonal) than they presently are (see Chapter 2). Polar ice, if present, was reduced. Sea levels were higher than today. Dinosaur faunas were more endemic than ever before.

The Late Jurassic has been called the Golden Age of Dinosaurs.[4] With the Middle Jurassic as poorly known as it is, dinosaurs appear to have sprung into high diversity in the Late Jurassic. Many of the dinosaurs that we know and love were Late Jurassic in age. Somehow that special Late Jurassic blend of high sea levels, equable climates, small brains, and massive size epitomizes dinosaur stereotypes and exerts a fascination on humans. Many *were* large – gigantic sauropods (*Brachiosaurus, Diplodocus, Camarasaurus*, among others) as well as theropods that reached upward of 16 m – but many were not (for example, *Compsognathus*). It was during the Late Jurassic that (comparatively tiny) avialans like *Archaeopteryx* appeared. Moreover, this was the time of stegosaurs, early iguanodontians, avialans, and even a few ankylosaurs. By Late Jurassic time, dinosaurs had consolidated their dominance of terrestrial vertebrate faunas.

Cretaceous (145 Ma to 66.1 Ma)

The Early Cretaceous (145 Ma to 100.5 Ma) was a time of enhanced global tectonic activity. With this came increased continental separation, as well as greater amounts of CO_2 in the atmosphere, producing "greenhouse" climates. Climates from the Early through mid Cretaceous (to about 96 Ma) were therefore warmer and more equable than today (see Chapter 2).

Who enjoyed these balmy conditions? Representatives of groups including all of our old friends from earlier times as well as a number of new groups of dinosaurs made their appearance. The Early Cretaceous marks the rise of the largest representatives of euornithopods. Ankylosaurs also became a significant presence among herbivores of the Early Cretaceous times, as did the earliest ceratopsians.

Moreover, the balance of the faunas seems to have changed. During the Late Jurassic, sauropods and stegosaurs were the major large herbivores, with ornithopods represented primarily by smaller members of the group. Now, in the Early Cretaceous (and, in fact, throughout the Cretaceous), euornithopods begin to make their mark. Sauropods continued their success story, and stegosaurs were still present (barely), but evidently on their way out. Was the spectacular Cretaceous ascendency of Ornithopoda due to the feeding innovations developed by the group? The parallel success of

---

[4] If only because Late Cretaceous dinosaurs weren't fully appreciated in the latter part of the 1800s when people began to imagine a "golden age" for dinosaurs!

ceratopsians in Late Cretaceous time, and the independent invention by that group of similar feeding innovations, suggest that sophisticated chewing didn't hurt. Then, too, the Early Cretaceous witnessed a revolution in small carnivorous theropods, most notably the deinonychosaurids and troodontids of both North America and Asia.

During the Late Cretaceous (100.5 Ma to 66.1 Ma; see Figure 2.7), never before in their history had Dinosauria been so diverse, so numerous, and so incredible. Something happened; there was a bump in diversity without parallel in dinosaur history.[5] The mid to Late Cretaceous boasted the beefiest terrestrial carnivores in the history of the world (North America and Asia – tyrannosaurids; South America and Africa – carcharodontosaurids), a host of sickle-clawed brainy (*and* brawny) killers worthy of any nightmare, and herds of horned herbivores, honking hadrosaurids, ambling ankylosaurs, dome-heads, therezinosaurs, the truly titanic titanosaurs...

Climate seems not to have affected diversity. In fact, although diversity increased, from the mid Cretaceous time onward, seasonality gently increased. This occurred at the same time as a marine regression, which undoubtedly played a role in the destabilization of climate. At the very end of the Mesozoic, there is no evidence for a sudden drop in temperatures, or any significant modification of climate; there was, however, a temporary increase in global temperatures just before the Cretaceous–Paleogene (Tertiary) boundary. This has been attributed to the effects of Deccan volcanism (see Chapter 17).

Endemism

Southern continents tended to maintain the veteran Old Guard: a large variety of sauropods (but some extraordinary new and large versions of the Old Guard; Bigger! Better!), some ornithopods and ankylosaurs, and, in South America, the abelisaurid theropods, a group representative of a somewhat primitive theropod stock. In northern continents, however, new, very different faunas appeared. Among herbivores, sauropods were still present, although oddly absent in northern North America. Lots of new creatures roamed, including potentially migrating herds of pachycephalosaurs, ceratopsids, and hadrosaurids. Europe, by contrast, consisted of an island archipelago, which dramatically increased endemism in the Old World.

Finally there is the magnificent diversity of Late Cretaceous theropods. Nothing shaped quite like tyrannosaurids had ever been seen, or has existed since. Yet, the Late Cretaceous theropod story might be better told in the diversity of smaller forms: oviraptorosaurs, alvarezsaurids, dromaeosaurids, ornithomimosaurs, troodontids, and therizinosaurs.

Across the Bering Straits?

North America and Asia share a rich Late Cretaceous record, including ceratopsians, tyrannosaurids, and ornithomimosaurs. Because of this, there may have been multiple migrations of herds of dinosaurs across a Bering land bridge throughout much of the Late Cretaceous (see Figure 11.35), just as humans and other Ice Age mammals are thought to have migrated to North America from Asia many tens of millions of years later.

---

[5] This conclusion, that dinosaur diversity increased throughout the Late Cretaceous, was challenged in 2016 in a controversial paper by Sakamoto and colleagues. They argued that during the last 50 million years of the Cretaceous, the rate of origination of dinosaur groups was exceeded by the rate of their extinction; that, therefore, dinosaurs were a group aiming towards extinction from 50 million years before the end of the Cretaceous, until its conclusion, the Cretaceous–Paleogene mass extinction. Most paleontologists would counter, however, that the diversity of dinosaurs was never greater than in the Late Cretaceous. What then, to conclude about the Sakamoto *et al.* study? See: Sakamoto, M., Benton, M. J., and Venditti, C. 2016. Dinosaurs in decline tens of millions of years before their final extinction. *Proceedings of the National Academy of Sciences*, 113 (18), 5036–5040. www.pnas.org/cgi/doi/10.1073/pnas.1521478113

And then, in the earliest Cenozoic, it was over. Just like that. While one of the great enigmas of dinosaur paleontology has historically been how the animals went extinct, the problem has begun to yield to concentrated study over the past 35 years. We'll save that story for Chapter 17.

After the Ball is Over

With the end of the Cretaceous, nonavian dinosaurs disappeared from Earth forever, and it definitely *was* the end of an Era. Mammals, well entrenched as the dominant terrestrial vertebrates in the Cenozoic, would be no more likely to give up their place in Tertiary ecosystems to dinosaurs than dinosaurs had been likely throughout the approximately 164 million years of their incumbency to give up *their* place to mammals!

Or so the story is told by we mammals. But remember: in today's world, some 66 million years after the nonavian dinosaurs supposedly became extinct, there are just under 5500 species of mammals. By contrast, there are somewhere around 10 000 species of avian dinosaurs (birds) alive today. Can it really be said that we've left the Age of Dinosaurs and entered the Age of Mammals?

Plants and Dinosaurian Herbivores

As in most extant terrestrial mammalian communities, the majority of dinosaurs were herbivorous. If dinosaurs were numerous enough, and their impact on terrestrial ecosystems was important enough, so the thinking goes, there ought to be some relationship between herbivorous dinosaur evolution and plants. To address this proposition, we begin by looking at plants.

Plants

Most paleobotanists – people who study extinct plants – recognize two major groupings of Mesozoic plants. The first is a nonmonophyletic cluster of plants including ferns, lycopods, and sphenopsids (Figure 15.7). All of these plants tend to be low-growing and primitive, but, like most land plants, they are vascular; that is, they possess specialized tissues that conduct water and nutrients throughout the plant.

The second major grouping of plants consists of gymnosperms and angiosperms. Together these two groups are united by the diagnostic character of possessing a seed (see Figure 15.8, inset). Seeds are ultimately nutrient-bearing pods apparently developed for the dissemination of gametes. Gymnosperms are today best known as pines and cypress, and a lesser-known but Mesozoic-ly important group known as cycadophytes: plants with large, pineapple-like trunks and bunches of leaves springing out of their tops. Angiosperms, the flowering plants, today consist of magnolias, maples, grasses, roses, and orchids, among many other groups (Figure 15.8).

Several qualities distinguish these plant groups. In general, Mesozoic gymnosperms tended to be of three types: conifers, cycadophytes, and ginkgoes. Conifers – epitomized, for example, by pines – were very tall and woody plants. They had relatively little nutritive value pound for pound, possessing coarse thick bark and cellulose-rich leaves. The modern representatives of these plants tend to secrete a variety of ill-tasting or poisonous compounds as a strategy to discourage their consumption; there is no reason to suppose that their Mesozoic counterparts were any different.

Figure 15.7. Representative ferns, lycopods, and sphenopsids from the Mesozoic. Club moss: (1) *Pleuromeia* (Early Triassic). Ferns: (2) *Matonidium* (Jurassic–Cretaceous); (3) *Onychiopsis* (Jurassic–Early Cretaceous); (4) *Anomopteris* (Middle–Late Triassic); (5) Osmundaceae (Late Paleozoic–Recent), (6) Tree fern (Jurassic). Sphenopsids: (7) *Equisetum* (Late Paleozoic–Recent); (8) *Neocalamites* (Triassic–Lower Jurassic); (9) *Schizoneura* (Late Paleozoic–Jurassic).

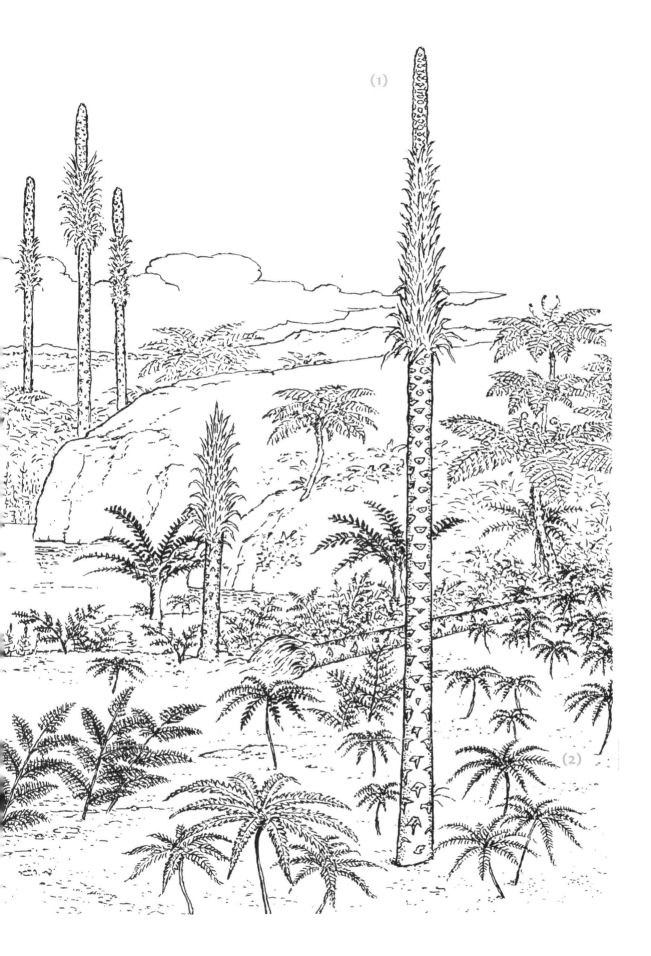

Figure 15.8. Representative cycads, ginkgoes, gymnosperms, and angiosperms from the Mesozoic. (1) Cycadeoids or bennettitaleans, as they are sometimes called (Triassic–Cretaceous); (2 and 5) *Williamsoniella* spp. (Triassic–Jurassic); (3) *Wielandiella* (Jurassic); (4) *Williamsoniella sewardiana* (Jurassic). Ginkgo: (6) *Ginkgoites* (Triassic–Recent). Conifers: (7) *Sequoia* (mid Cretaceous–Recent); (8) *Araucaria* (Late Triassic–Recent); (9) *Pagiophyllum* (Triassic–Cretaceous). Angiosperms: (10) Magnoliaceae (magnolias; still small-flowered in the early days of their appearance on Earth; Cretaceous?–Recent); (11) Nymphaeaceae (water lilies; Late Cretaceous–Recent). *Inset*: Seed (dicot) in cross-section. The cotyledons, shoot apex, root apex, and suspensor are all parts of the embryonic plant. The endosperm is a food source for the embryo as it develops, and the seed coat protects the embryo and its food source

(9)

(6)

(10)

(11)

(3)

Shoot apex

Seed coat

Root apex

Suspensor

Endosperm

Cotyledons

Cycadophytes, on the other hand, tended to be fleshier and softer, with perhaps more nutritive value. Ginkgoes would also have been plants available for dinosaur consumption, and circumstantial evidence suggests they too were eaten by Mesozoic herbivorous dinosaurs (Figure 15.8).

Flowering plants evolved an entirely different approach to life from gymnosperms. Far from discouraging herbivores from consuming them, they evolved a variety of strategies to actively seduce herbivores into consuming them: bright tasty flowers, fruits with tough seeds that can survive a trip through a digestive tract. Consumption by herbivores in the case of angiosperms appears to be a strategy for seed dispersal, not the destruction of the plant.

Dinosaurs and Plants

Figure 15.9 compares the record of Late Triassic through Late Cretaceous plant diversity with that of dinosaurian herbivores. The lower part of the figure gives approximations of the global composition of plants through the time of the dinosaurs. The upper part of the figure is divided into various groups of herbivorous dinosaurs.

Plants

In terms of plants, Figure 15.9 shows some key patterns. Lycopods, seed ferns, sphenopsids, and ferns decrease in global abundance during the Late Triassic interval. From then until the end of the Mesozoic, they constitute a roughly constant proportion of the world's floras. Not so with the gymnosperms, which dramatically increase their proportion of the total global flora during the Late Triassic. And it is clear that much of that increase is taken up by conifers, which constitute around 50 percent of the world's total floras throughout the rest of the Mesozoic.

Our best guess is that angiosperms first evolved in the very early part of the Cretaceous; however, it was during mid Cretaceous times that they underwent a tremendous evolutionary burst. The uniquely efficient angiosperm seed dispersal mechanisms afforded by flowers and their various dispersing agents – animal fur, insects, feces, you-name-it! – were (and are) unparalleled in the botanical world, and consequently flowering plants have literally blossomed as no other group of plants has.

Coevolution

It cannot be purely by chance that the rise of tall coniferous forests is coincident with the appearance on Earth of the world's first tall herbivores: prosauropods (and later sauropods). Here we see the possibility of coevolution, the evolution of one group affecting – and even effecting – the evolution of another. In this case, it's plants and dinosaurs: were those tall prosauropods favored by natural selection that could take advantage of comparatively succulent leaves at the tops of conifers? Alternatively, were conifers that were particularly tall favored by natural selection in response to the increasing height of prosauropods? Which is cause and which is effect is something we'll likely never know.

The figure also reveals another potentially compelling relationship. The rise of the angiosperms occurs at approximately the same time as several major radiations of dinosaur groups. Did these groups – ceratopsians, pachycephalosaurs, hadrosaurids, and late-evolved ankylosaurs – somehow take advantage of angiosperms as a food source and diversify? Is this a clue to what these dinosaurs were eating (Box 15.3)?

It appears that, during the middle of the Mesozoic, few vertebrates fed very *selectively* upon the relatively slow-growing conifers, cycads, and ginkgoes that formed the majority of terrestrial floras. Instead, it has been suggested, dinosaur feeding consisted of low browsing and was rather

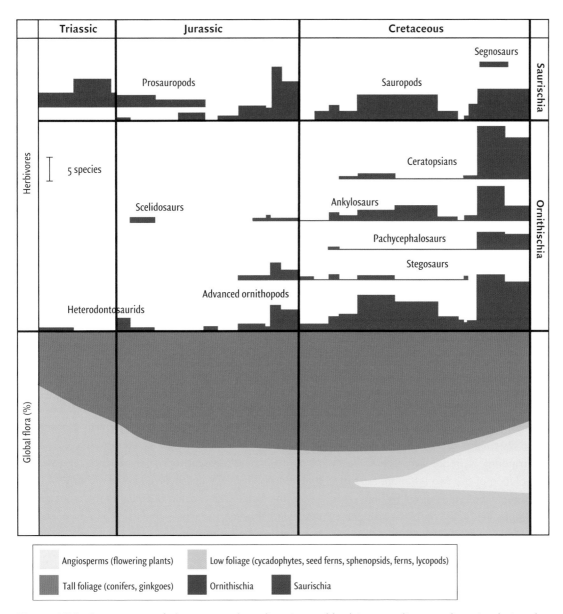

Figure 15.9. Comparison of changes in plant diversity and herbivorous dinosaur diversity during the Late Triassic through Late Cretaceous time interval. The upper part of the diagram shows diversity of major groups of herbivorous dinosaurs through time. Comparison between this diagram and that shown in Figure 15.2 suggests that one of the most important things driving dinosaur evolution and diversity was the development of new (or improved) ways to exploit the environments in which they lived (see the text).

generalized (similar to the way a lawn-mower "grazes" over whatever is in its path). Because so many of these Mesozoic herbivores were also very large and may have lived in large herds, they likely cleared expansive areas, trampling, mangling, uprooting, and otherwise disturbing areas that otherwise might be colonized by plants.

Such low-level, generalized feeding and disturbances of habitats tended to emphasize fast growth in plants, but discouraged the establishment of seed-dispersal relationships between plants and

Box 15.3 Dinosaurs Invent Flowering Plants... or at Least Fuel Their Evolution?

In his 1986 popular book *Dinosaur Heresies*, R. T. Bakker proposed that dinosaurs "invented" flowering plants. The germ behind Bakker's hypothesis is that Late Jurassic herbivores, epitomized by sauropods, were essentially high-browsers, while Cretaceous herbivores, epitomized by ornithopods, ankylosaurs, and ceratopsians, were largely low-browsers. Bakker argued that Cretaceous low-browsers put tremendous selective pressures on existing plants, so that survival could occur only in those plants that could disseminate quickly, grow quickly, and reproduce quickly. Angiosperms, he argued, are uniquely equipped with those capabilities. In his scenario, Bakker has Cretaceous low-browsing dinosaurs eating virtually all the low shrubbery, and plants responding by developing a means by which animals simply couldn't keep up with the growth, reproduction, and dissemination of the plants. The idea, here, is that dinosaurs drove plant evolution.

Since Bakker's original proposal, comprehensive, quantitative treatments of this idea have been given by a variety of paleontologists in several publications.[a] These authors noted that although the coincidence of dinosaurs and angiosperms is impressive, actual data supporting the coevolution of the two are more limited. We have seen that the most obvious direct evidence of dinosaurian diets – "mummies" and coprolites – all point to a gymnosperm-rich diet. In fact, because angiosperms were likely rare in the Early Cretaceous, it is only in the Late Cretaceous that angiosperms were numerous enough to have been an important constituent of dinosaur diets.

The issue is more nuanced. While the dinosaurs and angiosperms appear to have coevolved in time, if one incorporates *space* as well as time, there appears to be a discordancy; where the angiosperms appear seems not to be where the dinosaurs are recorded as having radiated.

Other herbivores almost certainly had at least as great an impact on angiosperm evolution. Strikingly, insects underwent a major radiation in the mid Late Cretaceous, and a variety of flower-loving pollinating forms evolved at that time, including bees, wasps, butterflies, and flies of many types. If coincidence of appearance is the evidence for the herbivorous dinosaur–angiosperm relationship, then surely an equal or stronger argument can be made for the appearance of all these flower-loving insects co-evolving with angiosperms. Other potentially important influences were mammals and birds; however, in all of these cases, evidence for direct interactions is sparse.

Another important potential influence on angiosperm evolution has also been proposed: CO_2 in the atmosphere. Recall that atmospheric CO_2 was high during the mid Cretaceous; this also correlates strongly with the angiosperm radiation. As high levels of atmospheric CO_2 fuel plant growth and the turnover of generations, it is possible that atmospheric CO_2 was involved in the mix as well.

So, the simple relationship between herbivorous dinosaurs and flowering plants is in fact a mighty complex one – surely far more complex than we have even portrayed here; and as we have seen, the balance of the evidence appears to suggest that dinosaurs had very little to do with the rise and spread of flowering plants, whether we love them (dinosaurs and flowering plants) or not!

[a] Here are two samples:

Barrett, P. M. and Willis, K. J. 2001. Did dinosaurs invent flowers? Dinosaur-angiosperm coevolution revisited. *Biological Reviews*, **76**, 411–447.

Butler, R. J., Barrett, P. M., Penn, M. G., and Kenrick, P. 2010. Testing coevolutionary hypotheses over geological timescales: interactions between Cretaceous dinosaurs and plants. *Biological Journal of the Linnaean Society*, **100**, 1–15.

animals. Thus the picture of Mesozoic plant–herbivore interactions appears to be one in which (a) plants produced vast quantities of offspring to ensure the survival of the family line into the next generation and (b) herbivores took advantage of the rapidly and abundantly reproducing resource base to maintain their large populations of large individuals. Plant–herbivore coevolution during the Mesozoic appears to have been based on habitat disturbance, generalized feeding, and rapid growth and turnover among plants.

This still doesn't really demonstrate whether angiosperms and dinosaurs coevolved; that is, each affected the trajectory of the other's evolution. Where we can actually track food sources, surprisingly perhaps, the evidence does not argue strongly for dinosaur–plant coevolution (Box 15.3).

For example, "mummified" remains of hadrosaurids (*Edmontosaurus* and *Corythosaurus*; see Figure 12.7) do not show the remains of angiosperms in the digestive tract, but rather the remains of coniferous plants. Late Cretaceous coprolites, reliably attributed (by size!) to either ceratopsids or hadrosaurids, contained conifer fragments as well. If angiosperms were fueling this dinosaur radiation, where are the angiosperm pieces (at least, seeds) that we might hope to find?

Dinosaur chewing efficiency increased markedly through the latter part of the Mesozoic. This is not to say that nonchewing dinosaurs were in a state of decline; as Figure 15.9 shows, sauropods – for whom, in most cases, chewing was by and large a minimalist artform – were successful throughout the Cretaceous, blooming, as it were, in the waning years of the Mesozoic. Moreover, animals that indulged in rudimentary chewing – such as ankylosaurs and pachycephalosaurs – underwent strong evolutionary bursts during the latter part of the Cretaceous. Far less abundant were the sloth-like therizinosaurs (see Chapter 7). Yet, was there something about the rise of angiosperms that fueled their unusual radiation too? Measured by the chewing standards of their ornithopod brethren, these animals were mighty primitive. But, as the Late Cretaceous global abundance (excepting North America) of sauropods, particularly titanosaurs, has shown, chewing is not the only way to be a herbivore.

Still, ceratopsids and hadrosaurids – groups that elevated chewing to new heights – are characteristic of the Late Cretaceous radiation. Did advanced chewing mechanisms allow hadrosaurids and ceratopsids to take advantage of food resources not heretofore available to other dinosaurs? Perhaps, but the evidence suggests that indiscriminate food selection and sophisticated chewing might have been ways to extract every last nutrient out of the nutrient-poor gymnosperms, rather than a means to access low-hanging, mouth-watering, and nutritious angiosperm fruit.

To actively drive plant evolution would take a significant total biomass, perhaps more than the dinosaurs – for all their large size as individuals – could muster. Think *really* large total biomass: like perhaps insects! And even then, more might have been involved (see Box 15.3). So overall, despite the coincidence of a variety of herbivorous dinosaurs and the rise of angiosperms, it is by no means clear that dinosaurs fueled the angiosperm radiation – even if they nibbled an apple a day.

The Late Cretaceous can rightly lay claim to being the "Golden Age of Dinosaurs"; as most dinosaurs of the time were herbivorous, it seems that, at a minimum, they themselves flowered during the flowering of the Mesozoic.

SUMMARY

Here we look at the overall sweep of nonavian dinosaur evolution. Factoring in time intervals of a poor geological record in which preservation is artificially low, dinosaurs as a group increased markedly in number and diversity, particularly during the Late Jurassic through latest Cretaceous time interval. This increase is attributable to ceratopsian and ornithopod herbivores, and theropods.

The global pattern of dinosaur evolution from the Late Triassic to the Late Cretaceous is one of generally increasing endemism, likely attributable to the increasing separation of continental masses. Late Triassic and Early Jurassic dinosaur faunas shared their terrestrial world with a variety of other vertebrates; and global vertebrate faunas were relatively homogeneous. Distinct among all the Late Triassic and Early Jurassic vertebrates, however, dinosaurian herbivores were the first to be able to reach, and thus add to their diets, tall foliage.

By Middle Jurassic time, dinosaurs likely consolidated their dominance in the terrestrial realm, even though terrestrial deposits from this time interval are comparatively rare. This fact is especially unfortunate for understanding the details of dinosaur evolution, since many of the groups that became so abundant and diverse in the Cretaceous had their roots in the Middle Jurassic. The Late Jurassic has been called, probably incorrectly, the "Golden Age of Dinosaurs," with the abundance of many familiar forms including very large theropods and sauropods.

Although the dinosaurs likely didn't drive Mesozoic plant evolution, it is probable that, as plants evolved effective methods for dispersal and colonization, dinosaurs hitched a ride, increasing markedly in number and diversity as they took advantage of the radiation of vascular plants.

The Cretaceous was a truly astounding time in dinosaur evolution. Aside from the wholesale dominance of new forms (in particular, ornithopods and ceratopsians, as well as a wide range of theropods), many of the spectacular adaptations that we've seen, such as advanced chewing, evolved in the Cretaceous. A driving force in all this evolutionary ferment may have been the rise of flowering plants; yet what we know of dinosaur diets suggests that the fibrous gymnosperms constituted the bulk of the nutrition.

SELECTED READINGS

Bakker, R. T. 1986. *The Dinosaur Heresies*. William Morrow and Company, New York, 481 pages.

Carrano, M. 2012. Dinosaurian faunas of the later Mesozoic. In Brett-Surman, M. K., Holtz, T. R., and Farlow, J. O. (eds.) *The Compete Dinosaur*, 2nd edn. Indiana University Press, Bloomington, IN, pp. 1003–1026.

Chin, K. 2012. What did dinosaurs eat: coprolites and other direct evidence of dinosaur diets. In Brett-Surman, M. K., Holtz, T. R., and Farlow, J. O. (eds.) *The Compete Dinosaur*, 2nd edn. Indiana University Press, Bloomington, IN, pp. 589–601.

Dodson, P. 1990. Counting dinosaurs. *Proceedings of the National Academy of Sciences*, **87**, 7608–7612.

Fastovsky, D. E. 2000. Dinosaur architectural adaptations for a gymnosperm-dominated world. In Gastaldo, R. A. and DiMichele, W. A. (eds.) *Phanerozic Terrestrial Ecosystems*. The Paleontological Society Papers, vol. 6, pp. 183–207.

Fastovsky, D. E., Huang, Y., Hsu, J., *et al.* 2004. The shape of Mesozoic dinosaur richness. *Geology*, **32**, 877–880.

Sues, H.-D. 2012. Early Mesozoic continental tetrapods and faunal changes. In Brett-Surman, M. K., Holtz, T. R., and Farlow, J. O. (eds.) *The Compete Dinosaur*, 2nd edn. Indiana University Press, Bloomington, IN, pp. 988–1002.

Tiffney, B. H. 1989. Plant life in the age of dinosaurs. *Short Courses in Paleontology*, **2**, 34–47.

Tiffney, B. 2012. Land plants as a source of food and environment in the Age of Dinosaurs. In Brett-Surman, M. K., Holtz, T. R., and Farlow, J. O. (eds.) *The Compete Dinosaur*, 2nd edn. Indiana University Press, Bloomington, IN, pp. 569–587.

Wang, S. C. and Dodson, P. 2006. Estimating the diversity of dinosaurs. *Proceedings of the National Academy of Sciences*, **103**, 13 601–13 605.

Weishampel, D. B. and Norman, D. B. 1989. Vertebrate herbivory in the Mesozoic: jaws, plants, and evolutionary metrics. *Geological Society of America Special Paper*, no. 238, pp. 87–100.

Wing, S. and Tiffney, B. H. 1987. The reciprocal interaction of angiosperm evolution and tetrapod herbivory. *Review of Palaeobotany and Palynology*, **50**, 179–210.

TOPIC QUESTIONS

1. What is meant by the words conifer, coevolution, cycadophyte, angiosperm, diversity, and endemism?
2. What is the general pattern of dinosaur diversity through time? When were dinosaurs at their most diverse? When were they at their least diverse?

3. On what continents are the most dinosaurs found? Why might this be?
4. Describe climatic conditions throughout the Mesozoic.
5. Describe the degree to which the continents were separated throughout the Mesozoic. When were the continents most like they are now? When were they least like they are now?
6. What is the relationship between endemism and the distribution of continents? Why is this so?
7. Describe the general outlines of plant evolution through the Mesozoic.
8. What is the general relationship between dinosaur diversity and plant evolution through the Mesozoic?
9. What is the relationship between herbivore chewing specializations and plant diversity in the Mesozoic?
10. Drawing on material from other chapters, can you think of highly evolved behaviors that appear to be related to plant diversity?
11. What kinds of evolutionary changes characterize Theropoda in the context of plant diversity and increased ornithischian diversity in the Mesozoic?

A History of Dinosaur Paleontology Through the Ideas of Dinosaur Paleontologists

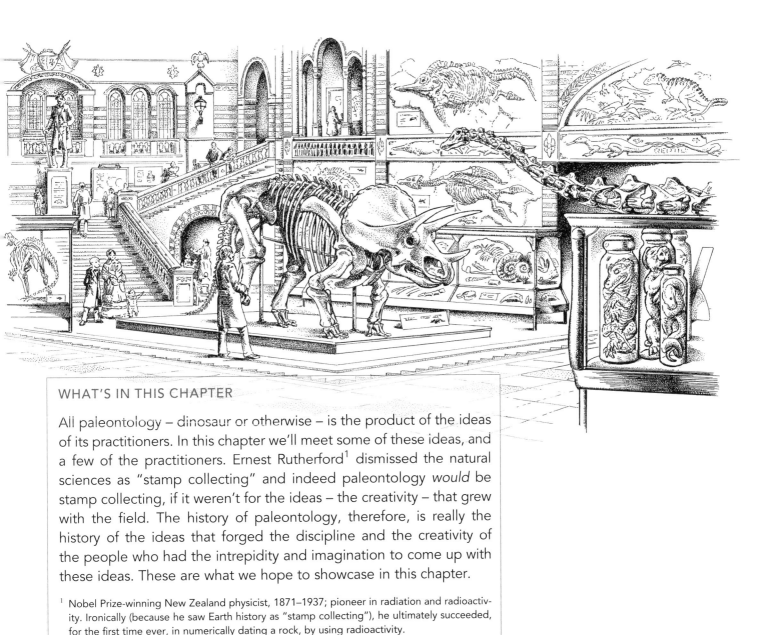

WHAT'S IN THIS CHAPTER

All paleontology – dinosaur or otherwise – is the product of the ideas of its practitioners. In this chapter we'll meet some of these ideas, and a few of the practitioners. Ernest Rutherford[1] dismissed the natural sciences as "stamp collecting" and indeed paleontology *would* be stamp collecting, if it weren't for the ideas – the creativity – that grew with the field. The history of paleontology, therefore, is really the history of the ideas that forged the discipline and the creativity of the people who had the intrepidity and imagination to come up with these ideas. These are what we hope to showcase in this chapter.

[1] Nobel Prize-winning New Zealand physicist, 1871–1937; pioneer in radiation and radioactivity. Ironically (because he saw Earth history as "stamp collecting"), he ultimately succeeded, for the first time ever, in numerically dating a rock, by using radioactivity.

Figure 16.1. The Mantells. (a) Mary Ann (Woodhouse) Mantell (1795-1847), Gideon's wife, whose sharp eyes actually spotted the tooth fossil (inset). Gideon became fixated on dinosaurs; found and described many more, and built a museum to showcase them. His monomania ultimately cost him his medical practice and his wife. (b) Gideon Mantell (1790–1852), the first westerner who recognized nonavian dinosaurs for what they were.

In the Beginning

Tradition usually identifies the beginning of dinosaur paleontology as 1822, when Englishwoman Mary Ann Mantell found large teeth along a Sussex country lane (Figure 16.1a). She brought them home to her physician-husband, Gideon Mantell (Figure 16.1b), who was something of a naturalist, and the discovery baffled him. The teeth looked very much like those of the living herbivorous lizard *Iguana*, but were ominously much, much bigger (Figure 16.2).

The Mantells may have been the first to interpret fossils meaningfully in a Western scientific context, but they weren't the first humans to see dinosaur fossils. Fossils of all types have been remarked upon for as long as there have been humans.

An earlier example of humans trying to interpret fossils comes from Adrienne Mayor, classical folklorist and historian of science, who has reconstructed the origin of the legend of the griffin, sharp-beaked, winged, four-legged creatures whose mythology was known across all of Europe and Asia (Figure 16.3). Her idea is that traders as early as the seventh century BCE, along ancient gold-trading caravan routes stretching from Europe through central Asia encountered abundant, beautifully preserved fossils of *Protoceratops* (see Chapter 11), whose strange (to them) combination of beak, frill, and limbs were explained as the mythical griffin's beak, wings, and legs. The richness of the Asian deposits was revealed more than a thousand years later in the American Museum of Natural History's Central Asiatic fossil Expeditions of the 1920s (Box 16.1).

Figure 16.2. The Mantells' *Iguanodon* teeth.

Seventeenth and Eighteenth Centuries

With a few exceptions, the birth of the Western scientific tradition is generally reckoned to have occurred in association with the seventeenth-and eighteenth-century intellectual revolution called the Enlightenment. The Enlightenment brought with it a number of scientific conclusions important to our story, including:

- The Earth is not static, that is, it has changed through time.
- The Earth of is of great antiquity (its age was not well understood until the mid twentieth century).
- The sequence of the rock record reveals the history of the Earth.
- Fossils are the remains of once-living organisms.
- Organisms on Earth were not static, they too had clearly changed through time, some in startling ways.

The Enlightenment represents the beginning of western science as it is currently understood, and it was during this time that people began to attempt to explore the natural world using observation and logic. As fossils began to accumulate in natural scientists' collections, it became clear that the Earth had been dramatically different in the past. An early attempt to come to grips with what appeared inexplicable was the first proper description of a dinosaur fossil, in this case the lower end of a theropod thigh (likely *Megalosaurus*) from Oxfordshire, England. As the bone was large, it was interpreted

(a)

(b)

Figure 16.3. (a) A griffin. (b) *Protoceratops,* as mounted at the American Museum of Natural History in New York.

Box 16.1 Indiana Jones and the Central Asiatic Expeditions of the American Museum of Natural History

There he is, in the middle of the remote, rugged, Mongolian desert: high leather riding boots, riding pants, broad-brimmed felt hat, leather-holstered sidearm hanging from a glittering ammunition belt. He carries a rifle and knows how to use it. Nobody else dresses like him, but then nobody else is Roy Chapman Andrews, the leader of the American Museum's Central Asiatic Expeditions to Mongolia in the 1920s. Fifty years later, he will be the inspiration, it is most plausibly rumored, for Indiana Jones (Figure B16.1.1).

His idea was simple: to search in Mongolia for fossils revealing the origin of humans. The logistics of the expedition were complex: Dodge cars, resupplied by a caravan of camels, would bear the brunt of exploring the Gobi Desert, the huge desert that forms the vast southern section of Mongolia and northern China (see Figure 1.9).

No ancient humans were found, but Andrews' crews brought back dinosaur fossils, so rich and unprecedented, that the origin of humans was quietly forgotten. Among the most famous dinosaur finds were *Protoceratops* (the species name of this famous dinosaur is *andrewsi*) and *Oviraptor* eggs – the first time that dinosaur eggs were ever found. Other finds included *Velociraptor* and a group of tiny Mesozoic mammals (still the rarest of the rare). It was only in 1992 that a specimen of *Mononykus*, collected by Andrews' scientists in the 1920s, was finally correctly identified (see Figure 7.14). All in all, it was quite a haul.

Figure B16.1.1. Roy Chapman Andrews (1884–1960), explorer, adventurer, and leader of what, in those days of unabashed imperialism, he revealingly called "The New Conquest of Central Asia."

by the Reverend Dr. Robert Plot in 1677 to have been the end of a thigh bone of an antediluvian (pre-Biblical Flood) giant – man or beast (Figure 16.4).[2]

The Nineteenth Century Through the Mid Twentieth Century

Dinosaurs in the Victorian Age

Mantell's discovery turned him into the western world's first true dinosaur junkie. But he was hardly alone. Perhaps it had to do with the Victorian penchant for collections and museums, perhaps it was just the novelty of the beasts being uncovered, but Victorian England was dino crazy. In 1824, the

---

[2] Stranger still, in 1763, Richard Brooke redrew the specimen in a publication on the uses of various natural objects (including fossils) in medicine. The specimen appeared to Brooke to preserve a giant's testicles; hence, his Latin description identified the fossil as "*scrotum humanum.*" It has been suggested that, tongue firmly in cheek, the rule of priority in the Linnaean classification (see Chapter 3) dictates that this first dinosaur bone should be referred to a genus *Scrotum*, species *humanum.*

natural historian William Buckland (1784–1856) described a jaw fragment with a single recurved, serrated tooth as *Megalosaurus*. This was the first formally named dinosaur,[3] now known to be a theropod, but thought by Buckland to be a rather large lizard. In 1842, English anatomist Sir Richard Owen (Box 16.2) invented the term Dinosauria, reflecting a growing realization among Victorian natural historians that a group of large extinct reptiles had once populated the Earth.

As the number of dinosaur discoveries continued to increase, Victorians immortalized their conception of dinosaurs with a variety of images and sculptures. The dinosaurs were reconstructed as large, heavy-set quadrupeds, the most famous of which were created life-sized in plaster and tile by an English sculptor, Benjamin Waterhouse Hawkins (1807–1889), on the occasion of the reopening of the Crystal Palace in 1854. Sureally, on New Year's Eve, 1853, Owen and Waterhouse Hawkins hosted a dinner for England's best and brightest *inside* the unfinished sculpture of *Iguanodon* (Figures 16.5 and 16.6).

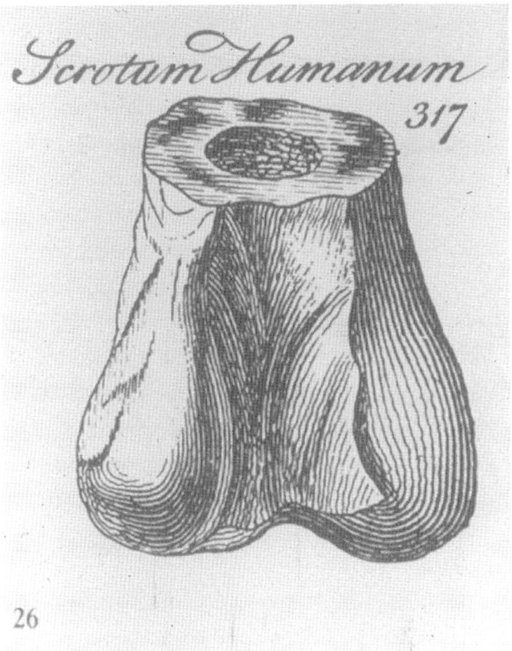

From its 1842 inception onward, membership of Owen's Dinosauria grew by leaps and bounds. Much of the attention was devoted to basic collecting and description, asking questions such as "What is this creature? A new genus? A species of an existing genus? Maybe even a new family?". This otherwise-healthy penchant for discovery, description, and naming reached absurd levels during the latter half of the century, fueled by the

Figure 16.4. Robert Plot's drawing of the lower end of a *Megalosaurus* (?) thigh bone. Dr. Plot was the first Professor of Chemistry at the University of Oxford. For explanation of the Latin inscription, see footnote 2.

extraordinarily rich fossil beds of the North American West and the not-so-healthy competition between Yale's O. C. Marsh and the Philadelphia Academy's E. D. Cope (Box 16.3).

Within 30 years of the establishment of Dinosauria, there was a revolution in scientists' conceptions of how dinosaurs looked, driven by remarkable finds such as a complete hadrosaurid from New Jersey (1858) and 33 complete *Iguanodon* skeletons, recovered from coal seams outside of the town of Bernissart, Belgium (1877–1878; Figure 12.2; Box 16.4). In the hands of imaginative, skilled paleontologists such as J. Leidy (the hadrosaurid) and L. Dollo (the *Iguanodon* specimens), dinosaurs were transformed from overfed, bear-like lumbering lizards to something more terrifying, unimaginable, and wonderful than anybody could have ever invented.

Dinosaurs Divided

And they were *different*. Not just from living animals, but also from each other. This was duly noted by Harry Govier Seeley, vertebrate paleontologist at Cambridge University, and Friedrich von Huene, dean of German dinosaur paleontology at the University of Tübingen, both of whom recognized the fundamental division in Dinosauria between Ornithischia and Saurischia (Figure 16.7).

That dinosaurs had two different types of pelvis implied to Seeley that the ancestry of Ornithischia and Saurischia was to be found separately and more deeply among primitive archosaurs, within a now-abandoned group called "Thecodontia" (see below). Therefore, Seeley's dinosaurs were not, had he known the term, monophyletic (Figure 16.8).

The perception that dinosaurs were at least diphyletic (that is, having two separate origins) continued well past the middle of the twentieth century. Most paleontologists, until even the early

---

[3] Mantell's *Iguanodon*, discovered in 1822, was not formally named and described by him until 1825.

Box 16.2 **Sir Richard Owen: Brilliance and Darkness**

Richard Owen (Figure B16.2.1), the man who invented the term "Dinosauria" and Victorian England's premier anatomist, was among the most powerful and influential scientists of his time. His personality was at once insightful, irascible, politically astute, ruthless, and condescending, and it would not be going too far to call him a liar. He looked like he came directly from Central Casting for the part of a Victorian serial killer (he was worthy of a poster appearance for a 1980s punk rock band)! He also was a brilliant scientist, capably reconstructing unimaginable (and unimagined) extinct organisms from their bones, most notably some of the first dinosaurs known to the western world, and New Zealand's recently extinct giant bird, the Moa.

By the age of 21, Owen was hired by the Royal College of Surgeons in London to assist in the curation of the Hunterian Collection, a collection of biological oddities and medical curiosities amassed by John Hunter, a famous London surgeon. Hunter's notes had been destroyed in a fire, and so the daunting job was to organize, identify, and catalog disorganized drawers of biological detritus. Owen nailed it, using clever inferences and his encyclopedic knowledge of comparative anatomy.

He became a lecturer in comparative anatomy, publishing scholarly tomes on organisms ranging from

Figure B16.2.1. Sir Richard Owen (1804–1892), eventually of the British Museum of Natural History, prodigious nineteenth-century English anatomist and the man who bequeathed to us the term "Dinosauria."

1980s, thought that dinosaurs had at least two and likely three or four, separate origins within "thecodonts." Certainly, saurischians and ornithischians must have had separate origins; after all, their hip structure was different. And among saurischians, surely sauropods and theropods had separate origins; after all, they *look* so different. And finally, among ornithischians, ankylosaur ancestry was also often sought separately within some thecodontian group (see Chapters 5 and 8 for discussions of "thecodonts").

And what about Owen's bold suggestion that these dinosaurs were endothermic? Well before the turn of the century – and despite some early advocacy of it by other natural scientists – it was largely forgotten, the victim of the "fact" that dinosaurs were reptiles, and reptiles are cold-blooded.[4] In 1953, Roy Chapman Andrews evocatively described *T. rex*'s meal in cold-blooded – literally and figuratively – terms, reflecting the prevailing ideas of the time:

> Then it [*Tyrannosaurus*] settles to the feast. Huge chunks of warm flesh, torn from the Duckbill's body, slide down the cave-like throat … The King's stomach is full to bursting. Walking slowly to the jungle, he stretches out beneath a palm tree … For days, or perhaps a

---

[4] A few paleontologists, notably G. R. Wieland of Yale University, shared a vision of some kind of dinosaur homeothermy.

the living chambered cephalopod *Nautilus* to the first description of the newly discovered *Archaeopteryx* (Chapter 8). It was Owen who first described the exotic South American fossils that Charles Darwin brought back with him from his voyage on the *Beagle*, and, naturally enough, it was Owen who made the connection between the still-fragmental and isolated bits of fossil material that seemed to represent unimaginably large – and thus terrifying – ancient reptiles.

By 1842, enough of dinosaurs was known for Owen to invent a new term: Dinosauria (*deino* – terrible; *sauros* – lizard). The charter members of the group were *Iguanodon* (an ornithopod), *Megalosaurus* (a theropod), and *Hylaeosaurus* (an ankylosaur). Presciently, Owen's initial idea of Dinosauria was that its members were endotherms like mammals and birds,[a] a conclusion based upon the now-discredited idea that Mesozoic air was somehow thinner than modern air: right idea, wrong reasons! Owen balked at the idea that organic evolution (as it was understood before Darwin) was a kind of linear process that ran from quite simple to more complex. Owen thought that by demonstrating that an ancient group of organisms had modern levels of complexity, he would successfully undermine the notion of evolution. He was on the wrong side of that battle!

Ultimately, he took a job as superintendent of the natural history collections in the British Museum. He had a stunning vision for the collections – that they would be available to all, and advocated their separation from the rest of the British Museum, viewing natural history as deserving its own museum. Seventy-one years after he died, The Natural History Museum, today a public museum much as Owen had envisioned it, finally separated from the British Museum in 1963.

---

[a] He presciently wrote, "The dinosaurs, having the same thoracic structure as the Crocodiles, may be concluded to have possessed a four-chambered heart; and, from their superior adaptation to terrestrial life, to have enjoyed the function of such a highly organized centre of circulation... more nearly approaching that which now characterizes the warm-blooded Vertebrata [i.e., mammals and birds]... A too-cautious observer would, perhaps, have shrunk from such speculations..." (Owen, 1842, p. 204).

---

week, he lies motionless in a death-like sleep. When his stomach is empty, he gets to his feet and goes to kill again. That is his life – killing, eating, and sleeping.[5]

Ironically – for how our views have changed! – this description would have appeared stranger to a nineteenth-century paleontologist than to most paleontologists 70 years later. In retrospect, it seems puzzling that thoughtful scientists could have so meticulously described the bones and studied the relationships, yet with hardly any thought assumed "reptilian" ectothermy for dinosaurs for so long. Yet a look at publications through this period suggests that dinosaur metabolism rarely crossed their minds. Such is the strength – or the inertia – of ideas.

Dinosaurs in the First Half of the Twentieth Century

The first 60 years of the twentieth century brought about an expansion and consolidation of our basic understanding of dinosaurs and their diversity. Collecting, describing, and naming were the game, and our understanding of fundamental dinosaur morphology and diversity was dragged into a modern framework. In North America in the early years of the twentieth century, spectacular

---

[5] Andrews, R. C. 1953. *All About Dinosaurs*. Random House, New York, pp. 64–67.

Figure 16.5. Sculptor Waterhouse Hawkins' life-sized iguanodons on the Crystal Palace grounds, Sydenham, south London, England, still enchanting visitors after 160 years. Inset: A modern reconstruction of *Iguanodon*.

collections were made by Charles H. Sternberg and his sons along the Red Deer River in Alberta, Canada. There he and his crews floated along the river in a mobile field camp, swatting mosquitoes and black flies, and harvesting Upper Cretaceous dinosaurs from the sandstones and mudstones exposed in its banks. Even more spectacular were the efforts of the American Museum of Natural History's redoubtable Barnum Brown (Box 16.5), who, one summer in 1902, unearthed a large theropod that his sponsor, H. F. Osborne, dubbed *Tyrannosaurus rex*. For exotic and ill-fated, however, German scientists seem to have had particularly poignant tales; in spite of this, their work led to the discovery of legendary beasts like *Spinosaurus* and *Brachiosaurus* (Box 16.6).

Dinosaur discoveries continued at a rapid rate, new names proliferated, skilled descriptions of the new material were written, but, from the standpoint of *ideas*, the field had largely stagnated. These discoveries were carefully collected and described by a host of extraordinarily fine, dedicated paleontologists, including (in addition to those mentioned above) W. Granger, C. W. Gilmore, J. B. Hatcher, L. M. Lambe, A. F. de Lapparent, R. S. Lull, W. D. Matthew, A. K. Rozhdestvensky, R. M. Sternberg, and C. C. Young. Each of these remarkable people made important contributions, and the full story of each would fill a book as long as this. It's really a shame that space keeps us from highlighting their lives and work. Yet no account of dinosaur paleontologists should omit the

Figure 16.6. Contemporary lithograph of the New Year's Eve, 1853, dinner inside of Waterhouse Hawkins' model of *Iguanodon*.

brilliant Franz Baron Nopcsa – paleontologist, Albanian nationalist, polyglot, and spy for the Austro-Hungarian Empire in World War I (Box 16.7).

Dinosaurs Before *Jurassic Park* (BJP): the Second Half of the Twentieth Century (Most of it Anyway!)

The 1960s and early 1970s are rightly known for social revolution, and paleontology was affected as well. It was then that the field of paleobiology (see also Chapters 13 and 14) came into its own. The people we mention here, like all the paleontologists we have mentioned in this chapter, might best be thought of as representatives of the biggest generation: Baby-boomers. All of these specialists, in a sense arbitrary exemplars, stand in for the many, highly accomplished vertebrate paleontologists produced in that generation.

The Dinosaur Renaissance

Yale University paleontologist J. H. Ostrom's 1969 description of *Deinonychus anthirropus* was the catalyzing event for the paleobiological revolution (Figure 16.9). Here was a predatory dinosaur (see Figure 5.15) obviously built for extremely high levels of activity; its skeletal design simply made no sense otherwise. Ostrom doubted that such levels of activity were likely in an animal with the metabolism of a crocodile, and he argued for the possibility that *Deinonychus* might have been an endotherm.

In 1974, Ostrom published an exacting study of the earliest-known bird, our old friend *Archaeopteryx*. Reviving the ideas of T. H. Huxley, a contemporary of Charles Darwin, Ostrom concluded that a close relationship between dinosaurs and birds was inescapable. "All available evidence," he wrote, "indicates unequivocally that *Archaeopteryx* evolved from a small coelurosaurian dinosaur and that modern birds are surviving dinosaur descendants." This also suggested that dinosaur physiology should be considered more along bird than crocodilian lines. With these two

Box 16.3 Dinosaur Wars in the Nineteenth Century: Boxer versus Puncher

One of the strangest episodes in the history of paleontology was the extraordinarily nasty and personal rivalry between late-nineteenth-century paleontologists Edward Drinker Cope and Othniel Charles Marsh (Figure B16.3.1). In many respects, it was a boxer versus puncher confrontation: the mercurial, brilliant, highly strung Cope versus the steady, capable, bureaucratic Marsh. Their rivalry resulted in what has been called the "Golden Age of Paleontology," a time when the richness of the dinosaur faunas from western North America first became apparent – when the likes of *Allosaurus*, *Apatosaurus*, and *Stegosaurus* were first uncovered and brought to the world's attention.

Cope was a prodigy; one of the very few in the history of paleontology. By the age of 18, he had published a paper on salamander classification. By 24, he became a Professor of Zoology at Haverford College, Philadelphia. Blessed with independent means, within four years he had moved into "retirement" (at the grand old age of 28) to be near Cretaceous fossil quarries in New Jersey. He quickly became closely associated with the Philadelphia Academy of Sciences, where he amassed a tremendous collection of fossil bones which he named and rushed into print at a phenomenal rate (during his life he published over 1400 works). He was capable of tremendous insight, made his

Figure B16.3.1. The two paleontologists responsible for the Great North American Dinosaur Rush of the late nineteenth century. (a) Edward Drinker Cope (1840–1897) of the Philadelphia Academy of Sciences; and (b) Othniel Charles Marsh (1831–1899) of the Yale Peabody Museum of Natural History.

papers, Ostrom rewrote the book on both dinosaur physiology *and* bird origins, and suddenly dinosaurs got a *lot* more interesting to a lot more people. The "Dinosaur Renaissance" had begun!

That same year (as we saw in Chapter 14), Robert T. Bakker (see below, Figure 16.10), Ostrom's student, and Peter M. Galton, an ornithischian specialist later based at the University of Bridgeport, Connecticut, proposed to remove dinosaurs from "Reptilia" and establish them and birds as a new Class of vertebrates, Dinosauria. The basis for this proposal was the "key advancements of endothermy and high exercise metabolism." In the end, however, the idea didn't stick because the authors failed to sufficiently demonstrate any relationship between dinosaurs and birds other than the assertion that they shared an endothermic metabolism – a point that was heatedly debated (Chapter 14), but ultimately left unresolved. Time has proven this idea resilient, and although perhaps the details of the proposal were flawed, the recognition that dinosaurs weren't overblown reptiles (in share of mistakes, and was girded with the kind of pride that did not admit to errors.

Marsh, nine years Cope's senior, was rather the opposite, with the exception that he, too, eventually rushed his discoveries into print almost as fast as he made them (some thought faster) and that he, too, did not dwell upon his mistakes. Marsh's own career started off inauspiciously; with no particular direction, he reasoned that if he performed well at school he could obtain financial support from a rich uncle, George Peabody. This turned out to be perhaps the most significant insight in Marsh's life: Marsh persuaded Peabody to underwrite a natural history museum at Yale (which to this day exists as the Yale Peabody Museum of Natural History), and, while he (Peabody) was at it, an endowed chair for Marsh at the Museum.

At first, there was no obvious acrimony, but this changed when Marsh apparently hijacked one of Cope's New Jersey collectors right out from under him. Suddenly, the fossils started going to Marsh instead of Cope. Then, in 1870, Cope showed Marsh a reconstruction of a plesiosaur, a long-necked, flippered, marine reptile. The fossil was unusual to say the least, and Cope proclaimed his findings in the *Transactions of the American Philosophical Society*. Marsh detected at least part of the reason why the fossil was so unusual: the head was on the wrong end (the vertebrae were reversed). Moreover, he had the bad manners to point this out. Cope, while admitting no error, attempted to buy up all the copies of the journal. Marsh kept his.

Cope sought revenge in the form of correcting something that Marsh had done. The rivalry ignited, and the battle between the two spilled out into the great western fossil deposits of the Morrison Formation. Both hired collectors to obtain fossils, the collectors ran armed camps (for protection against each other's poaching), and, between about 1870 and 1890, east-bound trains continually ran plaster jackets back to New Haven (Connecticut) and Philadelphia (Pennsylvania). There Marsh and Cope rushed their discoveries into print, usually with new names. The competition between the two was fierce, as each sought to out-science the other. Discoveries (and replies) were published in newspapers as well as scholarly journals, lending a carnival atmosphere to the debate. Because Philadelphia and New Haven were not that far apart by rail, it was possible for one of the men to hear the other lecture on a new discovery, and then rush home that night and describe it and claim it for himself.

Both Cope and Marsh eventually aged and, in Cope's case, his private finances dwindled. Moreover, a new generation of paleontologists arose that rejected the Cope–Marsh approach, believing, not unreasonably, that it had caused more harm than good. Both men ended their lives with somewhat tarnished reputations. History has viewed the thing a bit more dispassionately, and it is fair to state that the result ultimately was an extraordinary number of spectacular finds and a nomenclature nightmare that is still being disentangled (see Desmond, 1975). An interesting and unusual account of the Cope–Marsh feud was published as a graphic history (Ottaviani *et al.*, 2005).

the Linnaean sense), that birds are dinosaurs, and that dinosaurs as a group are far more central to tetrapod phylogeny than had been previously recognized, are all ideas that have come to be widely accepted; if now almost 50 years after the concept was first proposed!

In the meantime, the debate over dinosaur endothermy climaxed with the 1980 publication of an American Association for the Advancement of Science special volume that covered the 1978 proceedings of a symposium devoted solely to dinosaur thermoregulation (please see Selected Readings at the end of this chapter). Most authors seemed to lean toward, at a minimum, some kind of homeothermy (see Chapter 14). And while modern research on dinosaur metabolism has lost some of the contentiousness, it continues apace (Chapter 14) with the idea that although dinosaurs were not ectothermic in the crocodile sense, one size of metabolism surely does not fit all.

Box 16.4 Louis Dollo and the Beasts of Bernissart

Louis Antoine Marie Joseph Dollo (Figure B16.4.1), a Belgian paleontologist with a name almost as luxuriant as his moustache, gave us our first true picture of dinosaurs.

In 1878, commercial coal miners identified fossil bone some 322 m (1056') below ground, which was immediately brought to the attention of the Brussels museum and to Dollo in particular. This occurrence of bone turned into a treasure trove of more than 30 articulated skeletons of the Early Cretaceous ornithopod called *Iguanodon* (Figure B16.4.2; see also Figure 12.2).

Dollo devoted himself to understanding the anatomy and function of these extinct forms in ways that had not been possible before. He sorted the *Iguanodon* material into two species by successively eliminating different sources of skeletal variation. He used the disparity between forelimb and hindlimb length, the development of ossified tendons across the back, and footprints to establish bipedality in *Iguanodon*. And he outlined new approaches to reconstructing the jaw systems of numerous dinosaurs including *Iguanodon*, putting them into their comparative context with living vertebrates. In doing so, Dollo turned paleontological attention to what he called "ethological paleontology" – the study of behavior and environment of extinct organisms – which Othenio Abel, an Austrian paleontologist, termed paleobiology in 1912.

Other than his work on *Iguanodon*, best known is Dollo's Law of Irreversible Evolution. This biological principle, which Dollo formulated in 1893, argued that evolution is not a reversible process: Structures eliminated during the course of evolution cannot themselves reappear in the same form within a given lineage of organisms.

Figure B16.4.1. Louis Dollo (1857–1931), the Belgian paleontologist of the Musée Royal de Sciences Naturelles Belgiques, who, along with Joseph Leidy, Professor of Anatomy at the University of Pennsylvania, first understood the shapes of dinosaurs.

Figure B16.4.2. *Iguanodon*, the great beast of Bernissart, Belgium as currently displayed at the Musée Royal d'Histoire Naturelle de Belgique. In Figure 12.2 and here, we show the Bernissart *Iguanodons*, as originally mounted by Dollo. They're looking a tad upright by modern interpretations, but they were an astounding leap forward from the Crystal Palace figures of Waterhouse Hawkins (Figure 16.5), only 25 years earlier!

(a)

(b)

Figure 16.7. (a) is Cambridge University's Harry G. Seeley (1839–1909) and (b) Friedrich von Huene (1875–1969), University of Tübingen.

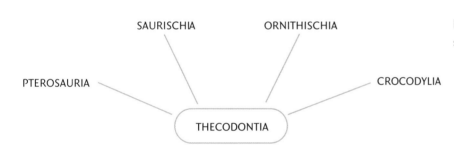

Figure 16.8. Seeley's evolutionary scenario of the origin of dinosaurs.

Phylogenetic Systematics Enters the Fray

Amid all of this intellectual ferment, yet another revolution was not-so-quietly taking place. This was the *cladistic revolution* (see Chapter 3). Phylogenetic systematics was not so new (although not nearly as old as the idea of endothermic dinosaurs); the basics had first been articulated by a German entomologist, Willi Hennig, in 1950 (Figure 16.11). English translations of Hennig's ideas appeared in 1966 and again in 1979. Hennig's great insight was, as we've seen in Chapter 3, to develop a scientific (testable) method whereby relationship can be inferred from anatomy.

The method was initially not widely appreciated, and the results were a bit shocking to people trained in the traditional Linnaean classification system (see, for example, Box 4.1), with the result that between 1966 and 1990 (or thereabouts), this approach engendered considerable controversy. By the 1990s, computer algorithms that allowed cladograms to be generated from large and complex datasets had come into common usage, and cladograms, thanks to vociferous support from scientists at the American Museum of Natural History and the Natural History Museum (London) became a ubiquitous and powerful tool for deciphering the relationships of both living and extinct organisms (including dinosaurs). It was cladistic methods that cemented the recognition of monophyly as the key ingredient in understanding relationships, and in so doing, highlighted the great monophyletic groups that characterize this book: Dinosauria; Ornithischia; Saurischia; and the many, many,

Box 16.5 "Mr Bones"

There have been dinosaur collectors; there have even been extraordinary dinosaur collectors, and then, in a league quite by himself, there was Barnum Brown (Figure B16.5.1). Born in 1873, and named after the then-popular circus showman P. T. Barnum, Brown virtually single-handedly turned the American Museum of Natural History (AMNH) from a place with not a dinosaur on the premises to perhaps the world's greatest dinosaur collection. Its great hall of Cretaceous dinosaurs has been described as a monument to his accomplishments.

He began his career in 1897 as a field assistant at the AMNH, first going out to the fossil grounds of Wyoming where he discovered the still-productive Bone Cabin Quarry, a site so rich that local ranchers built an entire cabin out of fossil bones. After three years of collecting, 35 tons of fossil bones were sent by train back to the AMNH, including what eventually became the largest mounted specimen of its time, the AMNH's magnificent "brontosaurus" mount (see Figure B9.1.1a).

in the early 1900s, Brown began to prospect in Hell Creek badlands of eastern Montana. There in 1902 he found the first of two magnificently preserved *Tyrannosaurus rex*; the first ever discovered. The thing

Figure B16.5.1. The well-dressed Barnum Brown (1873–1963), collector with the American Museum of Natural History, showing unequaled field couture.

groups on nodes of the cladograms that show dinosaur relationships as they are currently understood. Its influence can be felt in almost every aspect of the study of dinosaurs, and it is not an exaggeration to state that virtually every aspect of the book you're currently holding is grounded in cladistic methods.

Birds are Living Dinosaurs

Although the possibility of birds as dinosaurs goes back as far as T. H. Huxley, it is hard to imagine the resistance to the idea when Ostrom first proposed a bird–dinosaur relationship. It just didn't seem possible when dinosaurs were slow, dim-witted, and cold-blooded; and Bakker and Galton's (1974) proposal, based as it was upon the very shaky (at the time) evidence of dinosaur endothermy, was just too far out there. Neither Ostrom, nor Bakker and Galton, attempted a cladistic treatment of their hypotheses.

All that changed, however, in 1986, with Yale paleontologist and herpetologist Jacques Gauthier's (b. 1951) now-classic paper on saurischian monophyly, in which he addressed Ostrom's observations was preserved in a calcite-hardened sandstone concretion, and, aside from the dynamite necessary to free it from the hillside in which it was found, he had to cut a road to carry the massive blocks out to the nearest railroad for shipment to New York (Figure B16.5.2).

Brown's third major venue for fossil collecting was the Red Deer River in Alberta, Canada. There, taking a page from C. H. Sternberg's playbook, he fitted a barge with a canvas tent and prospected along the shores of the river. Here he collected trainloads of fossils, including the beautiful hadrosaur specimens for which the AMNH is justifiably renowned.

By the early 1930s, Brown had cut a funding deal with the Sinclair Oil Company (whose logo, not coincidently, was, and is, a green sauropod) and expanded his collecting efforts to fossils other than dinosaurs and locales other than western North America. His discovery in 1934 of the Jurassic Howe Quarry bonebed was another highwater mark in a career full of them; more than 20 dinosaurs represented by 4000 bones.

Figure B16.5.2. A 1985 photograph of the site of Barnum Brown's first *T. rex* discovery. The arrow points to the remains of the wagon trail cut by Barnum Brown into the hillside to remove the massive blocks containing the fossil.

from a cladistic viewpoint (Figure 16.12). The cladistic analysis didn't show that birds come from dinosaurs; it showed that birds *are* dinosaurs. Here, though, with the added rigor of phylogenetic systematics, it wasn't an amorphous question of opinion, as it had been in the past; birds had to be dinosaurs because they shared the diagnostic characters of dinosaurs, no more and no less. For most paleontologists as well as ornithologists, it was a done deal; the cladograms diagnosed the relationship after which, there was nothing left to say.

Our discussion of birds as dinosaurs would not be complete without at least a few words about feathers. Once the possibility of dinosaur endothermy became a question (as we have seen, this was in the late 1960s and early 1970s), it seemed that dinosaurs might need insulation, and it seemed obvious to many paleontologists that such insulation might come from feathers or proto-feathers. Thus, many of us predicted, we ought to find feathers on nonflying dinosaurs, on the assumption that feathers were initially more about endothermy than about flight. But as we have seen, feathers probably weren't initially about endothermy, but most likely about display or earlier, touch; however, we also have considerable evidence that they were coopted for insulation before they were

Box 16.6 Tales of Two Germans

In the latter part of the nineteenth and early twentieth centuries, imperialism ran high throughout Europe, and new colonies around the globe provided opportunities for paleontologists to look to new places for fossils. Germany was no exception to this, and thus it came to pass that some very distinguished German paleontological work was carried out by very distinguished German paleontologists in locations far removed from the Fatherland.

Saharan Dinosaurs

Ernst Stromer (Figure B16.6.1) was a Bavarian nobleman,[a] and professor at the University of Munich, who ran two trips to rich Saharan localities Wadi el Natrun and the Fayum Oasis (Egypt) after trips there in 1901 and 1902. In the latter months of 1910 and into 1911, still bitten by the Sahara Bug, Stromer organized yet another camel-supported expedition to the western Sahara Desert, ultimately aiming at the Baharia Depression, a bone-dry, large (about 1100 square miles) desolate swath of the western Sahara. The expedition was successful and, although Stromer returned to Munich in 1911, his crews continued to collect in the Baharia Depression region until 1914 and the outbreak of World War I.

Global events overtook paleontology, however, and getting the fossils out of Egypt (for study) became a nightmare. Egypt was a British "protectorate" (read, "colony"); Stromer was German; politics and the World War, of course, shut down both the expeditions and the removal of the fossils from Egypt. It took eight years but finally in 1922 Stromer received the specimens that he and his crews had collected 11 years before. The fossils came battered and in bad shape, but at least they came. Stromer began a publishing streak that ultimately led to descriptions of *Aegyptosaurus* (a sauropod), *Bahariasaurus* (a large theropod), *Carcharodontosaurus* (a very large theropod; see Figure 7.5c), and *Spinosaurus* (the largest carnivorous dinosaur that ever lived; see Figures 6.24 and 7.4d), and in so doing, he solidified his reputation as one

Figure B16.6.1. Ernst Freiherr Stromer von Reichenbach (1870–1952), aristocratic Professor of Paleontology at the University of Munich, and discoverer of a many dinosaurs, including *Spinosaurus*, and other vertebrates in the Cretaceous beds of the Baharia Depression, western Sahara, Egypt.

of the world's most careful, accomplished, paleontologists. Rapidly he obtained a professorship at the University of Munich as well as a directorship of the Bavarian State Collection of Paleontology and Historical Geology, which he maintained until his mandatory retirement (because of age) in 1937.

But by then national events were overtaking paleontology in Germany. Stromer refused to join the rising Nazi Party (whose *heimat* was in Munich); much worse, he denounced Hitler's regime and visibly maintained his

---

[a] The Germany of much of its history was a collection of separate states, and was finally united by Otto von Bismarck, in the latter part of the nineteenth century.

Jewish friendships. As a member of the aristocracy, he was largely untouchable, but as Hitler plunged Germany (and the rest of the world) into World War II, the Nazis found a way to extract their literal pound of flesh: Stromer's three boys were sent to war; two to the Russian front. Nobody viewed the Russian front as a recipe for longevity, and in fact one was killed in 1941 and the second disappeared into Russia in 1944. The third son was killed on the western front in 1945, two weeks before Germany capitulated. Stromer paid dearly for his principles.

Meanwhile, in 1940, the State Collection was taken over by Karl Beurlen, a paleontologist with the correct political outlook. Beurlen was an ardent supporter of *der Führer*, and simply refused to believe that the Third Reich could be defeated. And because he was sure of its invincibility, he also refused – despite Stromer's repeated warnings – to protect the valuable treasures of the State Collection by removing them (like so much else of value) to underground mines, such as the salt mines in nearby Salzburg.

And so it was that as April 24 turned to April 25, 1944, 16 speedy Royal Air Force deHavilland Mosquitoes with marker flares, and 244 bomb-laden Avro Lancasters, turned the Munich train station and surrounding buildings into utter rubble. Unfortunately, housed in one of those surrounding buildings was the Bavarian State Collection of Paleontology and Historical Geology. Stromer's fossils, most famously, *Spinosaurus*, were obliterated. Effectively all that was left were Stromer's meticulous notes and some detailed black and white photographs.

The end was bittersweet. On May 5, 1950, Stromer's middle son miraculously reappeared, literally walking out of the forest into his parents' ancestral home: having refused to make poison gas for Stalin, he spent the intervening six years since his disappearance at the Russian front in Siberian gulags. Two years later, Stromer died.[b]

Spinosaurus itself had to wait until well into the 2000s to be rediscovered. In 2008, then University of Chicago paleontology graduate student Nizar Ibrahim, working in the

western Sahara (inspired in part by Stromer's work!) was approached by a man with some fossil bones, one of which Ibrahim recognized as possibly belonging to *Spinosaurus*. Five years later, he heard of a newly acquired specimen at the Natural History Museum of Milan that someone thought might be *Spinosaurus*. The bones in Milan looked promising and, strikingly, had the same coloration as the ones Ibrahim remembered being shown by the man in the western Sahara. Unfortunately, the origin of these bones in Milan was uncertain; they were thought to come from somewhere (but where?) in Morocco. Incredibly – and serendipitously – Ibrahim managed to track down the man who in 2008 had first shown him the bones, and persuaded him to take Ibrahim and a crew to the site where he had found the bones. The bones of 2008 and the Milanese fossils actually were from the same *specimen*! *Spinosaurus* had been rediscovered at last.

Tendaguru!

Tendaguru, located in the hinterland of Tanzania on the eastern coast of Africa and today monotonously formed of broad plateaus blanketed by dense torn trees and tall grass thick with tse-tse flies, was formerly the site of perhaps the greatest paleontological expedition ever assembled, and much – thousands of millennia – before that it was the place where dinosaurs came to die.

Let's go back to 1907, when Tanzania was part of colonial German East Africa. The fossil wealth of Tendaguru was first discovered in 1907 by an engineer working for the Lindi Prospecting Company. Word spread quickly, ultimately to Professor Eberhard Fraas, a vertebrate paleontologist from the Staatliches Museum für Naturkunde in Stuttgart, who happened to be visiting the region. So excited was he at the prospect of collecting dinosaurs after his visit to Tendaguru that he took specimens back to Stuttgart (including what was eventually to be called *Janenschia*) and more especially started drumming up interest among other German researchers to continue field work in the area.

Wilhelm von Branca, director of the Humboldt Museum für Naturkunde in Berlin, seized upon the opportunity presented to him by Fraas. Yet before mounting an expedition of the kind demanded by Tendaguru, Branca had to

[b] The full story of Ernst Stromer is given in: Nothdurft, W. (with J. Smith). 2002. *The Lost Dinosaurs of Egypt*. Random House, NY, 242 pages.

tackle the problem of its financial backing. Ultimately, he received more than 200 000 deutschmarks – a fortune for the time – from the Akademie der Wissenschaften in Berlin, the Gesellschaft Naturforschender Freunde, the city of Berlin, the German Imperial Government, and almost a hundred private citizens.

With money, material, and supplies in hand, the Humboldt Museum expedition set off for Tendaguru in 1909. Under the leadership of mustached and jaunty Werner Janensch (Figure B16.6.2) for three of the next four seasons (Hans Reck took charge in the fourth season), it was possibly the greatest dinosaur collecting effort in the

Figure B16.6.2. Werner Janensch (1878–1969), paleontologist at the Humboldt Museum, and the driving force behind the extraordinarily successful excavations at Tendaguru, Tanzania.

modified and coopted for flight. We know this because of the extraordinary discoveries first appearing in the mid 1990s in Liaoning Province, China, and, as much as anybody, because of the work of Xu Xing, Chinese paleontologist *extraordinaire* at the Institute of Vertebrate Palaeontology and Palaeoanthropology (IVPP), Beijing, China (Figure 16.13). Xu has been associated with formal descriptions of many new genera of feathered dinosaurs. There is no question that the feathered, nonflying nonavian dinosaurs from China were utterly game-changing (see Chapters 7 and 8).

More Baby-Boomers in the Game

As we have seen, the 1970s were a time of intellectual ferment in paleontology, and for the dinosaur-loving public, nobody embodied those fresh winds of change more than Robert T. Bakker (b. 1945; Figure 16.10a). In a field historically known for bookish obscurity, Bakker was a ubiquitous media presence; so much so that he was recognizably caricatured as Dr Robert Burke in *Jurassic Park II*. Bearded, long-haired, and dressed for battle in his field-ready best[6] (belying

---

[6] Including an iconic battered hat that pre-dated that of Indiana Jones – but not that of Roy Chapman Andrews!

history of paleontology. The first season involved nearly 200 workers, mostly natives, laboring in the hot sun as they dug huge bones out of the ground. During the second season, there were 400 workers and in the third and fourth seasons 500 workers. By the end of the expedition's efforts, some 10 square kilometers of area was covered with huge pits.

But it got complicated. Many of the native workers brought their families with them, transforming the dinosaur quarries at Tendaguru into a populous village of upward of 900 people. With so many people, water and food became a severe problem. Not available locally, water had to be brought in, carried on the heads and backs of porters. And with the vast quantities of food that had to be obtained for workers and their families, and the pay for work carried out in the field, it is not surprising that the funds amassed by Branca disappeared rapidly.

Still, the rewards were great indeed. Over the first three seasons, some 4300 jackets were carried back to the seaport of Lindi – a four-day walk away and a trip made 5400 times there and back by native workers, each with the fossils balanced on his or her head – to be shipped from there to Berlin.

Overall, work at Tendaguru involved 225 000 person-days and yielded nearly 100 articulated skeletons and hundreds of isolated bones. When finally unpacked and studied, what a treasure-trove: in addition to ornithischians (*Kentrosaurus*, *Dryosaurus*) and theropods (*Elaphrosaurus*), and a pterosaur as well, the Tendaguru expeditions claimed two new kinds of sauropod (*Tornieria* and *Dicraeosaurus*), but also new material of *Barosaurus* and the finest specimen of "*Brachiosaurus*" (now called *Giraffatitan*) ever found – now mounted and peering into the fourth floor balcony of the Humboldt Museum für Naturkunde.

The Humboldt Museum never went back to Tendaguru after 1912. In 1914, World War I erupted and, with the Treaty of Versailles, German East Africa became British East Africa. This shift in European colonialism brought British paleontologists to Tendaguru in 1924, under the direction of W. E. Cutler. His team, from the British Museum (Natural History), hoped to enlarge the quarried area and retrieve some of the left-over spoils from the German effort. From 1924 to 1929, the British expedition found more of the kinds of dinosaurs discovered earlier, but suffered some severe health problems including malaria, from which Cutler died in 1925. There has been no significant paleontological effort at Tendaguru since.

degrees from Yale and Harvard), Bakker was filled with amazing ideas about birds, dinosaurs, and their world. A highly competent prose stylist (described in *Harper's Magazine* as "by far the most gifted writer in his profession"[7]) and a talented illustrator, Bakker was distinctive, articulate, and out to change ideas.

Equally informal of manner and no less intellectually endowed, is Bakker's contemporary, John R. "Jack" Horner (b. 1946; Figure 16.14). For much of his career, Horner lacked formal advanced degrees (although he now has an honorary doctorate); yet, it would be hard to find a more creative (he is a MacArthur "genius award" recipient) and accomplished paleontologist. As we have seen, it was Horner who first recognized extended parental care in *Maiasaura*; it has been Horner who has, most recently, led the field in understanding bone histology and its relationship to dinosaur growth and metabolism; it was in Horner's laboratory that proteins from *T. rex* were first extracted (see Chapter 13).

Horner's laconic, plain-spoken manner, reputed to be the inspiration for Dr. Alan Grant in the film *Jurassic Park* (he was the paleontological consultant on the film), belies a canny, sophisticated

---

[7] Silverberg, R. 1981. Beastly debates. *Harper's Magazine*, October, 1981, pp. 68–78.

Box 16.7 Franz Baron Nopcsa: Politics, Dinosaurs, and Espionage

When you visit Schönbrunn Palace, the Versailles-like summer home of the Austrian monarchy back in the heady days of the *fin-de-siècle* Austro-Hungarian Empire, go to the apartments of Emperor Franz Josef (1830–1916), and there you will find a sign referring to a paleontologist: Franz Baron Nopcsa (Figure B16.7.1). Only he is there in another role: diplomat; his career as a paleontologist was doubtless beneath imperial notice!.

Yet, there was never anyone quite like Franz Baron Nopcsa. He was one of the first paleontologists who saw to it that dinosaurs were interpreted in their full biological

Figure B16.7.1. Franz Baron Nopcsa (1877–1933), Transylvanian nobleman, patriot, spy, and paleontologist.

context. From him, we've learned about the unusual dinosaur fauna from Transylvania, that part of western Romania where his noble family's estate was located. This skeletal material formed the mainstay of Nopcsa's research, including soft tissue reconstruction and its relevance to jaw mechanisms, paleoecological reconstructions of the region as a Late Cretaceous island, and the evolution of the dwarf dinosaurs that lived on the island. He also published extensively on the early evolution of birds from small predatory dinosaurs, the origin of new evolutionary conditions due to disease, the pituitary gland and large size in dinosaurs, and the relationship between bone histology, growth rates, and thermoregulation among dinosaurs. A polyglot, Nopcsa published not only in German, but also regularly in English, Hungarian, and French.

Remarkable though these achievements were, they were conducted against a background of scientific and political involvement in the founding of the state of Albania. Nopcsa was captivated by the geography and people of this stark, yet beautiful land of the western Balkans. He began working there in 1906 and by the end of his career had published some still-current monographs on the geography, geology, and ethnography of Albania and its people. As part of his interest in Albanian nationalism, Nopcsa became a spy during the first and second Balkan wars, and continued his spy work in Romania during World War I.

Despite his international activities, Nopcsa was a private man. He lived most of his life in Vienna, except for two years as Director of the Hungarian Geological Survey in Budapest. Living with him was his secretary, friend, and lover, Bajazid Elmas Doda, an Albanian he met in 1906. Transylvania was ceded to Romania after World War I and the Nopcsa estate was lost. Thereafter, Nopcsa's mental health declined and early in the morning of April 15, 1933, he dosed Bajazid's tea with sleeping powder and then shot him. Going into his work room, Nopcsa wrote a suicide note and then killed himself.

approach to almost all he undertakes, whether it be building the Museum of the Rockies (Bozeman, Montana, USA) into a world-class paleontological museum and an important center for cutting-edge histological research (both living and fossil), or publishing well over 100 research papers on virtually every subject pertaining to Dinosauria.

Figure 16.9. John H. Ostrom (1928–2005), Yale University, the brilliant paleontologist whose ideas ignited the modern era of dinosaur research.

Figure 16.10. (a) Robert T. Bakker, then at Yale University, and (b) Peter M. Galton, University of Bridgeport, who first proposed that Linnaeus' venerable "Class" Aves be subsumed within an enlarged, new "Class Dinosauria."

Somewhat younger than Horner and Bakker is the Curator of the Division of Paleontology at the American Museum of Natural History (New York), Mark A. Norell (b. 1957; Figure 16.15). Norell's studies have been wide-ranging, including contributing to the development of the concept of "ghost taxa" or "ghost lineages"(see Box 15.2), the discovery and description of the unusual theropod *Mononykus* (see Figure 7.14), the discovery of the first embryo of a theropod dinosaur, and the first clear "proof" that dinosaurs nested on their eggs (*Oviraptor*; see Figure 6.30). Norell has led

Figure 16.11. Willi Hennig (1913–1976) of the Deutsches Entomologisches Institut, the German entomologist who was the father of cladistic analysis (phylogenetic systematics).

Figure 16.12. Yale University's Professor Jacques A. Gauthier, who carried out the seminal cladistic studies of bird origins and dinosaur monophyly.

Figure 16.13. Professor Xu Xing, of the Institute of Vertebrate Palaeontology and Palaeoanthropology, Beijing, China, one of the most prolific living paleontologists, and the man without whom feathered theropods would likely still be just a neat hypothesis.

Figure 16.14. Paleontologist *extraordinaire* Jack Horner, doing what he does best.

expeditions to the Gobi Desert for the past 18 years, during the course of which he discovered and described the dinosaurs *Shuvuuia*, *Apsaravis*, *Byronosaurus*, and *Achillonychus*, among other vertebrate fossils. He has even cowritten several award-winning books, including *Discovering Dinosaurs* (1995) and *Unearthing the Dragon* (2005).

Figure 16.15. Mark Norell, Curator of Paleontology at the American Museum of Natural History (New York).

Figure 16.16. Professor Paul Sereno in the waning sun of the Sahara Desert, digging up something not too dinosaur-like. He writes of this picture, "I [fell] prey to digging up 100 fossil humans in the Sahara – they got in the way of my dinosaurs!".

Our brief sampling of some of dinosaur paleontology's not-so-young Young Turks would surely be incomplete without the University of Chicago's redoubtable Paul Sereno (b. 1957; see Figure 16.16). Handsome and dashing, he was, after all, one of *People* magazine's "50 Most Beautiful People," *Newsweek*'s "100 People to Watch for the Next Millennium," and *Esquire* magazine's "100 Best People in the World." Reflecting his vigor/age ratio, did we not see him gracing the cover of the magazine of the American Association of Retired People [AARP]? – though he can hardly be said to have retired! Sereno has made it his business to travel to exotic locales (Egypt, Niger, Morocco, Argentina, China), collect, and then describe exotic fossils. Therein lies quite the list, including *Afrovenator abakensis* (a large theropod), *Carcharodontosaurus saharicus* (another large theropod, possibly larger than *Tyrannosaurus*), *Deltadromeus agilis and Rugops primus* (two more large theropods!), *Rajasaurus narmadensis* (a large, crested theropod), *Herrerasaurus ischigualastensis* and *Eoraptor lunensis* (two of the most primitive dinosaurs known; Figure 5.7a), *Jobaria tiguidensis* (a sauropod of over 20 m), the 13 m dinosaur-munching Saharan crocodile *Sarcosuchus imperator*, and the piscivorous large theropod *Suchomimus tenerensis*. Sereno, unsurprisingly, is a coauthor on the full description of the newly reconstructed (and surely rethought) *Spinosaurus* (see Figure 6.24; Box 16.6).

Other Big Changes BJP: Extinction of (Nonavian) Dinosauria

Our trip through the most recent ideas about dinosaur paleontology would not be complete without another thoroughly radical idea: extinction by asteroid impact (see Chapter 17). Since H. F. Osborn's time, the extinction of the dinosaurs was viewed by paleontologists as a gradual process of dwindling diversity, beginning well before the end of the Cretaceous. The prevailing view before 1980 was cleverly and succinctly summarized by University of California (Berkeley) paleontologist, the late W. A. Clemens, Jr., and colleagues who wrote (with apologies to T. S. Eliot):

This is the way Cretaceous life ended
This is the way Cretaceous life ended
This is the way Cretaceous life ended
Not abruptly but extended.[8]

Clemens and colleagues' article about the dinosaur extinction was aptly entitled "Out with a whimper, not a bang."

The revolution came in the form of the 1980 hypothesis of Alvarez and colleagues that an asteroid came from outer space, smashed into the planet, and ultimately reset global ecosystems for all time (we'll elaborate the details of this in Chapter 17). What a strange, wonderful, and terrifying idea – that extraterrestrial events are critical forces shaping the history of life on Earth. The conceptual revolution provoked by this vision, however, extended far beyond the deaths of a few dinosaurs, and reverberated throughout the geosciences.[9] For one thing, it took a global-system-based view of events: that the dinosaur extinction was not an isolated event pertaining only to dinosaurs (and thus dinosaur-based explanations for the extinction were sufficient), but rather it was a global biotic catastrophe with global physical ramifications. As such, it shined a (for some, unwelcome) spotlight into vertebrate paleontology, heretofore, a small, select community of scientists in large part talking exclusively to each other rather than to the scientific community as a whole.

At a deeper level, it provoked a fundamental revolution in the geosciences. Whereas the prevailing view in the geosciences, dating back to early-nineteenth-century geologist and Darwin contemporary Charles Lyell (1797–1875), had been that events on Earth occurred gradually over potentially immense time scales, the asteroid hypothesis catalyzed the widespread recognition that many Earth-shaping processes – as significant as asteroids, or as miniscule as a local flood – really occur on relatively short time scales; indeed, they might be said to occur, in a geological sense, instantaneously. Ultimately, this view has revolutionized the fields of paleontology and geology. The asteroid and its aftermath are the subjects of our final chapter.

Dinosaurs After *Jurassic Park* – the Late 1990s

Coming on the heels of the Dinosaur Renaissance and all the rethinking of dinosaur biology that had occurred in the 1970s, 1980s, and early 1990s, the movie *Jurassic Park* was a "perfect storm" for the study of dinosaur paleontology. Two generations (Gen X and Millennials; see below) were inspired by the images in that first movie and those that followed, with the result that there are more active, professional paleontologists working today than ever before in the history of the discipline, and the insights derived from the field have become particularly important, not just in the field of evolutionary biology, but also as they relate to climate science. Yet, paleontology is today at a crossroads, and calls for scientists with new skill sets required that would have been utterly foreign to paleontologists of a generation ago.

It has become very clear (see Gauthier, Kluge, and Rowe, 1988, in Selected Readings), that the present-day biota does not give us a complete enough picture of the past record of evolution.

---

[8] Clemens, W. A., Jr., Archibald, J. D., and Hickey, L. J. 1981. Out with a whimper, not a bang. *Paleobiology*, **7**, 297–298.
[9] See Powell, J. L. 1998. *Night Comes to the Cretaceous*. W. H. Freeman and Company, New York, 250 pages.

Deeper insights are afforded by the fossil record; among the many examples of this, how would we ever know that birds are dinosaurs if we only knew the living representatives and close relatives of Dinosauria?

Yet, modern evolutionary biology has acquired powerful new tools for unraveling the course of evolution, including: (a) molecular evolution, in which the molecular difference between two organisms is measured to determine how distantly or closely they are related (see Chapter 8); (b) evolution and development (or evo-devo), in which sophisticated genetic and embryological studies are revealing the way the organisms evolve new features and produce diversity; and (c) molecular clocks, in which the rate of molecular evolution can be used to date the time that two living organisms first diverged from a single (long-extinct) ancestor (see Chapter 8). In each of these cases, deep insights into the fossil record are necessary to optimize the results. Mary Schweitzer's (b. 1955; Figure 16.17) work with dinosaur soft tissues (see Chapter 13) and blood cells would hardly have been what paleontologists expected to do 50 years ago; but we would surely call her work paleontology! So paleontology's contributions continue to be on the forefront; yet, modern paleon-

Figure 16.17. Professor Mary Schweitzer, paleobiologist at North Carolina State University and preeminent discoverer of soft tissues from non-avian dinosaurs, including collagen fibers, red blood cells, and blood vessels.

tologists need a far greater sophistication and specialized training – more than was ever even imagined by the old timers – in the biosciences generally and in molecular genetics and embryology in particular.

We have also seen how many modern paleontologists now use geochemical tools unthinkable a generation ago: stable and unstable isotopes, as well as trace element compositions and geochemical signatures. Another important tool widely employed by paleontologists since the 1980s is statistics. Paleontologists deal with sparse datasets; incomplete datasets; data that are of variable quantity and quality; and all of these kinds of data require somewhat specialized mathematical treatments. An important part of a well-trained paleontologist's problem-solving arsenal, therefore, is statistical training; both frequentist (or classical) and Bayesian skills are called for. Finally, with the large databases required for phylogenetic systematics as well as for statistics, some fluency with computers is essential. At a minimum, active paleontologists must be ready to use the many published programs commercially (and noncommercially) available, but ideally some ability to program in one of several commonly used languages such as R or Python is very desirable. Such skills allow researchers to tailor the computer to their particular needs.

All of these skills – molecular biology; geochemistry; statistics; computer programming – are a far cry from the old-style collect-and-describe game of a generation ago. Of course, they haven't eclipsed it; they're just more possibilities in the professional toolbox.

Paleontologists have a unique perspective to offer the world about global climate change. It is clear that the Earth has gone through many episodes of global warming associated with greenhouse conditions, not the least of which was, as we've seen, during the mid Cretaceous. What was that world like? More importantly, how did the biota respond to that climate and to climate change in general? Since the Earth has experimented with greenhouse conditions in the past, it behooves us to consider how it and its biota responded to these events – information that may give us insights about what we may expect in the future. In this case, advanced expertise in geochemistry and computer-based climate modeling could serve as an important part of the preparation of future paleontologists.

So the days of extracting, describing, and naming dinosaurs are certainly still with us; it is clear from Benton's data (Figure I.1) that there are lots of dinosaurs to find; lots of unexpected discoveries to be made; Adelphi University vertebrate paleontologist Michael D'Emic estimated that in 2018, new species of dinosaur were named at the rate of three each month! Yet, the significance of these finds – in terms of evolutionary biology, paleogeography, paleoecology, functional morphology, and other paleobiological issues – can no longer be ignored. But in a way, such questions are more interesting, and the stakes far greater. It's a brave, new, and exciting world out there.

Today: Young Turks and Old Turkeys

"It's funny," University of Oregon paleopedologist Greg Retallack once said, "how quickly today's 'Young Turks' become tomorrow's old turkeys." The young, fire-breathing revolutionaries themselves become advocates of established dogma, as, even in their own lifetimes, their ideas are challenged and swept aside by a righteous, new, young, aggressive generation of scientists. Since the generation of Baby-boomer paleontologists most active in the 1970s and early to mid 1990s, two generations of scientists, many inspired in the "formative years" by the *Jurassic Park* (s), have made their marks: the "Gen-X" generation, approaching middle-age and at the height of their careers, and Millennials, currently just starting to build their accomplishments in unprecedented numbers. Again, the scientists we describe here represent – stand in for – the many, highly accomplished vertebrate paleontologists produced by these two generations.

The Gen-X generation has had no shortage of exciting talented paleontologists. We start with the US National Museum of Natural History's (Smithsonian Institution) Curator of Dinosauria Matthew T. Carrano (Figure 16.18). Carrano has wide-ranging interests, from collecting dinosaurs in the grand tradition, to large-scale evolutionary (macroevolutionary) patterns in the fossil record (such as we discuss in Chapter 15). Carrano is no stranger to exotic fieldwork, and has tromped around the wilds of Nigeria, Madagascar, and South America.

Coming at things from a different end is the University of Colorado's Karen Chin (Figure 6.19), who has made a unique and significant contribution studying coprolites and trace fossils. Coprolites remain the most direct means of interpreting what dinosaurs ate, assuming one can infer the source. Chin has done so with exceptional creativity and precision, and with such data,

Figure 16.18. Smithsonian curator and dinosaur specialist Matt Carrano, framed by a toothy Late Cretaceous friend.

Figure 6.19. Professor Karen Chin, who has built a unique field-based career in paleoecology (the ecology of ancient ecosystems) sitting in the Kaiparowitz Formation of Utah, USA, studying a (very) fossilized dinosaur coprolite.

she has been able to reconstruct dinosaur-based ecosystems with unprecedented resolution.

Another highly productive member of this generation, verging on being a Millennial, is the University of Edinburgh's affable Stephen Brusatte (Figure 16.20). Brusatte is an American; trained in the United States, although he obtained his MS (and his wife!), in Bristol, UK, working with Baby-boomer M. J. Benton (see below). Brusatte has made important contributions in dinosaur paleoecology, behavior, dinosaur origins *and* extinction, Mesozoic mammal evolution, functional morphology, phylogeny, all kinds of nondinosaurian archosaurs (phytosaurs, pterosaurs, crocodylians. . .).

It is no mistake that Brusatte came through the Benton lab at the University of Bristol in his career. There Benton (Figure 16.21) and colleagues have built what is arguably the one of the best evolutionary biology (including paleobiology) programs in the world. Benton himself is a prodigious dinosaur paleontologist whose work can be found throughout pages of this book; in 2019 he published *Dinosaurs Rediscovered*, which highlights the extraordinary contributions of his program to dinosaur research. Also at Bristol, and approaching dinosaurs from a slightly different perspective, is Professor Emily Rayfield (Figure 16.22), who has applied biomechanical engineering, computer programming, and CT scanning skills to the finite element analysis of bones, fossil (see Figure 6.16) and recent. These techniques allow her to obtain a quantitative understanding, based upon stresses undergone by their bones, of how animals – including dinosaurs – would have functioned in life.

Richard J. Butler, like Brusatte, is an extraordinarily productive researcher (Figure 16.20). His interests involve the early evolution of dinosaurs, as well as the vertebrate recovery following Permo-Triassic mass extinction (the largest extinction in the history of Earth), in archosaur systematics and functional anatomy, and in the nature of – and patterns in – the fossil record. Butler has carried out fieldwork in the Triassic of Argentina, South Africa, Portugal, Lithuania, and Poland.

Anusuya Chinsamy-Turan (Figure 16.23) is a productive and important South African paleontologist, now a Professor of Biology at the University of Capetown, who has done breakthrough work in dinosaur bone histology. It was largely through her studies that people were able to understand the significance of bone remodeling, develop means of determining the age of dinosaurs, and correlate morphology with various growth stages in dinosaurs lives. Beyond an impressive catalog of peer-reviewed scientific papers, Chinsamy-Turan has authored four books (three technical and one popular), and has been the recipient of a number of distinguished awards, from the "President's Award" of the National Research Foundation of South

Figure 16.20. Paleontologists Professor Richard J. Butler (left) and Professor Stephen Brusatte (right) looking for dinosaurs deep in red Triassic claystones of Lithuania.

Figure 16.21. Professor Michael Benton, the British dinosaur paleontologist whose work has dominated dinosaur research for many years. Benton commonly applies sophisticated statistical approaches to reveal underlying patterns that shaped dinosaur history.

Africa (1995), to the World Academy of Science "Sub-Saharan Prize for the Public Understanding and Popuarization of Science" (2013).

Paul Barrett (Figure 16.24), along with the standard taxonomic revisions and descriptions that are the bread and butter of the field (you'll note that Barrett was a coauthor of the very nonstandard contribution in which newly formed Ornithoscelida and Saurischa were proposed as substitutes for the venerable [and flawed?] but fundamental taxa Ornithischia and Saurischia), is interested in macroevolutionary patterns and mechanisms; that is, large-scale controls on dinosaur diversity and morphology. He has carried out extensive fieldwork in the UK, China, South Africa, and Australia and Antarctica. He, like many colleagues in his generation, has been particularly successful in applying formidable statistical skills to their research.

The University of Texas' highly accomplished Julia Clarke (Figure 16.25) is a graduate of Yale University with Jacques Gauthier (see Figure 16.12). Clarke's career has been built around early bird evolution; a timely choice, given the revolution that our understanding of bird evolution has undergone in the past. Clarke is interested in bird diversity, systematics, and function. With the new-found ability to identify colors in feathers, she has been a pioneer in the reconstruction of plumage coloration in nonavian dinosaurs. Her work has taken her to New Zealand, Antarctica, Mongolia, Argentina, the United States, and China. It was she that found the first representative of modern birds from the Mesozoic, the Cretaceous *Vegavis* (Figure 8.13).

The University of Calgary's Darla Zelenitsky (Figure 16.26) is particularly interested in dinosaur paleobiology, including senses, reproduction, and locomotion. She has published scientific papers on a broad range of dinosaur paleobiology, including the evolution of nesting, wings, and smell, and done extensive amounts of fieldwork, particularly in her "backyard," which just happens to be Dinosaur Provincial Park, Alberta, CA.

Some workers specialize in particular groups, rather than techniques applied across groups. One such worker would be Professor Kristi Curry Rogers, of Macalester College (Figure 16.27), who, although she is of course interested in all things dinosaurian, has nonetheless carved out an important and enviable reputation for working on perhaps the most difficult of dinosaurs: sauropods, whose extraordinary size has made the logistics of study challenging, to say the least.

Another highly accomplished dinosaur worker from the youngest generation of paleontologists is Lindsay Zanno, Assistant Professor of Biological Sciences at North Carolina State University, and Head of Paleontology, North Carolina State Museum (Figure 16.28). As a veteran field worker, she

Figure 16.22. Professor Emily Rayfield, Benton's colleague at the University of Bristol. Remembering that the shape of bones is related to the stresses they undergo in life, Rayfield has used finite element modeling to identify stresses undergone by the bones, which in turn has allowed her to quantitatively reconstruct dinosaur (and other extinct vertebrate) function.

Figure 16.23. Professor Anusuya Chinsamy-Turan, South African paleontologist and paleohistologist, whose work generated a revolution in understanding what the word "lifetime" meant to a dinosaur.

Figure 16.24. Professor Paul Barrett, getting his pants dirty in the field, excavating the partial skeleton of an early sauropodomorph in the Late Triassic Lower Elliot Formation near Lady Grey, Eastern Cape, South Africa.

Figure 16.25. University of Texas paleontologist Professor Julia Clarke, doing what she loves in the wilds of Argentina.

Figure 16.26. Professor Darla Zelenitsky, gluing some eggshell fragments in under the hot Canadian sun, before extracting them.

Figure 16.27. Professor Kristi Curry Rogers and her team, excavating a sauropod bone bed in the Maevarano Formation, Upper Cretaceous, Madagascar.

Figure 16.28. Professor Lindsay Zanno delicately manipulating a rock saw in the Triassic wilds of North Carolina, extracting the bones of a phytosaur (Figure 5.3a) from rock matrix she describes as "back-breakingly hard."

Figure 16.29. Professor Sterling Nesbitt doing field-work "always," as he bemoans, "in the Triassic!" The time was evidently well spent; Nesbitt is one of the world's preeminent authorities on Triassic vertebrates.

has worked globally, but has lately focused on the Cretaceous of the Western Interior of North America, where the combination of high dinosaur diversity, faunal turnovers, and greenhouse conditions has permitted rich insights into Late Cretaceous dinosaurs and ecosystems. Zanno has had a remarkably productive publishing career, including over 100 scientific papers, and a non-paleonotological children's book, *The Fall Ball.*

When it comes to the Triassic – Early Jurassic interval, that time in Earth history when non-dinosaur dinosauromorphs ceded the field to true dinosaurs – perhaps the most authoritative specialist in the world is Dr. Sterling Nesbitt, Assistant Professor of Geosciences at Virginia Polytechnic Institute and State University (Virginia Tech) (Figure 16.29). Paleontology, as we have seen, is not a field for prodigies such as one might find in music, but Nesbitt, we can vouch from personal experience, sounded like a PhD-bearing paleontologist while still in high school! Since then he has carried out exciting research that has taken him to Tanzania, Zambia, South Africa, Madagascar, Argentina, Mongolia, and of course all over United States. While we think of him particularly knowledgeable in the Late Triassic of the American Southwest, specialists who have worked with him elsewhere – such as Tanzania and Zambia – doubtless think of him as *the* authority in the Triassic of those places. With just under 100 technical publications and the discovery of many new genera, including the 3' tall tyrannosauroid *Suskityrannus*, under his belt,[10] Nesbitt has been one of the most promising, and productive, of the most recent generation of young paleontologist professionals.

Our utterly incomplete list of fresh, exciting paleontologists must not omit the prodigious Jingmai O'Connor (Figure 16.30),"paleontologista" (her term), the youngest Professor ever hired at China's

---

[10] He found *Suskityrannus* at 16, as a junior in high school!

IVPP. O'Connor is a Mesozoic bird specialist, and that background, in combination with the discoveries of Mesozoic birds in China, has placed the right person, with the right skills and talent to study these faunas, in the right place. The results have been spectacular, all the more impressive since O'Connor, aside from taking the lead in describing extraordinary fossils, asks broader, more complex paleoecological questions in an attempt to understand ancient behavior and ecosystems of early Aves, in particular, that of the Jehol biota described in these pages (see Chapter 8). In recognition of O'Connor's extraordinary accomplishments, the US-based Paleontological Society awarded her its highly prestigious and sought-after Schuchert Award[11] in 2019.

All of these workers – from Anusaya Chinsamy to Darla Zeletinsky – and so many more that we simply do not have room to highlight, are distinguished for the breadth of their interests, for their energy, and for the intelligence and creativity that they have applied to the questions they are asking. They all seek and have obtained extensive independent funding, and have ultimately built their careers on the love of, and enthusiasm for, paleontology and Earth history. Vertebrate paleontology is a career that is actively and aggressively pursued, rather than a default into which one falls.

Figure 16.30. "Paleontologista" Professor Jingmai O'Connor (and enantiornithine friend *Longipteryx*), seen at the IVPP, Beijing, China, where she works with the world's finest collection of Mesozoic feathered dinosaurs, nonavian and avian.

And like most other active professional paleontologists, the paleontologists we've highlighted here know the history of their chosen field and, to a person, each has consciously inherited the mantle from the grand (and not so grand) old paleontologists who preceded them, beginning with Mantell's Mantle, so to speak. They would all agree with Isaac Newton's famous epigram, "If I have seen further than others, it is by standing upon the shoulders of giants."

SUMMARY

Paleontology is a human endeavor, and like all human endeavors, ideas have changed as the context in which those ideas developed has changed. Our earliest suggestion that humans may have seen and taken note of dinosaur fossils comes from the recognition that mythical creatures may have been inspired by observations of very large or unfamiliar-looking dinosaur fossils.

Paleontology as a science began in the Enlightenment with the recognition that observation in combination with logic and rational thinking could reveal truths about the natural world. The earliest dinosaur fossil explicitly identified as something quite unlike anything alive today was found in 1822; within 40 years, not only was a variety of extinct animals recognized, but a relative geological time scale had been constructed. It remains valid in its essentials to this very day. In 1842, English anatomist Richard Owen established the word "Dinosauria" for an extinct group of reptiles, partly to demonstrate that organisms had not evolved. His concept of dinosaurs was one of bulky, elephantine quadrupeds, with a mammal – like metabolism.

[11] The Schuchert Award is given each year to a paleontologist under 40, whose accomplishments suggest unusual promise in that person's future career. O'Connor was an obvious choice.

The revolutionary idea that other worlds had existed that were very different from our own gained broad currency. At the same time, Dinosauria became far better known, and its members seemed so disparate that they were divided into two groups (Ornithischia and Saurischia) and thought to have had separate origins within early archosaurs (which were all lumped together as a group called "Thecodontia"). The rise of dinosaurs was interpreted in Darwinian terms as the competitive success of the superior forms (dinosaurs) over inferior forms (primitive archosaurs and advanced, nonmammalian synapsids).

The first 70 or so years of the twentieth century were largely about collecting, describing, and enhancing knowledge of the different forms of dinosaurs. This abruptly changed with John Ostrom's 1969–1970 interpretation of *Deinonychus* as endothermic and his 1974 reevaluation of *Archaeopteryx* as a theropod. These revolutionary views came as the field of paleontology was revitalized as paleobiology, and as phylogenetic systematics came to be recognized as a truly scientific way of inferring relationships among even extinct forms. Phylogenetic systematics demonstrated the fundamental monophyly of Dinosauria (also affirming, parenthetically, the monophyly of Saurischia and Ornithischia), destroyed "Thecodontia," and clearly showed that birds are dinosaurs.

When, in 1980, it was postulated that an asteroid caused the end-Cretaceous extinction (which obliterated the nonavian dinosaurs), paleontologists came to recognize that extraterrestrial events can have a profound effect on earthly events; that extinctions can occur regardless of how well adapted a particular group is. It was at this time that a reevaluation of the rise of Dinosauria was carried out, in which the success of the group was no longer ascribed to competitive superiority, but rather to extinctions that had liberated ecospace for dinosaurs to colonize.

Today, dinosaurs continue to be studied in a variety of ways: through the discovery and description of new forms; as biological entities functioning in ancient ecosystems; via analyses of the large-scale evolution of the group; through histological and even molecular analyses; and from the standpoint of evolution and development.

SELECTED READINGS

Andrews, R. C. 1929. *Ends of the Earth*. G. P. Putnam's Sons, New York, 355 pages.

Bakker, R. T. 1986. *Dinosaur Heresies*. William Morrow, New York, 481 pages.

Bakker, R. T. and Galton, P. M. 1974. Dinosaur monophyly and a new class of vertebrates. *Nature*, 248, 168–172.

Benton, M. J. 1984. Dinosaurs' lucky break. *Natural History*, 93(6), 54–59.

Benton, M. J. 2019. *The Dinosaurs Rediscovered*. Thames & Hudson, London, 319 pages.

Brusatte, S. L. 2012. *Dinosaur Paleobiology*. Wiley-Blackwell, UK, 322 pages.

Brusatte, S. L. 2018. *The Rise and Fall of Dinosaurs*. William Morrow, New York, 404 pages.

Bryson, B. 2003. *A Short History of Nearly Everything*. Broadway Books, New York, 544 pages.

Cadbury, D. 2000. *The Dinosaur Hunters*. Fourth Estate, London, 374 pages.

Cadbury, D. 2001. *Terrible Lizard*. Henry Holt and Company, New York, 384 pages.

Desmond, A. 1975. *The Hot-Blooded Dinosaurs*. The Dial Press, New York, 238 pages.

Desmond, A. 1982. *Archetypes and Ancestors*. University of Chicago Press, Chicago, IL, 287 pages.

Dingus, L. and Norell, M. A. 2010. *Barnum Brown. The Man who Discovered Tyrannosaurus rex*. University of California Press, Berkeley, CA, 368 pages.

Gauthier, J. A. 1986. Saurischian monophyly and the origin of birds. In Padian, K. (ed.) *The Origin of Birds and the Evolution of Flight*. Memoirs of the California Academy of Sciences no. 8, San Francisco, pp. 1–56.

Gauthier, J. A., Kluge, A. G., and Rowe, T. 1988. Amniote phylogeny and the importance of fossils. *Cladistics*, 4, 105–209.

Hennig, W. 1979. *Phylogenetic Systematics*, translation by D. D. Davis and R. Zangerl. University of Illinois Press, Urbana, IL, 263 pages.

Lessem, D. 1992. *The Kings of Creation.* Simon and Schuster, New York, 367 pages.

Mayor, A. 2000. *The First Fossil Hunters.* Princeton University Press, Princeton, NJ, 361 pages.

Nothdurft, W. (with Smith, J.) 2002. *The Lost Dinosaurs of Egypt.* Random House, New York, p. 242.

Ottaviani, J., Cannon, Z., Petosky, S., Cannon, C., and Schultz, M. 2005. *Bone Sharps, Cowboys, and Thunder Lizards: A Tale of Edward Drinker Cope, Othniel Charles Marsh, and the Gilded Age of Paleontology.* GT Labs, New York, 168 pages.

Preston, D. J. 1986. *Dinosaurs in the Attic: An Excursion into the American Museum of Natural History.* St. Martin's Press, New York, 244 pages.

Shubin, N. 2008. *Your Inner Fish: A Journey Into the 3.5-Billion-Year History of the Human Body.* Pantheon Books, New York, 230 pages.

Sternberg, C. H. 1985. *Hunting Dinosaurs in the Bad Lands of the Red Deer River, Alberta, Canada.* NeWest Press, Edmonton, Alberta, 235 pages.

Thomas, R. D. K. and Olson, E. C. (eds.) 1980. *A Cold Look at the Warm-Blooded Dinosaurs.* AAAS Selected Symposium no. 28, Washington, DC, 514 pages.

TOPIC QUESTIONS

1. Who invented the term "Dinosauria?" What was his idea about dinosaur metabolism?
2. What discoveries helped nineteenth-century paleontologists get a more modern idea of what dinosaurs looked like?
3. Describe a dinosaur as imagined by Victorian paleontologists. What changed our ideas from the Victorian conception?
4. Describe the kinds of vertebrate faunas that existed just before dinosaurs became the most abundant and diverse terrestrial vertebrates.
5. What were the dramatic "revolutions" that changed the face of paleontology after the 1960s?
6. Cite an example of how early twentieth-century imperialism moved the science of paleontology forward.
7. Give two examples of how cladograms changed *paleontological* thinking.
8. In what fundamental ways does the legend of the griffin differ from our understanding of *Protoceratops*?
9. How did the Cope–Marsh competition help the field of paleontology? How did it hurt it?
10. Explain the difference between a cladistic view of dinosaur origins and H. G. Seeley's understanding of the origins of dinosaurs.

The Cretaceous–Paleogene Extinction

The Frill is Gone

> **WHAT'S IN THIS CHAPTER**
>
> Perhaps you noticed that nonavian dinosaurs, the subjects of this book, aren't with us anymore? And so, not unreasonably, you might ask, "What happened to the dinosaurs?". The answer to that simple question, friends, is not so simple, and so the objective of this chapter is to look into what we know about the dinosaur extinction.

How Important Were the Deaths of a Few Dinosaurs?

The mass extinction to which the nonavian dinosaurs finally succumbed after thriving for about 164 million years on Earth is called the **Cretaceous–Paleogene extinction**, and is commonly abbreviated as **K-Pg**.[1] The K-Pg mass extinction (Box 17.1) involved much more than just dinosaurs. Among the "highlights" were:

- an approximately 10 km asteroid collided with Earth;
- the great cycles of nutrients that formed complex food webs in the world's oceans were temporarily disrupted and likely, at least partially, shut down;
- many marine and terrestrial animals, as well as many plants, went extinct;
- landscapes were deforested;
- epic tsunamis rolled across ocean basins; and
- wildfires likely raged across the continents.

By comparison with that, how important were the deaths of a few dinosaurs?

Asteroid Impact!

Whatever else is truth, the Cretaceous ended – and the Paleogene began – with a real bang. Here's how we learned about it. In the late 1970s, geologist Walter Alvarez and a team of coworkers (Figure 17.1) were studying K-Pg marine outcrops now exposed on land near a town called Gubbio, in Italy. They were struck by the fact that the lower half of the Gubbio rock exposure is composed of a rock made up entirely of thin beds of the microscopically sized shells of *Cretaceous* marine organisms. The upper half of the exposure was almost exclusively made of thin beds of the microscopic shells of *Paleogene* marine organisms. Between the two was a thin (2–3 cm) layer of clay, obviously the K-Pg boundary.

Analyses showed that the clay layer contained unusually high concentrations of **iridium**, a rare, platinum-group metal.[2] Instead of the expected amount at the Earth's surface, about 0.3 parts per billion (ppb), the iridium content was a whopping 10 ppb (parts per billion) at Gubbio. So the **iridium anomaly**, as it came to be called, contained about 30 times as much iridium as Alvarez and his coworkers had expected to find (Figure 17.2).

Iridium is normally found at the Earth's surface in very low concentrations, but it is found in higher concentrations in the core of the Earth and from **extraterrestrial** sources; that is, from outer space. Given that, the Alvarez group determined that the source of the iridium had to be extraterrestrial. The deal was sealed when they found iridium anomalies at two other K-Pg sites, one in Denmark and the other in New Zealand. Science often proceeds via brilliant intuition, and with exactly that (and little else), they concluded that, at 66.1 Ma,[3] a large **bolide**, that is, a fiery impacting body such as an asteroid, had to have smacked into Earth, delivering the iridium and, coincidently,

---

[1] The "K" comes from the German word *Kreide*, or chalk, because the Cretaceous was first identified at the chalk cliffs of Dover (England; the word "Cretaceous" is based upon the Latin for chalk, or *creta*). In older nomenclature, the boundary is sometimes called "K/T" for Cretaceous/Tertiary. The Tertiary is an outdated term that included the Paleogene.

[2] It is a common misconception that iridium metal is toxic and deadly. In fact, like its chemical relatives gold and platinum, it is quite unreactive. For those with significant disposable incomes, boutique fountain pens and watches made with iridium are available. Your bank account might go extinct if you buy them, however.

[3] The precise date was not actually known at the time; people thought it was around 65 Ma. It took almost 30 years before the high-precision date we report here – 66.1 Ma – was obtained.

Figure 17.1. The team of University of California (Berkeley) scientists who first successfully proposed the theory of an asteroid impact at the K-Pg boundary. Left to right: geochemists Helen V. Michel and Frank Asaro, geologist Walter Alvarez, and physicist Luis Alvarez.

causing the K-Pg mass extinction. Luis Alvarez, Nobel Prize-winning physicist, and a member of the team, described the relationship between an asteroid impact and the iridium layer in this way:

> When the asteroid hit, it threw up a great cloud of dust that quickly encircled the globe. It is now seen worldwide, typically as a clay layer a few centimeters thick in which we see a relatively high concentration of the element iridium – this element is very abundant in meteorites, and presumably in asteroids, but is very rare on earth. The evidence that we have is largely from chemical analyses of the material in this clay layer. In fact, meteoric iridium content is more than that of crustal material by nearly a factor of 10^4. So, if something does hit the earth from the outside, you can detect it because of this great enhancement. Iridium is depleted in the earth's crust relative to normal solar system material because when the earth heated up [during its formation] and the molten iron sank to form the core, it "scrubbed out" [i.e., removed] the platinum group elements in an alloying process and took them "downstairs" [to the core]. (Alvarez, 1983, p. 627)

Because the three sites are distributed around the globe, Alvarez and coworkers calculated that the asteroid had to have been about 10 km (about 6 miles) in diameter to spread an iridium dust layer globally. Really, it was more of a good idea than much else; but at least one geoscientist has since called it "the most important idea in geology."[4]

In the intervening years, a tremendous amount of work has been done to explore the possibility of an asteroid impact at 66.1 Ma. Most importantly, the number of K-Pg sites with anomalous concentrations of iridium at the boundary has reached well over 100 (Figure 17.3). Moreover, the iridium anomaly was discovered on land (Figure 17.4) as well as in ocean sediments, affirming that it is a global phenomenon.

---

[4] UC Berkeley plate tectonicist M. A. Richards. A lot of people would disagree; certainly plate tectonics, which explains so much more than continental drift, is a contender for the prize. Our personal preference would perhaps be the gift of geological time. Nonetheless, the 1980 Alvarez paper provoked a revolution in the geosciences that extended far past the K-Pg boundary. A user-friendly treatment of this is found in Powell, J. L. 1998. *Night Comes to the Cretaceous*. W. H. Freeman and Company, New York, 250 pages.

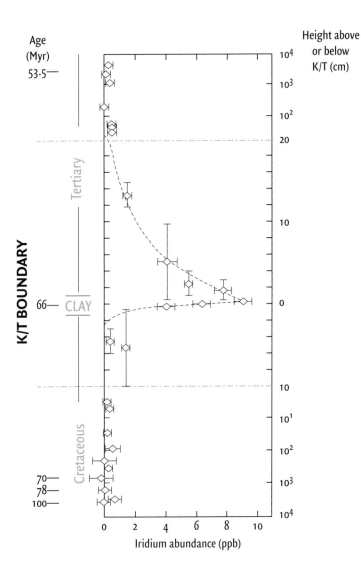

Figure 17.2. The iridium (Ir) anomaly at Gubbio, Italy. The amount of Ir increases dramatically at the clay layer to 9 ppb (parts per billion), and then decreases gradually above it, returning to a background count of about 1 ppb. On the right are numbers representing the thickness of the rock outcrop; on the left the time intervals (in millions of years (Myr)) and rock types are identified. Note that the vertical scale is *linear* close to the K-Pg boundary, but *logarithmic* away from the boundary, to show results well above and well below the boundary.

Shocked quartz and microtektites also came to be recognized as part of the fingerprint left by the asteroid. "Shocked quartz" is the name given to quartz that has been placed under such pressure that its molecular structure becomes deformed (Figure 17.5). It is now recognized that the kind of pressure that can cause such deformation could only be generated by impacts; indeed, shocked quartz is now known from many different impact sites, and has come to be recognized as a diagnostic criterion for meteor impacts.

Microtektites are small, droplet-shaped blobs of silica-rich glass. They represent material thrown up into the atmosphere in a gaseous state due to the tremendous energy released when a meteor strikes the Earth. Quick cooling occurs while they're still airborne and they liquify and then plummet down on Earth as a rain of solid, glassy blobs (Figure 17.6).

The "Smoking Gun"

As early as 1981, a bowl-shaped structure 180 km in diameter, buried under many meters of more recent sediments, was reported from the Yucatan Peninsula of Mexico, in the region near the town of

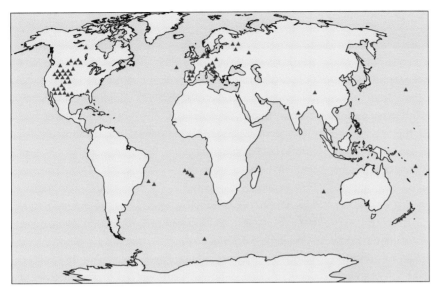

Figure 17.3. More than 100 K-Pg iridium anomalies are known around the world, about a third of which are shown here. Paleontologist W. A. Clemens, Jr., used to call it the "McDonalds" diagram because it showed "over [100] anomalies served!"

Figure 17.4. The iridium-bearing clay layer in Montana, USA; one of the first terrestrial localities where anomalous concentrations of iridium were discovered (they had heretofore been marine localities). Lines bracket thin Cretaceous–Paleogene boundary claystone. The sediments below the boundary claystone preserve dinosaurs; none has been found above the boundary claystone. To the left of the image are visible deposits of a small channel that carved into and eroded the claystone.

Box 17.1 Extinction

Paleontologists generally divide extinctions into two categories. The first is the so-called **background extinctions**, isolated extinctions of species that occur in an ongoing fashion. The second type is called **mass extinctions**. The latter certainly have caught the media and public's attention, and they appear to be something *qualitatively* as well as quantitatively different than background extinctions.

Background Extinctions

Although background extinctions are less glamorous than mass extinctions, they are essential to biotic turnover: University of Tennessee paleobiologist M. L. McKinney has estimated that as much as 95 percent of all extinctions can be accounted for by background extinctions. Isolated species disappear from a variety of causes, including out-competition, depletion of resources in a habitat, changes in climate, the growth or weathering of a mountain range, river channel migration, the eruption of a volcano, the drying of a lake, the spraying of a pesticide, or the destruction of a forest, grassland, or wetland habitat.

Dinosaur populations had a species-turnover rate of around 2 million years per species. This means that each species lasted about 2 million years, before a new one appeared and the old one disappeared.[a] Although some dinosaur extinctions coincided with earlier mass extinction events (such as those at the Triassic–Jurassic and Cretaceous–Paleogene boundaries), most dinosaurs fell prey to background extinctions. By far the majority of favorite and famous dinosaurs – *Maiasaura*, *Dilophosaurus*, *Protoceratops*, *Deinocheirus*, *Styracosaurus*, *Velociraptor*, *Iguanodon*, *Ouranosaurus*, *Allosaurus* (to name a tiny fraction) – were the victims of background extinctions. The ultimate dinosaur extinction didn't wipe out the total number of species accumulated over some 164 million years, it killed only the latest-evolved representatives of the group (see Figure 15.1).

Mass Extinctions

Mass extinctions involve *large numbers* of *species* and *many types* of species undergoing *global extinction* in a *geologically short* period of time. None of these has a truly precise definition, because there are no fixed rules for mass extinctions. Indeed, how do we know that there even were mass extinction "events" and how can we recognize them? A compilation of invertebrate extinctions through

[a] This is a simple statistical average for all of Dinosauria. It was not necessary for an older species to disappear before its descendant appeared.

Chicxulub (translated approximately as "devil's tail"; Figure 17.7). Ten years later, drill cores taken through the structure revealed shocked quartz: Chicxulub was a buried impact structure.

At about the same time, an approximately 1 m thick sequence of glass was discovered in Haiti, suggesting that the source of the glass had to be somewhere relatively nearby. Its chemical composition was shown to be the same as the composition of the rocks that make up the Chicxulub structure.

The pieces really started falling into place. Several years earlier (1988) evidence of a tsunami in K-Pg deposits in the Gulf Coast region of Texas had been reported. The Chicxulub site was well situated to produce the tidal wave deposits recognized in the sedimentary record. Finally, the Chicxulub structure has been dated at 66.1 Ma, the time of the K-Pg boundary.

Further study of Chicxulub below the surface of the Earth, using sophisticated geophysical techniques, showed a bull's-eye pattern with a circular peak and large concentric rings around it, representing topography preserved in buried rocks below the surface (Figure 17.8). Interestingly enough, the northwest part of the outermost ring is broken through. The distinctive ring pattern time (Figure B17.1.1) shows that, although extinctions characterize all periods (it is these that are termed background extinctions), there are intervals of time in which extinction levels are significantly elevated above background levels. Such intervals are said to contain the mass extinctions. Fifteen such intervals are recognized, of which five clearly towered above the others (Figure B17.1.1).

The 15 mass extinctions are classified into "minor," "intermediate," and "major" mass extinctions, on the basis of the amount of extinction that took place above background. In the entire history of life, only one extinction qualifies as "major"; that is, the Permian–Triassic (commonly called **Permo-Triassic**) extinction, 251 Ma. The remaining four of the Big Five – including "dinosaur extinction" (which, as we have seen, was far more than just dinosaurs) – are considered to have been "intermediate." The rest are considered "minor," although undoubtedly not to the organisms that succumbed during them.

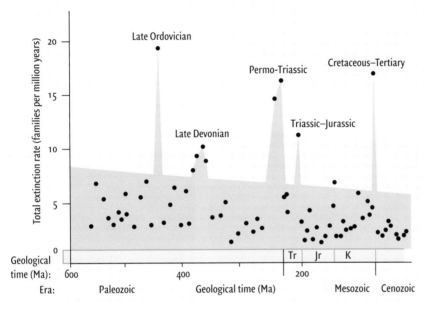

Figure B17.1.1. A compilation of extinction through time, taken from the work of D. M. Raup and J. J. Sepkoski. The most significant were the Late Ordovician (445 Ma to 444 Ma), the Late Devonian (around 373 Ma to 359 Ma), the Permo-Triassic (251 Ma), the Triassic–Jurassic (201 Ma), and the Cretaceous–Paleogene (66.1 Ma). A sixth mass extinction may soon join the list of the "most significant": the one we're in the middle of right now! Tr, Triassic; J, Jurassic, K, Cretaceous.

suggests that a large asteroid, 10 to 15 km in diameter,[5] approached Earth from the southeast at a low angle of about 30°.[6] The distribution of iridium, shocked quartz, and microtektites across the Western Interior of North America (north and west of the crater) reinforce the idea of a low-angle, directional impact (Figure 17.9).

What did the asteroid do to Earth when it struck? Numerous scenarios were proposed, most of them inspired by postnuclear apocalyptic visions. Of these, only a few remain current.

- *Blockage of sunlight.* It was initially hypothesized that sunlight would have been blocked from the Earth for about three to four months. This would have caused a cessation of photosynthesis and a short-term temperature decrease (now called an impact winter).

---

[5] Recall that previous estimates of size were based upon the global distribution of the impact ejecta; this estimate was based upon the morphology of the impact site.

[6] Some recent models have proposed as high an angle as 60°.

Figure 17.5. Shocked quartz from the terrestrial K-Pg boundary in eastern Montana. The etched angled lines across the face of a grain of quartz represent a failure of the crystal lattice along known crystallographic directions within the mineral. Grain is 70 μm across (1 μm = 10^{-6} m).

Figure 17.6. Microtektites. (a) In-situ tektites from the K-Pg boundary at Dogie Creek, Wyoming; (b) a single isolated microtektite.

- *Infra-red radiation pulse.* An 8 km asteroid, travelling at many times the speed of sound, and then being instantaneously brought to a complete stop would have to release the kinetic energy it possessed.[7] It has been proposed that tremendous amounts of energy in the form of infra-red

---

[7] D. S. Robertson and colleagues (2004) succinctly described it this way: "an asteroid 10–15 km in diameter, having a mass of ~1–4 × 10^{15} kg, arriving at tens of kilometers per second at an angle of perhaps 45° to Earth's surface, that produced a collapsed transient cavity ~80–100 km in diameter and a multi-ring basin 170–200 km in diameter on the Yucatán peninsula of Mexico." (Robertson, D. S. *et al.* 2004. *GSA Bulletin,* 116, 761.)

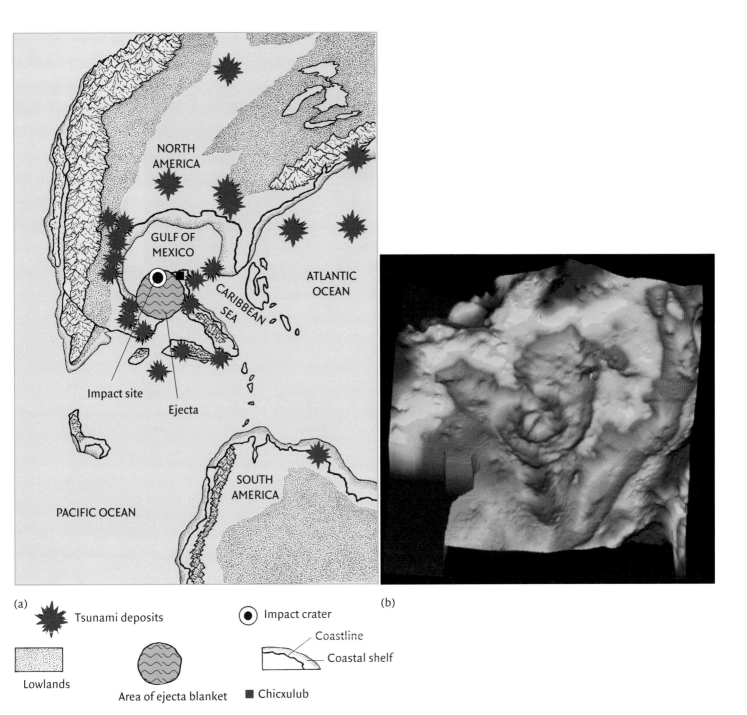

(a)

Tsunami deposits

Lowlands

Area of ejecta blanket

(b)

Impact crater

Coastline

Coastal shelf

Chicxulub

Figure 17.7. (a) Paleogeographic map of the ground-zero region for the K-Pg asteroid, Yucatan Peninsula, Mexico. The geography of the region as we know it today is contrasted against the geography of the region at 66.1 Ma. The pink-brown color represents the land masses during latest Cretaceous time; the blue shows modern the continental margins, with the modern exposed North and South American continents outlined with a thicker line. (b) Three-dimensional geophysical reconstruction of the remnants of the Chicxulub crater. A gravimeter measures subsurface changes in gravitational attraction of rocks under the town of Chicxulub. These variations in gravitational attraction show a large-scale bull's-eye pattern of concentric rings, diagnostic of a meteor impact. North is toward the top of the page.

Figure 17.8. Reconstruction of an asteroid impact with Earth. Planetary geologist P. H. Schultz and geobiologist S. L. D'Hondt suggest that the asteroid struck Earth at an angle of about 30°, coming from the southeast.

radiation and heat must have been released immediately upon impact. The initial global heat release at ground zero might have been 50 to 150 times as much as the energy of the sun as it normally strikes Earth. One group of scientists likened this radiation at the Earth's surface to an oven left on the broil.

- *Global wildfires.* With so much instantaneous heat production, fires might have broken out spontaneously around the globe. Soot-rich horizons from K-Pg sites in Europe, North America, and New Zealand have been identified, in which the amount of the element carbon (the soot) was enriched between 100 and 10 000 times over background. The soot has been attributed to wildfires, perhaps resulting from the infra-red heat pulse.

All of these catastrophic effects were short term, which means that they affected the globe for days, months, or at most a few years. In a longer-term sense, that is, on ecological time scales (10^4 to 10^6 years), climates were little affected by the asteroid impact. What we know of climates in the latest Cretaceous suggests that, considered overall, they did not differ significantly from those in the early Paleogene.

Volcanic Eruptions

The Earth wasn't completely peaceful – it never is – before the asteroid impact. In fact, in India, there was an unusual and very powerful volcanic episode going on – the **Deccan Traps**. Deccan volcanism wasn't your basic Mt. Vesuvius–Krakatoa experience – short-term explosive eruption sending a geyser-like column of hot ash miles into the stratosphere, followed by red-tinged sunsets from all the ash particles floating in the atmosphere – no, it was something potentially far deadlier.

The Deccan Traps were a series of continental **flood basalts**: multiple episodes of hot basaltic lava pouring through fissures in the Earth's crust. No big explosion, no huge dust cloud, but as the molten lava poured out over the Earth's surface, the pressure that the lavas were under when buried deep within the Earth, was released, and gases – certainly CO_2, SO_2, NO_2, and methane, among many

Figure 17.9. Latest Cretaceous basalt flows of the Deccan Traps, exposed on the western part of the Indian subcontinent.

others – were emitted into the atmosphere. Ultimately the Deccan Traps produced more than about 1.3 million km^3 of basalt, leaving a very palpable record on the Indian subcontinent (Figure 17.9).

The Deccan Traps are composed of a series of closely timed (in a geological sense) events, the most important of which contributed a basaltic plateau some 3000 m thick on the western Indian subcontinent, timed around the K-Pg boundary. But how close (in time) to the K-Pg boundary was the massive outpouring of basalt? Here, things get a bit murky. Two important studies were published simultaneously in the prestigious journal *Science*, each led by veteran K-Pg geochronologists (scientists who do the kind of numerical dating described in Chapter 2). But the results, and thus their conclusions, did not agree. One study, led by Princeton's Blair Schoene, used U-Pb dating to identify four high-volume eruptive time intervals. They identified the K-Pg boundary by using published dates from Chicxulub. As these authors saw it, the two highest eruption rates occurred just before (around 65.9 Ma) and after (around 66.04 Ma) the Chicxulub K-Pg date (with a third major outpouring occurring just a bit later, at about 66.17 Ma), and they thus concluded that both Deccan volcanism and the asteroid impact had to be involved in the extinction.

The other study team, led by UC Berkeley's Courtney Sprain, came to a somewhat different conclusion. First they found the K-Pg boundary in the Deccan Traps using ^{40}Ar– ^{39}Ar dating. Then they determined that, given the position in time of that date, 90 percent of the volume of the Deccan

Traps erupted within less than a million years, and 75 percent of that total eruptive volume occurred after the K-Pg boundary. In short, they did not interpret the K-Pg boundary eruption as the extinction-causing force that Schoene and colleagues interpreted it to be.

But what do we know about the extinction, itself? Who went extinct, and how quickly did this occur? For that, we turn to the biotic record of the K-Pg boundary.

Biological Record of the Latest Cretaceous

Like any good murder mystery, no amount of volcanoes; bolides; hot climates; cold climates; natural (or unnatural) catastrophes – or any other causes – can explain *any* extinction until we understand the anatomy of the extinction itself. What went extinct? When did those extinction(s) occur? Was there one or were there many? Were they fast or were they slow? Who survived and who did not? These things were very poorly known at the time of the Alvarez hypothesis, which made their conclusion that much more radical. Now, with many years under our belts of research in this mass extinction, we have far better answers than we had in 1980.

Oceans

Continental Seas and Shelves

Because the shallow seas that covered large expanses of the continents receded before the K-Pg boundary, very few shallow marine deposits are preserved that record the last 2 to 3 million years of the Cretaceous. And because many groups of organisms lived and died in shallow continental seas and shelves, we lack data for such groups.

How well or badly fishes and sharks fared remains largely conjectural, although it is apparent that, where sharks, skates, and rays are understood (at this point, only in North America), there was a significant extinction.

The whale- and dolphin-like marine diapsids called **ichthyosaurs** (Figure 17.10a) are known to have disappeared well before the K-Pg boundary. Not so in the case of marine-adapted lizards called **mosasaurs** (Figure 17.10b). Recent work suggests that these went extinct geologically abruptly, at the end of the Cretaceous. More equivocal is the record of **plesiosaurs**, the long-necked, Loch Ness-type, fish-eating diapsids of the Jurassic and Cretaceous (Figure 17.10c), for whom there are, Loch Ness notwithstanding, no credible post K-Pg records.

Among fossil invertebrates, perhaps the most famous group is the **ammonites** (Figure 17.10d). Ammonites lived right up to the K-Pg boundary, before finally going extinct. Another important group of invertebrates are the bivalves. Careful study has shown that, with one exception (which went extinct much earlier), 63 percent of all bivalves went extinct sometime within the last 10 million years of the Cretaceous. The record is, unfortunately, not yet more precise than this, but it does show that the extinction took place without regard for latitude: bivalves in temperate regions were just as likely to go extinct as those in the tropics.

Marine Microorganisms

Because of the richness of their fossil record, **foraminifera**, marine microscopic, shelled, single-celled organisms that are either **planktonic** (living in the water column) or **benthic** (living within sediments) have dominated discussions of K-Pg boundary events (Figure 17.11). **Micropaleontologists** studying foraminifera have shown persuasively, since as early as the late 1970s (and in many studies thereafter, including the observations of the Alvarez team at Gubbio) that the planktonic foraminiferal extinction was abrupt, with only a few species crossing the boundary into the Paleocene.

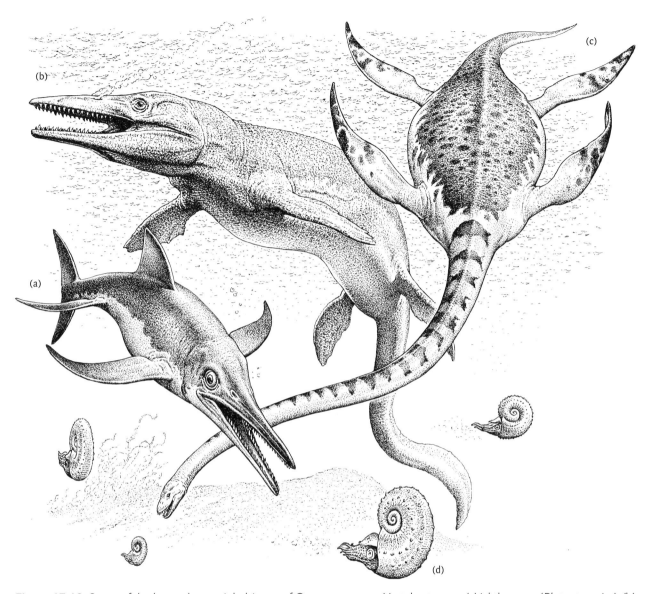

Figure 17.10. Some of the better-known inhabitants of Cretaceous seas. Vertebrates are: (a) ichthyosaur (*Platypterygius*), (b) mosasaur (*Tylosaurus*), and (c) plesiosaur (*Elasmosaurus*). (d) The shelled, tentacled invertebrates toward the bottom of the drawing are cephalopod mollusks called ammonites.

An entirely different group of marine microorganisms, calcareous nanofossils, also shows an abrupt extinction. It is safe to say that most (but not all!) paleontologists working with marine microorganisms are inclined toward a catastrophic view of the extinction.

The "Strangelove" Ocean
Some of the most exciting results came from a series of studies of K-Pg oceanic primary production; that is, the amount of organic matter synthesized by organisms from inorganic materials and sunlight.

At the K-Pg boundary there was observed a rapid and complete breakdown in nutrient cycling between surface and deep waters, to less than 10 percent of what it had been. For some oceanographers, this signaled that the world's oceans were effectively all but dead. For the next 1.5 million years, nutrient cycling remained at levels well below those preceding the original drop.[8]

Obviously, nutrient cycling is fundamental to oceanic health. With oceans covering 75 percent of the Earth's surface (or even more during the many high sea levels experienced during Earth history), it would not be an exaggeration to state that Earth's marine and terrestrial ecosystems are dependent upon these great cycles. In short, dead oceans could have threatened the whole biosphere.

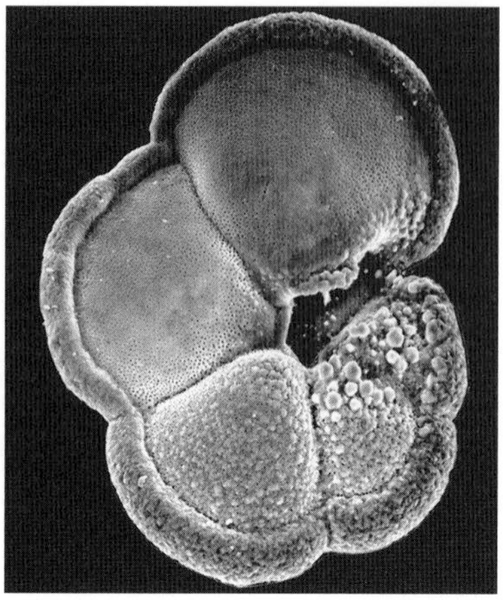

Figure 17.11. The carbonate shell of a modern planktonic (free-swimming) foraminifer, *Globorotalia menardii*. The long dimension is 0.75 mm.

Terrestrial Record

For better or worse, virtually all of what we know of the K-Pg boundary on land also comes from the Western Interior of North America (Figure 17.12). There, several well-studied, complete sections have provided the best insights available into the dynamics of the extinction.

Plants

The plant fossil record in the Western Interior has two major components, a palynoflora (spores and pollen) and a macroflora (the visible remains of plants, especially leaves; Figure 17.13). After 15 years of intensive scrutiny, both records agree nicely with each other and both records indicate that a major extinction occurred geologically instantaneously at the K-Pg boundary.

Interestingly, pollen that is typical of early Paleocene time does not immediately follow the extinction of the Cretaceous pollen. Instead there is a high concentration of fern spores just after the iridium anomaly, suggesting that, immediately after the extinction of the Cretaceous plants, there was a "bloom" of fern growth, interpreted to be a pioneer community growing on a devastated postimpact landscape. Within a short time, the fern flora gave way to a more diverse angiosperm flora characteristic of the early Paleocene. The North American plant record, however, is clear: the pollen shows a 30 percent extinction right at the K-Pg boundary.

Outside North America, an interesting southern high-latitude flora is known from New Zealand. There, an abrupt pollen and spore extinction, as well as the fern spike, are also known. In short, the pollen record suggests that the terrestrial K-Pg boundary was characterized by global deforestation.

The megafloral record based upon 25 000 plant specimens for the Western Interior of North America shows that while some environmental changes caused extinctions earlier than the K-Pg boundary, a 57 percent extinction took place precisely at the boundary, exactly correlated with the pollen extinction and iridium anomaly. If slightly more time is included in the analysis, the extinction climbs to a whopping 78 percent!

Which number is right? Probably neither, and the true extinction number is likely somewhere between the two. Although 57 percent went extinct directly at the boundary, some of those that *appear* to have gone extinct before the boundary might have actually gone extinct right at the boundary as well, but were not preserved. Still, these are large collections from many localities, and so the expectation is that the real extinction was closer to 57 percent than to 78 percent.

---

[8] The moribund K-Pg (or K/T) oceans were called "Strangelove" oceans after Dr. Strangelove, a brittle, grotesque character, played by Peter Sellers in the eponymous black comedy film, who was untroubled by the possibility of a scorched, post-nuclear world.

Figure 17.12. The K-Pg boundary in eastern Montana, USA. The boundary is midway up the butte, right at the dotted line. Below is the dinosaur-bearing latest Cretaceous Hell Creek Formation; above is the Paleogene Fort Union Formation. No dinosaurs have ever been found in the Fort Union Formation. The Hell Creek–Fort Union is the world's most-studied K-Pg boundary sequence.

The extinction of more than 57 percent of the known Cretaceous angiosperms from the Western Interior of North America suggests that, as suspected, the fern "bloom" may have been a response to the absence of flowering plants that would normally have occupied and colonized the ecosystem.

Vertebrates

Vertebrates suffered their fair share of extinction at the K-Pg boundary: the poster children are of course dinosaurs. But as we've said, lots of animals went extinct, and some of these are actually far better suited than dinosaurs to understanding what might have happened.

Patterns of survivorship, that is, who made it across the boundary and who did not, obviously might shed light upon events 66.1 Ma. And indeed, these data were extracted from the K-Pg vertebrate fossil record of the US Western Interior in 1990 by paleontologists J. D. Archibald and L. J. Bryant. Overall, an important pattern emerged: organisms that lived in aquatic environments (that is, rivers and lakes) showed up to 90 percent survival, whereas organisms living on land

Figure 17.13. Plant fossils. (a) Late Cretaceous leaf. The leaf is from an angiosperm that became extinct at the K-Pg boundary. The specimen is from just outside Marmarth, North Dakota, USA. (b) Pollen grains belonging to the genera *Proteacidites* (1) and *Aquilapollenites* (2), both important genera in measuring the moment of the terrestrial K-Pg extinction. *Proteacidites* is about 30 μm across; *Aquilapollenites* is about 50 μm.

showed as little as a 10 percent survivorship. Thus the extinction seems not to have drastically affected aquatic organisms such as fishes, turtles, and crocodiles, but apparently wreaked havoc among terrestrial organisms such as mammals and, of course, dinosaurs. Several other survivorship patterns, not as statistically robust, also appeared in the data: small vertebrates are favored over large vertebrates, ectotherms over endotherms, and nonamniotes over amniotes (Figure 17.14).

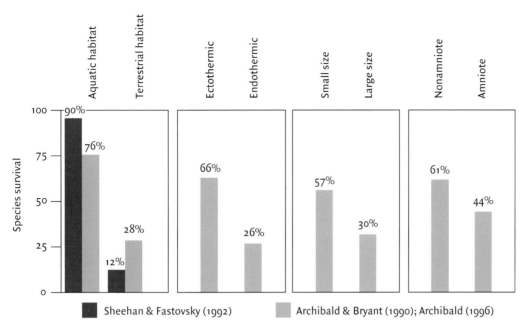

Figure 17.14. Patterns of survivorship at the K-Pg boundary as reconstructed by J. D. Archibald (1996). The study suggests that aquatic habitat, ectothermy, small size, and the absence of an amnion were qualities that statistically facilitated survival across the K-Pg boundary. Of these, aquatic habitat may have been the most important; in a separate publication, P. M. Sheehan and D. E. Fastovsky reconstructed the aquatic and land-dwelling survivorship pattern as even more extreme than that proposed by Archibald, with land-dwelling organisms showing only a 12 percent survivorship, but aquatic organisms showing 90 percent survivorship (shown in red).

Two other recent studies have corroborated these results: a 2014 study of Western Interior amphibians, such as salamanders, showed a decline only in the last 400 kyr[9] of the Cretaceous. Snakes and lizards, fully terrestrial animals, evidently also underwent a major, abrupt extinction.

Mammals

Mammals were alive and well in the Late Cretaceous, having first appeared on Earth during the Late Triassic. Mammals turn out to be extremely useful fossils: their small teeth preserve well and show a lot of morphology, allowing us to precisely identify them, obtain insights into aspects of their diet and behavior, and use them biostratigraphically.

In general, as we have seen (Chapter 14), they were small and possibly nocturnal throughout the "first two-thirds of mammalian history," as paleomammalogist W. A. Clemens, Sr., trenchantly put it. By the end of the Cretaceous, however, they were undergoing some important changes, as recorded in the Western Interior of North America. University of Washington paleontologist G. P. Wilson has reconstructed the pattern of their extinction.

Some 500 kyr before the boundary, Wilson recognized a first episode of faunal change, involving decreases in body size and possible migration out of the region. A second wave of change occurred around 200 kyr before the boundary; this consisted of a series of apparent stepwise disappearances of fossil mammals. He correlates these extinctions with short-term temperature changes identified from the plant record, although he is unable to rule out the possibility that they are artifacts of the Signor–Lipps effect (Box 17.2). Whichever the case, the upshot was significant: mammals show in

---

[9] 1 kyr = 1000 years.

Box 17.2 Getting Fooled by the Fossil Record: the Signor–Lipps Effect

A key question in any reconstruction of an extinction event is how fast or slowly it occurred. But it turns out that this can be very tricky to ascertain, in part due to a problem called the "Signor–Lipps effect," named after paleobiologists Phil Signor and Jere Lipps, who first articulated the problem.

We begin with a thought experiment. We take a handful of various coins, and distribute them on a table (Figure B17.2.1a). If they are randomly distributed, the diversity of coins (pennies, nickels, dimes, and quarters) is unchanging all over the table. Now, we draw an arbitrary line right through the center, as in Figure B17.2.1b. That line cuts across the full diversity of coins; that is, because the coins are randomly distributed on the table, anywhere one puts that line should reflect the same diversity (all coins represented).

Let's suppose the coins represent fossils, and the different types of coins represent different types of fossil animals. Now, think stratigraphically: imagine that the bottom of the pictures in Figures B17.2.1a and b is older, and the top of the pictures is younger; it can be said that the diversity of fossil animals does not change throughout the full amount of "time" represented by the table (from bottom to top). And, where the line is arbitrarily drawn, the diversity of fossil animals (represented by coins) is still constant (because nothing has changed).

Now, let's say that the arbitrary line actually represents an extinction boundary (Figure B17.2.2; E is of course the same line as in Figure B17.2.1b; here, however, it is labeled "E" and represents an "extinction boundary" in the figure; thus the "extinct" coins above the boundary are greyed). So while there are no fossil animals above the boundary (they go extinct), the diversity of fossil animals leading up to the boundary is still unchanged and constant. The "extinction," therefore is *abrupt*; diversity is unchanged preceding the "extinction" (the

(a)

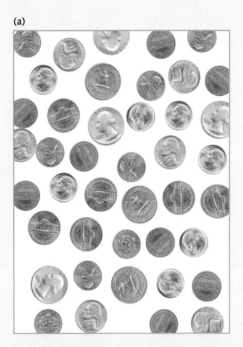

Figure B17.2.1a. A diversity of coins (pennies, nickels, dimes, and quarters) representing the diversity of a fauna or flora. The coins are randomly distributed.

(b)

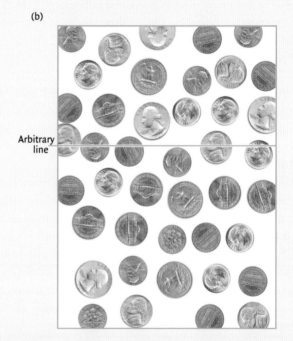

Arbitrary line

Figure B17.2.1b. Here is the same random diversity of coins, but with an arbitrary line drawn across that same random distribution of coins/taxa.

arbitrary line), and then after it, nothing is present (e.g., survives).

You are the paleontologist, though, and you'd like to reconstruct this extinction – and if these coins were fossils, you would have no idea how they are distributed through time. So you ask the basic question: was the extinction abrupt or gradual?

If it is *gradual*, you correctly assume, the diversity of fossils (coins) will decrease as you approach the extinction boundary. So you look at successively smaller intervals approaching the boundary, and what you discover is that as you get closer and closer to the boundary, the diversity of fossils (coins) starts decreasing: first one goes "extinct," then another, and finally, as you get quite close to the boundary, only one type of fossil is left (Figure B17.2.2). The extinction of the fossils, you therefore conclude, is stepwise through time; while you started with four different types of coins (representing fossils) in the stratigraphically lowest parts of the section, by the time you get to the boundary, there is only one type of coin present. So that looks like a gradual extinction, with the fossils (again, the coins) going extinct in stepwise fashion.

This is the Signor–Lipps effect. You drew an arbitrary line across the full diversity of fossils (coins), and so you know that diversity did *not* decrease. Yet, as you attempted to reconstruct the pattern of extinction, diversity appeared to decrease as you approached the boundary – as you expected that it would if the extinction were not abrupt, but rather stepwise (gradual). You've been fooled!

So we return the question of the mammal extinction: as the K-Pg boundary is approached, mammal diversity decreases in the last few hundred thousand years. Should we take this as a stepwise (gradual) extinction, or is this just an artifact of the Signor–Lipps effect, and the extinction was actually abrupt? G. P. Wilson, studying the patterns of mammal extinction at the K-Pg boundary, and fully aware of the Signor–Lipps effect, was rightly unsure of which was the signal that he had observed.

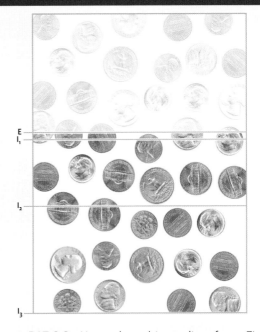

Figure B17.2.2. Here, the arbitrary line from Figure B17.2.1b represents an extinction boundary (E) if we think about this stratigraphically. The coins in this exercise represent a fauna and/or floral and they are visible in the figure to show that it is the same image as Figure B17.2.1b. However, they are grayed above E to indicate that they would no longer be there if there was actually an extinction. Between I_3 and the "extinction" E, the largest interval of time, the full diversity is represented. Between I_2 and E, a smaller interval, closer to the boundary, there is a slight decrease in diversity. Between I_1 and E, the smallest interval – and closest to the boundary – there is yet a further decrease in diversity. But, since we know that diversity is actually unchanged (we have not altered our original distribution of coins/taxa), the apparent decrease in diversity as the boundary is approached is an artifact of how we study, and not truly a gradual or stepwise extinction. As we saw in Figure B17.2.1b, diversity did not truly change as the "extinction boundary" is approached; the extinction, therefore, is abrupt (even it if looked gradual as an artificial byproduct of our flawed method of trying to understand it).

Box 17.3 Dinosaurs: All Wrong for Mass Extinctions

What are some of the problems with reconstructing changes in dinosaur populations over time? For one thing, dinosaurs are, by comparison with foraminifera for example, large beasts and, more importantly, not particularly common.[a] For this reason, the possibility of developing a statistically meaningful database is impractical, and rigorous studies of dinosaur populations are very hard to carry out. Just counting dinosaurs can be difficult. Mostly, you don't find complete specimens, and adjustments have to be made. For example, if you happen to find three vertebrae at a particular site, they might be from one, or two, or three individuals. The only way to be sure that they belong to a single individual is to find them articulated. Suppose they are not; then one must speak of **minimum numbers of individuals**, in which case the three vertebrae would be said to represent one individual: that would be the minimum number of individual dinosaurs that could have produced the three vertebrae. On the other hand, if one found two left femora, then the minimum number of individuals represented would be two.

It would be nice to use all the specimens that have been collected in the past 170 years of dinosaur studies in a study of changes in dinosaur diversity. Unfortunately, dinosaur specimens have been collected in the past generally because they are either beautiful or rare; hardly criteria for ensuring that an accurate census of dinosaur populations has been performed. So any study that really is designed to get an accurate census of dinosaur abundance or diversity at the end of the Cretaceous must begin by counting specimens in the field, which is a labor-, time-, and cost-intensive proposition.

Then, of course, the taxonomic level at which to count dinosaurs can create problems. Suppose that two specimens are found; one is clearly a hadrosaurid and the other is an indeterminate ornithischian. The indeterminate ornithischian might be a hadrosaurid, in which case we should count two hadrosaurids. But then again it might not (because its identity is indeterminate), in which case calling it a hadrosaurid would give us more hadrosaurids in our survey than actually existed. On the other hand, calling both specimens "ornithischians" is quite correct, but not very informative, if we hope to track the survivorship patterns of *different types* of dinosaurs.

Finally, within the sediments themselves, problems of correlation exist. Suppose that, in Montana, we record the last (highest level) dinosaur in the Jordan area and then record the last (highest) dinosaur in the Glendive area, about 150 km away from Jordan. Can these two dinosaurs be said to have died at the same time? How could one possibly know? Suppose that in fact these dinosaurs died 200 years apart. An interval of 200 years, viewed from a vantage point of 66 million years, is literally a snap of the fingers. Yet 200 years is a long time when one is considering an instantaneous global catastrophe that ideally is measured in milliseconds.

[a] How rare are dinosaurs in this part of the world? Of course, we cannot know the density of dinosaurs within the rocks, but their surface density was calculated by sedimentologist P. White and colleagues, using the Sheehan *et al.* database (Fastovsky and Sheehan, 1997, p. 527). They reported, "White and Fastovsky calculated that 0.000056 dinosaurs are preserved per m² of outcrop. Considered more realistically, in a statistical sense one must search a 5 m wide path of exposed rock that is 4 km long to find a single dinosaur fragment identifiable to family level (or lower)." (Fastovsky, D. E. and Sheehan, P. M. 1997. Demythicizing dinosaur extinctions at the Cretaceous–Tertiary boundary. In Wolberg, D. L., Stump, E., and Rosenberg, G. D. (eds.) *Dinofest International*. Academy of Natural Sciences, Philadelphia, p. 527.)

total a 75 percent extinction in this region. But the question of stepwise extinction versus Signor–Lipps artifact looms over these data; thus, we will return to them a bit later in this chapter when we try to figure out what it all means.

Dinosaurs

Nonavian dinosaurs are difficult animals to study (Box 17.3) and for many years, no scientific study of dinosaurs at the K-Pg boundary was ever carried out. Inexplicably, although no data were ever published to show this, it was long thought that dinosaurs died off gradually, beginning about 10 million years before the boundary.

While that is clearly not correct, the exact pattern of how they went out remains controversial.[10] While the overall trajectory of dinosaur diversity through time is ever-increasing (see Figure 15.2), data obtained in 2012 suggest that, globally, dinosaurs may have experienced a decline in diversity several million years before the K-Pg boundary. Indeed, it may have depended upon who you were and what you did: theropods may not have declined, while ceratopsians and hadrosaurs may have. These fluctuations in their biodiversity, however, were no greater than many that had occurred in the preceding 164 million years or so of dinosaurs on Earth. Nobody has identified anything that makes this particular fluctuation in dinosaur populations unique.

Decline in biodiversity or no, we can be sure that there were very much more than just a few dinosaurs hanging on right before the end of the Cretaceous. Indeed, nonavian dinosaurs were still a very diverse, abundant, and important terrestrial group of vertebrates within less than 500 kyr of the K-Pg boundary. How quickly, then, were they eliminated from Earth?

In the late 1980s and through the 1990s, field-based studies were finally designed and carried out to determine the rate of dinosaur extinction. All of them took place in the American West: two on what had been a low-lying coastal plain in what is now eastern Montana and western North Dakota, and one in an intermontane basin in what is now Wyoming, which formed in the ancestors of the present-day Rocky Mountains, rising to the west (Figure 17.15). The three studies all concluded the same thing: the dinosaur extinction was geologically abrupt.

Two studies in the Hell Creek Formation (Figure 17.12), a rock unit that represents ancient coastal plains in what is now eastern Montana and western North Dakota, were quantitative censuses of dinosaur diversity during the last 1.5 million years of the Cretaceous. One looked at the ecological diversity; that is, the proportion of the total dinosaur population taken by each of eight families of dinosaurs. The other counted genera of all vertebrates through the last 1.5 million years of the Cretaceous in that region, looking for changes in either abundance or diversity. Both demonstrated that, within about 160 000 years of the K-Pg boundary, neither ecological diversity nor abundance and generic diversity changed (Figure 17.16).

The last of the three studies, carried out in what was an ancestral Rocky Mountain intermontane basin (Figure 17.15), utilized an approach very similar to the coastal plain study of vertebrate genera described above. And the results from that study were much the same as those from the other studies: the extinction of the dinosaurs was geologically abrupt. Major extinctions occurred in most groups, but particularly in dinosaurs and mammals (Figure 17.17). A key point, however, is that none of these studies can distinguish whether the extinction took every day, or whether it took only the last minute, of that last 150 000 years of the Cretaceous.

In summary, the very limited data from the Western Interior of the USA strongly indicate an abrupt end for the nonavian dinosaurs. Only time and much further study will enable us to integrate other dinosaur-bearing localities from around the world into what is already known of North America.

The study of *avian* dinosaurs brings its own challenges, because bird fossils are so rare. We know that this group suffered very significant extinctions; all Cretaceous toothed birds – four major groups of archaic birds – went extinct at or very near the K-Pg boundary. We also know that some of the modern groups of birds – for example, ducks and their relatives – had begun to evolve before the close of the Cretaceous, and obviously survived whatever killed off their archaic relatives (see Figure 8.13). The survival of the modern groups may have been tied into an ability to fly and seek refuge, or it may have been dumb luck. One recent (2018) study ties

[10] See, for example, Chapter 15, footnote 4.

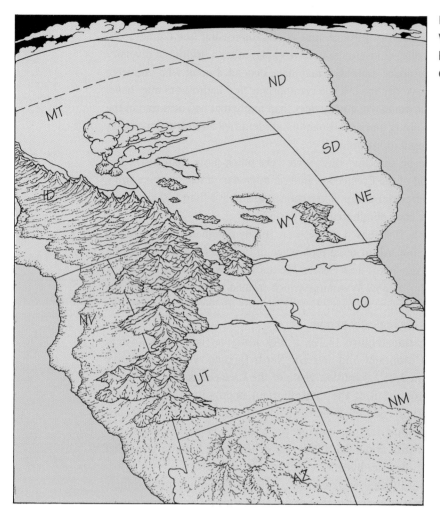

Figure 17.15. Paleogeography of the Western Interior of the USA, as it would have looked during Late Cretaceous time.

survival to whether birds reproduced on the ground, or in trees; ground-dwelling birds had much higher survivorship than the tree-dwellers. The driving cause for this pattern may be something that we have not yet landed upon.

Extinction Hypotheses

Testability and Parsimony

Any hypothesis that purports to explain the extinction of the dinosaurs at the K-Pg boundary must contain the basic prerequisites for a viable, scientific theory (see Part I and Chapter 3):

(1) *The hypothesis must be testable.* A scientifically meaningful hypothesis must be *testable*; that is, it must have predictable, observable consequences. Without testability, there is no way to falsify a hypothesis and, in the absence of falsifiability, we are then considering **belief systems** (perspectives or viewpoints that require neither deductive inferences from observation, nor logical support, for their credibility), rather than scientific hypotheses. If an event occurred and left no traces that could be observed (by whatever means available), science is simply not an appropriate tool to investigate the event.

(2) *The hypothesis must be parsimonious.* By this we mean that it must explain as many of the events as possible. If each step of the event (or events) requires an additional ad-hoc explanation, our hypotheses lose strength. They are strongest when the explanation that explains the most observations is used. If K-Pg boundary events can be explained by a single hypothesis, that is the most parsimonious hypothesis, and it has a good chance of being correct.

Extinction Hypotheses

In Table 17.1 we present about 80 years of serious, published proposals designed to explain the extinction of the dinosaurs (although see Box 17.4). The majority was published within the past 40 years. Consider each; you don't need to be a professional paleontologist to reject most of them, for most fail to meet the twin criteria for science enumerated above.

Parsimony tells us that any theory that purports to explain K-Pg events in a meaningful way must also explain as many as possible of the other events associated with the boundary. With that in mind, the hypothesis that an asteroid impact caused the events at the K-Pg boundary becomes an interesting and plausible hypothesis.

Does the Idea that an Asteroid Impact Caused the K-Pg Extinctions have Predictable Consequences?

Clearly, the answer to the above question is "Yes." Firstly, if the asteroid produced global consequences, evidence for it should be visible globally. After 40 years and hundreds of thousands of person-hours of research, the evidence for global influence of the K-Pg boundary asteroid impact is undeniable (see Figure 17.3). And it delivers on the promise of predictable consequences in terms of the extinction.

In the case of the bivalves and plants, the fact that the extinctions took place regardless of latitude is strong evidence that those extinctions were due to a global effect, which was apparently unrelated to climate. Had climate, for example, been involved as a causal agent, one might expect to see latitudinal changes in the patterns of extinction, but as we have seen such is not the case.

Other evidence comes from the rate at which the extinction took place. If the asteroid really caused the extinction, the event should have been what, in 1981, W. A. Clemens (after W. S. Gilbert) dubbed a "short, sharp shock." Patterns of *gradual* extinction would falsify the asteroid impact as a causal agent, whereas patterns of *abrupt* or *catastrophic* extinction would support the hypothesis.

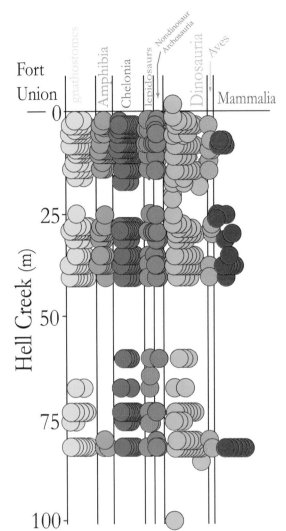

Figure 17.16. Evidence for sudden extinction of the dinosaurs. The vertical axis shows meters through the Hell Creek Formation, the uppermost unit in the Western Interior of the USA. "0" is the K-Pg boundary. The horizontal axis shows various vertebrate groups (including dinosaurs) that are found within the Hell Creek. Virtually all vertebrate groups are present throughout the thickness of the Hell Creek; there is no gradual decrease in the groups as the boundary is approached. The data indicate that the extinction of the dinosaurs and other vertebrates at 65.5 Ma was geologically instantaneous.

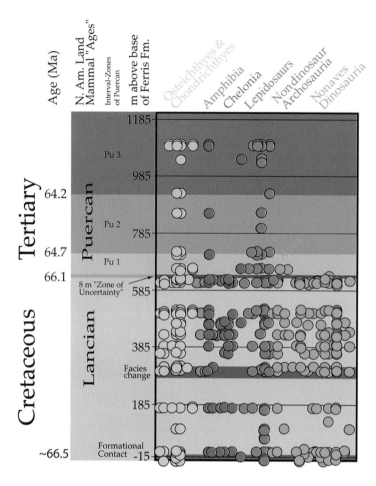

Figure 17.17. Evidence for sudden extinction of the dinosaurs in intermontane basin sedimentary deposits of the Late Cretaceous of the US Western Interior, as demonstrated by Lillegraven and Eberle (1999). The vertical axis shows meters through the K-Pg Ferris Formation and time (Ma). The horizontal axis shows various vertebrate groups (including dinosaurs) that were found in the study. Virtually all vertebrate groups are present below the boundary; there is no gradual decrease in the groups as the boundary is approached. After the boundary, diversity is drastically reduced. The data indicate that the extinction of the dinosaurs and other vertebrates 66.1 Ma was geologically instantaneous.

Recovery

Catastrophic events tend to leave a distinctive mark: organisms that first colonize deserted **ecospace** tend to **speciate** rapidly, to be rather small, and to adopt **generalist** lifestyles (rather than developing a highly specialized behavior such as exclusively meat-eating or herbivorous behaviors). Such organisms are termed **disaster biota**s and are known in vertebrate, invertebrate, and plant communities.

At the K-Pg boundary, recall that in the plant realm, the initial colonizing flora was a short-lived growth of ferns. These have been interpreted as a disaster flora developing in a disrupted and unstable landscape.

A 2019 study by Denver Museum of Natural History paleontologist T. R. Lyson and colleagues contains the best data on the terrestrial recovery known, and this work suggests that the postasteroid radiation of mammals may have occurred rapidly, in as little as 1 Ma to 1.5 Ma. The mammals that first appeared in the earliest stages of the recovery were generalists; perhaps somewhat mouse-like (although surely not rodents!). They appear to have speciated rapidly, doubling their diversity in the first 100 000 years after the impact, evolving **specializations** (such as herbivores and carnivores) within 300 000 years of the impact, and achieving a range of sizes by 700 000 years after the impact. The mammalian pattern of postimpact evolution is the pattern of a disaster fauna that came through a catastrophic event and radiated in deserted ecospace (Figure 17.18).

Table 17.1 Proposed causes for the extinction of the dinosaurs (based in part upon Benton, 1990).

I Proposed biotic causes

A *Medical problems*

   (a) Slipped disks in the vertebral column causing dinosaur debilitation

   (b) Hormone problems

       (1) Overactive pituitary glands leading to bizarre and non-adaptive growths

       (2) Hormonal problems leading to eggshells that were too thin, causing them to collapse in on themselves in a gooey mess

   (c) Decrease in sexual activity

   (d) Blindness due to cataracts

   (e) A variety of diseases, including arthritis, infections, and bone fractures

   (f) Biting insects carrying diseases that did dinosaurs in over hundreds of thousands to millions of years

   (g) Epidemics leaving no trace but wholesale destruction

   (h) Parasites leaving no trace but wholesale destruction

   (i) Change in ratio of DNA to cell nucleus causing scrambled genetics

   (j) General stupidity

B *Racial senescence*

   This is the idea, no longer given much credence, that entire lineages grow old and become "senile," much as individuals do. Thus, in this way of thinking, late-appearing species would not be as robust and viable as species that appeared during the early and middle stages of a lineage. The idea behind this was that the dinosaurs as a lineage simply got old and the last-living members of the group were not competitive for this reason

C *Biotic interactions*

   (a) Competition with other animals, especially mammals, which may have out-competed dinosaurs for niches, or perhaps ate their eggs

   (b) Overpredation by carnosaurs (who presumably ate themselves out of existence)

   (c) Floral changes

       (1) Loss of marsh vegetation (presumably the single most important source of food)

       (2) Increase in deforestation (leading to loss of dinosaur habitats)

       (3) General decrease in the availability of plants for food with subsequent dinosaur starvation

       (4) The evolution in plants of substances poisonous to dinosaurs

       (5) The loss from plants of minerals essential to dinosaur growth

II Proposed physical causes

A *Atmospheric causes*

   (a) Climate became too hot so they fried

   (b) Climate became too cold so they froze

   (c) Climate became too wet so they got waterlogged

   (d) Climate became too dry so they desiccated

   (e) Excessive amounts of oxygen in the atmosphere caused:

       (1) changes in atmospheric pressure and/or atmospheric composition that proved fatal; or

       (2) global wildfires that burned up the dinosaurs

   (f) Low levels of CO_2 removed the "breathing stimulus" of endothermic dinosaurs

   (g) High levels of CO_2 asphyxiated dinosaur embryos

   (h) Volcanic emissions (dust, CO_2, rare earth elements) poisoned dinosaurs one way or another

B *Oceanic and geomorphic causes*

   (a) Marine regression produced loss of habitats

Table 17.1 (*cont.*)

(b) Swamp and lake habitats were drained

(c) Stagnant oceans produced untenable conditions on land

(d) Spillover into the world's oceans of Arctic waters that had formerly been restricted to polar regions, and subsequent climatic cooling

(e) The separation of Antarctica and South America, causing cool waters to enter the world's oceans from the south, modifying world climates

(f) Reduced topographical relief and loss of habitats

C *Other*

(a) Fluctuations in gravitational constants leading to indeterminate ills for the dinosaurs

(b) Shift in the Earth's rotational poles leading to indeterminate ills for the dinosaurs

(c) Extraction of the Moon from the Pacific Basin perturbing dinosaur life as it had been known for 140 million years (!)

(d) Poisoning by uranium from Earth's soils

D *Extraterrestrial causes*

(a) Increasing entropy leading to loss of large life forms

(b) Sunspots modifying climates in some destructive way

(c) Cosmic radiation and high levels of ultraviolet radiation causing mutations

(d) Destruction of the ozone layer, causing (c)

(e) Ionizing radiation as in (c)

(f) Electromagnetic radiation and cosmic rays from the explosion of a supernova

(g) Interstellar dust cloud

(h) Oscillations about the galactic plane leading to indeterminate ills for the dinosaurs

(i) Impact of an asteroid (for mechanisms, see the text)

Recent phylogenies based upon the rates of molecular evolution have suggested that modern mammals' roots are to be found within the Cretaceous, showing that the mammalian radiation that characterized the Paleogene was actually well underway during the latest Cretaceous. In fact, the far-distant ancestors of modern mammals were likely scurrying around during the Late Cretaceous, but the rapid species turnovers of the earliest Paleogene disaster faunas show the clear mark of a catastrophic event.

Does the Asteroid Impact Hypothesis Explain all the Data?

In fact, there does appear to be a correlation between extinction selectivity and the asteroid as a causal agent in the extinctions. Those marine creatures that suffered the most extinctions were those that depended directly upon primary productivity for their food source. Such creatures included not only the planktonic foraminifera and other planktonic marine microorganisms, but also ammonites, other cephalopods, and a variety of mollusks. On the other hand, organisms that not only depended on primary productivity but could also survive on detritus, that is, the scavenged remains of other organisms, fared consistently better. In marine deposits, detritus-feeders were apparently less affected by the extinction.

In the terrestrial realm, the strong selectivity between land-dwelling and aquatic tetrapod survival (see Figure 17.14) correlates with feeding strategy: aquatic vertebrates tend to utilize detritus as a major source of nutrients, while land-dwelling vertebrates are far more dependent upon primary productivity.

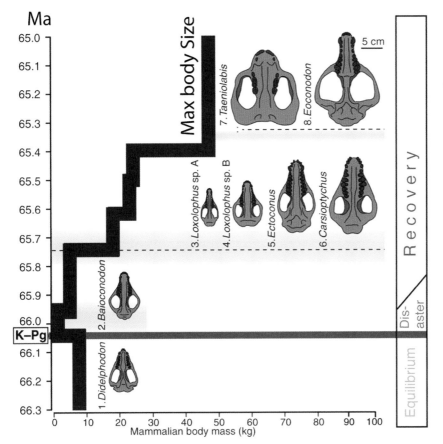

Figure 17.18. The anatomy of the earliest Paleocene radiation of mammals after the K-Pg extinction boundary. On the vertical axis is time, with extraordinarily high-precision dates. A clear increase in the diversity and estimated body mass of mammal species is visible in the first few 100 000 years after the K-Pg impact. Redrawn from Lyson, T. R. *et al.* 2019. Exceptional continental record of biotic recovery after the Cretaceous–Paleogene mass extinction, *Science*, **366** (6468), 977–983.

In this scenario, the tetrapods that survived the K-Pg boundary were primarily aquatic detritus-feeders. This is because river and lake systems can serve as a repository for detrital material, and organisms that live in such environments and can utilize this resource were protected against short-term drops in primary productivity. They may also have been protected from the strong infra-red radiation pulse, as well as the wildfires. In short, then, a lot of evidence points to an abrupt extinction.

Other Hypotheses

So there is a striking coincidence between the timing of the asteroid impact, and these biological events. Moreover, recent high-precision numerical dating shows that, within a very few tens of thousands of years,[11] the asteroid impact and the biological extinction occurred at the same time. In fact, although it was not as well understood then as now, this remarkable coincidence of events is what first suggested to the Alvarezes that maybe the asteroid and the extinction were paired as cause and effect.

However, there is another remarkable coincidence that nags almost as compellingly as the asteroid. For many years, French volcanologist Vincent Courtillot has documented the relationship between episodes of continental flood basalt (CFB) volcanism and mass extinctions. The relationship that he observed is really quite striking; each continental flood basalt event since the Permian seems to have been associated with a mass extinction (Figure 17.19).

The K-Pg iconic CFB is the Deccan Traps, that ginormous outpouring of continental flood basalts in India described earlier in this chapter (see Figure 17.9). These bracketed the K-Pg boundary,

---

[11] An astounding achievement, considering that these events occurred tens of *millions* of years ago.

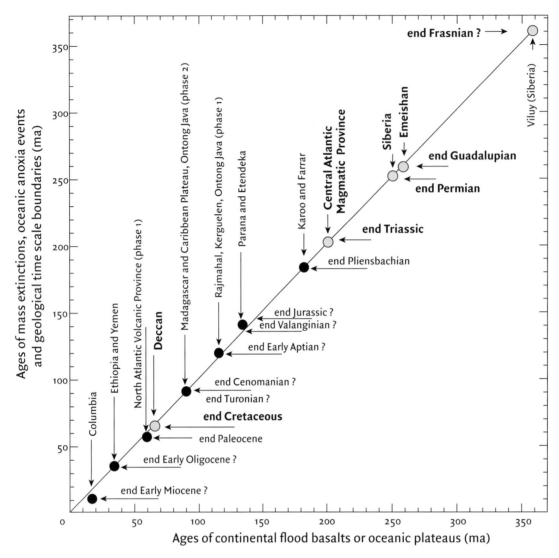

Figure 17.19. Comparison of the ages of continental flood basalt events and mass extinction events. Light-blue colored dots represent the most important mass extinction events. The straight line indicates excellent correlation between the ages of the basalts and the timing of the mass extinction events. Redrawn from: Courtillot. V. 1995. *La Vie en Catastrophes*, Librairie Arthème Fayard, Paris, 278 pages.

beginning at about 67.5 Ma, and occurring in four major eruptive phases until about 64.7 Ma. While work done in 2015 suggested that the major eruptive phase (the middle one) took place within a period of less than 750 kyr (66.3 Ma to 65.5 Ma), bracketing the K-Pg boundary we have seen (above) that the most recent studies do not agree on the timing of the middle basalt outpouring: one study has two eruptive pulses closely bracketing the K-Pg extinction, and the other suggests that the K-Pg extinction postdates the vast preponderance of the Deccan CFB events.

How would CFBs do the extinction deed? Two models have been proposed: In the first, CO_2 (carbon dioxide) from the volcanos would get into the atmosphere, causing greenhouse conditions, which would in turn impede oceanic circulation, leading to a host of nasty consequences, such as the release into the atmosphere of hydrogen sulfates, in turn producing H_2SO_4 (sulfuric acid), effectively acidifying and poisoning the planet. The other model is no less grim: again, sulfur in the form of SO_2 (sulfur dioxide) would be injected into the stratosphere via the volcanism, and produce H_2SO_4

aerosols, which would cause a variety of nasty consequences, including acid rain, atmospheric, and ecosystemic poisoning.

These models have quite differing consequences. The CO_2 release model, because it requires the slowing of oceanic circulation by atmospheric (greenhouse) warming, would occur over long-ish time scales, likely measured in thousands, or even tens of thousands of years. But our best data suggest that the K-Pg extinction likely occurred on shorter time scales. That takes us to the sulfur dioxide atmospheric injection model which, because it depends upon rates of atmospheric circulation that occur on much shorter time scales than tens of thousands of years, seems more attractive. Yet, the kind of volcanism that characterized the Deccan Traps, CFB volcanism, is not generally associated with atmospheric injections; the volcanic plumes never reach the stratosphere.

Other aspects of the K-Pg extinction do not clearly implicate flood basalts: the global fires, the selectivity of the extinction, the evidence for physical disruption at the boundary globally, and chemical signals around the Gulf of Mexico region that match those of the impact crater, and of course the presence at the K-Pg boundary globally of an extraterrestrial signature, namely iridium, clays derived from impact glass, and shocked quartz. Moreover, volcanic events implicated in other mass extinctions have left a global record. By contrast to this and to the asteroid, there is no global record of the Deccan volcanic event. For all these reasons, then, although the CFB idea may be intuitively attractive, it simply does not fit best the known data profile that characterizes K-Pg events (including the extinctions).

In 2015, tectonicist M. A. Richards and a host of coauthors including Walter Alvarez (of the 1980 hypothesis fame), and four years later, Schoene and colleagues in 2019, proposed something of a compromise: perhaps, they suggested, the asteroid impact provoked the main episode of Deccan volcanism, and the two events together caused the extinction. The possibility of this occurring could be demonstrated on theoretical grounds; so, conceptually, it is not out of the question. This hybrid extinction mechanism takes us into the world of multiple-cause hypotheses, which, although they may explain the data more mechanistically, run the risk of not being parsimonious.

Multiple Causes

It's been said that "complex extinctions require complex causes," and a number of workers have suggested that a unique suite (the fabled "perfect storm") of events fortuitously coalesced at the K-Pg boundary to produce the K-Pg mass extinction. These events include the asteroid impact, Deccan volcanism, sea-level fall, global climatic deterioration through warming just before the K-Pg boundary, competition among various groups… in fact, something of a laundry list of what is known about boundary events. In this case, every known physical event is imputed to have caused "weakened" ecosystems, and made them more susceptible to extinction. An inclusive compromise is attractive in politics, but is it science? Does it better explain the data?

Well, not exactly. Ultimately in this way of thinking, each K-Pg event is ascribed a role in the extinctions; ultimately, therefore, any and all K-Pg events must have been involved in the extinctions. Because of this, the hypothesis is unfalsifiable: if all K-Pg events are involved in the extinction, there is no way to eliminate any particular event. If we can't know which event – because none can be absolutely eliminated – it really says nothing new about the K-Pg boundary: the hypothesis is so general that it ultimately explains nothing.

Ironically (and sadly), the multicause hypothesis *could* be correct: a unique suite of events uniquely occurring at the K-Pg boundary could have uniquely caused this mass extinction. The scenario could go like this: the asteroid impact, itself unimaginably destructive, could have provoked the Deccan Traps, which released various gases (CO_2, NO_2, SO_2) into the atmosphere, none of which likely improved life for the Earth's biota.

Box 17.4 The Real Reason the Dinosaurs Became Extinct

Not every published hypothesis has been serious. In 1964, for example, E. Baldwin suggested that the dinosaurs died of constipation. His reasoning went as follows. Toward the end of the Cretaceous, there was a restriction in the distribution of certain plants containing natural laxative oils necessary for dinosaur regularity. As the plants became geographically restricted, those unfortunate dinosaurs living in places where the necessary plants no longer existed acquired stopped plumbing and died hard deaths. The same year, humorist W. Cuppy noted that "the Age of Reptiles ended because it had gone on long enough and it was all a mistake in the first place," a view with which many characters in the *Jurassic Park* series would have probably agreed.

The November 1981 issue of the *National Lampoon* offered its explanation, entitled "Sin in the Sediment." The Christian right was the target:

> It's pretty obvious if you just examine the remains of the dinosaurs . . . Dig down into older sediments and you'll see that the dinosaurs were pretty well off until the end of the Mesozoic. They were decent, moral creatures, just going about their daily business. But look at the end of the Mesozoic and you begin to see evidence of stunning moral decline. Bones of wives and children all alone, with the philandering husband's bones nowhere in sight. Heaps of fossilized, unhatched, aborted dinosaur eggs. Males and females of different species living together in unnatural defiance of biblical law. Researchers have even excavated entire orgies – hundreds of animals with their bones intertwined in lewd positions. Immorality was rampant.

In 1983, sedimentary geologist R. H. Dott Jr. published a short note in which he vented his frustrations with the pollen season, suggesting that it was pollen in the atmosphere that killed the dinosaurs. He called his contribution "Itching Eyes and Dinosaur Demise."

The issues raised by the *National Lampoon* were compelling enough to again be raised in 1988 by the *Journal of Irreproducible Results*. There, L. J. Blincoe developed a then-new hypothesis about the "fighting dinosaurs" specimen (see Figure 6.23):

> A thorough but cursory review of fossil specimens. . . has revealed a unique fossil found in the Cretaceous "beds" of Mongolia in 1971. The fossil featured two different species of dinosaur, one a saurischian carnivore (*Velociraptor*), the other an ornithischian herbivore [sic] (*Protoceratops*), in close association at the moment of their deaths. Prejudiced by their preconceived notions of dinosaur behavior, paleontologists have almost unanimously interpreted this find as evidence of a life and death struggle [*see also Figure 6.23*]
>
> . . . However, an alternative theory has now been developed which not only explains this unusual fossil, but also answers the riddle of the dinosaurs' disappearance. Quite simply, when their lives were ended by sudden catastrophe, these two creatures were locked together . . . in a passionate embrace. They were, in fact, prehistoric lovers.

The implications of this startling interpretation are clear: dinosaurs engaged in trans-species sexual activity. In doing so they wasted their procreative energy on evolutionarily pointless copulation that

In the marine realm, the combination of volcanism and an asteroid would have darkened skies, leading to inhibition of photosynthesis and a drop in primary productivity of phytoplankton. Food chains, ultimately dependent upon that primary productivity, would have collapsed.

In the terrestrial realm, perhaps a few other things had already undermined existing ecosystems: some extra competition among mammalian species, some latest Cretaceous temperature increases; but for whatever the reasons, populations were not replenished, and dinosaur-based ecosystems became unusually susceptible to collapse. The heat pulse and forest fires induced by the asteroid, as well as the darkened skies produced by the asteroid and the Deccan Traps would have then destroyed the standing vegetation (the primary production on land), and shut down food chains that were

Box 17.4 *cont'd*

Figure B17.4.1. O'Donnell's take on the cause of extinction of dinosaurs.

resulted in either no offspring or, perhaps on rare occasions, in bizarre, sterile mutations (the fossil record is replete with candidates for this later category.[a]

For ultimate cause(s) of the extinction, however, we think O'Donnell's perspective published in the *New Yorker* says it all (Figure B17.4.1).

[a] Blincoe, L. J. 1988. *Journal of Irreproducible Results*, **33**, 24.

primary production- and not detritus-based. Aquatic organisms would have been protected by their wet habitats, and their survival potential would thus have been enhanced.

Alternatively, the same scenario might have occurred *without* the Deccan Traps provoking the asteroid. In this case, the Deccan Traps themselves might have weakened marine and terrestrial ecosystems,[12] and the asteroid would have then provided the *coup de grâce* to these staggering ecosystems.

[12] By any number of mechanisms: darkened skies decreasing photosynthesis; an atmosphere terminally polluted with various gases, many of which could have come back to Earth as acid rain; temperature increases precipitated by those same greenhouse gases in the atmosphere. Multicause hypotheses are truly multicaused!

This idea, of weakened ecosystems made particularly vulnerable to an asteroid impact, has been called the "press-pulse" hypothesis; ecosystems were stressed (pressed) by a variety of potential stressors; and then, in a more susceptible state zapped (pulsed) by the asteroid impact. Unfortunately, to date, this hypothesis has little support in the known data; instead, it appears unparsimonious, and therefore as science, its significance is undermined.

So What Happened at the K-Pg Boundary?

It seems to us that, in the K-Pg mass extinction, the Deccan Traps were superfluous. Deccan volcanism has no global signature; the timing of events (so compelling to some) is suspect; what, then is the compelling case for the volcanism in the first place? Why isn't the asteroid impact – for which a comprehensive model of extinction has been developed and tested, and in which the timing is demonstrably coincident with the extinctions – enough? Among supporters of anything other than the asteroid alone, is there some unarticulated, intuitive distrust of an extraterrestrial event controlling life on Earth?

Scientists who view the asteroid impact as the cause of K-Pg events generally envision some kind of dramatic and short-term disturbance to the ecosystem. Such a disturbance was likely a dust cloud blocking sunlight for a few months, a pulse of infra-red radiation, and global wildfires. It may never be possible to know exactly which factor(s) did which deed(s), and trying to reconstruct it in so exact a manner may be stretching the resolution of the fossil record well past its capability. Whatever the disturbance to the terrestrial ecosystem, it seems to have had, at the very least, a deadly effect on global primary productivity, which in turn seems to have decimated organisms solely dependent upon primary productivity, and ultimately shattered complex food webs. Aquatic habitats evidently provided some measure of protection, and thus in the terrestrial realm, detritus-feeders survived. Dinosaurs, being nonaquatic terrestrial animals, ultimately succumbed. But whatever cause or combination of causes produced the K-Pg extinction, the absence of nonavian dinosaurs after their approximately 164 million years of terrestrial importance was the event that allowed the mammals to diversify and occupy the place in the global ecosystem that they presently hold.

SUMMARY

Although there has been considerable speculation about what happened to the nonavian dinosaurs at 66.1 Ma, any meaningful hypothesis must operate within the bounds of science: it must be testable and it must explain as much of the data as possible. Given those constraints, the single-cause hypothesis matches what is known about the K-Pg extinction: the hypothesis that an asteroid hit the Earth and killed many organisms, including the nonavian dinosaurs. Multicause hypotheses may be correct, but are weaker, because what is known of the extinction does not require further causes than an asteroid impact.

The evidence for a large (10 km in diameter) asteroid striking the Earth at 66.1 Ma in the Yucatan region of Mexico is now extensive, and virtually incontrovertible. Its immediate effects are likely to have been, globally, blockage of sunlight for three to four months and the propagation of an energy pulse and associated wildfires in the terrestrial realm as well as, locally, tidal waves and severe environmental disruption. Foremost among the various biotic consequences of these events is that the great nutrient cycles that characterize healthy oceans were severely disrupted.

The asteroid impact is consistent with the limited amount that is known of the dinosaur extinction, which is that in the Western Interior of North America, at least (the only place in the world in which rates of dinosaur extinction have been studied), the dinosaur extinction was as instantaneous as we are able to resolve at a distance of 66.1 Ma.

The patterns of terrestrial vertebrate survivorship are relatively well understood. Analyses show that vertebrates not in aquatic ecosystems were most susceptible to extinction; other inferences, less robust, suggest that larger animals, endotherms, and amniotes also were less likely to survive than smaller animals, ectotherms, and anamniotes. Nonavian dinosaurs were large organisms, and clearly dependent upon primary productivity.

One explanation for the survivorship of organisms living in aquatic food webs is that these are generally not as dependent upon primary productivity as those in land-based food chains. Because primary productivity was clearly perturbed at the K-Pg boundary, organisms in aquatic food webs were likely protected both by dietary preference and by the fact that the aquatic realm provided a refuge from the physical catastrophes caused by the asteroid impact.

The recovery took as little as a few thousand years in the oceans, and was well underway by 1.5 million years on land.

SELECTED READINGS

Alvarez, L. W. 1983. Experimental evidence that an asteroid impact led to the extinction of many species 65 Myr ago. *Proceedings of the National Academy of Sciences*, **80**, 627–642.

Alvarez, L. W., Alvarez, W., Asaro, F., and Michel, H. V. 1980. Extraterrestrial cause for the Cretaceous–Tertiary extinction. *Science*, **208**, 1095–1108.

Alvarez, W. 1997. *T. rex and the Crater of Doom*. Princeton University Press, Princeton, NJ, 185 pages.

Archibald, J. D. 1996. *Dinosaur Extinction and the End of an Era*. Columbia University Press, New York, 237 pages.

Archibald, J. D. 2011. *Extinction and Radiation: How the Fall of the Dinosaurs Led to the Rise of Mammals*. Johns Hopkins University Press, Baltimore, MD, 108 pages.

Benton, M. J. 1990. Scientific methodologies in collision the history of the study of the extinction of the dinosaurs. *Evolutionary Biology*, **24**, 371–400.

Fastovsky, D. E. and Bercovici, A. 2016. The Hell Creek Formation and its contribution to the Cretaceous–Paleogene extinction: A short primer. *Cretaceous Research*, **57**, 368–390.

Frankel, C. 1999. *The End of the Dinosaurs – Chicxulub Crater and Mass Extinctions*. Cambridge University Press, New York, 223 pages.

Hartman, J., Johnson, K. R., and Nichols, D. J. (eds.) 2002. The Hell Creek Formation and the Cretaceous–Tertiary Boundary in the Northern Great Plains. Geological Society of America Special Paper no. 361, 520 pages.

Koeberl, C. and MacLeod, K. G. (eds.) 2002. Catastrophic Events and Mass Extinctions: Impacts and Beyond. Geological Society of America Special Paper no. 356, 746 pages.

Lillegraven, J. A. and Eberle, J. J. 1999. Vertebrate faunal changes through Lanican and Puercan time in southern Wyoming. *Journal of Paleontology*, **73**, 691–710.

Lyson, T. R., Miller, I. M., Bercovici, A. D., *et al.* 2019. Exceptional continental record of biotic recovery after the Cretaceous–Paleogene mass extinction. *Science*, **366** (6468), 977–983.

Maas, M. C. and Krause, D. W. 1994. Mammalian turnover and community structure in the Paleocene of North America. *Historical Biology*, **8**, 91–128.

Powell, J. L. 1998. *Night Comes to the Cretaceous*. W. H. Freeman and Company, New York, 250 pages.

Richards, M. A., Alvarez, W., Self, S., *et al.* 2015. Triggering of the largest Deccan eruptions by the Chicxulub impact. *Geological Society of America Bulletin*, published online on April 30, 2015. doi:10.1130/B31167.1

Robertson, D. S., McKenna, M. C., Toon, O. B., Hope, S., and Lillegraven, J. A. 2004. Survival in the first hours of the Cenozoic. *Geological Society of America Bulletin*, **116**, 760–768. doi: 10.1130/B25402.1

Ryder, G., Fastovsky, D. E., and Gartner, S. (eds.) 1996. *The Cretaceous–Tertiary Event and Other Catastrophes in Earth History*. Geological Society of America Special Paper no. 307, 569 pages.

Sharpton, V. L. and Ward, P. D. (eds.) 1990. *Global Catastrophes in Earth History: An Interdisciplinary Conference on Impacts, Volcanism, and Mass Mortality.* Geological Society of America Special Paper no. 247, 631 pages.

Schoene, B., Eddy, M. P., Samperton, K. M., *et al.* 2019. U-Pb constraints on pulsed eruption of the Deccan Traps across the end-Cretaceous mass extinction. *Science,* **363,** 862–866.

Schoene, B., Samperton, K. M., Eddy, M. P., *et al.* 2015. U-Pb geochronology of the Deccan Traps and relation to the end-Cretaceous mass extinction. *Science,* **347,** 182–184. doi: 10.1126/science.aaa0118

Silver, L. T. and Schultz, P. H. (eds.) 1982. *Geological Implications of Large Asteroid and Comets on Earth.* Geological Society of America Special Paper no. 190, 528 pages.

Sprain, C. J., Renne, P. R., Vanderkluysen, L., *et al.* 2019. The eruptive tempo of Decca volcanism in relation to the Cretaceous–Paleogene boundary: *Science,* **363,** 866–870.

Wilson, G. P., Clemens, W. A., Horner, J. R., and Hartman, J. H. (eds.) 2014. *Through the End of the Cretaceous in the Type Locality of the Hell Creek Formation in Montana and Adjacent Areas.* Geological Society of America Special Paper 503, 392 pages.

TOPIC QUESTIONS

1. What do the words regression, detritus, iridium anomaly, shocked quartz, microtektites, Chicxulub, ichthyosaurs, plesiosaurs, ammonites, mosasaurs, foraminifera, planktonic, and benthic refer to?
2. What are primary production and disaster faunas?
3. What is meant by nutrient cycling? Why is that important?
4. What is the K-Pg boundary? When was it?
5. What kinds of physical events took place at the K-Pg boundary?
6. Describe the biotic extinctions that took place at the K-Pg boundary.
7. Describe the results of the studies that concluded that the dinosaurs died abruptly.
8. What is meant by "geologically abrupt" when speaking of the extinction of the dinosaurs?
9. Describe the kinds of physical events that might have occurred when the asteroid hit the Earth.
10. Choose four extinction hypotheses from Table 17.1 and evaluate them in terms of the criteria for a viable scientific theory.

GLOSSARY

Using This Glossary

The goal of this Glossary is to help to clarify language and images that may be unfamiliar to students who happen to live in the Recent (us!). Below, therefore, follows a complete listing of all the words highlighted in this textbook, as well as other words of relevance to the subject of dinosaur paleontology. Readers are provided with the chapter in which the word is highlighted; in some cases, however, readers are also referred to places where the concept(s) embodied by the word is treated. Finally, in a few cases, readers are referred to a figure as well. A relatively few words have no cross-reference, implying that the ideas they represent recur throughout the book.

A

Acetabulum. Hip socket. *See* Chapter 4.

Acromion (acromial) process. A broad and plate-like flange on the forward surface of the shoulder blade. *See* Chapter 10.

Adenosine diphosphate. *See* ADP. *See also* Chapter 14.

Adenosine triphosphate. *See* ATP. *See also* Chapter 14.

ADP (adenosine diphosphate). A molecule involved in the energy production of a cell. It is produced when ATP breaks down to release energy. *See* Chapter 14.

Advanced. In an evolutionary context, shared or derived (or specific), with reference to characters. *See* Chapter 3.

Aerobic. A type of metabolism involving a complex series of oxidation steps through the citric acid cycle. *See* Appendix 14.1.

Akinetic. *See* Kinetic. *See also* Chapter 12.

Allometry. The condition in which, as the size of organisms changes, their proportions change as well. For example, if an ant were scaled up to the size of a 747 airplane, its features – body, head, legs, etc. – would no longer have the same proportions relative to each other. *See* Chapter 13.

Altricial. Pertaining to organisms that are born relatively underdeveloped, requiring significant parental attention for survival. *See* Chapter 9.

Alvarezsauridae. An unusual group of theropods equipped with stout carpometacarpus-like features. Their phylogenetic status is currently poorly understood. *See* Chapter 7.

Alveoli (singular alveolus). Sac-like anatomical structures. *See* Chapter 14.

Ammonites. Once-abundant, now-extinct, shelled cephalopod (related to modern-day squids and octopuses) mollusks. *See* Chapter 17.

Amnion. A membrane in some vertebrate eggs that contributes to the retention of fluids within the egg. *See* Chapter 4.

Amniote. An organism bearing amniotic eggs. *See* Chapter 4.

Anaerobic. Without oxygen. *See* Chapter 14.

Analog. In anatomy, structures that perform in a similar fashion but have evolved independently. *See* Chapter 3.

Analogous. Adjectival form of analog. *See* Chapter 3.

Anamniote(s). An organism whose egg has no amnion. *See* Chapter 4.

Anapsid(s). The group that contains all amniotes with a completely covered skull roof. *See* Chapter 4.

Ancestral. In an evolutionary sense, relating to forebears. *See* Chapter 3.

Angiosperms. Flowering plants. *See* Chapters 10 and 15.

Ankylosauria. Armored quadrupedal thyreophorans (Ornithischia) whose bodies were encased in a bony pavement of osteoderms. *See* Chapter 10.

Ankylosauridae. Along with Nodosauridae, one of the two clades that make up Ankylosauria. *See* Chapter 10.

Antediluvian. Occurring before the Biblical flood. *See* Chapter 16.

Anterior. Pertaining to the head-bearing end of an organism. *See* Chapter 4.

Antorbital fenestra. An opening on the side of the skull, just ahead of the eye. This is a character that unites the clade Archosauria. *See* Chapter 4.

Arboreal. Pertaining to trees; as the arboreal hypothesis, referring to the idea that bird flight evolved by birds jumping out of trees. *See* Chapters 8 and 10.

Archosauria. A clade within Archosauromorpha. The living archosaurs include birds and crocodiles. *See* Chapters 4 and 10.

Archosauromorpha. The large clade of diapsids that includes the common ancestor of rhynchosaurs and archosaurs, and all its descendants. *See* Chapter 4.

Artifact. An artifact in science is a false discovery, occurring perhaps as a result of some aspect of the method that is being applied to address the particular scientific problem. Suppose you were looking for Y chromosomes in humans, but your sample was made up only of men. You would conclude that all humans had Y chromosomes, a false result that would be an artifact of your sampling method (only men). *See* Chapter 15.

Ascending process (of the astragalus). A wedge-shaped splint of bone on the astragalus that lies flat against the shin (between the tibia and fibula) and points upward. Diagnostic character of Theropoda. *See* Chapter 7.

Assemblage. In paleontology, a group of organisms. The term is used to refer to a collection of fossils in which it is not clear how accurately the collection reflects the complete, ancient formerly living community. *See* Chapters 2 and 14.

Asteroid. A large extraterrestrial body. *See* Chapters 16 and 17.

Astragalus. Along with the calcaneum, one of two upper bones in the vertebrate ankle. *See* Chapter 5.

Atom. The smallest particle of any element that still retains the properties of that element. *See* Chapter 2.

Atomic number. The number of protons (which equals the number of electrons) in an element. *See* Chapter 2.

ATP (adenosine triphosphate). A molecule synthesized by the body as a means to store energy. The energy is stored in the three phosphate bonds, one of which is then broken to release energy (ATP → ADP + inorganic phosphate + energy). *See* Chapter 14.

Avemetatarsalia. All ornithodirans, excluding Pterosauria. *See* Chapter 5.

Aves. Aves is the original Linnaean name for all living birds. It has since been expanded to include extinct ones as well. *See* Chapters 7 and 8.

Avetheropoda. That group of theropods defined as the most recent common ancestor of *Passer*, carnosaurs, and all its descendants. *See* Chapter 7.

Avialae. That group of theropods defined as the most recent common ancestor of *Passer*, *Archaeopteryx*, and all its descendants. *See* Chapter 7.

B

Background extinctions. Continually occurring, isolated extinctions of individual species. As distinct from mass extinctions. *See* Chapter 17.

Badlands. Extremely rough country, generally carved by rivers and floods. *See* Chapter 1.

Barb. Feather material radiating from the shaft of the feather. *See* Chapter 8.

Barbule. A small hook that links barbs together along the shaft of the feather. *See* Chapter 8.

Basal. Relatively low in the cladogram, that is, less derived than most members of a particular clade.

Beak. Sheaths of keratinized material covering the ends of the jaws (synonym: rhamphotheca). *See* Chapters 4 to 7.

Belief system. A conceptualization that is faith-based, knowledge that does not require the application of logic, reason, and/or observation. *See* Chapter 17.

Bennettitaleae. A group of extinct cone-bearing plants. *See* Chapter 15.

Benthic. With reference to the marine realm (oceans), living within sediments. *See* Chapter 17.

Bidirectional. Moving in two, generally opposite, directions. *See* Chapter 13.

Biogeography. Pertaining to the distribution of organisms in space. *See* Chapter 11.

Biomass. The sum total of the weights of organisms in the assemblage or community being studied. *See* Chapter 14.

Biosphere. A spherical zone, encompassing the entire Earth, which contains all its living organisms. The biosphere extends from up in the atmosphere to deep down into the Earth's crust.

Biostratigraphy. The study of the relationships in time among groups of organisms. *See* Chapter 2.

Biota. The sum total of all organisms that have populated the Earth.

Biotic or organismic evolution. As it is used today, the type of change through time of life first described by Charles Darwin in 1859, and studied ever since. *See* Appendix 3.1.

Body fossil. The type of fossil in which a part of an organism becomes buried and fossilized as opposed to trace fossil. *See* Chapter 1.

Bolide. A bright (via friction upon entry through the Earth's atmosphere) meteor. *See* Chapter 17.

Bone. The mineralized tissue (generally mineralized by sodium apatite) that forms the structural support (skeleton) and dermal armor in vertebrates.

Bone histology. The study of bone tissue. *See* Chapter 14.

Bonebeds. Relatively dense accumulation of bones of many individuals, generally composed of a very few kinds of organisms. *See* Chapters 6 and 11.

Brain endocasts. Internal casts of braincases; *see also* Endocast(s). *See* Chapters 6 and 11.

Braincase. Hollow bony box that houses the brain; located toward the upper, back part of the skull. *See* Chapter 4.

C

Cadence. In locomotion, the rate at which the feet hit the ground. *See* Chapter 10.

Calcareous nanofossils. Fossils of extremely small planktonic microorganisms. *See* Chapter 17.

Carapace. A shell or protective covering across the back of an animal. *See* Chapter 10.

Carbohydrates. A family of five- and six-carbon organic molecules whose chemical bonds, when broken, release energy. *See* Chapter 14.

Carpal. Wrist bone. *See* Chapter 4.

Carpometacarpus (plural carpometacarpi). Unique structure in all living, and in most ancient, birds, in which bones in the wrist and hand are fused. *See* Chapter 8.

Cast. Material filling up a mold. *See* Chapter 1.

Caudal. Referring to the tail. *See* Chapter 9.

Cellular respiration. The breakdown of carbohydrates through a regulated series of oxidizing reactions. *See* Chapter 14.

Cenozoic. An Era, lasting from 66.1 Ma to present. *See* Chapters 2, 13, and 15.

Centrum. The spool-shaped, lower portion of a vertebra, upon which the spinal cord and neural arch rest. *See* Chapter 4.

Cerapoda. The ornithischian clade of Ceratopsia + Pachycephalosauria + Ornithopoda. *See* Part III.

Ceratopsia. Beaked dinosaurs of Asia and North America who, together with Pachycephalosauria, make up Marginocephalia. *See* Chapter 11.

Cervical. Referring to the neck. *See* Chapters 4 and 9.

Character. An isolated or abstracted feature or characteristic of an organism. *See* Chapter 3.

Cheek. The muscular, fleshy organs along the side of the jaw that retain food in the mouth. *See* Part III.

Cheek teeth. Teeth that lie against the cheeks; in mammals, the cheek teeth consist of the premolars and molars. *See* Part III.

Choanae (singular, choana). Paired openings in the mouth, at the tip of the palate, connected to the external nares.

Clade. Group of organisms in which all members are more closely related to each other than they are to anything else. All members of a clade share a most recent common ancestor that is itself the most basal member of that clade. Synonymous with "monophyletic group." *See* Chapter 3.

Cladogram. A hierarchical, branching diagram that shows the distribution of shared, derived characters among selected organisms. *See* Chapter 3.

Clavicle. Collarbone. *See* Chapter 4.

Clumped isotopes. Clumped isotopes are a means of telling temperatures in various carbonate (CO_3 minerals). The idea in its simplest form is that the amount of ordering of the stable isotopes of oxygen and carbon (^{18}O and ^{13}C, respectively) is temperature dependent, meaning that if one measures the amount of ordering of these isotopes in carbonate molecules, (s)he can back calculate to the temperature at which the carbonate was formed. *See* Chapter 14.

Coelophysoidea. The theropod clade containing *Coelophysis* and its close relatives. *See* Chapters 6 and 7.

Coelurosauria. That group of theropods defined as the most recent common ancestor of *Passer*, tyrannosaurids, and all its descendants. *See* Chapter 7.

Coevolution. The idea that two organisms or groups of organisms may have evolved in response to one another. *See* Chapter 15.

Collect. To obtain fossils from the Earth. *See* Chapter 1.

Completeness. In paleontology, how well the fossil sample you recover represents the original sample you're trying to understand. If in life there were 32 different types of dinosaurs living, but only 16 types are recovered, then the sample is 50 percent complete. Completeness also has a geological meaning: sedimentation represents elapsed geological time, and how complete a rock sequence is depends upon how much of the total elapsed time is represented in the sedimentary sequence. *See* Chapter 15.

Computed tomography (CT). A medical scanning technique using computers to construct a two-dimensional image using X-ray imaging around an axis of rotation. It has proven to be an extremely successful means of resolving incompletely prepared fossils.

Continental drift. The movement of the continents during Earth history.

Continental effects. The effect on climate exerted by continental masses. *See* Chapter 2.

Contour feathers. Pennaceous feathers that generally cover the body.

Convergent. In anatomy, pertaining to the independent invention (and thus, duplication) of a structure or feature in two lineages. The streamlined shape of whales, fish, and ichthyosaurs is a famous example of convergent evolution. *See* Chapter 9.

Coprolite. Fossilized feces. *See* Chapters 1 and 9.

Coracoid. The lower (and more central) of two elements of the shoulder girdle (the upper being the scapula). *See* Chapter 4.

Coronoid process. A bony enlargement or process at the back of the lower jaw for the attachment of jaw-closing musculature. *See* Part III.

Cretaceous. A Period lasting from 146 Ma to 66.1 Ma.

Cretaceous–Paleogene extinction. The event, 66.1 million years ago at the Cretaceous–Paleogene (Tertiary) boundary, in which all the nonavian dinosaurs, as well as many other terrestrial vertebrates, became extinct. *See* Chapter 17.

Crop (cropping). To cut short; in anatomy, to bite off. *See* Part III.

Crurotarsi (adjectival form "crurotarsal"). A clade of archosaurs including crocodilians and their close relatives, identified by the morphology of the ankle. *See* Chapter 4.

Curate. In paleontology, to incorporate, preserve, and catalog specimens into museum collections. *See* Chapter 1.

Cursorial. Pertaining to running; as the cursorial hypothesis, referring to the idea that bird flight evolved by birds running along the ground. *See* Chapters 4, 8, and 10.

Cycadophyte. A bulbous, fleshy type of gymnosperm. *See* Chapter 15.

D

Deccan Traps. Interbedded volcanic and sedimentary rocks in western and central India of Cretaceous–Tertiary age. *See* Chapter 17.

Definition. In phylogenetic systematics, as distinguished from diagnosis, definition outlines the evolutionary group to which one is referring, not by identifying it with a diagnostic character, but by delineating the extreme margins of the group in question (its so-called "crown"), and then identifying that group as the common ancestor of the two margins, and all descendants of that ancestor (see Chapter 3). By way of example, suppose we wanted to define Hadrosauridae. Looking at the cladogram (Figure 12.21), we could *diagnose* the group by the characters listed at (1). But, we could also *define* hadrosaurids as the common ancestor of *Telmatosaurus* and *Lambeosaurus* (the margin members of the crown group), and all of that ancestor's descendants. These would naturally encompass all hadrosaurids, and nothing

else. We might not even know the animal that was the ancestor, or be familiar with all of its descendants, but the cladogram would allow us to recognize any fossil as either belonging, or not belonging, to Hadrosauridae.

Deltoid crest. A large process at the head of the humerus. *See* Chapter 8.

Dense Haversian bone. A type of Haversian bone in which the canals and their rims are very closely packed. *See* Chapter 13.

Dental battery. A cluster of closely packed cheek teeth in the upper and lower jaws, whose shearing or grinding motion is used to masticate plant matter. *See* Chapter 11.

Denticles. Small bumps or protuberances generally associated with teeth. *See* Chapters 6 and 7.

Derived. In an evolutionary context, pertaining to characters that uniquely apply to a particular group and thus are regarded as having been "invented" by that group during the course of its evolutionary history. *See* Chapter 3.

Determinate. As applied to growth, meaning that growth stops after a certain "adult" size is reached; after which the animal continues to live its life getting no larger. Humans, for example, have determinate growth. *See* Chapter 13.

Detritus. Loose particulate rock, mineral, or organic matter; debris. *See* Chapter 17.

Developmental biology. The study of how organisms grow from a fertilized egg to an adult. *See* Chapter 16.

Diagnostic. In phylogeny, a feature that uniquely pertains to a group of organisms. Diagnostic features permit the identification of groups of organisms because, uniquely, all members of that group possess the feature (and ideally all other groups do not). *See* Chapter 3.

Diapsida. The large clade of amniotes that includes the common ancestor of lepidosauromorphs and archosaurs, and all its descendants. *See* Chapter 4.

Diastem(a). A gap. *See* Part III.

Digitigrade. In anatomy, a position assumed by the foot when the animal is standing, in which the ball of the foot is held high off the ground and the weight rests on the ends of the toes. Opposite of plantigrade. *See* Chapter 8.

Dinosauria. A clade of ornithodiran archosaurs. *See* Chapter 4.

Dinosauriformes. A clade, closer to Dinosauria than Dinosauromorpha, that includes just the very closest relatives of Dinosauria, including Marasuchus and the silesaurids, as well as Dinosauria itself.

Diphyletic. Meaning derived from two separate (monophyletic) groups. In evolution, we look for monophyletic groups, as these show evolutionary relationship. Diphyletic groups can't show much of anything. *See* Chapter 16.

Disarticulated. Dismembered. *See* Chapter 1.

Disaster biota. Organisms that colonize a landscape immediately after an ecological disaster. Disaster biotas tend to have three characteristics: small size, high rates of speciation, and generalist life strategies. *See* Chapter 17.

Display. Messaging or signaling by organisms, aimed at other organisms, including members of the same species. Such signaling is commonly carried out by coloration, behavior, vocalizations, features on the head such as horns, wattles, or crests, or combinations of some or all of these. Of the many messages that might be conveyed by display, possession of territory or spoils, mating suitability, and position in a social hierarchy are among the most common.

Distal. In anatomy, in the direction away from the central part (or core) of the animal. *See* Chapters 3, 7, and 12.

Diversity. The variety of organisms; the number of *kinds* of organisms. *See* Chapter 15.

DNA hybridization. A molecular biological technique that measures the difference between two comparable strands of DNA. *See* Chapters 8 and 11.

Dominant. Dominant in biology commonly refers to the most numerically abundant; e.g., one might say that insects are the dominant animals in terrestrial ecosystems because they are the most abundant (common). Dominance also has a social meaning, referring the ability of one animal to impose its will – whatever that may be – upon another. *See* Chapter 15.

Down (adjectival form "downy"). A bushy, fluffy, type of feather in which barbules and vanes are not well developed, used for insulation. *See* Chapter 5.

Dynamic similarity. A conversion factor that "equalizes" the stride rates of vertebrates of different sizes and proportions, so that speed of locomotion can be calculated. *See* Chapter 13.

E

Ecological diversity. The proportion of an ecosystem that is occupied by a particular lifestyle, such as feeding type or mode of locomotion. For a simple example, one might study an ecosystem by dividing it into herbivores, carnivores, and omnivores. *See* Chapter 17.

Ecospace. Niches that can characterize an ecosystem. Simple categories of ecospace, for example, include "carnivore," "herbivore," and "scavenger." More refined categories could include "grazing" versus "browsing" herbivores, and large and small carnivores. *See* Chapter 17.

Ectotherm. Organism that regulates its temperature (and thus metabolic rate) using an external source of energy (heat). Such an organism is said to be ectothermic. The opposite of endotherm. *See* Chapter 14.

Edentulous. Lacking teeth. *See* Chapter 6.

Electron. A negatively charged subatomic particle. Electrons reside in clouds around the nucleus of an atom. *See* Chapter 2.

Element. In chemistry, an atom distinguished by the number of protons in its nucleus; in anatomy, discrete part of the skeleton, that is, an individual bone. *See* Chapters 2 and 4.

Enantiornithes. A group of sparrow-like, relatively common Mesozoic avialians. *See* Chapter 8.

Encephalization. A reference to the proportional size of the head to the body. An animal that is highly encephalized has a proportionately large head (and usually, but not necessarily, a large brain is implied); *see also* Encephalization quotient (EQ). *See* Chapter 13.

Encephalization quotient (EQ). An estimate based on brain size and body weight, designed to determine the intelligence of an extinct organism, relative to a living organism whose intelligence can be ascertained. *See* Chapter 13.

Endemic. An organism or fauna is said to be endemic to a region when it is restricted to that region. *See* Chapter 15.

Endemism. The property of being endemic. *See* Chapter 15.

Endocast(s) (*endo* - within). A cast obtained from an internal anatomical region. The word is generally applied to casts of braincases, in which a mold is taken of the internal surface of the braincase (where the brain would have resided), and the cast resulting from that mold would give at least a partial sense of the shape of the animal's brain. *See* Chapter 11.

Endosymbionts. Organisms that live within another organism in a mutually beneficial relationship. *See* Chapter 9.

Endotherm. Organism that regulates its temperature (and thus, metabolic rate) using an internal source of energy. Such an organism is said to be endothermic. The opposite of ectotherm. *See* Chapter 14.

Enlightenment. The Enlightenment was a Western European seventeenth- and eighteenth-century political, social, and philosophical movement that, among other qualities, emphasized the primacy of logic, reason, and observation for understanding the natural world. *See* Chapter 16.

Epeiric. Referring to marine waters (seas) covering continental crust. An epeiric sea is a sea that covers part (or all) of a continent, such as, for example, the North Sea, which sits atop some of the crust of the European continent.

Epicontinental sea. Relatively shallow (at most, a few hundred meters) marine water covering a continent (synonym: epeiric sea). *See* Chapter 2.

Epoch (plural Epochs). An interval of time, shorter than a Period, but longer than a Stage. The duration of any given Epoch varies, but they generally tend to extend for tens of millions of years.

Era. A very large block of geological time (hundreds of millions of years long), composed of Periods. *See* Chapter 2.

Erect stance. In anatomy, the condition in which the legs lie parasagittal to (alongside) the body and do not extend laterally from it. *See* Chapter 4.

Euornithopoda. A monophyletic group containing the more derived members of Ornithopoda. *See* Chapter 12.

Eurypoda. The ornithischian clade Stegosauria + Ankylosauria. *See* Chapter 10.

Eustatic. Global. *See* Chapter 2.

Evo-devo (evolution and development). The coupling of the sciences of evolutionary biology and developmental biology to attempt to understand how new features in organisms have evolved. *See* Chapter 16.

Evolution. In biology, descent with modification. *See* Appendix 3.1.

Evolutionary trees. "Trees of life," that purport to show the evolution of life, through time, showing which organisms gave rise to which other organisms. Trees of life are untestable, and are therefore scenarios, rather than scientific hypotheses.

Exhumed. Disinterred; brought to the Earth's surface after having been buried.

Extant phylogenetic bracketing. The use of closely related living forms to understand the anatomy of extinct forms (see Box 6.1).

Extraterrestrial. From outer space. *See* Chapter 17.

F

Fauna(s). A group of animals presumed to live together within a region and/or within a particular time interval.

Femur. The upper bone in the hindlimb (thigh bone). *See* Chapter 4.

Fibula. The smaller of the two lower leg bones in the hindlimb; the bone that lies alongside the shin bone (tibia). *See* Chapters 4 and 10.

Fit. Adjectival form of fitness. Fitness is the degree to which natural selection acts upon an organism to assure the presence of its genetic makeup in the succeeding (descendant) generation. *See* Chapter 3.

Flight feather. Elongate feather with well-developed, asymmetrical vanes; usually associated with flight. *See* Chapters 5 and 10.

Flood basalts. Episodic silica-poor lava flows from fissures in the Earth's crust. *See* Chapter 17.

Flux. A measure of change; rate of discharge times volume.

Footplate. The expanded bony area present at the distal of many theropod pubes. *See also* pubis.

Footprint. Trace fossil left by the feet of vertebrates. *See* Chapter 1.

Foramen magnum. The opening at the base of the braincase through which the spinal cord travels to connect to the brain. *See* Chapters 4 and 5.

Foraminifera (singular foraminifer). Single-celled, shell-bearing organisms that live in the oceans. *See* Chapter 17.

Fossil. Technically, anything buried; generally refers to the buried remains of organisms. *See* Chapter 1.

Fossilization. The process of becoming a fossil. *See* Chapter 1.

Fossorial. Burrowing. *See* Chapter 12.

Fractionation. The process of separation of the components in a mixture (for example, liquids, gases, solids, and isotopes), in which the separation takes place during a phase transition (changing from one phase [liquid, gas, or solid] to a different phase [liquid, gas, or solid]). A familiar example of fractionation is distillation, in which the alcohol and water components of a mixture of water and alcohol are separated in the process of turning the mixture from a liquid to a gas (and then returning the separated components back into their liquid phases. *See* Chapter 14.

Frill. In ceratopsians, a sheet of bone extending dorsally and rearward from the back of the skull, made up of the parietal and squamosal bones. *See* Chapter 11.

Furcula. Fused clavicles (collarbones); the "wish-bone" in birds and most nonavian theropods. *See* Chapters 7 and 10.

G

Gamete(s). A mature sex cell; either male or female. *See* Chapter 13.

Gastralia. Belly ribs. *See* Chapter 8.

Gastrolith. Smoothly polished stone in the stomach, used for grinding plant matter. *See* Chapter 5.

Genasauria. The ornithischian cheek-bearing clade of Thyreophora + Cerapoda. *See* Part III.

General. In phylogenetic reconstruction, referring to a character that is nondiagnostic of a group; in this context, synonymous with primitive. *See* Chapter 3.

Generalist. In ecology, unspecialized. Basically willing and able to eat anything to stay alive. In terms of diet, at least, humans are generalists; a meat-eater, like a panther, would be considered a specialist. *See* Chapter 17.

Generic. Adjective from the word genus, the second-smallest grouping in the Linnaean classification (a genus [plural genera] is composed of species, the smallest formal category in the Linnaean classification). *See* Chapter 4.

Genotype. The total genetic makeup of an organism. *See* Chapter 3.

Geochemistry. Chemistry particularly as related to geological problems. *See* Chapter 16.

Geochronology. The science of determining the age (in years) of the Earth. Adjectival form: geochronological. Practitioners of geochronology are, not surprisingly, geochronologists. *See* Chapter 2.

Ghost lineage. Lineage of organisms for which there is no physical record (but whose existence can be inferred). *See* Chapter 15.

Gigantism. A tendency towards very large (gigantic) size. *See* Chapter 7.

Gigantothermy. Modified mass homeothermy, which mixes large size with low metabolic rates and control of circulation to peripheral tissues. *See* Chapter 14.

Girdle(s). Something that encircles; in this case attachments for the limbs, that partially encircle the trunk. *See* Chapter 4.

Gizzard. A muscular chamber just in front of the glandular part of the stomach. *See* Chapter 6.

Glycogen. A complex, carbohydrate-based molecule used by the body for energy storage. *See* Chapter 10.

Gnathostome. A vertebrate with a jaw (formal term: Gnathostomata). *See* Chapter 4.

Gondwana. A southern supercontinent comprising present-day Australia, Africa, South America, and Antarctica. *See* Chapter 2.

Grade. A group of organisms that have achieved a similar level of evolutionary advancement. Distinguished from "clade," in which the members constitute a monophyletic group. One might think of the largest theropods as a grade: large, bipedal carnivorous creatures that evolved this morphology independently (i.e., are not a clade). *See* Chapter 8.

Gregarious. Highly socialized; regularly moving in flocks or herds. *See* Chapters 5 to 8 and Chapter 10.

Growth trajectory (plural trajectories). A growth trajectory is the sequence of growth stages through which an organism passes as it grows. For humans, it could be something like: baby → toddler → child → adolescent → young adult → mature adult → middle-aged → senior. *See* Chapter 13.

Gut. Stomach, intestine, and bowels. *See* Chapter 9.

Gymnosperms. A group of seed-bearing, nonflowering plants, including pines and cypress. Gymnosperms are not a monophyletic grouping, unless angiosperms are also included within the group. *See* Chapters 9 and 14.

H

Hadrosauridae (hadrosaurids). The derived iguanodontian family comprising duck-billed dinosaurs. Defined as the most recent common ancestor of *Telmatosaurus* and *Lambeosaurus* and all its descendants. *See* Chapter 12.

Half-life. The amount of time that it takes for 50 percent of a volume of unstable isotope to decay. *See* Chapter 2.

Hard part. In paleontology, all hard tissues, including bones, teeth, beaks, and claws. Hard parts tend to be preserved more readily than soft tissues. *See* Chapter 1.

Hatch mark. In a cladogram, the bar next to a node that indicates which derived characters pertain to that node.

Haversian canal. In bone histology, a canal composed of secondary bone. *See* Chapter 13.

Heat capacity. The amount of heat (energy) required raise an object of a particular mass, a given temperature (usually, 1 °C).

Hemal. Referring to blood. In bone anatomy, hemal arch. A vertebral process straddling the caudal artery and vein on the ventral side of the vertebral column and pointing ventrally (oriented opposite to the neural arch). *See* Chapter 4.

Hierarchy. As applied here, the ordering of objects, organisms, and categories by rank. The military and the clergy are both excellent examples of hierarchies; in these, rank is a reflection of power and, one hopes, accomplishment. Another hierarchical system is money, which is ordered by value. *See* Chapter 3.

Histology. The study of any tissues. *See also* Bone histology.

Homeotherm. Organism whose core temperature remains constant. *See* Chapter 14.

Homologous. Two features are homologous when they can be traced back to a single structure in a common ancestor. *See* Chapter 3.

Horn core. The horn in many animals, cows and sheep being two familiar examples, are constructed of a bony central part, the core, covered by a layer, or sheath, made of keratin. Ceratopsian dinosaurs had this type of horn. *See* Figure 11.21.

Hornlets. Small horns. *See* Chapters 6 and 11.

Humerus. The upper arm bone. *See* Chapter 4 (especially Figure 4.5).

Hyoid bone(s). Paired elongate bones in the throat that form a support for the tongue. *See* Chapter 10.

Hypothesis of relationship. A hypothesis about how closely or distantly organisms are related. A cladogram is a hypothesis of relationship. *See* Chapter 3.

I

Ichnofossil. Impression, burrow, track, or other modification of the substrate by organisms. *See* Chapter 1.

Ichthyosaurs. Dolphin-like marine reptiles of the Mesozoic. *See* Figure 17.10a.

Ilium. The uppermost of three bones that make up the pelvis. *See* Chapter 4.

Impact ejecta. The material thrown up when an asteroid strikes the Earth. *See* Chapter 17.

Impact winter. The idea that a temporary period of climatic cooling caused by sunlight blockage by ejecta would follow an impact. Based on climatic data from large volcanic eruptions such as Krakatoa. *See* Chapter 17.

Indeterminate. As applied to growth, meaning that growth continues as long as the animal lives. Crocodiles, for example, have indeterminate growth, and continue growing in size throughout their lives. *See* Chapter 13.

Interspecific. Among different species. *See* Chapter 11.

Intraspecific. Within the same species. *See* Chapters 5, 6, 10, and 11.

Iridium (Ir). A nontoxic, platinum-group metal, rare at the Earth's surface. *See* Chapter 17.

Iridium anomaly. High concentrations of iridium in a single place; the name comes from the idea that since Ir levels at the Earth's surface are generally low, high concentrations would be considered unexpected (anomalous). *See* Chapter 17.

Ischium. The most posterior of three bones that make up the pelvis. *See* Chapter 4.

Isotopes. In chemistry, elements that have the same atomic number but different mass numbers. *See* Chapter 2.

J

Jacket. In paleontology, a rigid, protective covering placed around a fossil, so that it can be moved safely out of the field. Commonly made up of strips of burlap soaked in plaster. *See* Figure 1.11.

Jurassic. A Period lasting from 201 Ma to 145 Ma.

K

K-strategy. The evolutionary strategy of having few offspring, which are cared for by the parents. The symbol *K* stands for the carrying capacity of the environment. *See* Chapter 9.

K-Pg (boundary). Common abbreviation for that moment in time, 65.5 Ma, which marks the boundary between the Cretaceous and the Paleogene (Tertiary). *See* Chapter 17.

Keel. A flange or sheet of bone, as in the keeled sternum of birds; named for its resemblance to the keel on a sailboat. *See* Chapter 10.

Keratin. A protein that forms the basis of nails, horns, hooves, feathers, and hair. *See* Part III.

Kinetic. With reference to skull anatomy, movement between bones of the skull. *See* Chapter 12.

L

Lactic acid. An organic acid produced as a byproduct of muscle exertion; $CH_3CH(OH)COOH$. *See* Chapter 14.

Lagerstätte (plural Lagerstätten). A fossil-rich locality, or group of fossil-rich localities, in which the preservation is extraordinary. Such localities might contain, along with bones and teeth, skin impressions, organic matter suggestive of the original soft tissue of the organ, the fossils of soft-bodied creatures such as insects, and the last meals (stomach contents) of the fossils that have been preserved. *See* Chapter 7.

LAGs. *See* Lines of arrested growth. *See also* Chapter 13.

Lambeosaurines. The hollow-crested hadrosaurid dinosaurs. *See* Chapter 12.

Land bridge. A corridor of land between two continents that allows the passage of organisms from one continent to the other. Such corridors have occasionally existed, for example, between South America and North America, as well as between North America and Asia. *See* Chapter 15.

Laurasia. A northern supercontinent. *See* Chapter 2.

Lepidosauromorpha. One of the two major clades of diapsid reptiles; the other clade is Archosauromorpha. *See* Chapter 4.

Lines of arrested growth (LAGs). Lines that are inferred to represent times of nongrowth, visible in the cross-section of bones. *See* Chapter 13.

Lithostratigraphy. The general study of all rock relationships. *See* Chapter 2.

Locality. A location; in paleontology, the place where a fossil is – or fossils are – found.

Lower temporal fenestra. The lower opening of the skull just behind the eye; *see* Temporal fenestra. *See also* Chapter 4.

M

Macroflora. In paleontology, plant remains that are clearly visible to the naked eye, such as leaves, stems, and fruits. Distinguished from a microflora, which generally consists of fossil pollen, and is virtually impossible to distinguish without a hand lens or microscope. *See* Chapter 17.

Mandible. The lower jaw. *See* Chapter 4.

Mandibular foramen. A fenestra in the posterior portion of the mandible. *See* Part III.

Marginocephalia. The clade of dinosaurs that includes the most recent common ancestor of pachycephalosaurs and ceratopsians and all of its descendants. *See* Chapter 11.

Mass extinctions. Global and geologically rapid extinctions of many kinds, and large numbers, of species. *See* Chapter 17.

Matrix. In paleontology, the rock that surrounds fossil bone. *See* Chapter 1.

Megaflora (adjective megafloral). The visible remains of plants, especially leaves. *See* Chapter 17.

Melanosomes. Microscopic dermal organelles that are used to produce, store, and facilitate the transportation of melanin, the pigment responsible for dark coloration a well as for protection from UV light (from the Sun).

Mesotarsal. A linear type of ankle in which hinge motion in a fore–aft direction occurs between the upper ankle bones (the astragalus and calcaneum) and the rest of the foot. *See* Chapter 4.

MDT. *See* Minimal divergence time. *See also* Chapter 15.

Mesozoic. An Era lasting from 251.9 Ma to 66.1 Ma. *See* Chapter 2.

Metabolism. The sum of the physical and chemical processes in an organism. *See* Chapter 14.

Metacarpal. Bone in the palm of the hand. *See* Chapter 4.

Metapodial. A general name for metacarpals and metatarsals. *See* Chapter 4.

Metatarsal. Bone in the sole of the foot. *See* Chapter 4.

Micropaleontology. The study of microscopic organisms, such as marine plankton. A person who specializes in micropaleontology is a *micropaleontologist*. *See* Chapter 17.

Microtektite. A small, droplet-shaped blob of silica-rich glass thought to have derived from impact ejecta. *See* Chapter 17.

Minimal divergence time (MDT). The minimal amount of time missing between the two descendant species and their common ancestor; calculated by comparing phylogeny and age of fossils. *See* Chapter 15.

Minimum number of individuals (MNI). A technique for estimating how many individual organisms are represented in a locality. If we find two left thigh bones (among other elements), then we know that at least two different individuals are represented. If we found six theropod teeth, however, we would not know whether they came from one or more than one (up to six) different individuals; the *least* number of individuals that could have produced those six teeth is one. *See* Chapter 17.

Mold. Ichnofossil that consists of the impression of an original fossil. *See* Chapter 1.

Molecular clock. The use of the rates of mutation in certain molecules to determine how long ago two organisms diverged from a common ancestor. *See* Chapters 8, 11, and 14.

Molecular evolution. The idea that molecules can evolve at particular rates. *See* Chapter 16.

Monofilamentous. A primitive feather type, consisting of a single thread or filament, constructed of a hollow cylinder of protein. Monofilamentous feathers would have looked a bit like hair, and are seen on many dinosaurs, from *Psittacosaurus* to *Yutyrannus*.

Monophyletic group. A group of organisms that has a single ancestor and contains all of the descendants of this unique ancestor (synonymous with "clade" and "natural group"). *See* Chapter 3.

Monospecific. Pertaining to, or containing, a single species.

Monotypic. Of one type; generally, composed of one species. *See* Chapter 12.

Morphology. The study of shape. *See* Chapter 3.

Mosasaurs. Late Cretaceous marine-adapted lizards. *See* Chapter 17.

N

Naris (plural nares). Opening in the skull for the nostrils. *See* Chapter 4.

Natural selection. The process by which certain members of a population are more effectively able to assure the representation of their genes in the succeeding generation (e.g., the descendant generation). *See* Chapter 3.

Neoceratosauria. The theropod clade *Coelophysis*, *Ceratosaurus*, and near relatives. *See* Chapters 6 and 7.

Neonates (*neo* - new; *natus* - born). Newly born (hatched, in the case of dinosaurs). *See* Chapter 9.

Neornithes. As used here, modern birds, denoted by a variety of anatomical features, most famously, loss of teeth. *See* Chapter 8.

Neornithischia. Genosaurian ornithischians, consisting of the most recent common ancestor of *Agilisaurus* and ornithopods, and all its descendants. *See* Part III.

Neotheropoda. That group of theropods defined as the most recent common ancestor of *Passer*, *Coelophysis*, and all its descendants. *See* Chapter 7.

Neural arch. A piece of bone that straddles the spinal cord; generally with a central process that rises dorsally. *See* Chapters 4 (especially Figure 4.5) and 10.

Neutron. Electrically neutral subatomic particle that resides in the nucleus of the atom. *See* Chapter 2.

Node. A bifurcation or two-way split point in a phylogenetic diagram (cladogram). *See* Chapter 3.

Nodosauridae. Along with Ankylosauridae, one of the clades making up Ankylosauria. *See* Chapter 10.

Nonavian. Nonbird (Aves – birds); recalling that birds are dinosaurs, nonavian dinosaurs would be those dinosaurs which are not birds.

Nonavian dinosaurs. All dinosaurs *except* birds. *See* Chapters 1, 9, and 10.

Nondiagnostic. In phylogeny, a feature that does not uniquely pertain to a group of organisms. Nondiagnostic features do not permit the identification of a particular group of organisms because members outside of that group also possess the feature. *See* Chapter 3.

Notochord. An internal rod of cellular material that, primitively at least, ran longitudinally down the backs of all chordates. May be thought of as a precursor to the vertebral column. *See* Chapter 4.

Nuchal ligament. An elastic ligament running dorsally in the neck from the back of the head to a posterior cervical vertebra. In sauropods, it likely helped to support the head. *See* Chapter 8.

Nucleus. Central core of an atom or a cell. *See* Chapter 2.

Nutrient cycling. As used here, the movement of nutrients from the shallow surface waters to the bottom of the ocean. *See* Chapter 17.

Nutrient turnover. The amount of nutrients that pass through the system; in the case of bone development, the amount of metabolic activity associated with bone growth. *See* Chapter 14.

O

Obligate biped. Tetrapod that must walk or run on its hind legs. *See* Chapters 5, 9, and 15.

Occipital condyle. A knob of bone at the back of the skull with which the vertebral column articulates. *See* Chapter 4.

Occiput. The back of the skull. *See* Chapter 11.

Occlusion (verb to occlude). Contact between upper and lower teeth; necessary for chewing. Teeth that contact each other between the upper and lower jaws are said to occlude. *See* Part III.

Olfactory bulbs. An enlarged part of the brain that deals with the sense of smell. *See* Chapter 10.

Ontogeny. Biological development of the individual; the growth trajectory from embryo to adult. *See* Chapters 6 and 11.

Opisthopubic. The condition in which at least part of the pubis has rotated backward to lie close to, and parallel with, the ischium. *See* Part III.

Orbit. Eye socket. *See* Chapter 4.

Ornithischia. One of the two monophyletic groups comprising Dinosauria. *See* Chapter 5.

Ornithodira. The common ancestor of pterosaurs and dinosaurs, and all its descendants. *See* Chapter 4.

Ornithopoda. Biped and quadrupedal herbivorous ornithischian dinosaurs. *See* Chapters 11 and 12.

Ornithoscelida. Part of the new (2017) proposed classification of Dinosauria; Ornithoscelida includes Theropoda and Ornithischia. According to this classification, the other great clade of Dinosauria would be Sauropodomorpha and Herrerasauridae (here considered theropods). The 2017 classification contrasts dramatically with the current classification, dating back to 1887, in which Dinosauria is thought to be made up of two clades: Saurischia (containing Theropoda and Sauropodomorpha) and Ornithischia. *See* Chapter 5.

Ornithothoraces. A group of Mesozoic birds including Aves. *See* Chapter 8.

Ornithurae. Hesperornithiformes, Ichthyornithiformes, and Aves. *See* Chapter 8.

Ornithuromorpha. The lineage of ornithothoracean birds, including Aves. *See* Chapter 8.

Ossified. Having turned to bone (usually from cartilage). *See* Chapter 10.

Ossified tendons. Tendons that have ossified (turned to bone; *see* ossified). *See* Chapter 12.

Osteichthyes. Bony fishes that include ray-finned and lobe-finned gnathostomes. *See* Chapter 4.

Osteoderm. Bone within the skin; may be small nodule, plate, or a pavement of bony dermal armor. *See* Chapters 9 and 10.

Oxidation. Bonding of oxygen. *See* Chapter 14.

P

Pachycephalosauria. Dome-headed ornithischians of North America and Asia who, together with Ceratopsia, make up Marginocephalia. *See* Chapter 11.

Palate. The part of the skull that separates the nasal cavity (for breathing) from the oral cavity (for eating); usually strengthened by a paired series of bones. *See* Chapter 4.

Paleobiology. The discipline of paleontology, arising in the 1970s, that actively sought to understand the biology of fossil organisms. *See* Chapters 13 and 14.

Paleobotany. The study of ancient plants. A person who studies ancient plants is called a *paleobotanist*. *See* Chapter 15.

Paleoclimate. Ancient climate. *See* Chapter 2.

Paleoenvironment. Ancient environment.

Paleontology. The study of ancient life; distinguished from anthropology, which is the study of humans, and archaeology, which is the study of past civilizations. A *paleontologist* is someone who studies paleontology.

Paleozoic. An Era lasting from 541.0 Ma to 251.9 Ma. *See* Chapter 2.

Palpebral. A rod-like bone that crosses the upper part of the eye socket. *See* Chapter 5.

Palynoflora. Spores and pollen. *See* Chapter 17.

Pangaea. The mother of all supercontinents, formed from the union of all present-day continents. *See* Chapter 2.

Parasagittal stance. Stance in which the legs are in the same plane as the plane of symmetry of the body; they do not extend out from it. *See* Chapters 4 and 15.

Parascapular spine(s). An enlarged spine over the shoulder. *See* Chapter 10.

Parsimony. A principle that states that the simplest explanation that explains the greatest number of observations is preferred to more complex explanations. *See* Chapter 3.

Patellar groove. A groove at the distal end of the femur to accommodate the patella (knee cap). *See* Chapter 8.

Pectoral girdle. The bones of the shoulder; the attachment site of the forelimbs. *See* Chapter 4.

Pectoralis (muscle). The muscle that drives the wing's powerstroke in bird flight. *See* Chapter 8.

Pedestal. In paleontology, a pillar of matrix underneath the fossil. *See* Figure 1.11.

Peer-reviewed. As a part of the publication of scientific papers, experts familiar with the subject under study (i.e., *peers*) are asked to read manuscripts critically – to *review* them – to determine whether or not the work merits publication.

Pelvic girdle. The bones of the hips; the attachment site of the hindlimbs. *See* Chapter 4.

Perforate acetabulum. A hole in the hip socket; a diagnostic character of Dinosauria. *See* Figure 4.11.

Period. Subdivision of an Era, consisting of tens of millions of years. *See* Chapter 2 (especially Figure 2.4).

Permineralization. The geological process in which the spaces in fossil bones become filled with a mineral. *See* Chapter 1.

Permo-Triassic. Relating to events at the Permian/Triassic boundary (251.9 Ma). *See* Chapter 17.

Phalanx (plural phalanges). Small bone of the fingers and toes that allows flexibility. *See* Chapter 4.

Phanerozoic. The time interval from 541.0 Ma to 0 Ma; the most recent Eon in Earth history. For more details about the history of the Earth. *See* Chapter 2.

Phenotype. The physical appearance and features of an organism. *See* Chapter 3.

Photosynthesis. The process by which organisms use energy from the Sun to produce complex molecules for nutrition. *See* Chapter 17.

Phylogenetic. Pertaining to phylogeny. *See* Chapter 3.

Phylogenetic systematics. The method of determining organismic relationships that uses parsimony to select among competing hierarchical distributions of shared, derived characters (that is, cladograms). *See* Chapter 3.

Phylogenetic tree. Another name for a cladogram.

Phylogeny. The study of the fundamental genealogical connections among organisms. *See* Chapter 3.

Phylum. A grouping of organisms whose makeup is supposed to connote a very significant level of organization shared by all of its members. *See* Chapter 3.

Phytosaur. Long-snouted, aquatic, fish-eating member of Crurotarsi. *See* Figure 5.3.

Piscivorous. Fish eating (*pisces* – fish; *vorous* – eating).

Planktonic (or planktic). Living in the water column. *See* Chapter 17.

Plantigrade. A foot position in which the bottom of the foot (the tissue below the metatarsals) lies flat on the ground. Opposite of digitigrade. *See* Chapter 8.

Plate tectonics. The study of the movement of the geological plates; its history, processes, and consequences. *See* Appendix 2.2; *also* tectonics.

Plesiosaurs. Long-necked fish-eating reptiles with large flippers that inhabited Mesozoic seas. *See* Figure 17.10.

Pleurocoel. A well-marked excavation on the sides of a vertebra. *See* Chapter 8.

Pleurokinesis. Mobility of the upper jaw. *See* Chapter 12.

Pneumatic. Having air sacs or sinuses; the state of having such is called pneumaticity. *See* Chapter 8.

Pneumatic foramina. Openings for air sacs to enter the internal bone cavities. *See* Chapters 8 and 10.

Poikilotherm. Organism whose core temperature fluctuates. *See* Chapter 14.

Precocial. The condition in which the young are rather adult-like in their behavior. *See* Chapter 9.

Predator / prey biomass ratios. The ratio of the total estimated weight of predators to the total estimated weight of their prey in a particular ecosystem. *See* Chapter 14.

Predentary. The bone that caps the front of the lower jaws in all ornithischians. *See* Part III.

Preparation (prep) laboratory. Where preparation takes place. *See* Figure 1.12.

Preparators. Highly skilled individuals who prepare fossil material.

Prepare. In paleontology, the process of freeing fossils from the rocks or sediment which encase them (see also Matrix), so that the fossils can be studied. Preparation can be the most time-consuming process in the study of fossils. *See* Chapter 1.

Prepubic process. A flange of the pubis that points toward the head of the animal. *See* Chapters 5, 10, and 12.

Primary bone. Bone tissue that was deposited or laid down first. *See* Chapter 13.

Primary production (or productivity). The sum total of organic matter synthesized by organisms from inorganic materials and sunlight. *See* Chapter 17.

Primitive. *See* Ancestral. *See also* Chapter 3.

Process. In relation to anatomy, part of a bone that is commonly ridge-, knob-, or blade-shaped and sticks out from the main body of the bone. *See* Chapter 4.

Productive. With reference to a fossil locality: rich; lots of collectable fossils. With reference to the biosphere. *See* Productivity.

Productivity. The amount of biological activity in an ecosystem. *See* Chapter 1.

Prosauropoda. Late Triassic and Early Jurassic saurischian dinosaurs; the world's first high-browsing herbivores. Once thought to be ancestral to Sauropoda, now believed to share a common ancestor with sauropods. *See* Chapter 9, especially Figure 9.2, for diagnostic characters.

Prospect(ing). To hunt for fossils. *See* Chapter 1.

Proton. Electronically charged (+1) subatomic particle that resides in the nucleus of the atom. *See* Chapter 2.

Proximal. In anatomy, in the direction toward the central part (or core) of the animal.

Proxy (plural proxies). In science, when something allows us to infer information about something else. For example, high blood pressure could be seen as a potential proxy for heart disease. The term comes from voting, when a person would send another person to vote on his behalf. The person physically voting would be the proxy for the person who sent him (or her). *See* Chapter 14.

Pterosauria. Flying ornithodiran archosaurs, closely related to (but not included within) dinosaurs. Uniquely, the wing was supported by an extraordinarily elongate digit IV. *See* Chapter 4.

Pubis (plural pubes). One of the three bones that make up the pelvic girdle. *See* Chapter 4.

Pull of the Recent. The inescapable fact that as we get closer and closer to the Recent, fossil biotas become better and better known. *See* Chapter 15.

Pygostyle. A small, compact, pointed structure made of fused tail bones in birds. *See* Chapter 8.

R

r-strategy. The evolutionary strategy where organisms have lots of offspring and no parental care. The letter *r* is obtained from the symbol *r* for growth rate. *See* Chapter 9.

Radius. One of the two lower arm bones; the other is the ulna. *See* Chapter 4.

Rarefaction. A statistical technique that allows comparison of two differently sized samples. Suppose you wanted to have all the different kinds of American coins in your hands: pennies, nickels, dimes, quarters, half-dollars, and dollars. What are the chances that the first six coins you pick up would have one of each of these? Very very low. But, if you have a beer keg full of coins, your chances would be much better. The rarefaction curve allows you to predict what the likely coin distribution might be if you only had half a beer keg full of coins. Or if you had only a beer glass full of coins. The beer glass likely would not have examples of all the coins; the half-keg would be more likely to have at least one example of each coin, and the full keg would be even more likely to have at least one example of each coin. Arranging those samples along a curve of volume vs. different types of coins would allow you to predict how many different kinds of coins you might find for different volumes of loose change. *See* Chapter 15.

Recent. The time interval that encompasses 13 000 years ago to the present. *See* Chapters 2 and 15.

Recovery. The process of repopulating an ecosystem after an extinction. *See* Part IV.

Recurved. Curved backward, like a scimitar. *See* Chapters 6 and 9.

Regression. Retreating of seas due to lowering of sea level. *See* Chapter 15.

Relationship. The phylogenetic closeness of two organisms; that is, the genealogical nearness or distance of their most recent common ancestor. *See* Chapter 3.

Relative dating. The type of geological dating that, although not providing ages in years before present, provides ages relative to other strata or assemblages of organisms. *See* Chapter 2.

Remodel. In bone histology, to resorb or dissolve primary bone and deposit secondary bone. *See* Chapter 13.

Renewable resources. Resources that can be synthesized or manufactured, either naturally or artificially. *See* Chapter 1.

Replace. To exchange the original mineral with another mineral. *See* Chapter 1.

Reptilia. The old Linnaean category for turtles, lizards, snakes, and crocodiles. Reptilia as formulated by Linnaeus and as commonly used is not monophyletic; only the addition of birds to these four groups constitutes a monophyletic group. *See* Chapter 4.

Respiratory turbinate. A thin, convoluted or complexly folded sheet of bone located in the nasal cavities of living endothermic vertebrates. *See* Chapter 14.

Retroverted. The pelvis is said to be "retroverted" when the pubis points downwards or posteriorly. *See* Chapter 7.

Rhamphotheca. Cornified covering on the upper and lower jaws (for example, a beak). *See* Part III and Chapter 6.

Robust. (1) In the context of hypothesis testing, a hypothesis is said to be robust when it has survived repeated tests; that is, despite meaningful attempts, it has failed to be falsified. (2) In anatomy, strong and stout.

Rock. An aggregate of minerals.

Rostral. Referring generally to the rostrum, or snout region of the skull; in ceratopsians, a unique, diagnostic bone at the tip of the snout on the skull. *See* Chapters 9 and 11.

Rostral bone. A unique bone on the front of the snout of ceratopsians, giving the upper jaws of these dinosaurs a parrot-like profile. *See* Chapter 11.

S

Sacrum. The part of the backbone where the hip bones attach. *See* Chapter 4.

Sarcopterygii. Lobe-finned fish. *See* Chapter 4.

Saurischia. One of the two monophyletic groups that Dinosauria comprises; the other is Ornithischia. *See* Chapter 5 and Part III.

Sauropoda. Monophyletic group of long-necked, long-tailed, quadrupedal herbivorous saurischians. *See* Chapter 9, especially Figure 9.20, for diagnostic characters.

Scapula (plural scapulae). The shoulder blade. *See* Chapters 4 (especially Figure 4.5) and 10.

Scenario. An untestable story that may or may not be correct.

Sclerotic ring. A ring of bony plates that support the eyeball within the skull. *See* Figure 4.6 and Chapter 12.

Scute(s). Keratinized (horny) or bony plate. *See* Part III.

Seasonality. Highly marked seasons. *See* Chapter 2.

Secondarily evolve. To re-evolve a feature. *See* Chapters 5 and 15.

Secondary bone. Bone deposited in the form of Haversian canals. *See* Chapter 13.

Secondary palate. A shelf of bone, above the palate, over which air can be directed so that air is not mixed with the food during chewing. All mammals and crocodiles have secondary palates; some turtles do as well. *See* Chapters 4 and 10.

Sedimentary rock. A rock that generally represents the lithification, or hardening, of sediment. *See* Chapter 1.

Sedimentologist. Someone who studies sedimentary rocks and processes. *See* Chapter 1.

Seed. A capsule that contains gametes, as well as nutrients. These are generally encased in a protective pod. *See* Chapter 15.

Semi-lunate carpal. A distinctive, half-moon-shaped bone in the wrist. *See* Chapter 9.

Sexual dimorphism. Size, shape, and behavioral differences between sexes *See* Chapters 6 and 10.

Sexual selection. Selection not between all of the individuals within a species, but between members of a single sex. *See* Chapter 6.

Shaft. (1) The hollow main vane of a feather. (2) The title and eponymous lead male character in the first and most famous of the 1970s "blaxploitation" films. *See* Chapter 8.

Shocked quartz. Quartz that has been placed under such pressure that the crystal lattice becomes compressed and distorted; correctly termed "impact metamorphism." *See* Chapter 17.

Sigmoidal. Having an "S" shape. *See* Chapter 11.

Sinus. A cavity. *See* Chapter 6.

Skeleton. The supporting part of any organism. In vertebrates, the skeleton is internal and consists of tissue hardened by mineral deposits (sodium apatite). Such tissue is called "bone." *See* Chapter 4.

Skin impressions. When the skin of an organism presses against a soft substrate, such as clay or mud, it can leave a cast or imprint of itself. When that cast is preserved, we can see the preserved imprint years, even millions of years later: the skin impression. *See* Chapter 1.

Skull. That part of the vertebrate skeleton that houses the brain, special sense organs, nasal cavity, and oral cavity. *See* Figure 4.6.

Skull roof. The bones that cover the top of the braincase. *See* Chapter 4 (especially Figures 4.6 and 4.9).

Soft tissue. In vertebrates, all of the body parts except bones, teeth, beaks, and claws. *See* Chapter 1.

Specialization. In biology, the idea that an organism is adapted for a particular circumstance (for example, diet, climate, ecosystem, a particular host (if it is a parasite), or any other aspect of its existence). Such an organism is termed a specialist. *See* Chapter 17.

Speciate. Evolve new species; diversify. *See* Chapter 17.

Species-specific. Appying only to a particular species (the one under discussion). *See* Chapter 10.

Specific. (1) Diagnostic of a monophyletic group; uniquely evolved. (2) Adjectival form of the word "species," the smallest formal category in the Linnaean classification. Operationally (among living organisms), if two creatures breed and produce viable offspring, they are generally said to be the same species. *See* Chapter 4.

Sphenopsid(s). A primitive type of vascular plant. *See* Figure 15.7.

Sprawling stance. Stance in which the upper parts of the arms and legs splay out approximately horizontally from the body. *See* Box 4.3.

Stable isotope. An isotope that does not spontaneously decay. *See* Chapter 14.

Stapes. The middle-ear bone that transmits sound (vibrations) from the tympanic membrane to a hole in the side of the braincase (allowing auditory nerves of the brain to sense vibration). *See* Chapter 4.

Stegosauria. Thyreophorans (Ornithischia) with paired rows of bony osteoderms running along the back. *See* Chapter 10.

Sternum. The breastbone. *See* Chapters 4 and 10.

Strata. Layers of rock. *See* Figure 2.2.

Stratigraphy. The study of the relationships of strata and the fossils they contain. *See* Chapter 2.

Subatomic. Smaller than atom-sized. *See* Chapter 2.

Substitution. In molecular biology, the notion that one of the bases in the pairs that compose a DNA molecule can be substituted for another base or base-pair. *See* Chapters 8 and 11.

Superposition. The geological principle in which the oldest rocks are found at the bottom of a stack of strata and the youngest rocks are found at the top. *See* Chapter 2.

Supracoracoideus (muscle). The muscle that drives the wing's recovery stroke in bird flight. *See* Chapter 8.

Survivorship. The pattern of survival measured against extinction. *See* Chapter 17.

Synapsida. The large clade of amniotes, including mammals, diagnosed by a single, lower temporal opening. *See* Chapter 4.

Synsacrum. A single unit of fused bones consisting of the sacral vertebrae. *See* Chapter 10.

T

Tarsal. Ankle bone. *See* Chapter 4.

Tarsometatarsus. The name for the three metatarsals fused together with some of the ankle bones. *See* Chapter 10.

Taxon (plural taxa). A group of organisms, designated by a name, of any rank within the biotic hierarchy. *See* Chapter 4.

Tectonics (adjectival form tectonic). That branch of geology dealing with the development and form of the surficial and near-surficial parts of the Earth. *See* Chapter 2.

Temnospondyl. Belonging to the group Temnospondyli, large, carnivorous largely Paleozoic amphibians, whose behavior must have been something like that of a crocodile. *See* Chapter 14 (especially Figure 14.7).

Temporal. (1) Referring to time. (2) In anatomy, the side region of the skull, in the position of the brain case. *See* Chapters 2 and 4.

Temporal fenestra (plural fenestrae). Openings in the temporal region of the skull. *See* Chapter 4.

Terrestrial. Referring to land; that is, not marine. *See* Chapter 1.

Testable. With reference to scientific hypotheses, the construction of a hypothesis that can be logically considered and potentially rejected because it can be falsified. An example of testable hypothesis would be: *All mammals have fur.* The test would consist of checking all mammals and determining if they have fur. The hypothesis would be rejected if a mammal was found that had, for example, feathers instead of fur. A hypothesis that is not testable is not appropriate for science.

Testable hypothesis. A hypothesis that makes predictions that can be compared and assessed by observations in the natural world. *See* Chapter 3.

Tetanurae. Literally, "stiff tails"; the clade of theropods with tails stiffened as a result of overlapping caudal zygapophyses. *See* Chapter 9.

Tethys Ocean (adjectival form Tethyan). The ancient water mass that eventually became today's Atlantic Ocean. Named after the Titan Goddess of the Ocean (Greek), *Tethys. See* Chapter 2.

Tetrapoda. A monophyletic group of vertebrates primitively bearing four limbs. *See* Chapters 3 and 4.

Thecodontia. A paraphyletic taxon that at one time was used to unite the separate ancestors of crocodilians, pterosaurs, dinosaurs, and birds. *See* Chapters 5 and 10.

Therapsida. The clade of synapsids that includes mammals, some of their close relatives, and all of their most recent common ancestors. *See* Chapters 13 and 15.

Thermoregulation. Control of body temperature. *See* Chapter 10.

Theropoda. Extremely diverse clade of bipedal, primarily carnivorous saurischian dinosaurs.

Thorax. In vertebrates, the part of the body between the neck and abdomen. *See* Chapter 8.

Thyreophora. Armor-bearing ornithischians; stegosaurs, ankylosaurs, and their close relatives. *See* Part III and Chapter 5.

Tibia. One of the two lower bones in the tetrapod hindlimb; the other is the fibula. *See* Chapters 4 and 10.

Trace fossil. Impressions in sediment left by an organism. *See* Chapter 1.

Trachea. The windpipe. *See* Chapter 13.

Trackway. Group of aligned footprints left as an organism walks. *See* Chapter 1.

Triassic. A Period lasting from 251.9 Ma to 201.3 Ma. *See* Chapter 2.

Triosseal foramen. The hole formed by three bones, the coracoid, furcula, and scapula, through which the tendon for the supracoracoideus connects to the humerus for the wing recovery stroke in birds. *See* Chapter 8.

Tubercles. Small bumps or protuberances. *See* Chapter 11.

Turn (a fossil). To separate the fossil from the surrounding rock at the base of the pedestal and to rotate it 180°. *See* Figure 1.11.

Tympanic membrane. Eardrum. *See* Chapter 4.

U

Ulna. One of the two lower arm bones; the other is the radius. *See* Chapter 4.

Ungual phalange. An outermost bone of the fingers and toes. *See* Chapter 4.

Unidirectional. Moving in a single direction. *See* Chapter 13.

Unstable isotope. An isotope that spontaneously decays from an energy configuration that is not stable to one that is more stable. *See* Chapter 2.

Upper temporal fenestra. The opening in the skull roof above the lower temporal fenestra; *see* Temporal fenestra. *See* also Chapter 4.

V

Vane. The sheet of feather material that extends away from either side of the shaft. *See* Chapter 8.

Vascular. Pertaining to vessels that conduct fluids. In animals, a region that is vascular has a lot of blood vessels; a vascular plant is a plant that has tubes (xylem and phloem) that conduct water and nutrients. *See* Chapter 15.

Vertebrae. The repeated structures that compose the backbone and that, along with the limbs, support the rest of the body. *See* Chapter 4.

Vertebrata. The group of all animals containing vertebrae. *See* Chapter 4.

Vestigial. A feature, present in the ancestors of an organism, which remains in the descendants, but commonly in a highly reduced, and nonfunctional state. A famous example of a vestigial feature is the rudimentary pelvis and small femur still present in the body cavity of some modern whales, presumed to be a relic of the hind legs of its tetrapod ancestor.

W

Weathering. The physical or chemical breaking down of earthly materials (for example, minerals, rocks, and bones). *See* Chapter 1.

Z

Ziphodont. Thin, scimitar-like teeth, serrated on the edges, and curved towards the back of the mouth.

Zygapophysis (plural zygapophyses). A fore-and-aft projection from the neural arches (of vertebrae). *See* Chapters 7 and 9.

INDEX OF SUBJECTS

INDEX OF GENERA